Natural Heritage

NATURAL HERITAGE DIVERSITY

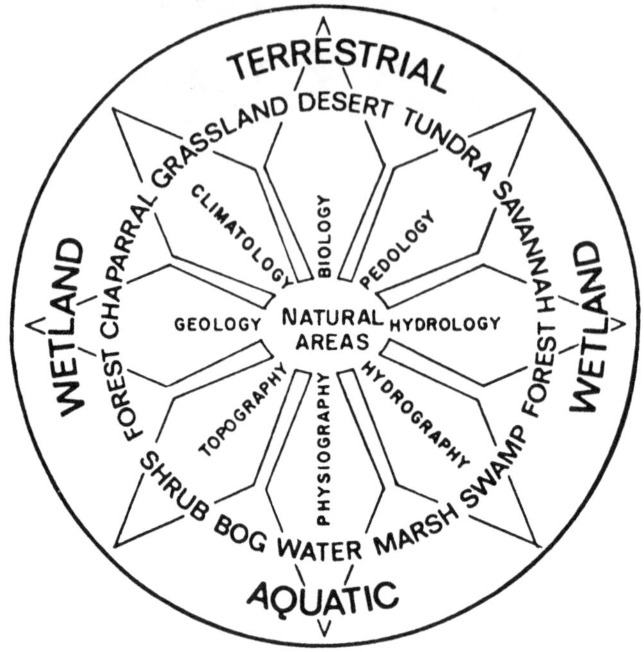

BIOSPHERE

Hierarchical arrangement of ecosystems: Natural Areas, Biomes, Ecosystems, and the component composition of the Biosphere

Natural Heritage

Classification, Inventory, and Information

by Albert E. Radford
Deborah Kay Strady Otte   Lee J. Otte
Jimmy R. Massey   Paul D. Whitson
and Contributors

The University of North Carolina Press *Chapel Hill*

Library of Congress Cataloging in Publication Data

Main entry under title:

Natural heritage.

Bibliography: p.
Includes index.
1. Natural areas—United States. I. Radford, Albert E.
QH76.N284 333.95'0973 80-23087
ISBN 978-0-8078-6752-5

Contents

Part I. Ecological Diversity Classification, Inventory, and Information

Part II. Natural Area Reports

Part III. Natural Heritage Resource Information

Part IV. Field Inventory and Research Procedures

Tables

Part I

Part II

Part III

Part IV

Maps and Figures

Maps

Part I

Part II

Part IV

Figures

Part I

Part II

Part III

Part IV

Prologue

This book was written primarily for diversity inventory use in natural areas by conservationists, heritage workers, and students of flora and fauna and for habitat analysis of special sites by advanced students of systematics, ecology, and species biology. A system of classification for ecological diversity and inventory procedures was developed as a basis for effective and efficient use of resources and field time. The classification system was designed as a usable and useful tool, of practical and theoretical relevance, for natural heritage preservation and field training of students.

Our **natural heritage** includes the full spectrum of ecological diversity and embraces all of **natural diversity**, which is comprised of "plant and animal species that have evolved over the last 600 million years and the discrete types of terrestrial and aquatic communities and ecosystems in which these species live." Our **national natural heritage** should have "representative examples of the full array of discrete types of terrestrial and aquatic communities, geologic features, landforms, and habitats of native plant and animal species that may be eliminated without deliberate protection" conserved in carefully selected natural areas.*

A comprehensive approach to the study of ecological diversity in our natural heritage involves (1) observing, identifying, and learning the distinguishing and diagnostic characteristics of taxa, populations, and communities; (2) analyzing the habitat for factors that control or limit populations and biotic assemblages; (3) synthesizing the biotic, abiotic, spatial, and temporal characteristics for ecological characterizations of floras, faunas, species, and communities; and (4) hypothesizing and theorizing as to origin, migration, and evolution of taxa and ecosystems. This approach to the study of diversity is fundamental to the determination of significance in the preservation of our heritage and certainly adds insight to natural resource inventory. Comprehensive training for the development of diversity perspective by the systematist, ecologist, and evolutionist must be based on studies using inclusive ecological diversity classification and information systems.

This book, especially Part I on diversity classification and inventory, is the outgrowth of an effort to produce a pedagogically sound classification and information systems base for a holistic approach to field studies in the plant systematics program at the University of North Carolina at Chapel Hill. The use and application of these systems in diversity inventory and analysis were intended to make floristic studies more sophisticated and scientifically valuable, monographic studies more effectively documented, biosystematic and evolutionary research more soundly based, ecological diversity analyses and syntheses more meaningful and definitive, and conservation decisions more realistic and significant. The information and data obtained from the literature search, from field observations, and from limited analyses while using these systems provide a fundamental background for working out analytical procedures and experimental design for advanced research on the more challenging aspects of field-based problems and for making wiser decisions on matters of conservation and preservation issues.

We firmly believe that on a nationwide scale the use and application of these systems will produce an information and data base that will be consistent, comprehensive, and comparable for

1. Habitat descriptions
2. Ecological characterizations
3. Population status inventories
4. Threatened and endangered species studies
5. Natural area diversity summaries
6. Natural area acquisition decisions
7. Management recommendations
8. Effective resource uses
9. Environmental impact studies

We are equally convinced that this type of information and data base has scientific and social relevance, gives perspective to the formulation and addition of predictive parameters to resource analysis, and in general provides a sound basis for decision making on scientific-social issues.

This presentation is a convenient source of information, procedures, bibliographies, and glossaries for natural heritage workers and a basic reference for field

*See definitions of natural heritage, natural diversity, and natural area in Part III A and treatment of natural and ecological diversity in the summary statement at the end of Chapter 2.

courses, environmental program training, conservation curricula, and natural resource studies. The intrinsic value of this book is in its use as an interdisciplinary training manual for field analysis of habitat, inventory of natural diversity, and characterization of ecological diversity.

This treatise may be considered a pioneer text in the field of **ecosystematics**—the systematics of ecological diversity—which is comprised of the classification, identification, nomenclature, and description of components and elements of ecological diversity; the study of the relationships between biotic and abiotic diversity; and the determination of the significance of those relationships. A knowledge of ecosystematics, including an understanding of the significance of diversity in its many aspects, is requisite to the preservation of our natural heritage.

These classification and information systems represent a holistic approach to field study. Ideally, students and heritage workers using this publication at the reconnaissance, basic, and population status inventory levels should have a background in plant and animal identification, soils classification, physical geology, climatology, map and aerial photo interpretation, and general ecology. Few students or professional heritage staff members have this basic background, thus most will have to learn one or more aspects of diversity inventory on the job. It is imperative that heritage workers and students be sufficiently experienced to know when to obtain professional help in the acquisition of diversity information and data.

Graduate and advanced undergraduate students involved in the development of these systems over the past six years have been challenged by the holistic diversity approach. The disciplined and responsible use of a broad ecological diversity classification has enabled them to gain a basic understanding of natural diversity and species-habitat relationships at the population, community, and ecosystem levels. Natural area diversity inventory is education in action. Inventory reports provide products for immediate and direct use in academic studies, conservation, and resource management.

This publication is organized into four parts: Part I deals with ecological diversity classification, inventory, and species information systems; Part II with natural area reports, which illustrate the application of the classification system and inventory procedures to a wide variety of habitats for a variety of purposes; Part III with natural heritage resource information; and Part IV with notes, suggestions, and worksheets on inventory procedures for heritage workers and students.

The ecological diversity classification in Part I has been designed for inventory use on the local to national levels; for habitat analysis and ecological characterization of the individual organism, population, community, and ecosystem; for identification of significant species, communities, habitats, and abiotic features; and for the provision of information basic to decisions on land classification, habitat management, natural area acquisition, and resource use and protection.

The classification system can be used at one to seven orders or levels of resolution, from system to population, and in one to seven components, from biological to physiographic. The orders and components used depend upon the individual interest, training, and expertise of the investigator, as well as upon time and resources available for the work. Different hierarchical levels of the various components can be combined in many ways to achieve desired goals or purposes.

The basic rationale and philosophy for the development of the classification for ecological diversity is presented in Chapter 1, along with the general organization of the system. Comments on the relationships of the ecological diversity classification to major ecosystems, habitat, identification, maps and mapping, inventory, and site-species relationships are included to show the utility and the potential for application of the system. The classification system is explained in its entirety, from system to population level, in Chapter 2. The inventory treatment in Chapter 3 is a "how to use" presentation of the classification system for academic and heritage conservation purposes.

The species information system, Chapter 4, was contributed by our colleagues J. R. Massey and P. D. Whitson, who together have developed over the past three years a comprehensive program for aiding in the acquisition of pertinent information necessary for an understanding of species of special concern, especially endangered and threatened vascular plants. Their program consists of four units or levels of information: Species General Information; Species Population, Habitat, and Threat Inventory Information; Species Biology Information; and Environmental Factor Analysis. To justify its inclusion in this volume, a brief explanation of our perception of the relationship of special species information to natural area endeavors and information seems appropriate.

The close relationship of the two efforts emerged as we developed and discussed the basic concepts of information acquisition, classification, storage, retrieval, and evaluation. It became apparent that an exchange or sharing of information could be profitable to each effort and that the most valuable link occurred at the Species Population, Habitat, and Threat Inventory level of information, as seen by the following comparison:

Natural Diversity Information

Natural Area Information	*Special Species Information*
Natural area checklists	Special species checklists
Natural area general information	Species general information
Ecological diversity classification	Species population, habitat, and threat inventory
Natural area status information	Species biology status information

The lists above suggest that information acquired in both inventory efforts can provide a much clearer understanding of the whole natural diversity picture for an area. To provide natural area workers with a better knowledge of the kinds of general plant information the species specialists desire, we have also included the Massey and Whitson first unit. It is our hope that some of this information may be supplied in the future by natural area workers as a result of observations made during the conduct of normal activities. A close evaluation of their second unit will reveal the exchangeable information and will underscore the dual value it can possess. A sound research proposal that utilizes the information presented in the two units, but directs it toward a species biology study, is presented in Part IV, Section F. The preservation of special species will frequently require detailed information of this genre to establish sound management programs.

The natural area reports in Part II are included to show the use of the system in ecological diversity inventory and in analysis of a variety of sites. The diversity presented covers terrestrial, wetland, and aquatic ecosystems; successional and climax communities; and endangered, threatened, endemic, disjunct, and restricted species. The communities range from upland forests and savannahs to salt marshes, grass dunes, and submerged aquatic beds. The natural areas encompass the regionally to locally significant. The ecological diversity supporting data for the reports extend from ecosystematics class data through undergraduate honors research to several doctoral programs and from a limited field data base to wide literature documentation. Even though the reports are based on natural areas in North Carolina, the diversity covered exhibits the potential for the use of the system in the remainder of the nation.

The treatments in Part III are presented to make the text more useful to academic and heritage workers. A specialized terminology has evolved in relation to conservation and heritage work, so the editors thought a glossary of these words and phrases in Section A would be timely and pertinent. State heritage programs are introduced in Section B. Registries for natural areas are now being developed or recently have been developed, and an example in Section C is included for general information. More heritage resource information is given in Sections D–H.

The notes and worksheets in Part IV, A–E, are included as suggestions for efficient and cost-effective inventory. A new classification for archeology is presented in Section E. A sound research proposal on species biology in Section F incorporates the use of the classification and information systems described in this book.

New initiatives have been stimulated, ideas promoted, concepts matured, inventory procedures introduced, systems presentations modified, and criteria selected for the determination of significance as a result of the senior editor's personal experience with the natural heritage programs of The Nature Conservancy and of the state of North Carolina, with the natural landmarks program of the National Park Service (now with the Heritage Conservation and Recreation Service), in the application of the system and inventory procedures to a study of the Dan River Basin for the U.S. Army Corps of Engineers, in natural areas program advising for the North Carolina Department of Natural Resources and Community Development, in the development of the ecological diversity classification for the Heritage Conservation and Recreation Service, in the publications on threatened and endangered species for the North Carolina State Museum of Natural History and the New York Botanical Garden, in the definition and circumscription of wetlands for the Environmental Protection Agency and the Environmental Defense Fund, in the review and discussions of the classifications for wetlands of the Fish and Wildlife Service, and in the great conservation effort of the North Carolina Nature Conservancy. These experiences have been of inestimable value in the evolution of this manuscript.

The authors hope that the natural heritage classification and information systems presented in this publication (listed below) will be of benefit to individuals, agencies, and organizations that have a desire to understand ecological diversity, an interest in habitat, and a fundamental concern for the conservation of our natural heritage.

Classification Systems
1. Ecological diversity
2. Natural heritage inventory
3. Criteria for determination of significance
4. Natural areas, management, and protection priorities
5. Rare species
6. Archeology

Information Systems
1. Ecological characterization
2. Diversity summary
3. Lowest common denominator element file system (an introduction only)
4. Species general information
5. Population status summary
6. Documentation system

Albert E. Radford

In 1789 Baron Alexander von Humboldt stated: "There is a harmony in nature, an invisible harmony. All natural objects are interrelated. We cannot study rocks by themselves. . . . Find a certain type of soil and a certain type of plant and you will find a certain type of rock. And it is the same with plants. . . . And as for human beings: Are we not related to—perhaps it is better to say influenced by—our natural surroundings?" [In Buol et al., 1980]

THE FUTURE OF MANKIND IS DEPENDENT

upon

CONSERVATION

of

GENETIC RESOURCES

in

DYNAMIC NATURAL ECOLOGICAL SETTINGS

"CONGENDYNECS"

Acknowledgments

We are indebted to our colleagues, Jimmy R. Massey and Paul D. Whitson, for their observations and suggestions on the many versions of the manuscript. The editors are particularly grateful to Julie A. Smith for her professional excellence in the editing and typing of the manuscript, to Laurie S. Radford for her patience and thoroughness in an editorial review of Part I and the discussions in Part II, and to Karen H. Hildebrandt and Linda Naylor for their indispensable aid in the typing of the manuscript. We are grateful to Marion S. Seiler for her design of the biosphere logo and to Betsy Birkner and Susan Sizemore for their preparation of the diagrams and figures in the text. The senior editor and author will always be obligated to his students in the ecosystematics classes for their helpful comments, criticism, corrections, and patience in the development of the many phases of the ecological diversity classification and inventory procedures.

We are especially indebted to the following organizations and agencies: The Nature Conservancy (TNC) for permission to use information on its state heritage programs and lowest common denominator element files; the North Carolina Nature Conservancy (NCNC) for support in the preparation of natural area reports in North Carolina; the Heritage Conservation and Recreation Service (HCRS) for preparation of the natural heritage glossary; the North Carolina State Museum of Natural History (NCMNH) for permission to extract information from "Vascular Plants" (Hardin and Committee, 1977); the North Carolina Natural Heritage Program (NCNHP) for permission to use inventory worksheets and reports and the "Registry of Natural Areas"; the Corps of Engineers of the U.S. Army (CEUSA) for two field data reconnaissance reports; The Highlands Biological Station for its support of several graduate students during the summers of 1974 and 1975 in a natural areas study of the southern Appalachians in which earlier versions of the diversity classification were applied and tested; the Illinois Department of Conservation (Ill.) for permission to print its appendices, as modified by John White, on boundaries, grading of natural quality, and procedures for detecting disturbances in natural areas; the New York Botanical Garden for permission to use Tables 17, 18, 19, 22, 23 from Geographical Data Organization for Rare Plant Conservation; and the New England Botanical Club for the use of Tables 16, 22, 23, which appeared in Rhodora 82.

Furthermore, we appreciate the reviews of the manuscript by John White (a contributor), Timothy Nifong, Timothy Atkinson, Loyal Mehrhoff III, and C. Ritchie Bell of the University of North Carolina at Chapel Hill, Terry Huffman of the U.S. Army Engineers Waterways Experiment Station at Vicksburg, Julie Moore of the North Carolina Natural Heritage Program, Robert E. Jenkins of The Nature Conservancy, Alan S. Weakley of the Coastal Zone Resources Division at Wilmington, and John Taggart of the Division of State Parks in the North Carolina Department of Natural Resources and Community Development.

The authors are indebted to the George R. Cooley Fund and the Department of Botany of the University of North Carolina at Chapel Hill for absorbing all preparation costs for manuscript copy—typing, preparation of figures, indexing, and photocopying. We are particularly grateful to George R. Cooley for publication support. All royalties from this publication will be deposited in the George R. Cooley Fund for the support of heritage work and taxonomic research.

Contributors

The editors have tried to integrate and coordinate all of the presentations in the text. The ultimate responsibility for content and accuracy, however, belongs to each author. The contributions are as follows by part, chapter, report, or section as appropriate. Reports excerpted from previous publications or prepared by an agency are indicated by the acronym (see acknowledgments) of the agency granting permission for use.

Prologue. Radford

I. 1. Radford, D. Otte, L. Otte
2. Radford, D. Otte, L. Otte
3. Radford, D. Otte, L. Otte
4. Massey, Whitson

II. A. L. Otte
B. Weakley
C. Taggart
D. L. Otte
E. Lynch
F. Weakley, Dickerson
G. Weakley, Dickerson

III. A. HCRS
B. TNC
C. NCNHP
D. TNC
E. NCMNH
F. White
G. White
H. White

IV. Introduction. Radford, D. Otte, L. Otte
A. Radford, D. Otte, L. Otte
B. Radford, D. Otte, L. Otte, NCNHP
C. Massey, Whitson
D. Radford
E. Schneider, Dittmar
F. Henifin

Epilogue. Radford

Bibliography. D. Otte

Albert E. Radford
Editor

Deborah Kay Strady Otte
Associate Editor

Lee J. Otte
Associate Editor

Chapel Hill, North Carolina
31 May 1980

Part I

Ecological Diversity Classification, Inventory, and Information

Part I deals with ecological diversity classification, inventory, and information. These classification and information systems have been designed as a basis for inventory use on the local to national levels; for habitat analysis and ecological characterization of the individual organism, population, community, and ecosystem; for identification of significant species, communities, habitats, and abiotic features; and for the provision of information basic to decisions on land classification, habitat management, natural area acquisition, and resource use and protection.

1

Ecological Diversity Classification

Ecological diversity is the foundation of our natural heritage. A holistic, comprehensive classification of the diversity features in our environment is essential to the basic inventory of species, communities, and habitats in that heritage. A system of classification of ecological diversity is a fundamental requirement for consistent and comparable descriptions of natural areas and sites. Classification is basic to the identification of elements of diversity in the landscape and is essential to habitat analysis and synthesis for populations and communities. A comprehensive ecological diversity classification system is necessary for perspective in ecosystem analysis and categorization. Such a classification is also fundamental to studies in ecology, taxonomy, evolution, and resource management.

Any conservation effort must be based upon a thorough survey of biotic and abiotic features. All types of communities, from the pioneer to the climax, developed during time over different rock and water types, should be included in the representative site samples of an area. The successional communities, the topoedaphic climaxes, and the continua should be part of the master ecosystem study. Biogenesis must be integrated with pedogenesis in explaining the present and past development of species and communities; climatogenesis and phylogenesis must be coupled with succession and soil formation to explain the present composition and distribution of biotic assemblages. We must try to conserve the total diversity of species in as broad a range of habitats as possible within the different climates in each province in order to understand the origin, migration, and evolution of species, floras, and faunas as well as the productivity and composition of present communities. Sound conservation depends upon a comprehensive classification of ecological diversity.

The natural diversity for any province or area includes: (a) vegetation (with animal dependents and/or communities), (b) climate, (c) soils, (d) geology, (e) hydrology, and (f) topography. All of these components of the habitat are interacting but conceptually independent systems that compose the ecosystems, biomes, and natural areas. Vegetation (composition, distribution, and development) is dependent upon climate, soils, geology, and topography acting through time. Climate (microclimate) is dependent upon vegetation, soils, geology, and topography. Soil (composition, distribution, and development) is dependent upon vegetation, climate, geology, and topography acting through time. Geology (structures and formations as well as sedimentary rocks) is dependent upon vegetation, climate, soils, topography, and time. Topography (landforms and features, structures, and development) is dependent upon vegetation, climate, soils, geology, and time. These interrelated components are related to the various habitat types and must be major parts of any classification scheme for ecological diversity.

Who knows which component of our habitat is more significant than another in the maintenance of species diversity or in the origin, migration, and evolution of species, faunas, floras, and communities? Who understands the role of species diversity in maintaining habitat diversity? Who knows which elements of the biotic or abiotic habitat will be of great natural resource value to man? These fundamental questions can be answered only if we preserve total species/habitat diversity in carefully selected natural areas. Identification of that total species/habitat diversity depends upon the development of the best possible classification of that diversity.

Many systems of classification have been devised for components and elements of diversity. The reasons for development of those systems, however, have been mapping, productivity, cover, and the like, rather than diversity specifically. No commonly used system has as its goal the delineation and delimitation of ecological diversity. Most systems are hybrids, combining various factors inconsistently. These earlier systems, hybrid or otherwise, can be used in part or whole to construct a comprehensive classification system for natural diversity.

The purpose of classification is to arrange elements, components, objects, or taxa in a way that gives the greatest possible command of knowledge, makes the most efficient and effective use of information, and leads most directly to the acquisition of more data, information, and knowledge.

We believe that the hierarchical component approach, as we have developed it, has enabled us to accomplish most effectively the goals for the classification of natural diversity (see Section I, the Biosphere Figure on the cover, and Table 1). We adopted this approach for the following reasons:

1. The establishment of separate ecological component classes permits the description of natural areas in terms of independently variable factors. This will enhance the recognition of varying ecological character combinations that leads to identification of maximum ecological diversity.

2. The component approach to diversity classification can be used easily as a basis for consistent, comprehensive, and comparable habitat analysis for populations, communities, and ecosystems.

3. The hierarchical component approach enables the developers of this system to utilize the professional expertise from each of the respective disciplines and to take the greatest advantage of the diversity knowledge in those fields gained through years of experience.

4. An inventory of ecological diversity within a state or region can be evaluated effectively for thoroughness using the component classification system, for example, in checking the inventory of pioneer, transient, and climax communities in relation to each soil series, rock type, landform feature, and others within each climatic regime in a physiographic province.

5. Many systems of classification presently in use can be included easily in this system: Society of American Foresters' cover types and the Braun-Blanquet system (slightly modified) at the type level in the biological component, the Köppen climatic classification at the system level in the climate hierarchy, the Soil Conservation Service soil classification at all levels, traditional rock classification at various levels in geology, the wetlands classification at several levels in the hydrology hierarchy, and the Heritage Conservation and Recreation Service geological classification at several levels in the topographic component.

6. In an open-ended system, analysis can be done as thoroughly as the time and experience of the investigator will permit. Appropriate data or information can be added efficiently whenever available and changed or deleted at the proper hierarchical level for each component when necessary.

I. Objectives, Principles, and Use of the System

A. Objectives of a Natural Diversity Classification System

At present no means exists for the identification of the full array of diversity in our natural heritage. Nearly all classification systems dealing with the subject invariably tend to homogenize diversity, usually inconsistently mixing distinctly different factors, such as vegetation, climate, or soils. The resulting composite classes inherently negate the recognition of maximum diversity.

Our system has internal consistency and comparability and the potential for effective and efficient compatibility with world, national, regional, and state classifications used in the inventory of natural elements. This system provides a classification that is holistic and comprehensive. Our primary objective has been to develop a system of classification that will (1) form the framework for a comprehensive survey of ecological diversity; (2) provide the classification base for an inventory of biotic, pedologic, geologic, topographic, and hydrologic features as well as natural areas and sites; and (3) serve as the basis for inventory and identification of all types of successional communities, all types of mono- and topoedaphic climaxes, and all typical as well as rare, endangered, threatened, relict, or disjunct species, communities, and ecosystems.

B. Guiding Principles for the Development of a Natural Diversity Classification System

The principles used in the development of the classification system are included to help the user understand the conceptual basis for the system and the application of the classification in inventory and to provide herewith a device for evaluating the effectiveness of the system. These guiding principles are:

1. That the system be comprehensive enough to encompass all elements of ecological diversity.

2. That a hierarchical component classification system is most useful for efficient handling of information.

3. That the classification system have precisely defined classes and elements so that information will be treated consistently and comparably.

4. That the existing classification systems of various scientific disciplines be utilized as much as possible.

5. That the classification system be compatible with other pertinent systems as much as practical.

6. That the fundamentals of classification be used as rigorously as possible.

7. That the system be open ended and capable of being expanded and changed.

8. That each species is selectively and uniquely adapted to its habitat.

9. That species diversity is related to habitat diversity within an area.

10. That habitat diversity is related to the diversity of climate, soils, geology, hydrology, topography, and biology within an area.

11. That species assemblages are recurring combinations under similar habitat conditions within an area at a given time.

12. That species assemblages are the result of the interaction of species and habitat diversity in an area through time.

C. Inventory Use of a Natural Diversity Classification System

An inventory of the species, communities, and habitat diversity (natural diversity) in natural areas and sites—using a standardized classification system—will provide the baseline data and fundamental information for the following efforts:

1. Conservation of species, community, and habitat diversity.

2. Ecological characterization of species, communities, and habitats.

3. Formulation of a predictive system for occurrence and distribution of species, communities, and habitats (what is where under which conditions).

4. Development of perspective in species biology studies (endangered and threatened), community analyses, and habitat significance.

5. Interpretation of the origin, migration, and evolution of species, floras, faunas, and communities.

6. Foundation for research in applied and advanced problems in many fields of endeavor (hydrology, pedology, habitat cover, food productivity, community ecology, and others).

7. Stimulation of production of integrative classifications of many kinds (such as map systems, habitat productivity, trout stream catch).

8. Decisions on land use, impact evaluation, and management problems of many types.

9. Establishment of priorities through environmental analysis for formulation of land use policies, land classification, and management programs.

10. Establishment of priorities through natural area basic inventory analyses for natural area acquisition and resource protection.

D. General Comments on Development of the System

The development of this system of diversity classification is based on inventory experience gained during the past six years in studying some 200 natural areas from Florida to New Jersey. These sites represent most of the plant biotic systems; the estuarine, lacustrine, riverine, and palustrine wetland systems; the Alfisols, Entisols, Histosols, Inceptisols, Spodosols, and Ultisols among the soil orders; the basic, acid, calcareous, ferruginous, carbonaceous, salinaceous, and siliceous rock classes; and a wide variety of climatic, topographic, and hydrologic features in at least eight distinct physiographic regions. Most of the data for inventory reports have been collected by members of plant ecosystematics classes at the University of North Carolina at Chapel Hill and by natural area research associates sponsored by The Highlands Biological Station.

For many ideas and suggestions we are indebted to a panel that was convened in Washington, D.C., 20–24 March 1978, under the auspices of the Heritage Conservation and Recreation Service of the U.S. Department of the Interior for the development of a national classification system for ecological diversity. Many parts of the draft report made by the panel have been incorporated into this presentation, directly or indirectly.

Pertinent documentation for the system is contained in the bibliography at the end of the book. Some of the background resources used in the development of this classification system are:

Vegetation: Fosberg, Küchler, Dansereau physiognomic systems; Braun-Blanquet floristic approach; Mueller-Dombois and Ellenberg physiognomic-ecological classification; Raunkiaer plant life forms; Clements monoclimax; Society of American Foresters' cover types; Daubenmire forest vegetation; Whittaker botanical review

National Park Service National Natural Landmark

Reports and Theme Studies: Southern California; Blue Ridge Region; Piedmont Region; Coastal Plain Region

State Nature Preserves Classification and Habitat Type Inventories: California; Illinois; Wisconsin; West Virginia; North Carolina; Tennessee

Component Classifications: Köppen climate; Soil Conservation Service soil taxonomy; Fish and Wildlife Service wetlands classification (Cowardin et al., 1977); Fennemen physiography; Heritage Conservation and Recreation Service geology panel classification system

Combination Classifications: Bailey ecoregions; *The National Atlas* (U.S. Department of the Interior, 1970); Otte systems comparison; Radford natural diversity.

Much work remains to be done on the system from a classification standpoint, particularly in the development of a glossary for the diversity elements. There is an absolute necessity for professional input into the component classifications. A great need exists for the incorporation of field experience at the state and local levels for the lower orders of the system, particularly the biological.

II. Organization of the System

The following are the major components and hierarchical levels in the Ecological Diversity Classification System. The components are indicated by a roman numeral and the levels by one or two capital letters. (See Table 1.)

Components	*Hierarchical Levels*
I. Biology	A. System
II. Climate	AA. Subsystem
III. Soils	B. Class
IV. Geology	BB. Subclass
V. Hydrology	C. Generitype
VI. Topography	CC. Type
VII. Physiography	D. Population

The element entries (see Part III B, D) for the levels of each diversity component are indicated by an arabic numeral, for example, I.A.*6* forb system; III.AA.*17* Andepts; IV.B.*8* ferruginous; V.BB.*7* semipermanently flooded. The "forb system" is the element entry at the System level of the biological component; "Andepts" is the element entry at the Subsystem level in soils; "ferruginous" is the element entry at the class level in the geologic component; "semipermanently flooded" is the element entry at the Subclass level in the hydrologic component. (See Tables 2–8.)

With this three-part code (roman numeral, letter, arabic numeral), or any similar code, it is possible to encode, store, and retrieve any element of diversity in the classification system. (See Tables 2–8 in Chapter 2 for summary of element entries for each hierarchical level for each component.)

The ecological components used in this classification system are considered traditional with the exception of physiography and topography. The inclusion of physiography as a major theme and the separation of topography from geology in the system require explanation.

Physiography is a regional approach to the study of natural areas and the environment, an approach basic to an understanding of the origin, migration, and evolution of species, floras, faunas, communities, and ecosystems. Application of the classification of the biological through the topographic (I–VI) components to inventory of a site results in an ecological characterization of the natural area, communities, and populations. A physiographic characterization gives a spatial and temporal description of the site. Spatial systems, exemplified by drainages and watersheds, mountain chains and ranges, enable investigators to determine possible migration routes for plants and animals; temporal systems, such as geological formations by era or period and successional schemes, help in deciphering the history of an area and in predicting its future. Both systems are required for an understanding of distributional and phylogenetic relationships of species and the present and future composition and habitat of communities. Physiography thus helps give perspective to the significance of natural areas, communities, and populations. Physiography as an integral component of the system also facilitates the search for natural areas and the geographic location of the sites.

The physiographic characterization of sites and natural areas provides the spatial and temporal references necessary for comparative studies of communities and ecosystems. The physiographic component classification represents a standardized approach to regionalization of spatial and temporal (historical) relationships of biotic assemblages, species, and habitats.

Gary et al. (1972, p. 745) give the following definitions for **topography**: (a) "the general configuration of a land surface or any part of the earth's surface, including its relief and the position of its natural and man-made features"; (b) "the nature or physical surface features of a region, considered collectively as to form; the features

Table 1. *Ecological Diversity Classification System*

CODE	HIERARCHY	COMPONENT	INCLUSION
		I. BIOLOGY	
I.A.	Biotic System		Biotic System
I.AA.	Biotic Subsystem		Biotic Subsystem
I.B.	Biotic Class		Community Cover Class
I.BB.	Biotic Subclass		Community Class
I.C.	Biotic Generitype		Community Cover Type
I.CC.	Biotic Type		Community Type
		II. CLIMATE	
II.A.	Climatic System		Climatic Regime
II.AA.	Climatic Subsystem		Climatic Subregime
II.B.	Climatic Class		Sectional Climate
II.BB.	Climatic Subclass		Local Climate
II.C.	Climatic Generitype		Natural Area Climatic Site Type
II.CC.	Climatic Type		Community Climatic Site Type
		III. SOILS	
III.A.	Soil System		Soil Order
III.AA.	Soil Subsystem		Soil Suborder
III.B.	Soil Class		Soil Great Group
III.BB.	Soil Subclass		Soil Subgroup
III.C.	Soil Generitype		Soil Family
III.CC.	Soil Type		Soil Series
		IV. GEOLOGY	
IV.A.	Geologic System		Rock System
IV.AA.	Geologic Subsystem		Rock Subsystem
IV.B.	Geologic Class		Rock-Sediment Chemistry
IV.BB.	Geologic Subclass		Rock-Sediment Occurrence
IV.C.	Geologic Generitype		Natural Area Geologic Site Type
IV.CC.	Geologic Type		Community Geologic Site Type
		V. HYDROLOGY	
V.A.	Hydrologic System		Hydrologic System
V.AA.	Hydrologic Subsystem		Hydrologic Subsystem
V.B.	Hydrologic Class		Water Chemistry
V.BB.	Hydrologic Subclass		Water Regime
V.C.	Hydrologic Generitype		Natural Area Hydrologic Site Type
V.CC.	Hydrologic Type		Community Hydrologic Site Type
		VI. TOPOGRAPHY	
VI.A.	Topographic System		Topographic System
VI.AA.	Topographic Subsystem		Topographic Subsystem
VI.B.	Topographic Class		Landscape
VI.BB.	Topographic Subclass		Landform
VI.C.	Topographic Generitype		Natural Area Topographic Site Type
VI.CC.	Topographic Type		Community Topographic Site Type
		VII. PHYSIOGRAPHY	
VII.A.	Physiographic System		Physiographic Region
VII.AA.	Physiographic Subsystem		Physiographic Province
VII.B.	Physiographic Class		Physiographic Section
VII.BB.	Physiographic Subclass		Local Landform
VII.C.	Physiographic Generitype		Natural Area Physiographic Site Type
VII.CC.	Physiographic Type		Community Physiographic Site Type
D.	Population level inventory for endangered, threatened, rare species. I.D. Biotic Associates, II.D. Population Climatic Site Type, III.D. Population Soil Site Type, IV.D. Population Geologic Site Type, V.D. Population Hydrologic Site Type, VI.D. Population Topographic Site Type, VII.D. Population Physiographic Site Type.		

revealed by the contour lines of a map"; (c) "the art or practice of accurately and graphically delineating in detail, as on a map or chart or by a model, selected natural and man-made surface features of a region." We define **geology** as the study of the local origin, chemical and mineral composition, classification, structure, and outcrop pattern of rocks, instead of the broad general definition usually employed—"the scientific study of the origin, history, and structure of the earth" (Morris, 1969).

We consider topography as a separate theme and the coequal of geology, rather than as simply a subdivision, as far as ecological influence is concerned. For example, topographic slope position, aspect, shelter, angle, profile, and surface patterns are critical in the distribution of populations, species, and communities within an area. Most of these slope characteristics have been traditional parts of habitat definitions for many years. Also, from a classification standpoint, it is much easier to deal with geology and topography as separate entities than it is to combine the two into one six-level scheme. Contrarily, we now regard hydrography (see the logo on the text cover) as a component of topography.

The decision to use six basic hierarchical levels was made because two very well worked-out six-level systems already exist: (1) plant and animal classification —division, class, order, family, genus, and species; and (2) soil taxonomy—order, suborder, great group, subgroup, family, and series. Use of six levels throughout the seven major themes was desired as a mnemonic device and for consistency and comparability. The six hierarchical levels become more analytical and technical with descent from the highest (A.) to the lowest (CC.) order.

Three of our levels—A., B., and C.—are quite distinctive, the level falling under each of these—AA., BB., and CC., respectively—being simply a more technical description of A., B., or C. For example, Topographic System (VI.A.) is a broad term describing regional topography (for example, plains), and Topographic Subsystem (VI.AA.) is a more detailed description of VI.A. (for example, flat plains or irregular plains with slight relief). Likewise, the Community Cover Type (I.C.) gives the dominant species of only the top layer of the community, whereas the Community Type (I.CC.) more specifically names all the layers of the community that have a dominant species. A conscious effort was made to follow the above pattern, but the goal was often not attainable in some of the seven themes, such as hydrology. Hopefully, further input from users of this system will aid in the solution of problems of hierarchical arrangement.

A seventh level (D.) is included as an optional level so that populations might be described in the system. The D. level is essentially an application of the CC. level to population rather than community, an application particularly useful in threatened and endangered species studies.

The Diversity Classification System is summarized in Table 1 and explained in detail in Chapter 2. This classification provides perspective, as well as position and rank, for the hierarchical elements in each component.

III. Major Ecosystems in the Biosphere: Aquatic, Wetland, Terrestrial

A fundamental assumption underlying the Ecological Diversity Classification System is that basically three main types of habitats and ecosystems exist in the biosphere: aquatic, wetland, and terrestrial. **Habitat** here includes all abiotic components that provide the environment for the biotic; **ecosystem** consists of all components, the biotic as well as abiotic. This fundamental assumption constitutes a major concept that is worthy of considerable attention and detailed explanation because it is (1) essential to understanding natural diversity; (2) fundamental to national inventory of our natural heritage; and (3) consequential to legal interpretation of the wetland laws with the potential for tremendous environmental and economic impact upon an area.

A continuum exists in habitats from the one extreme of terrestrial to the other extreme of aquatic, with wetlands lying in between and forming a transition zone whose boundaries often are difficult to draw. Problems arise in defining these three habitats because wetland habitats have not been sharply distinguished from terrestrial and aquatic. Sculthorpe (1967, p. 3) addresses this problem:

> In most climates there is a seasonal fluctuation of the water table. Habitats with standing water for most of the year may dry out completely in the summer whilst normally terrestrial soils may be flooded during a rainy season. At no time is there an abrupt change from land

to water, but rather a gradual transition from dry through waterlogged to submerged soils. The reversion of vascular plants to aquatic life has involved colonization of all these transitional habitats as well as the water itself, and some of the marginal sites that are periodically flooded have come to possess their own distinctive plant associations.

Because the wetland definition is the crux of the problem, most of the following discussion pertains to wetlands. An effort is made here to distinguish clearly the wetland habitat from the terrestrial and aquatic.

A. Wetland Habitat

Our concepts of wetland habitats basically follow those of Cowardin et al. (1976, 1977), from whom most of the following guidelines, definitions, and limits are taken. Cowardin et al. (1977) define **wetland** in a broad general sense as land where water is the dominant factor determining the nature of soil development and the types of plant and animal communities living in the soil and on its surface. They further characterize the concept of wetland with the following features: (1) elevation of water table with respect to the ground surface; (2) the duration of surface water; (3) the soil types that form under permanently or temporarily saturated conditions; (4) the various kinds of organisms that have become adapted to life in a "wet" environment; and (5) the presence of more soil moisture than is necessary to support the growth of most plants and, in fact, often causes severe physiological problems for most plants except hydrophytes, which are adapted for life in water or in periodically to permanently saturated soil.

More specifically, Cowardin et al. (1977) delimit two divisions within the general category of wetland: wetland and aquatic (deep-water) habitats. The latter type is discussed in Section B of this chapter. They define **wetland** as follows (Cowardin et al., 1977, pp. 4–5):

> Wetland is defined as land where the water table is at, near or above the land surface long enough to promote the formation of hydric soils or to support the growth of hydrophytes. In certain types of wetlands, vegetation is lacking and soils are poorly developed or absent as a result of frequent and drastic fluctuations of surface-water levels, wave action, water flow, turbidity or high concentrations of salts or other substances in the water or substrate. Such wetlands can be recognized by the presence of surface water or saturated substrate at some time during each year and their location within, or adjacent to, vegetated uplands or deep-water habitats.

The wetland habitat circumscription should include the biologic, pedologic, geologic, hydrologic, and topographic factors because the ecological diversity in wetlands directly relates to these components. We feel the Cowardin et al. (1977) characterization of wetlands needs amplification, and we have developed the wetland concept and definition in relation to each of the aforementioned major components.

Vegetation. Adaptations of vegetation to life in soil that is saturated or periodically inundated aid in delimiting wetlands. Radford (1979, pp. 3–4) discusses these adaptations in the following paragraphs.

> The general life cycle of plants has to be considered for the determination of adaptations to the wetland habitat. The phases of the life cycle are: "(1) reproduction—production of propagules; (2) dispersion—dissemination and distribution of propagules; (3) establishment—origination and settlement of a new, non-selfsufficient individual from a propagule; and (4) maintenance—sustention and interactions of a selfsufficient individual" (Massey & Whitson, 1978).
>
> Adaptive structures for the maintenance of plant species occurring in wetlands in intermittently exposed to semipermanently flooded (saturated) areas are pneumatophores (Bald Cypress), adventitious roots (Water Ash), buttressed or swollen stem bases (Water Tupelo), large lenticel areas on stems or surficial roots (Water Locust), short unbranched roots with long sheathing root caps (Swamp Privet), and aerenchyma and lacunar tissue (Water Hyacinth). Adaptive physiologies for the establishment of plant species in intermittently exposed areas are nondormant seeds, rapid germination and growth, e.g., Sycamore, Cottonwood, and River Birch on sand bars or spits at low water, and shallow root systems of the species, e.g., Green Ash and Overcup Oak, in seasonally flooded areas. An example of adaptive dispersion would be floating fruits, e.g.,Water Tupelo. Filamentous pollen is one reproductive adaptation in some of the aquatics occasionally found in a wetland situation, e.g., Pondweeds.
>
> Most species in saturated soils which are temporarily, briefly, occasionally, or rarely inundated do not have readily observable adaptations. The best indication of their adaptation is their abundance and dominance in a wetland habitat, a situation in which they are competitive in a natural environment. Many of these species, however, establish and maintain themselves in disturbed habitats; many become weeds in

> some upland fields and drained, disturbed bottomlands. These species remain best adapted, however, for competition in their natural wetland habitat.

The concept of wetland species must be expanded to include those woody and herbaceous species that may or may not normally reproduce during immersion but can tolerate, and often even are adapted to, periods of partial or total immersion or saturated soils. Some of these species may even occur in upland situations. Some examples include *Acer rubrum*, *Liquidambar styraciflua*, *Platanus occidentalis*, and *Quercus phellos*. These species might best be called hydro-mesophytic.

Perhaps of great assistance to us here is the concept of wetland based on vegetation and developed in a report by Tripp of the Environmental Defense Fund, Incorporated (1979, pp. 7–8). This report relies on a classification of bottomland hardwood species developed by Teskey and Hinckley (1977), after whom Tripp writes:

> *Very tolerant*. Trees which can withstand flooding for periods of two or more growing seasons. These species exhibit good adventitious or secondary root growth during this period.
>
> *Tolerant*. Trees which can withstand flooding for most of one growing season. Some new root development can be expected during this period.
>
> *Intermediately tolerant*. Species which are able to survive flooding for periods between one to three months during the growing season. The root system of these plants will produce few new roots or will be dormant during the flooded period.
>
> *Intolerant*. Species which cannot withstand flooding for short periods (1 month or less) during their growing season. The root systems die during this period.

Tripp (1979, pp. 8, 11) goes on to define wetland utilizing the above four classes:

> As a general rule, an area where the very tolerant species predominate should be classified as a wetland without the need for any detailed consideration of other system factors. On the other hand, an area in which the "intolerant" species predominate can presumptively be classified as a dry area or upland. Although the individual species listed as "tolerant" or "intermediately tolerant" are all well adapted to life in saturated soil conditions, some may be found, under varying soil and climatic conditions, in uplands. Thus, the presence of several trees of one such species in an area does not establish that the area is a wetland. However, larger plant communities or associations made up entirely or predominantly of such species would not be expected to be found in uplands; instead, such communities would be found in wet areas, which, due to soil, geological, flooding and precipitation factors, are periodically flooded or the soils of which are saturated or inundated during a significant portion of the year. Such plant communities are typically adapted to life in saturated soil conditions.
>
> Thus, our methodology adopts a powerful, although rebuttable, presumption that large areas in which tolerant and intermediately tolerant species predominate, with scattered or no intolerant species, are wetlands.

Therefore, a plant community may be considered wetland if the community as a whole has developed, regenerated, and maintained itself because of its ability to tolerate periodic inundation or soil saturation, even though some individuals may grow in upland situations.

The presence of species with visible structural adaptations facilitates the use of vegetation as an indicator of wetland. Species without easily recognizable adaptations present the problem. A prime example is those species that undergo shoot growth while the root zone of the soil is inundated or saturated. In these species anaerobic respiration occurs in the roots and indicates a physiological adaptation to the wetland habitat. These species typically inhabit wetlands flooded in the spring, the season during which shoot growth takes place. One can accomplish this type of biotic wetland determination either by checking flooding records or by examining the root zone of the soil—mottled soil suggests oxidizing/reducing conditions and consequently an anaerobic/aerobic root adaptation to wetlands by the species in question.

Soils. Soils in wetlands are those classified in the following suborders: Aquents, Aquepts, Aqualfs, Aquolls, Aquods, Aquults, Saprists, Fibrists, Hemists, and Fluvents. With the exception of some Fluvents, all of these are saturated during most of the growing season. Most of these have an aquic moisture regime.

In conjunction with these factors, a characteristic of soils that must be considered is mottling. **Mottled** means marked with spots of contrasting colors (Soil Survey Staff, 1975). Generally these spots represent iron complexes that have been through several oxidation and reduction sequences because of alternating periods of inundation and exposure (Whelan, unpublished, 1979). Mottles of wetland soils have chroma values of two or less. Soils without mottles that have chroma values of one or less are considered hydric or wetland. **Chroma** is the relative purity, strength, or saturation of a color; it is directly related to the dominance of the determining wavelength of the light and inversely related to grayness (Thompson and Troeh, 1973; Steila, 1976).

Geology. Geomorphic-topographic wetland areas (as described below) have several geologic features in common in addition to their shape and relative positioning in the landscape.

1. These areas act as sites of geologically temporary depocenters for water- and, infrequently, wind-transported clastic or chemical sediments and/or water-accumulated organic sediments that formed in place or were transported over only short distances.

2. Generally water flows through these areas, either above or below ground. Because of their topographic configuration, these areas function as sites of (a) temporary decrease in flow rate, (b) temporary storage of surface or near surface waters, (c) discharge of ground water, or (d) a combination of the above.

3. No limitation is set on the physical, chemical, or organic origin of the depression or lowland.

4. Wetlands are such that four mechanisms exist that can destroy the environment by alteration into either a terrestrial system or an aqueous system (outlined by Hinds, 1943): (a) drainage of the depression or lowland; (b) filling the depression or building up the lowland with sediment accumulations; (c) evaporation of the water within the area to "dry up" the system (these first three methods result in a relative lowering of local base level and could result in a terrestrial system); and (d) total flooding of the system could raise the water-air contact above the earth-air contact to such a height that it becomes an aqueous system. Other processes can destroy or create wetlands over geologic time, such as large-scale lowering of land masses or rises in sea level; however, these long-term factors generally do not produce results noticeable in one's lifetime.

Hydrology. Flooding and drainage must be considered in delimiting a wetland. Usually areas are intermittently exposed to intermittently flooded (as defined by Cowardin et al., 1977) or have aquic moisture regimes (as defined by Soil Survey Staff, 1975). In addition, soils are very poorly drained to moderately poorly drained (as defined by Soil Survey Staff, 1951).

Topography. Most wetlands occur in depressions or lowlands. A **topographic depression** is a "low-lying area completely surrounded by higher ground and has no natural outlet for surface drainage" (Gary et al., 1972). The term **lowlands**, according to Gary et al. (1972, p. 419), has three different definitions, all of which apply to this situation:

1. Lowlands is "a general term for low-lying land or an extensive region of low-land [elevationally low when compared to the local base level], especially near the coast and including the extended plains or country lying not far above tide level."

2. "The low and relatively level ground of a region, in contrast to the adjacent, higher country."

3. "A low or level tract of land along a water course."

Wetland vegetation typically inhabits features known as marshes, bogs, and swamps, which are found on levees, sloughs, low ridges, swales, old and new channels, meanders, slip-off slopes, bars, flats, terraces, depressions on plateaus, hills, and flats over basic or calcareous rock.

Summary Statement. Wetlands represent a transition zone of intermediate moisture conditions between those typical of the terrestrial or upland and aquatic habitats. Wetlands occur in topographic depressions or lowlands that range from intermittently exposed to intermittently flooded and characteristically are saturated or inundated during the shoot-growing season. Wetlands have very poorly drained to moderately poorly drained soils with aquic moisture regimes and mottles with chroma values of two or less or no mottles and chroma values of one or less. Wetland vegetation is dominated by species with structural and/or physiological (anaerobic-aerobic) adaptations to the edaphic, hydrologic, and topographic features typical of that habitat.

Because all of the above factors contribute to the formation of a wetland, this system gives all of them equal consideration. One can utilize one (for example, vegetation) or a combination of these habitat clues to detect the presence of a wetland and then execute a comprehensive analysis to verify the existence of a true wetland.

B. Aquatic Habitat

Aquatic (deep-water) habitats are "permanently flooded lands lying below the deep-water boundary of wetland" (Cowardin et al., 1977). In this aquatic situation the substrates are considered "not-soil" because the water is too deep to support emergent vegetation (Soil Survey Staff, 1975). The vegetation is submerged or free-floating.

Boundaries between wetland and aquatic habitats in the five hydrologic systems are as follows (Cowardin et al., 1977, p. 6):

1. Marine—"coincides with the elevation of the extreme low water of spring tide (ELWS); permanently flooded areas are considered deep-water [aquatic] habitats."

2. Estuarine—same as that given for marine.

3. Riverine—"lies at a depth of 2 m (6.7 ft) below low water" (this limit was selected because it represents the maximum depth to which emergent plants normally grow—Sculthorpe, 1967); "however, if shrubs or trees grow beyond this depth at any time, their deep-water edge is the boundary."

4. Lacustrine—same as that given for riverine.

5. Palustrine—same as that given for riverine.

C. Terrestrial Habitat

Terrestrial habitats lie at the opposite end of the moisture gradient. In a terrestrial system the soil is fairly to highly permeable, thus allowing regular diffusion of water and also good aeration in the root zone of plants. A terrestrial system ranges from a situation in which the soil remains wet a substantial length of time but retains oxygenating conditions because of continual flow of water (for example, a seepage area), to an intermediate situation in which the soil retains a good deal of rainwater and yet absorbs it readily in deeper layers (for example, mesic soil of a cove forest), and finally to the most xeric situation in which water is removed from the soil very rapidly (for example, soils in the Carolina Sand Hills). A terrestrial habitat can support **mesophytic** (growing under medium moisture conditions) or **xerophytic** (growing under low moisture conditions) vegetation, but not **hydro-mesophytic** (tolerant of or adapted to periods of immersion or saturated soils) or **hydrophytic** (standing and growing in water or normally saturated soils) vegetation.

The limits of wetland (as distinguished from terrestrial) are outlined by Cowardin et al. (1977, pp. 5–6): "1) the boundary between land with predominantly hydrophytic cover and land with predominantly mesophytic or xerophytic cover; 2) the boundary between soil that is predominantly hydric and soil that is predominantly nonhydric; or in the case of wetlands without vegetation or soils; 3) the boundary between land that is flooded or saturated at some time during years of normal precipitation and land that is not."

A Classification Justification

The addition of the major habitat (terrestrial, wetland, aquatic) as a modifier to the Biotic System (I.A.) in the Ecological Diversity Classification System (see Chapter 2, end of section B) violates the principles of classification. Nevertheless, we feel strongly about its usage in this manner, because the major habitat (1) represents a summary characterization of the component habitat elements not available elsewhere in the system; (2) gives immediate basic perspective to any ecological characterization or habitat analysis; and (3) immediately and easily places the system and community under investigation into a major, widely applied, national inventory category.

IV. Habitat

Habitat is usually defined as the natural environment of a plant or animal, its place of abode, or the kind of place to which an organism is adapted for its life, growth, and reproduction. Traditionally habitat has also been used to indicate the home environment of biotic assemblages of different hierarchical levels. Confusion in the use of the term arises from the application of the same environmental element to the different hierarchical levels of biotic assemblages and different hierarchical levels of environmental elements to a single level of a biotic assemblage or community. The major problem, however, in habitat description and analysis is the lack of a circumscribed and classified terminology. In order to render habitat descriptions and analyses more meaningful, a conscious effort has been made in the development of the Ecological Diversity Classification System to make the hierarchical levels of the biotic and abiotic components comparable and to produce a reference base in which the terms are classified and circumscribed for the habitat elements.

We have tried to include the environmental factors that seem critical to the growth and reproduction of individual organisms, populations, and Community Types under field conditions. The CC. and D. levels in the classification system, for instance, represent an analysis of the habitat for critical factors at those levels. An application of the classification to a field population or Community Type would result in a comprehensive description and analysis of the habitat for those biotic assemblages.

The abiotic environmental factors used in the system at the Community Type (CC.) and Population (D.) levels are summarized below. Any species within a community usually will have the Community Type as its biotic environment; any population will have its biotic associates as its biotic environment. **Biotic associates** are those extrapopulational species that shade the population, those that release exudate, leachate, or litter on the population, those that have root competition, and/or those that have symbiotic relations with the population.

Climate

Annual sunshine (temperature)
Annual precipitation
Freeze-free period
Average mean maximum temperature of hottest month
Average mean minimum temperature of coldest month
Average mean precipitation of driest month
Average mean precipitation of wettest month
Wind velocity and direction
Thermal belt
Frost pocket
Air drainage pattern
Late freeze pattern

Geology

Rock-sediment type
Rock-sediment weathering type
Rock-sediment fabric
  Fracture-jointing pattern
  Dip
  Porosity-Permeability
Rock-sediment particle size of constituent minerals
Rock-sediment consolidation
Rock-sediment exposure type
Rock-sediment exposure cover

Soils

Particle size
Mineralogy
Moisture
Temperature
Chemistry
Organic matter
Structure
Depth (Root system)
Color

Hydrology

Chemistry
Drainage
Regime
Color
Turbidity
Depth
Temperature
Velocity
Oxygen saturation
Dissolved solids and gases
Pollutant concentration

Topography

Landform type
Slope shelter
Slope aspect
Slope angle
Slope profile
Slope surface pattern
Slope position

A standardized approach to habitat description and analysis of pioneer to climax species and Community Types will help us in the determination of critical habitat for threatened and endangered populations and will provide consistent, comprehensive, and comparable baseline data for advanced analytical and experimental studies on our rare and/or economically valuable biotic assemblages. Much work must be done in standardizing habitat descriptions and analyses, but we feel that the use of this diversity classification system and format represents a big step forward. (See the Ecological Characterization examples by component in Chapter 2 and the Community Diversity Summary Ecological Characterizations in Part II.)

V. Identification

A classification of natural diversity is absolutely basic to the identification of diversity elements. Identification of diversity elements within a natural area is fundamental to the inventory of the site. Inventory and identification are applications of a classification system. Identification enables us to retrieve from the diversity classification system the facts that are pertinent to an inventory of a natural area.

Identification is usually synonymous with recognition. Recognition of objects, taxa, elements, and so on can be made by smell, touch, hearing, taste, or sight. Identification of elements of natural diversity, however, must be made by comparing the element to be identified with named, defined, and circumscribed element entries arranged hierarchically according to an established system of classification. This is done for comparability and consistency of treatment.

The traditional methods of identification, according to Massey (Radford et al., 1974), for the elements of diversity within a natural area include (1) expert determination, (2) recognition, (3) comparison, and (4) the use of keys or similar devices. Identification of the elements of diversity in information inventory is usually based on recognition or the use of keys; in reconnaissance inventory, the methods used are comparison and the use of keys; and in basic inventory, it is primarily expert determination. The most reliable method of identification of the elements of diversity is expert determination. In general, for reliability of determination, identification of elements in basic inventories of natural areas or sites should be made by experienced botanists, zoologists, taxonomists, ecologists, geologists, pedologists, and other specialists. General resources used in identification, or as information sources of the elements of diversity, are presented in Part IV of this book.

The absence of identification systems and/or experts will necessitate many hours of tedious comparison of the elements of diversity with the descriptions in pertinent literature before a determination can be made. Much time can be saved in identification by using this classification system properly, with a little help from the literature or from professionally trained individuals. For example, ten soil orders are listed as element entries in the diversity classification at the system level. Four have never been reported in the state of North Carolina, nor would their occurrences be expected, based on their circumscriptions. An individual determining soil orders in a natural area in North Carolina would have to decide among six, not ten. By the same elimination process, many topographic features, rock types, soil series, and so on could be removed from consideration for identification.

We must be able to identify reliably the elements of diversity in our natural sites in order to know what we have preserved and what we need to conserve. Identification in conjunction with a comprehensive classification system for natural diversity is a basic process in the inventory of our natural areas. The intelligent application of the classification system can make identification more efficient and can lead to an excellent inventory of our natural heritage.

VI. Inventory

Inventory of a natural area is the cataloging of the elements of natural diversity on the site. The elements of diversity need to be identified according to a classification system for ecological diversity.

Many products are derived from on-site inventory based on a holistic, comprehensive classification of ecological diversity. Natural Area Diversity Summaries provide diversity data and information for data banks or manual files, which can be used in master inventory. The bibliographies associated with the discussions represent in-depth studies of the important aspects of a site. The elements of significance furnish the essential ingredients for decision making on conservation, environmental impact, property acquisition, management recommendations, and protection suggestions. Community Diversity Summaries provide the data and information for Ecological Characterizations. Integrated elements of the classification system can be used in the formulation of predictive models. Basic inventory is a fundamental step in analytical, experimental, and theoretical work on population, community, and ecosystem problems. This basic on-site inventory phase of research provides publishable products useful to ecologists, systematists, evolutionists, planners, developers, managers, classification experts, and conservationists.

Ultimately, a national classification for ecological diversity should include a complete inventory for all of the elements for each hierarchical level of each component. At present the soils classification is the most complete and the biological classification the most incomplete. In time, circumscriptions will be available for each hierarchical element, as in class, order, family, genus, and species levels now used in regional floras of vascular plants. We have attempted to make the element summaries (Tables 2–8) in this classification as complete as possible in the higher categories and have given references to elements in the lower levels where available. The Community Type in the biological component is the least known scientifically and also has the poorest representation in the literature. Species Population Level inventory and analysis is essentially a new field of endeavor, which is basic to an understanding of endangered and threatened species biology and is fundamental to monographic work and to other types of systematic studies.

State heritage organizations can use this classification to check the thoroughness of their diversity survey, the number and types of elements conserved, and the predicted occurrence of elements not previously reported. The Natural Area Diversity Summaries and Master Species Presence Lists are designed to expedite master inventory of natural areas, counties, physiographic regions, and states. (See Chapter 3 for a much more comprehensive treatment of inventory.)

VII. Maps and Mapping

Component maps are fundamental to inventory application of diversity classification. Topographic maps, for example, show the size, shape, and distribution of features of the earth's surface. These features are usually classified in three groups: relief, drainage or water, and culture. This topographic map information is basic to all field inventory. Geologic maps illustrate another use for component maps. Maps with formations and component rock types facilitate location of pioneer-transient vegetation on particular rock types, for example, dolomite, quartzite, and shale.

Even though mapping and map production were not included in the objectives for developing the ecological diversity classification, the two activities logically coincide with utilization of the system. By combining land cover vegetation and one or more pedologic, geologic, or hydrologic components (III.–V.) at the System (A.) to Class (B.) levels, "ecoclass" maps of great value to land classification and use can be developed from satellite data and aerial photography. Meaningful maps for land use can be produced by relating vegetation Community Cover Classes (I.B.) and Classes (I.BB.) with comparable soil, geologic, or hydrologic classes within a physiographic area and climatic regime.

Note, however, that maps are dated tools or guides that are scaled, generalized representations of groups of specific features such as soils, geologic formations, and vegetation as to position and size. Because nature is ever changing through time and maps are generalized representations that are accurate only as to position and size according to scale, the ground truth of specific features should be checked frequently.

VIII. Site-Species Relationships

Site-species relationships are broadly interpreted here as relationships of populations, species, and communities to the biotic and abiotic components of the site each occupies. The study of these relationships involves an understanding of the developmental principles for ecological diversity, a knowledge of general biotic-abiotic relationships, a familiarity with fundamental ecosystem and classification concepts, and the training and ability to pose the proper field questions and to perceive past, present, and future dynamics within a natural area. The use of a comprehensive system of classification for ecological diversity is a basic requirement for the determination of site-species relationships.

Three groups of questions are presented for comprehension of site-species relationships. The concept questions (1) are designed primarily for site conspectus, the basic questions (2) for understanding site-species relationships, and the aids questions (3) for perception of population-habitat relationships—a "conspective, perspective, inspective" series.

A master set of element files covering the categories, I. Biology through VIII. Natural Communities, is introduced as fundamental helps in the determination of site-species relationships. Prior use of the information in the files can make field study of these relationships most cost-effective and efficient. These files can be a major part of any natural heritage inventory program and a convenient system for determining the progress of heritage conservation.

This treatment of site-species relationships represents a logical approach to the study of the spatial, relational, and holistic aspects of the subject within a temporal framework. This effort presumably articulates the intuition and intelligence of knowledgeable naturalists of the past for the improvement of field training for students of the future.

A. Site and Natural Area

Site is the position or location of any natural area; a specific plot of ground on which biotic and/or abiotic elements occur within a population or community; a smaller area within a natural area on which a community type, rock type, soil series, or other abiotic feature is located. Site in this section and in all ecological characterizations is used as the area or plot of ground within a natural area on which a particular biotic and/or abiotic feature occurs. Significant species-site relationships exist when a population or species is located on a single lithology, a single soil series, a single landform feature, or a single hydrologic type within a natural area. Site in the sense of position or location is used frequently in the general discussion of natural areas.

Natural area, as defined by the Heritage Conservation and Recreation Service* (HCRS) of the U.S. Department of the Interior, is "an area of land or water that either retains or has reestablished its natural character and values, although it need not be completely undisturbed, which provides scientific, recreational, or inspirational benefits." **Research natural areas** (HCRS) represent a "system of natural areas administered by the Federal Committee on Ecological Reserves that seeks to establish a representative array of natural ecosystems and their inherent processes as baseline areas for education and scientific research purposes." A research natural area is a type of natural area that is a representative example of a natural ecosystem. Not all natural areas, however, are research natural areas, because of size limitations, recreational and/or inspirational use, or lack of a natural ecosystem.

B. Site-species Relationships and Ecological Characterizations

Site-species relationships (biotic, abiotic, spatial, temporal) are described in terms of ecological characteristics. An **ecological characterization** of a population, species, or community type at a locality is a description of each in relation to the biotic and abiotic components of the site (such as biology, climate, soils, geology, hydrology, topography, and physiography). The habitat and environmental relationships of the species to the site are also expressed in terms of ecological characteristics.

An Ecological Characterization of a Species†

Rhododendron catawbiense Michx. (Purple Laurel) at a site on Round Bald on Roan Mountain, in Tennessee.

Vegetationally: Ericalean/Polytrichalean terrestrial shrub system with a closed shrub layer of tall, rhizomatous, evergreen shrubs, an open herb layer of scattered, mixed herbs, and a closed moss layer composed of tall turfs of mosses with erect branches. *Climatically*: Boreal, microthermal climate, cooler and wetter than the sectional climate, which is moderately cold and moist yearly and which has warm and moderately wet summers, moderately cold and moderately wet winters, and a short freeze-free period. *Pedologically*: Typic Haplumbrept soil. *Geologically*: Strongly folded, regionally metamorphosed, intermediate to basic hornblende gneiss exposed as abundant, scattered, fresh to highly weathered, pebble- and cobble-sized fragments covering about 50% of the community surface. *Hydrologically*: Well-drained, permanently exposed, dry-xeric, terrestrial system, wetted by fresh rains, high elevation fogs, and downslope drainage. *Topographically*: Open, west-facing, gravelly and irregular, moderately steep, concave portion of an overall more gentle lower slope of a wind gap on an east-west-running ridge on the upper slopes and crest of a water-eroded mountain in a region of high mountains. *Temporally and Spatially*: A seral stage of a lithosere on Round Bald, underlain by Precambrian Roan Gneiss, on the upper slopes and crest of Roan Mountain in the Southern Section of the Blue Ridge Province of the Appalachian Highlands.

*See Part III, Section A.

†See the characterizations at the end of each section in Chapter 2 and at the end of each community description in each of the natural area reports in Part II; also see the basic inventory in Chapter 3 as an example of population-habitat relationships.

Some of the questions that may come to mind as one reads the Purple Laurel ecological characterization are listed here:

1. Are all of the characteristics used in the Purple Laurel description necessary?
2. Should some of the characteristics be eliminated? Others added? Which? Why?
3. Which of the descriptive elements are critical to the maintenance and establishment of the species?
4. Can the critical habitat of the Purple Laurel be determined by the habitat analysis represented in the ecological characterization?
5. Does the characterization follow a standard format?
6. Why should the characterization be based on a multicomponent classification system for ecological diversity?
7. Why should site-species relationships be determined anyway? What is the significance of the effort?

Some of the fundamental reasons for the determination of site-species relationships using a diversity classification system (1, 2) are indicated below:

1. An environmental or habitat analysis of a site(s) is necessary for the determination of component elements or combinations of elements that affect and/or control the establishment, maintenance, reproduction, and dispersion of the species of concern; the determination of critical habitat for the species; and the effective and efficient management of the species.
2. An analysis of the site-species relationships for several sites can lead to the identification of elements that can be used in predicting the occurrence or distribution of the species of concern and to the discovery of clues that can be of significance in interpreting the origin, migration, and evolution of populations, species, and communities.

Ecological characterizations should be written according to a fundamental set of procedures based on a stable classification system for ecological diversity. Descriptive consistency, comparability, and comprehensiveness in ecological characterization require the development, adoption, or adaptation of a multicomponent classification system for ecological diversity and the establishment of standard report procedures for accuracy, completeness, and thoroughness of treatment. Meaningful analyses that lead to the identification of elements used in the prediction of species distribution and to the discovery of clues to the origin, migration, and evolution of populations, species, and communities have to be based on consistent, comparable, and comprehensive data.

C. Principles and Relationships

The basic principles of diversity underlying the study of site-species relationships under natural competitive conditions are:

1. Each species is adapted selectively and uniquely to its habitat.
2. Species diversity is related to habitat diversity within an area.
3. Habitat diversity is related to the diversity of climate, soils, geology, hydrology, topography, and biology within an area.
4. Species assemblages are recurring combinations under similar habitat conditions within an area at a given time.
5. Species assemblages are the result of the interaction of species and habitat diversity in an area through time.

The general biotic-abiotic relationships fundamental to an understanding of site-species relationships are the interacting, interrelated, but conceptually independent systems operating through time:

1. $O_V = f\,(C, G, R, S)\ t$
2. $C = f\,(O_V, R, S, G)\ t$ (Micro)
3. $G = f\,(O_V, C, R, S)\ t$ (Structure & sedimentary formations)
4. $S = f\,(C, O_V, G, R)\ t$
5. $R = f\,(C, G, O_V, S)\ t$

O_V = Organisms (vegetation); C = Climate; G = Parent material (rock-sediment); R = Relief or topography; S = Soils; t = time; f = function.

D. Fundamental Classification and Ecosystem Concepts

Some basic concepts required for perception of site-species relationships and for an understanding of the dynamics of an ecosystem are defined in the following treatment.

Classification. An arrangement of objects, taxa, and entities into groups horizontally (position) and/or vertically (rank).

System. An ordered and comprehensive assemblage of facts, principles, or elements in a field of knowledge.

Component. An element in something larger; a major constituent of the habitat or ecosystem, as geology, biology, and climatology.

Element. A fundamental, ultimate part of any class of diversity. Element in an ecological diversity classification system represents one or more types of each component at each hierarchical level, for example: Lakeland soil series is an element of the soils component (III.) at the CC. level (Soil Series), with Gilead, Rains, and McColl series as examples of other elements in that component at the CC. level; Longleaf Pine and White Oak represent elements in the biology component (I.) at the C. level (Community Cover Type).

Hierarchy. Any system of things or persons ranked one above the other; in each component within the diversity classification system an ascending series of levels or ranks with the one above being inclusive of all below.

Ecological diversity classification system. A systematic hierarchical arrangement of the elements of ecological diversity horizontally into major components (position) and vertically into levels (rank).

Diversity Class. All of the elements in a particular component at a given rank, as the 10,555 soil series at the CC. level in the soils component (III.).

Habitat. That part of the environment in which biotic and abiotic relationships and exchanges occur between organisms and the resources they utilize and produce (Dansereau, 1957).

Landscape. A section or portion of land on the surface of the earth on which biotic and abiotic features may be seen from a single viewpoint.

Bioscape. An assemblage of biotic communities in a particular landscape.

Soilscape. An assemblage of soil bodies in a particular landscape.

Lithoscape. An assemblage of lithologies in a particular landscape.

Hydroscape. An assemblage of moisture classes (or drainages) in a particular landscape.

Toposcape. An assemblage of topographic features in a particular landscape.

Landscape-forming processes and agents (also see Chapter 2, Section VII). Change occurs at sites through erosion and deposition caused by animals, plants, man, wind, water, and meteorite impact; vent and fissure eruption, ash fall and flow and caldera formation resulting from igneous activity; folding, faulting, subsidence, and uplift in tectonic deformation; and solution, precipitation, organic accumulation, and weathering related to water activity. (Processes include biogenesis, pedogenesis, topogenesis, and so on.)

Natural community. A group of organisms that are related to each other and their environment. Natural community can be at any scale from biome to microcommunity. A combination of soil moisture, major habitat, topography, and others, such as loess hill prairie, low shrub bog, brackish marsh, dry barren, dry upland forest. (From White, 1978, p. 316.)

Pattern. Arrangement, design, configuration.

Catena. A chain or connected series, as of soils derived from the same parent material, which change with position on a slope; a toposequence of soils.

Continuum. A continuous extent, series, or whole with a set of elements such that between any two there is a third, as hydric to xeric with mesic in between; the occurrence of populations or communities of organisms along a gradient such as moisture, nutrient, and altitude.

Discrete (population or community). A body or entity possessed of distinct or definite identity or individuality.

Ecotone. The transition zone between two different plant communities.

Gradient. The rate of change with respect to distance of a variable quantity, such as nutrient and moisture; constant change of a variable quantity with respect to distance.

Mosaic. A combination of diverse or discordant elements, as a quartz vein in a peridotite intrusion (acid and ultrabasic elements, respectively) or seepage areas on a dry hillside (wet and dry elements, respectively).

Sequence. An orderly arrangement of objects, taxa, or entities in space or time.

Biosequence. A set of biotic communities with characteristic sequential differences resulting from dynamic biotic or soil differences through time; biochronosequence.

Climatosequence. A set of soils or biotic communities with characteristic sequential differences caused by differences in temperature and/or moisture.

Pedosequence. A set of soils with characteristic sequential differences caused by differences in topography.

Lithosequence. A set of soils or biotic communities with characteristic sequential differences caused by differences in lithology.

Hydrosequence. A set of soils or biotic communities with characteristic sequential differences caused by differences in moisture or drainage classes.

Toposequence. A set of soils or biotic communities with characteristic sequential differences caused by differences in topography.

Sere. A group of plant communities that successively occupy the same area from pioneer to climax (sensu Clements).

Hydrosere. A sere in which the pioneers invade water.

Halosere. A sere in which the pioneers invade salt water or salt flats.

Psammosere. A sere in which the pioneers invade sand.

Pelosere. A sere in which the pioneers invade clay or mud.

Lithosere. A sere in which the pioneers invade rock.

Subsere. A sere in which the pioneers invade an area disturbed by man, such as pasture, old field.

Succession (generation and community cyclic). Changes of communities in time in a given area from pioneer to climax (sensu Clements).

Pioneer (species or communities). Plant invaders of an area previously uninhabited by plants or into an area disturbed by man.

Transient (species or communities). Later invaders into pioneer and subsequent communities prior to climax.

Climax (species or communities). The community that is in equilibrium with its environment in a given area; one that perpetuates itself indefinitely within a climatic regime on the best-drained landform and the most differentiated soil (sensu Clements). Climax species are those typically associated with climax communities.

E. Site-Conspectus Questions: "Conspective"

These questions are included to aid in understanding the site in relation to its surrounding environment, for example, its location in relation to various parts of the landscape, its position in relation to the types of sequences, seres, continua, and mosaics, and its relation to the present natural dynamics occurring in the area.

1. In which part of the landscape is the site located? In which part of the bioscape? The lithoscape? The soilscape? The hydroscape? The toposcape?
2. What is (are) the natural community(ies) of the site?
3. What are the continua at the site?
4. What are the obvious gradients at the site?
5. What are the abiotic mosaics at the site? The biotic mosaics?
6. What are the soil and/or biotic toposequences at the site? Lithosequences at the site? Climatosequences? Biosequences? Pedosequences? Hydrosequences?
7. Which seres are at the site? What stages of succession are present at the site?
8. What are the ecotones at the site?
9. What is the potential biotic climax at the site? (sensu Clements and topoedaphic.)
10. What are the controlling geomorphic processes at the site? Geologic processes? Biologic processes? Hydrologic processes? Soils development processes? (See Chapter 2.)

F. Basic Site-species Relationships Questions: "Perspective"

The ability to describe and/or determine site-species relationships is closely associated with an individual's understanding of observable phenomena. One cannot see and describe accurately what one does not understand. Perception of relationships comes with a knowledge of the dynamics of the ecosystem and a comprehension of the biotic and abiotic features within an area. The following questions are designed to aid in gaining perspective on the field site-species relationships.

1a. What are the seres in the natural area? In which sere does the species occur? At which stage of succession is the species found?

1b. What community cover types are in the natural area? In which community cover type does the species occur? Are associated community types present in the area? Is the species restricted to one community type?

2a. What are the climatic subregimes in the natural area? In which subregime does the species occur?

2b. What are the climatic features within the natural area? Is the species restricted to one climatic feature?

3a. What are the soil series within the natural area? Does the species occur on associated soil series? Is the species restricted to one soil series?

3b. In which soil layer are most of the roots of the species found? What are the texture class, the pH class, and the organic content of the root-bearing layer(s)?

4a. What are the rock occurrences and/or geologic formations within the natural area? On which occurrence and/or formation does the species occur? What are the rock type associations within the area?

4b. On or over which rock type does the species occur? Is the species restricted to one rock or mineral type?

5a. What are the water systems and regimes in the natural area? In which water regime does the species typically occur?

5b. What are the water chemistry types in the natural area? Is the species restricted to one moisture class within the water chemistry type?

6a. What are the landscape types within the natural area? What are the associated landform features? Does the species usually occur on one landform feature?

6b. On what type of slope profile does the species occur? What is the species position on the profile? Is the species restricted to a certain slope aspect, a specific slope shelter, or a slope surface relief feature?

7a. On which major range, plain, basin, drainage, or water body is the species found?

7b. Is the species restricted to a secondary ridge, plain, basin, water body, or tributary?

8a. In which natural community does the species occur?

8b. What is the species relationship to each of the following in a continuum and/or mosaic?

1. Sere
2. Moisture class
3. Temperature class
4. Wind
5. Light
6. Nutrient class
7. pH class
8. Texture class
9. Soil series
10. Rock occurrence
11. Geologic formation
12. Rock type
13. Mineral type or mineralogy
14. Water system
15. Water regime
16. Water chemistry
17. Drainage class
18. Water body
19. Topographic feature
20. Landform type
21. Slope profile
22. Slope position
23. Slope surface relief
24. Slope degree
25. Slope shelter
26. Slope aspect
27. Physiographic province
28. Range, plain, basin
29. Drainage or water body
30. Fire
31. Predator

G. Population-habitat Questions: "Inspective"

The fundamental site concepts are included primarily to enhance broad perspective for the site. The basic questions should facilitate perceptions of site-species relationships, and the next set of questions should lead to a better comprehension of population-habitat relationships.

1. What are the observable symbiotic phenomena at the site? Is the species being damaged by predators? Disease? Grazing? What are the parasites on the species? What are the mycorrhizal relations?

2. What are the wind patterns at the site? Is the species restricted to a northwesterly wind-ice shear zone? To a northeasterly salt spray zone?

3. What are the temperature gradients at the site? Is the species restricted to a thermal belt? Frost pocket? Cold air drainage?

4. What are the moisture gradients at the site? Is the species restricted to a fog belt? Rain shadow? Orographic precipitation zone?

5. What are the light conditions at the site? Is the species restricted to full sunlight? Full shade? Openings in the canopy? Partial shade?

6. What are the discordant soil texture classes at the site? pH classes? Moisture classes? Is the species restricted to one of these classes?

7. What are the discordant lithologies in the site? Are alteration products present? Is the species restricted to a discordant lithology or specific alteration product?

8. Does a mineralogical reaction series occur at the site? Is the species restricted to one part of the series, as olivine rather than biotite?

9. Do minor rock types occur in the geological formation? Is the species restricted to a minor lithology within the formation?

10. Is there much variation in rock structure, fabric, porosity, and permeability? Is the species restricted to one of these variants or combinations thereof?

11. Is microrelief due to animal burrows, crayfish holes, anthills, termite nests, gopher mounds? Is the species restricted to one of these topographic features?

12. What is the role of fire or disturbance at the site? Is the species part of a pyroclimax or does it follow fire as a pioneer or early transient? Is the species dependent upon disturbance for its survival?

13. What is the critical factor or combination of factors controlling the distribution of the species?

14. What is the critical habitat for the species?

15. What are the management recommendations for the species on the basis of your knowledge of its site-species relationships and critical habitat?

H. Master Set of Element Files

Ideally, natural heritage files should include all of the elements for each component, I. Biology to VIII. Natural Communities, to be used in preparation for the determination of site-species relationships in the field. The master element lists of biotic Community Types by geologic formations over landscape types in definitive water regimes, for instance, would be of inestimable value in natural heritage inventory, in the determination of site-species relationships, and in the preparation of environmental impact statements. The expertise and knowledge is now generally available for the compilation of fairly complete element lists for the components II. Climate through VII. Physiography within each state. Professional geologists, pedologists, hydrologists, geomorphologists, and climatologists, helped by geological and soil survey staffs, could prepare master lists of elements found or predicted within the state or physiographic region with little effort or cost. The biologists-ecologists-taxonomists still have a great deal of work to do before comprehensive lists are made for the biotic elements in most states. The natural community lists await the cooperation of biotic and abiotic specialists.

Professional heritage and agency workers along with academic specialists can do society a great service by actively aiding and abetting the preparation of the abiotic master element lists and stimulating biologists to get busy on their field work. In the meanwhile, the student of site-species and community relationships will have to continue to study maps, consult the literature, and visit professional agency and heritage staffs for element information prior to field work.

1. Biologic Element Files

a. Species of native plants and animals
b. Species of special concern (threatened and endangered)
c. Biotic community cover types
d. Biotic community types
e. Biotic community type associations
f. Biotic successional community associations
g. Breeding and feeding territories

The biotic element file entries (a through d) are self-explanatory; entries e through g require explanation (see Chapter 2, Section I). Community Type associations (e) represent the Community Types associated with a particular Community Cover Type. Northern Red Oak, for example, is a Community Cover Type found in the Southern Appalachians, which normally has the Community Type associates: N. Red Oak/Flame Azalea, N. Red Oak/Great Laurel, N. Red Oak/Purple Laurel, N. Red Oak/Hay Scented-New York Ferns, N. Red Oak/Beaked Hazelnut. The biotic successional community associations (f) represent the pioneer to climax communities found on a rock type within a province such as a granitic flatrock or a dolomite cliff, or in a major habitat such as a bog or salt marsh. A master list of nesting areas, bedding areas, and feeding ranges (g) should be prepared for each province within a state. Shad spawning areas, osprey nesting sites, and flounder bedding areas are examples of elements in this file.

2. Climatic Element Files

a. Climatic subregimes
b. Climatic features
c. Climatic feature associations

The climatic element file entries (a through b) are sufficiently explained in texts on climate (see Chapter 2, Section II). In large natural areas climatic feature associations (c), such as frost pockets and cold air drainages, orographic precipitation zones and rain shadows, thermal belts and temperature inversion areas, usually are associated with each other.

3. Soil Element Files

a. Soil series
b. Soil topographic associations
c. Soil development associations (geology and biology)
d. Soil lithologic associations

Soil Series (a) lists for each county or parish in the United States are usually available from the Soil Conservation Service (see Chapter 2, Section III). Soil topographic associations (b) are those Soil Series commonly found together on a topographic feature within a province, as the Cecil, Louisburg, and Appling series on a hillslope over granite. Soil development associations (c) are those commonly occurring together in a geologically dynamic area such as a floodplain. For example, the following developmental sequence is found on a floodplain in the Piedmont of North Carolina: Wehadkee series—Fine-loamy, mixed, nonacid, thermic Typic Fluvaquents; Chewacla series—Fine-loamy, mixed, thermic Fluvaquentic Dystrochrepts; Altavista series—Fine-loamy, mixed, thermic Aquic Hapludults. Another type of soil development association is a combination of biological succession and rock weathering, as pioneer to climax communities on granitic flatrock that would have a combination of Udorthents, Umbrepts, and Hapludults under the pioneer to climax communities. Soil lithologic associations (d) are those located on related rocks or commonly associated rocks within a formation. Examples include the Cecil series over mica gneiss, the Lloyd series over hornblende gneiss, the Davidson series over diorite within a limited area, and the White Store and Creedmoor soil series over fine sandstones, mudstones, and siltstones in the Triassic Newark Group.

4. Geologic Element Files

a. Rock occurrences
b. Geologic formations
c. Rock types
d. Rock type associations
e. Rock-sediment chemistry associations
f. Mineral types

Rock occurrences (a), rock types (c), and rock-sediment chemistry types (e) are treated in most general rock texts (see Chapter 2, Section IV). Geologic formations (b) are indicated on state geologic maps, and lists of mineral types (f) within a state can be obtained from the state geological survey. Rock type associations (d), as dolomite and chert in the Shady Dolomite and hornblende gneiss, hornblende schist, mica gneiss, and mica schist in a Hornblende Gneiss belt, can be obtained from the state geologic maps that have been prepared for most of the states. Frequently, discordant lithologies (e), such as a diabase dike (basic) in a Triassic siltstone bed (acid), can be of great significance in explaining site-species relationships. Occasionally, a concordant granitoid series (e), such as granite (acid), granodiorite (intermediate), and diorite (basic), occurs at a large site. This might account for a lithologic sequence distribution of populations.

5. Hydrologic Element Files

a. Water systems
b. Water regimes
c. Water chemistry types
d. Moisture classes and drainages
e. Topographic moisture class associations

Water systems (a), water regimes (b), water chemistry types (c), and moisture classes and drainages (d) have been listed in the wetlands classification (see Chapter 2, Section V). Moisture class and drainage sequences or associations (e) are usually found on topographic features or landscape types, as hydric to mesic and moderately well drained to very poorly drained on a floodplain.

6. Topographic Element Files

a. Landscape types
b. Landform type associations

Landscape types (a) are treated in most geomorphology texts (see Chapter 2, Section VI). Landform type associations (b) are the parts of a landscape type, as toeslope, footslope, backslope, shoulder, and ridgeslope of a hillslope and levee, terrace, oxbow, slipoff slope, undercut slope, terrace slope, and a bar of a floodplain.

7. Physiographic Element Files

a. Physiographic regions
b. Physiographic provinces
c. Physiographic sections
d. Major ranges, plains, basins
e. Major drainages, water bodies

All of the physiographic element files are treated appropriately in texts such as Fenneman (1931, 1938) (see Chapter 2, Section VII).

8. Natural Community Element Files

a. Natural community developmental associations (concordant)
b. Natural community mosaic associations (discordant)

Natural community developmental associations (a) are the successional communities (pioneer to climax) with associated soil developmental types (as Udorthents to Hapludults) over a rock type (as granite) on a landscape type (as a flatrock) with a moisture gradient (as xeric to mesic) and drainage sequence (as excessively drained or with excessive runoff to well drained) within a climatic subregime (as warm temperate, mesothermal) and physiographic province (as Piedmont).

Natural community mosaic associations (b) are those in which a mosaic of biotic communities occurs over a landscape type within a climatic subregime and physiographic province with a mosaic of soil types (as Udifluvents, Humaquepts, and Umbraquults on a floodplain) and/or a mosaic of lithologies (as basic and acid volcanic flow rock and granite on a slope).

The ultimate in the determination of site-species and community relationships is in the deciphering of the natural community mosaic. A better understanding of population variation, hybrid swarms, and presence of closely related species within a site will come with the proper interpretation of the natural community mosaic. In order to perceive and preserve ecological diversity, scientists should initiate the registering of concordant and discordant natural communities with their component parts and relationships.

I. Conclusions

The intelligent, informed discernment of the dynamics of biotic succession, soils genesis, rock formation, geomorphism, hydrologic changes, and climatic phenomena is basic to the conceptual, detailed interpretation of the significance of seres, gradients, continua, catenas, mosaics, and discrete elements in natural areas. A knowledge of these dynamics and concepts is a requirement for obtaining maximum natural diversity inventory information with the least effort and expense by natural heritage workers. The comprehension of these dynamics and concepts is indispensable to the perception of taxonomic, ecologic, and evolutionary relationships of species to sites by students and professional workers. The perception of taxonomic, ecologic, and evolutionary relationships of species to sites is an absolute must for effective and efficient management of populations and communities of special concern. An understanding of these dynamics and concepts is a necessity for the application of any system of classification of ecological diversity to site-species (population-community) relationships (see Chapter 2).

A knowledge of site-species relationships and a perception of ecologic, taxonomic, and evolutionary relationships of populations to habitats is essential to sound training in floristics, faunistics, forestry, and natural heritage work.

A systematic and standardized application of an ecological diversity classification (Chapter 2) and of inventory-report procedures (Chapter 3) is requisite for the effective and efficient determination of site-species and community relationships as well as for solutions to the perspective, basic, and perceptive questions posed. A comprehensive classified approach is also fundamental to the understanding of population-habitat relationships (Chapter 4).

IX. Our Heritage and Classification: A General Statement

Hopefully, the classification and information systems presented in this publication represent a significant advance in the progression toward

> the development of a sufficiently standardized classification, inventory, and data collection program from state to state to achieve an ever more comprehensive national overview of the status of natural ecological diversity in this country—in other words, a national picture of the existence, characteristics, conditions, status, location, numbers, and distribution of occurrences of the elements of natural ecological diversity. In addition to the obvious importance this will have in maximizing cost-effective allocation of ecological conservation resources in this country, it will also have a profound effect in influencing objective decisions in all kinds of land use and development siting determinations, and improving land management practices. The benefits of a standardized, comprehensive overview for guiding all kinds of disciplinary and interdisciplinary scientific research should be immeasurable, and the efficiency of research investment should be expanded many-fold by the accessibility and comprehensibility of previous research results, survey products, specimen collections, and other repositories. [Heritage Conservation and Recreation Service, 1978, pp. 14–15.]

2

The System of Classification

Chapter 2 treats each major component, or theme (I. Biology, II. Climate, III. Soils, IV. Geology, V. Hydrology, VI. Topography, and VII. Physiography), of the Diversity Classification System. Each component is discussed in its own section. Within each of these sections a general introduction is followed by explanations of the hierarchical levels (A., AA., B., BB., C., CC.) included in each component. In the explanation of each hierarchical level we attempt to furnish the following information: (1) the basis for the level—what is it?, (2) the manner in which the level name is composed, (3) the reasons for the inclusion of the level in the system, (4) whether the hierarchical elements are chosen from an established list or are newly composed, and (5) examples. We adhere to this format as closely as possible, but some sections needed further expansion (for example, Biology) and others required a less extensive discussion (for example, Climate). Because of the significance of level D. (Population Level), it is treated as a separate section in the chapter, following the discussion of the seven themes.

Excerpts from Community Diversity Summaries and Ecological Characterizations that resulted from habitat analysis using the Ecological Diversity Classification System come after the discussion of each theme. Lists of hierarchical elements under each hierarchical level in the major themes are placed in tables at the end of each section.

Many of the hierarchical elements are excerpted or modified from previously existing systems. For instance, all soil elements are from *Soil Taxonomy* (Soil Survey Staff, 1975), and Water Regimes (V.BB.) are from *Classification of Wetlands and Deep-Water Habitats of the United States (An Operational Draft)* (Cowardin et al., 1977). When the elements can be located in a reference, this is indicated in the discussion. When we have formulated the elements ourselves or have modified the elements greatly or when a reference is not readily available, the elements are defined in the text, for example, Hydrologic System (V.A.).

I. Biology

The primary emphasis of this system of classification is biology, which is, in effect, the pedestal for the remainder of the system. Therefore, it is requisite that the reader comprehend certain essential ideas outlined in this section of the chapter. Flora and fauna are influenced by the interaction of factors in the other vertical themes, and they in turn can influence these factors, for example, vegetation affects soil formation.

As one can easily see, the classification of animal and nonvascular plant communities is not well developed in this system and is usually delineated in terms of vegetational habitat types. If mosses or lichens are a conspicuous part of the community, however, they may be indicated in the community name. Professional input is needed in the development of animal and nonvascular plant ecological classification systems.

A. Basic Concepts

1. Site Selection

Certain underlying methodological concepts are fundamental to the understanding of the derivation and nomenclature of each level of the Biology theme. The most significant of these concepts is subjectivity in selecting sites (relatively "uniform" topoedaphic and microclimatic situations) to be analyzed, in drawing imaginary lines around communities, and in choosing sites for relevés and quarter-points. These subjective activities are performed while fully realizing that in nature such discontinuities do not exist so distinctly.

Webb (1954) says that we shall succeed in recognizing, naming, and describing plant communities only if we acknowledge the great difficulties inherent in the task. Webb (1954, p. 364) further states:

> The fact is that the pattern of variation shown by the distribution of species among quadrats of the earth's surface chosen at random hovers in a tantalizing manner between the continuous and the discontinuous. If variation were continuous and all possible combinations of species could be realized, then the science of plant-sociology would be impossible. If variation were discontinuous, with a finite and manageable number of combinations of species which could be realized, and sharp boundaries between the communities, then a satisfactory taxonomy of communities would have been agreed on long ago, and no problem would exist. What we find, however, is that variation is continuous, but in some regions sudden and in others very gradual, and that there is a series of often striking but never perfect correlations between the distribution of different pairs of species. We find, in consequence, that there is a recognizable but very imperfect predictability of the remaining species of a community when some of the principal ones have been ascertained.

As though with the same ideas in mind, Daubenmire (1966) remarks that undoubtedly vegetation presents a continuous variable by virtue of ecotones and that the argument hinges on the existence or absence of plateaulike areas exhibiting minor gradients separated by areas of steeper gradients, with the plateaulike areas being of sufficient similarity to warrant being designated as a class.

In addition, the classifier or the user of a classification system dealing with vegetation science must realize that the units or names employed are abstract entities and that natural communities do not consist of concrete, well-defined, "natural" units. "Classes are always, whatever the extent to which they are suggested or determined by characteristics of that which is classified, human creations, products of the classificatory process" (Whittaker, 1962).

For the purposes of this system, the subjective approach is better than a random selection of areas. We are seeking to delimit variation in natural diversity and to preserve representative examples of each type of diversity. Subjective choice is the only answer to as great a problem as delimiting and classifying natural diversity in the United States.

As with Braun-Blanquet's (1932) approach (a European system of vegetation study upon which much of our approach is based), this subjectivity is desired for the following reasons (Westhoff and van der Maarel, 1973): (1) a selection of relevés should effectively represent the variation in the vegetation under study; and (2) when classification is one of the purposes, a subjective selection allowing for uniformity of stands is desirable, that is, no obvious structural boundaries or variation are visible within the stand and the floristic composition is uniform. Whittaker (1962) makes the following statement, which applies equally to our approach, concerning the Braun-Blanquet method: "The objective is not so much classification of all possible stand combinations as these might be randomly sampled, as it is development of a most effective classification, representing most significant features of vegetational and environmental variation and interrelation, by a skilled practitioner relying on his judgment and experience. Subjective sample choice is quite consistent with the objectives and the rest of the system."

Egler (1977, p. viii) aptly describes the role of subjectivity in vegetation science:

> Technology reigns, as it now does in the arts, social areas, and other sciences—with predictable results. The mere accumulation of data, acquired by hordes of job-holding mathematic-worshipping technicians capable of activating a quantity-understanding computer is able to produce an elegant "hard" and "exact" science—or so it is said. There is no question but that the greatest brains in these subjects have not only made notable contributions, but they realize also the limitations of their methods, as well as the gross abuses of the "privates" in this technologic army. In the meantime, with the vigor and indifference of the superior mutant mongrel that it is, I do believe that Vegetation Science will yet emerge as a blue-blooded breed in its own right.
>
> I look upon this science as one essentially *interpretative* in nature, akin in certain respects to geomorphology, involving long spans of time. It is a science that requires judgment, intuition, background, understanding and experience. For such activity contemporary quantitative and mathematical methods may supply ornamental details of considerable importance, but they can never supply a fundamental conceptual framework for the growing science.

2. Stratification

Stratification, another concept basic to the system, is defined as vertical differentiation within a community —different species occur at various heights above the ground or depths below the water surface. Vegetation is analyzed by recording each species in each stratum by height classes. Definite height limits cannot be set for any stratum because each individual community differs in structure. Hence, we have defined four basic strata with the emphasis on plant habits:

(a) *Tree stratum(-ta)*—A **tree** may be defined as a tall, woody perennial plant usually with a single trunk (Radford et al., 1974). It may sometimes prove difficult to determine whether a plant is a tree or shrub (see definition below). If in doubt, consult a flora or manual (for example, *Manual of the Vascular Flora of the Carolinas* by Radford, Ahles, and Bell, 1968).

A forest will often have more than one tree stratum. In the eastern deciduous forests two strata commonly occur—a **canopy** (upper stratum) and a **subcanopy** (lower stratum). The **subcanopy** contains younger individuals of the canopy and mature trees of smaller species that normally do not reach canopy height, such as *Carpinus caroliniana*, *Cornus florida*, and *Oxydendrum arboreum*. Each tree stratum should be analyzed separately. See page 37 for a list of tree size classes.

(b) *Shrub stratum(-ta)*—A **shrub** may be defined as a much-branched woody perennial plant usually without a single trunk (Radford et al., 1974). This layer generally lies below the lowest tree layer. If different height classes in this layer are obvious, subdivisions may be indicated (see page 37 for a list of shrub size classes). An example is normal shrubs (1–2 m) having typical dwarf shrubs (10–30 cm) beneath them, as in the Community Type *Menziesia pilosa–Nemopanthus mucronata/Vaccinium angustifolium*.

Transgressives and woody vines are also analyzed in the shrub layer. A **transgressive** is an immature individual of a woody species. A **woody vine** is a plant with elongate, nonself-supporting, climbing or clambering, woody stems. If these vines compose a significant percentage of the community, a separate vine analysis table may be incorporated into the study site report.

(c) *Herb stratum(-ta)*—An **herb** may be defined as a plant with an annual above-ground stem. Herbs include herbaceous vines, forbs, graminoids, ferns, and fern allies. An **herbaceous vine** is an elongate, weak-stemmed, often climbing plant (such as *Dioscorea villosa*). A **forb** is any herbaceous plant other than grasses, sedges, or rushes. A **graminoid** is any herbaceous plant that is a grass (Poaceae), a sedge (Cyperaceae), or a rush (Juncaceae). A **fern** is any herbaceous plant that is a flowerless, seedless vascular plant of the class Filicopsida, characteristically reproducing by means of spores and having fronds (megaphylls). A **fern ally** is any herbaceous plant that is a flowerless, seedless vascular plant of the classes Lycopsida (clubmosses, spikemosses, quillworts) and Sphenopsida (horsetails), characteristically reproducing by means of spores and having very small, needlelike or scalelike leaves (or long, linear-subulate, grasslike leaves in quillworts) that are microphylls.

Various size class layers may be present within the herb stratum (see page 37 for a list of herb size classes). For example, in a fen Community Type *Juncus subcaudatus* (10–30 cm) overtops *Rhynchospora capitellata* (3–10 cm), which in turn overtops *R. alba* (3 cm).

(d) *Moss/lichen stratum(-ta)*—These nonvascular plants often form a layer under the herb layer. Ideally, one would like to name these layers as specifically as the vascular plant layers, but this requires expert recognition or identification, which often will not be available. If mosses and lichens form 50% cover and expert help is not available, a general name indicating the presence of mosses should be given, such as *Betula lutea*/Moss spp., *Eriophorum virginicum/Sphagnum* spp. In the reports provided in Part II, moss and lichen species are analyzed in the herb layer.

The majority of species positioning results from variations of light intensity, mostly because of light absorption by the plants themselves. An example (taken from Whittaker, 1975) of how light intensity decreases along a vertical gradient in a community may be found in a forest. The uppermost trees (canopy) of the community receive full sunlight and may absorb and scatter more than half of the sunlight energy. A lower layer of smaller trees (subcanopy) uses most of the remaining light. Thus, less than 10% sunlight (the spectral composition of the internal forest light now changed) penetrates the foliage of both tree layers. The shrub stratum photosynthesizes with this weaker light and further reduces the amount reaching the next layer, the herbs. Often only 1% to 5% incident sunlight supports the herb layer. A moss layer may occur beneath the herbs and may receive only a fraction of 1% incident sunlight.

Stratification forms a significant aspect of the system because understory species are often just as indicative as or more indicative than canopy species. Daubenmire (1976, p. 124) emphasizes that many aspects of vegetation are potentially significant as indicators:

> Concentrating attention on just the trees, or even just the dominants of the undergrowth cannot yield an ecologically refined classification of vegetation. *All* species must be studied quantitatively to learn which are useful indicators and whether or not their relative abundance is significant. Cajander (1926) concluded that in each community type "there exists a stock number of plant species, which are always or nearly always present and

from this stock number there leads an uninterrupted series through those which are often present down to those which are only very seldom present."

For example, *Quercus rubra* might be the dominant tree species in 10 different Community Types with various shrubs and herbs becoming dominant under certain microhabitat, especially topoedaphic, conditions. The preservation of different Community Types under various conditions is important for reasons previously stated. Systematic inventory is the only way to identify these Community Types under various conditions for preservation purposes. For instance, in one natural area inventoried, Panther Creek in Georgia, four previously undescribed mycetozoan species were discovered (Boufford and Wood, 1976). If the many different habitat types are not surveyed and preserved, many species may go undiscovered and much information remain lost.

3. Cover

Cover, the vertical projection of the crown or shoot area of a species to the ground surface expressed as a fraction or percent of a reference area (Mueller-Dombois and Ellenberg, 1974), employs three modifers: **closed**—50% or more cover, **open**—25% to 50% cover, and **sparse**—less than 25% cover. Cover is a significant feature in the biology component because it functions as the basis for nomenclature (this concept is elaborated upon below) and plays an important role in populations and communities. It is indicated not only in closed strata having a dominant or co-dominant (as delimited below in the nomenclature section) but also in all other community strata. This information, in conjunction with a physiognomic description including size, growth form, and duration, is denoted in the Community Ecological Characterization which appears at the end of each Community Diversity Summary of a study site report.

B. General Methods

In doing an inventory for evaluation purposes, each layer is analyzed. Techniques employed are the quarter-point method (Curtis and Cottam, 1962; Mueller-Dombois and Ellenberg, 1974) for the tree strata and the relevé method (Mueller-Dombois and Ellenberg, 1974) for the shrub, herb, and moss/lichen strata. These two methods are described generally below.

1. Quarter-point (Point-centered Quarter) Method

a. Field procedure

Amount of available time usually will determine the number of sample points and the method by which the sample points are selected. Reconnoiter the region to estimate the size of the Community Type area to be sampled. This size establishes the number of quarter-points that can be accomplished and the method by which the quarter-points can be selected. A compass line along which sample points are placed may be followed. If the study site is large enough, a line can be run in any direction. If, on the other hand, the study site is limited in area, the line must be run parallel to the orientation of the community. If a site is extremely confined, it may be necessary to discard the compass line and simply zigzag the sample points randomly across the community. Of course, accuracy increases with number of sample points.

One may determine the distance between sample points either by using a random numbers table as found in any general statistics book (for example, Schefler, 1969, pp. 210–13) or by setting a standard distance between points, for example, 50 m, 100 m, 200 m. The point at which the pacer stops is the center point of the quarter-point (sample plot). Divide the area into four quarters (quadrants) by visualizing two lines passing through the point—one line is the compass line and the other runs at a right angle to the first. Within each of the four quarters, select the tree closest to the center point, record its species name, measure and record the distance from the center point to the midpoint of the tree, and record the diameter at breast height (dbh). Note that distances and dbh's are recorded for four trees at each sample point, one tree in each quarter (quadrant). So if ten quarter-points are completed, a total of 40 trees will have been identified, measured, and recorded. See Figure 1 for a visual representation of the method.

Generally, one person paces, remains at the quarter-point as the center point, and records the data, while the other team member measures the distances and trees. A worksheet for quarter-points field data is provided on page 398.

b. Data computation

Formulas are used to perform data computations from the raw field data. A worksheet for recording computation results is provided on page 399. Data should be calculated in the following manner:

1. Convert dbh's to basal areas (circular area values)

by using a standard conversion table (as in Curtis and Cottam, 1962). **Basal area** is defined as the number of square feet of cross-sectional surface of individual tree species determined at breast height (4.5 feet) per unit area (Radford et al., 1974).

2. List the species on the data sheet.
3. Determine the total number of quarter-points or sample points in which each species occurred.
4. Determine the total number of trees represented by each species.
5. Determine the total basal area of each species.
6. Calculate the **relative frequency**. Note that the maximum value for a relative frequency index is 100.

$$\text{relative frequency} = \frac{\text{number of quarter-points in which a single species occurred}}{\text{total number of quarter-points in which all species occurred}} \times 100$$

7. Calculate **relative density**. Note that the maximum value for a relative density index is 100.

$$\text{relative density} = \frac{\text{number of individual trees of a single species}}{\text{total number of trees of all species}} \times 100$$

8. Calculate **relative dominance**. Note that the maximum value for a relative dominance index is 100.

$$\text{relative dominance} = \frac{\text{total basal area of a single species}}{\text{total basal area of all species}} \times 100$$

9. Calculate the **importance value** (I.V.). Note that the maximum value for an importance value index is 300 (100 + 100 + 100).

$$\text{importance value} = \text{relative frequency} + \text{relative density} + \text{relative dominance}$$

The importance value indicates the relative importance of each tree species in a community and therefore dictates which species is/are dominant/co-dominant.

An example of quarter-point data computation, taken from the *Plant Ecology Workbook* of Curtis and Cottam (1962), is presented below. The data are based on 20 quarter-points (sample points).

Species	*Number of Quarter-points in which Species Occurred*	*Number of Trees*	*Total Basal Area*	*Relative Frequency*	*Relative Density*	*Relative Dominance*	*Importance Value*
Sugar maple	17	47	2,490	17/41 = 41.5%	47/80 = 58.8%	2,490/6,022 = 41.3%	141.6
Hemlock	10	14	1,224	10/41 = 24.4	14/80 = 17.5	1,224/6,022 = 20.3	62.2
Basswood	8	13	838	8/41 = 19.5	13/80 = 16.2	838/6,022 = 13.9	49.6
Yellow birch	6	6	1,470	6/41 = 14.6	6/80 = 7.5	1,470/6,022 = 24.4	46.5
Total	41	80	6,022	100.0%	100.0%	99.9%	299.9

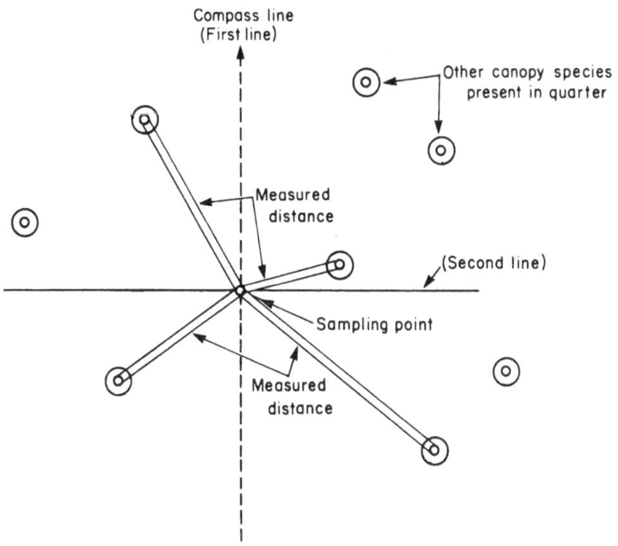

Figure 1. *Quarter-point method* (after Mueller-Dombois and Ellenberg, 1974).

2. Relevé Method

Relevé is defined by Mueller-Dombois and Ellenberg (1974) as the European equivalent to sample stand or vegetation sample. We use it specifically to indicate a square unit area—5 m × 5 m squares for the shrub strata and 1 m × 1 m squares for the herb and moss/lichen strata. The growth form, cover, and sociability for each species present in a relevé are determined. Because these three terms are often employed somewhat ambiguously, we define them in the following discussion.

Growth forms are defined as plant habits, the general appearance of plants. A few examples are: trees—bottle, tall succulent; shrubs and herbs—bulbous, reptant, rhizomatous; vines—root climbers, winding climbers; aquatics—free-floating thalloid, submergent scapose, emergent stoloniferous. For a complete list see Table 2, pages 37–38. The majority of growth form terms are found in Appendix A in Mueller-Dombois and Ellenberg (1974).

Cover, as circumscribed in our diversity classification system, is the estimated amount of horizontal space shaded or occupied by the foliage of a species. Cover values that represent the percentage of the land surface covered are defined in *Vascular Plant Systematics* (Radford et al., 1974, p. 314). A copy of this cover value table also appears on page 386 of this book. These values are determined separately for each species overlapping the relevé, regardless of where the individuals are rooted. The sums of cover values for a stand commonly exceed 100%, because the foliages of different species are interlaced and those of different heights are superimposed over the same area. Cover values are used as the basis for nomenclature throughout the Biology theme (explained in the section on nomenclature).

Sociability pertains to the gregariousness of individual organisms within populations (Radford et al., 1974) or to the relationship of individuals with each other. This classification ranges from plants occurring in pure populations, through intermediate situations in which plants occur in colonies or carpets, small patches or cushions, or tufts, to plants occurring singly (Radford et al., 1974, p. 312). Also see the sociability table on page 386 of this book. Using sociability one can determine the degree to which individuals of a species are grouped or the distribution pattern within a stand and thus can formulate a picture of the plant mosaic. One must realize that sociability values are not constant, for they are affected by habitat, competition with species with which the plant is associated, and the different forms of the plant as it ages. Even in the same locality the sociability of a species may change because habitat factors are variable and communities change, particularly through plant succession.

C. Nomenclature

Nomenclature for the biological hierarchies (I.A.–I.CC.) depends upon cover, which is estimated in this system by cover values in the shrub, herb, and moss/lichen strata and importance values in the tree strata. The name or entry at each level is based upon the condition that the Community Type (I.CC.) has a **dominant** or **co-dominants**, that is, one or few species present in high numbers and therefore of considerable biomass. A species **dominates** only if it makes up 50% or more of the cover or importance values or **co-dominates** only if it makes up 25% of more of the values; otherwise, a mixture of species predominates. These dominant species largely regulate the energy flow, strongly affect the environment of all other species in the community, and exert major controlling influence by virtue of their numbers, size, productivity, or other activities (Odum, 1971). Only under these conditions is a layer incorporated into the name of the Community Type (I.CC.) and hence into the other entries (I.A.–I.C.).

Exceptions to these rules do occur. In specialized habitats in which little vegetation occurs or in which plant form is such that the plants never provide much cover but very characteristic plants are present, a Community Type based on the most abundant species should still be composed. Exemplifying this is a coastal salt flat in which *Salicornia bigelovii* is the characteristic species.

Cover and importance values are used as the bases for nomenclature because they provide a fairly adequate measure of plant biomass. Plant biomass holds ecological significance (Mueller-Dombois and Ellenberg, 1974) because it (1) allows estimation of the capacity of vegetation to accumulate organic material if something is known about the developmental status of the community and its use as a food supply for animals; (2) has a major influence on the stand climate in terms of light and temperature; (3) influences the water relations through rainfall interception and transpiration rate per unit area; (4) is closely related to the volume of circulating nutrients in the ecosystem; and (5) is of direct importance to the animals associated with the vegetation because the plant biomass provides their shelter and food. An obvious advantage of utilizing cover as a measurement is that nearly all plant life forms, from trees to mosses, can be evaluated by the same parameter and therefore can be compared.

Biologic Hierarchies

Code	*Hierarchy*	*Inclusion*
I.A.	Biotic System	Biotic System
I.AA.	Biotic Subsystem	Biotic Subsystem
I.B.	Biotic Class	Community Cover Class
I.BB.	Biotic Subclass	Community Class
I.C.	Biotic Generitype	Community Cover Type
I.CC.	Biotic Type	Community Type

With these general concepts outlined, let us turn to the crux of the system—determination of entries I.A. through I.CC. Vegetational units utilized in the classification scheme are described below. In this section, it is easier to deal with the explanations of the hierarchical levels by starting with the lowest, most specific level (CC. Type) and progressing to the highest, most inclusive level (A. System). Please note that in the other themes (II. through VII.) it is done in the opposite manner.

Biotic Associates (I.D.). This is discussed on page 13 in Chapter 1 and page 90 in the last section of this chapter.

Community Type (I.CC.). Community Type (I.CC.) is based on quantitative field data for a biotic assemblage with a relatively uniform microclimatic and edaphic situation—it is a site-specific description. It is named by enumeration of dominants, layer by layer, and is based on 50% cover or importance value for a single dominant or 25% or more for co-dominants.

Stratified layers are separated by a slash, for example, *Quercus prinus/Kalmia latifolia*; co-dominants are separated by a dash, for example, *Pinus strobus–Tsuga canadensis*. A one-layered Community Type is denoted by a monomial, for example, *Phragmites communis*. Two-layered Community Types are indicated by a binomial, with the first name based on the upper layer and the second on the lower layer, for example, *Aesculus octandra/Athyrium pycnocarpon* and *Scirpus americanus/Sagittaria subulata*. A Community Type with three layers is designated by a trinomial, for example, *Tsuga canadensis/Rhododendron maximum/Mitchella repens*. If vines occur in two or more layers and have a total cover value of 50%, then the vine dominants are placed at the end of the name and separated by a double slash, for example, *Quercus prinus/Kalmia latifolia//Smilax rotundifolia*. Epiphytes with a large cover value are placed at the end of the name and are preceded by a triple slash, for example, *Nyssa aquatica/Spirodela polyrrhiza///Tillandsia usneoides*.

An additional aspect to consider in nomenclature in many communities is seasonality—with a change in season a change in the herbaceous layer dominants will undoubtedly occur. This can be denoted in the Community Type simply by listing each season's dominant(s) and by including an abbreviation for the season in parentheses following the species. These abbreviations are: vernal (V), aestival (Ae), autumnal (Au), and hibernal (H). An example would be: *Fagus grandifolia/Dicentra cucullaria* (V), *Eupatorium rugosum* (Au). Most of the time, investigators will not know the seasonal dominants because the analysis in all likelihood will be based on one or two site visits.

Often in very diverse, rich areas with a number of microhabitats no species will be dominant or co-dominant. Instead, a mixture of species will predominate. How do we handle the nomenclature in this situation? A detailed example of the cove hardwood forests illustrates the handling of such a situation.

A **cove hardwood community** is a very rich forest composed of some 25 or 30 tree species (not all in any one cove), of which six or eight may dominate in varying proportions in different stands (Braun, 1950). Usually a luxuriant herbaceous layer is present and contains a variety of ferns and geophytes. In the Southern Appalachians, where the following examples occur, these communities are generally limited to coves and lower north slopes.

In a cove hardwood stand often as many as eight tree species predominate in the canopy: *Tsuga canadensis*, *Halesia monticola*, *Aesculus octandra*, *Tilia hetero-*

phylla, *Acer saccharum*, *Betula lutea*, *Liriodendron tulipifera*, and *Fagus grandifolia* (Whittaker, 1956). Each of these forms 10% to 20% of the canopy in some stands, perhaps becoming locally more important in others. Together these species make up 80% or 90% of the canopy, dominating stands in different combinations, with the other 10% to 20% composed of various other species.

Species of subcanopy stature include *Magnolia tripetala* and *Carpinus caroliniana* at low elevations, *Magnolia fraseri*, *Cladrastis lutea*, *Ilex opaca*, and *Ostrya virginiana* at most elevations, and *Acer spicatum* and *Amelanchier laevis* at high elevations (Whittaker, 1956). Once again, these may occur in mixed combinations, or in specific localities one may be more dominant than the others.

The shrub layer is most often poorly developed, and no one shrub grouping is characteristic of the cove forests as a whole. Whittaker (1956) lists the following shrubs associated with the cove forests under particular conditions: (1) *Euonymus americanus* and *Lindera benzoin* in low-elevation stands; (2) *Cornus alternifolia*, *Viburnum alnifolium*, and *Ribes cynosbati* in high-elevation stands; (3) *Hydrangea arborescens* locally at all elevations; and (4) *Rhododendron maximum* along streams.

The herb layer of the cove forests is the richest layer in the mountains, particularly in the spring, with over a hundred species of forbs, graminoids, and ferns often reported present. As in the other strata certain herbs may form extensive cover and become dominant or co-dominant in local situations under specific conditions, for example, *Impatiens pallida* in a seepage area or *Monarda didyma*, *Rudbeckia laciniata*, or *Oxalis montana* at high-elevation sites.

With the structure of the cove forests and the cove species in mind, let us examine examples of Community Types (I.CC.) to see how the nomenclature works. If the stand is a typical cove hardwood stand in which all the strata are composed of mixtures of species with none dominating (that is, none have a 50% cover for dominance or 25% cover for co-dominance), then the Community Type would be Mixed cove hardwoods/Mixed cove hardwood transgressives/Mixed herbs. No shrub layer is indicated because generally in a cove hardwood community only scattered shrubs occur and hence do not form a layer. This name tells us that the diversity is great and that no dominants or co-dominants, as we define them, are present. Important species and any microhabitat preferences for species should be elaborated upon in any discussion concerning a study site.

On the other hand, when certain environmental parameters are extended over a large area, particular species adapted to these parameters tend to segregate and form distinctive Community Types. Let us consider some examples and discern how the nomenclature is affected. A clear example occurs at high-elevation sites, where the so-called northern hardwood species (*Acer saccharum*, *Aesculus octandra*, *Betula lutea*, *Tilia heterophylla*) increase in dominance (Braun, 1950). Along with the canopy species, species in the other layers also obtain dominance at high elevations (see examples above). Consequently, our Community Types acquire more specific meaning—"northern hardwoods" as the canopy name denotes more specific species than does "cove hardwoods." Moreover, segregates of these species (therefore, Community Types) appear in conjunction with various environmental factors. The following list will serve as examples of types under possible limiting environmental factors:

Betula lutea/*Acer spicatum*/*Ribes glandulosum*/Mixed tall ferns—over boulder fields

Betula lutea/*Ribes glandulosum*/*Impatiens pallida*/Moss spp.—over boulder fields with seepage areas

Acer saccharum/*Acer saccharum* transgressives/Mixed tall composites—basic soils

Aesculus octandra/*Eupatorium rugosum*—seepage areas

Segregates other than those of the northern hardwoods include the following examples:

Tsuga canadensis/*Rhododendron maximum*—along streams in deep, narrow, moist coves

Liriodendron tulipifera/*Lindera benzoin*/*Jeffersonia diphylla*—rich basic soils or alluvial deposits

Fagus grandifolia/*Fagus grandifolia* transgressives/*Epifagus virginiana*—silty loams with good drainage

When working at the Community Type level and dealing with a mixture of species, one should always try to modify "hardwoods" with a term denoting the specific type of hardwoods, such as southern, upland, bottomland, northern, cove. As a result, the term "mixed hardwoods" is reserved for the upper level Community Cover Class (I.B.), which, as is explained below, is of a more general and inclusive nature.

Another consideration in naming Community Types concerns the term "hardwoods." Traditionally, **hardwoods** are dicotyledonous tree species, as distinguished from softwoods, which are coniferous tree species. Departing slightly from tradition, we reserve the term **hardwoods** for dicotyledonous tree species other than oaks (*Quercus* spp.) or hickories (*Carya* spp.). For example, if a forest has a canopy layer composed of

Quercus spp., *Carya* spp., and other genera (such as *Liriodendron*, *Fraxinus*, *Liquidambar*), the canopy stratum in the Community Type would be "Mixed oaks–hickories–hardwoods." Because oaks and hickories characteristically compose a considerable portion of the eastern deciduous forest, they are deemed significant enough usually to merit recognition, particularly in a mixed situation.

Another situation to examine is the following set of data:

Species	*Importance Value*
Fagus grandifolia	120
Pinus taeda	60
Liquidambar styraciflua	46
Carya glabra	32
Ilex opaca	18
Quercus falcata	18
Acer saccharum ssp. *floridanum*	6

This is a case where one co-dominant (*Fagus grandifolia*) is present, and a mixture of other species, including a fairly high percentage of a pine (*Pinus taeda*), makes up the remainder of the canopy. The name cannot be (1) *Fagus grandifolia*–Mixed relict dune hardwoods because *Pinus* constitutes a major portion of the mixture and is not a hardwood; nor (2) *Fagus grandifolia*–Mixed relict dune hardwoods–*Pinus taeda* because this name would indicate that *Pinus* is a co-dominant with *Fagus* and the mixed hardwoods (the dash indicates co-dominance). The situation can be remedied with the use of the conjunction "and": *Fagus grandifolia*–Mixed relict dune hardwoods and *Pinus taeda*. Note that the term "hardwoods" is used here even though an oak and a hickory are present in the mixture, for they are not co-dominants in the community but only a minor percentage of the entire co-dominant mixture.

The Community Type level (I.CC.) in the classification scheme will certainly demonstrate and strongly emphasize biological diversity. Hopefully, one will become aware of the number of Community Types that must be preserved. We can no longer speak in terms of simply an American Beech (*Fagus grandifolia*) forest but must observe carefully, see relationships, and provide lists such as the following, which truly demonstrate natural diversity.

Fagus grandifolia/*Taxus canadensis*/*Cymophyllus fraseri*
Fagus grandifolia/*Rudbeckia laciniata*–*Eupatorium rugosum*
Fagus grandifolia/*Carex pensylvanica*
Fagus grandifolia/*Vaccinium erythrocarpum*
Fagus grandifolia/*Viburnum alnifolium*
Fagus grandifolia/Mixed herbs
Fagus grandifolia/*Viburnum acerifolium*
Fagus grandifolia/*Kalmia latifolia*
Fagus grandifolia/*Stewartia malacodendron*
Fagus grandifolia/*Ostrya virginiana*/*Gaylussacia frondosa*
Fagus grandifolia/*Athyrium pycnocarpon*
Fagus grandifolia/*Aster divaricatus*

Further selected examples appear in Table 2 (pp. 42–45).

Community Cover Type (I.C.). The Community Cover Type (I.C.) indicates the dominant (50% cover or importance value) or co-dominants (25% or more cover or importance values) of the top layer. This layer need not be the canopy, for in many instances a tree stratum does not exist. For example, on balds the top layer may be *Rhododendron minus* or *Kalmia latifolia*, and in a salt marsh the top layer may be *Juncus roemerianus* or *Salicornia bigelovii*. Examples of Community Cover Types from the Community Types given in the second paragraph under Community Type (I.CC., p. 30) are *Quercus prinus*, *Pinus strobus*–*Tsuga canadensis*, *Phragmites communis*, *Aesculus octandra*, *Scirpus americanus*, *Tsuga canadensis*, *Quercus prinus*, and *Nyssa aquatica*, respectively.

Naturally, a significant reason for erecting this level is its hierarchical inclusion of the Community Type. More important, however, it is included because of its similarity to other systems already extensively worked out, thereby allowing for efficient and effective communication and comparability with and relativity to the literature. For example, considerable compatibility exists with the Society of American Foresters' (1954; Buckman and Quintus, 1972) forest cover types (for example, *Pinus virginiana*, SAF-79; *Quercus virginiana*, SAF-89; *Pinus strobus*–*Tsuga canadensis*, SAF-22; *Pinus strobus*–*Quercus rubra* var. *borealis*–*Fraxinus americana*, SAF-20) and some of the Küchler (1964) vegetation types (Northern hardwoods, *Acer saccharum*–*Betula lutea*–*Fagus grandifolia*–*Tsuga canadensis*, Appalachian oak forest, *Quercus alba*–*Q. rubra*, Northern floodplain forest, *Populus deltoides*–*Salix nigra*–*Ulmus americana*).

The hierarchical elements in the Community Type (I.CC.) and Community Cover Type (I.C.) levels are therefore delimited by analysis and nomenclature in each particular community site, not by selection from a fixed listing of types. Those entries listed in Table 2, pages 40 to 42, are simply selected examples known to occur in nature.

Community Class (I.BB.). The Community Class (I.BB.) includes two features: Species Associates and

Taxonomic Order. The Community Class adds considerable value to the system by providing significant clues to the interpretation of the origin, migration, and evolution of species, floras, faunas, and communities. In addition, this level employs a carefully developed classification system recognized worldwide.

Species Associates are those species characteristically associated with the Community Type (I.CC.). Often only the top layer species associates will be listed. As in the Community Cover Type (I.C.), the top layer is not necessarily the canopy because frequently a tree stratum does not exist. If, however, the lower layers contain high fidelity, or indicator, species, then these species should be included in the associates. **Fidelity** pertains to the degree to which a species is restricted to a particular type of community (Radford et al., 1974) and comprises five classes, three of which concern the present matter: (1) **preferents**, found in several types of communities but predominantly in one type; (2) **selectives**, usually found in one type of community but occasionally found in others; and (3) **exclusives**, found in only one type of community, rarely in another. Quantitatively, in this system these species associates normally must represent at least 10% of the cover or importance value.

A species associate more than likely belongs to one or both of the following types of species—constant species or character species, two slightly modified Braun-Blanquet (1932) concepts that aid in understanding precisely what constitutes a species associate. A **constant species** occurs in a high percentage of the relevés or quarter-points but does not necessarily compose a great proportion of the cover or importance value. A **character species** shows a distinct maximum concentration (quantitatively and by presence) in a well-definable vegetation type and its associated abiotic factors. These two definitions partially follow those of Mueller-Dombois and Ellenberg (1974).

Mueller-Dombois and Ellenberg (1974) discuss in greater detail character species, which can exhibit three degrees of exclusiveness or fidelity in relation to a given vegetational unit and/or habitat: (1) **absolutely restricted** corresponds to the term "exclusives" defined above; (2) **strongly associated** corresponds to the term "selectives" defined above; and (3) **favorably associated** corresponds to the term "preferents" defined above. They further review the concept by defining the different geographical bases delimiting the diagnostic validity of character species: (1) **local character species** have a closely limited range of diagnostic validity (for example, applicable to a mountain valley, an island, or the surroundings of a village); (2) **territorial character species** have diagnostic validity for larger, naturally defined regions (for example, the Vienna basin, the Northwest German lowland); and (3) **absolute character species** have diagnostic validity without any geographic limitation (for example, endemics or ecologically strongly specialized plants in extreme habitats, such as certain alpine species, species of serpentine soils or granitic flatrocks).

The following exemplify the species associates concept (in these particular examples, a Community Type is given first as a reference point and then the list of associates—do not include the Community Type when actually completing I.BB. for the system). Indicate the layers (canopy, subcanopy, shrub, herb) in each entry.

Mixed northern hardwoods/Mixed tall forbs: CANOPY: *Fagus grandifolia, Betula lutea, Aesculus octandra*/SUBCANOPY: *Prunus pensylvanica*/HERB: *Solidago curtisii, Aster acuminatus, Eupatorium rugosum, Rudbeckia laciniata, Laportea canadensis*

Mixed bottomland trees: CANOPY: *Nyssa aquatica, Taxodium distichum, Populus heterophylla, Acer rubrum, Quercus hemisphaerica*

Mixed bottomland oaks–hardwoods: CANOPY: *Fraxinus pennsylvanica, Acer rubrum, Liquidambar styraciflua, Quercus michauxii, Q. hemisphaerica, Carya aquatica, Q. lyrata, Q. falcata* var. *pagodaefolia*

Picea rubens–Abies fraseri: CANOPY: *Betula lutea, Sorbus americana*

Abies fraseri: CANOPY: *Picea rubens, Betula lutea*/SHRUB: *Acer spicatum, Viburnum alnifolium*

Picea rubens: CANOPY: *Betula lutea*/HERB: *Oxalis acetosella*

Picea rubens: HERB: *Athyrium asplenioides, Dryopteris campyloptera, D. intermedia*

Mixed pocosin shrubs: SHRUB: *Zenobia pulverulenta, Cassandra calyculata, Kalmia angustifolia, Ilex coriacea, I. glabra, Lyonia ligustrina*

Mixed bay hardwoods: CANOPY: *Persea borbonia, Magnolia virginiana, Gordonia lasianthus, Nyssa sylvatica* var. *biflora, Acer rubrum* var. *tridens*

Pinus palustris/Quercus laevis: SHRUB: *Gaylussacia frondosa, G. dumosa*/ HERB: *Stipulicida setacea, Cnidoscolus stimulosus, Arenaria caroliniana, Aster linariifolius, Selaginella arenicola*

Mixed salt shrubs: SHRUB: *Ilex vomitoria, Baccharis halimifolia, Myrica cerifera, Iva frutescens*

Nymphaea odorata: HERB: *Nuphar luteum, Brasenia schreberi*

Mixed floating, nonanchored aquatics: HERB: *Lemna gibba, Spirodela polyrrhiza, Wolffia columbiana, Wolffiella floridana, Limnobium spongia, Azolla caroliniana*

Once again, as in the two previous levels, the hierarchical elements are determined through site observation and analysis, not by selection from a standardized list. Those entries appearing in Table 2, page 39, are selected examples. Some clues as to species commonly associated with particular communities can be gained by perusing the literature.

A few remarks may help explain the significance of the species associates concept. Quite simply, species associates represent the biological factors active in the total community environment—an integrated complex of numerous interrelated abiotic and biotic factors. In addition, because we arbitrarily have designated 50% and 25% for dominance and co-dominance limits, respectively, this species associates list gives a more realistic set of species comprising the Community Type in nature. From a practical standpoint, many situations are found in nature where a mixture of species constitutes the Community Type (as in many of the examples provided above)—no true dominant or co-dominants exist as we define these terms. Because the Community Type (I.CC.) in these cases consists of very general terminology (for example, mixed cove hardwoods), the investigator and the reader are enlightened considerably if the most important species are indicated in the Species Associates category. These species associates recur in the literature as well.

Additionally, a character species might consistently occupy the same habitat or restricted environment, whereas a species that dominates with regard to cover in a particular situation might not. It might be a generalist species that inhabits many habitats. Therefore, these character species facilitate determination of abiotic and biotic relationships on the species and community levels and provide the key to the identification of vegetational types and habitat factors. The concept thus becomes important when concerned with species and community occurrences and distributions (for example, disjuncts) and with community evolution and migration, particularly what replaces what.

The second Community Class feature is the **Taxonomic Order**—the order or family (for grasses and sedges) according to Cronquist (1968) or Foster and Gifford (1974) in which the dominants (50% cover or importance value) or co-dominants (25% or more cover or importance value) of each layer belong. Thus, in this level hierarchical elements are chosen from an established list (Table 2, pp. 39–40). Angiosperm families and orders are those of the Cronquist classification scheme (1968). A comprehensive list of the scheme is found on pages 617–40 of *Vascular Plant Systematics* (Radford et al., 1974). Gymnosperm, fern, and fern ally orders are those of Foster and Gifford (1974). The following examples are from the Community Type examples used above (p. 30): Poaceae, Sapindales–Malvales/Filicales, Cyperaceae/Alismatales, Coniferales/Ericales/Rubiales, Fagales/Ericales//Liliales, Cornales/Arales///Bromeliales.

With regard to format, the Taxonomic Order follows an em dash (—) placed after the Species Associates list. The following examples illustrate this (once again a Community Type is provided simply as a reference point).

(1) Community Type (I.CC.): *Pinus Palustris/Quercus laevis*

Community Class (I.BB.): SHRUB: *Gaylussacia frondosa, G. dumosa*/HERB: *Stipulicida setacea, Cnidoscolus stimulosus, Arenaria caroliniana, Aster linariifolius, Selaginella arenicola*—Coniferales/Fagales

(2) Community Type (I.CC.): *Abies fraseri*

Community Class (I.BB.): CANOPY: *Picea rubens, Betula lutea*/SHRUB: *Acer spicatum, Viburnum alnifolium*—Coniferales

Community Cover Class (I.B.). The Community Cover Class (I.B.) is composed of two categories: Generic Class and General Habitat Feature. As the name suggests, the **Generic Class** is based on the generic concept of botanical nomenclature and is indicated with a common generic name or, if a common name does not exist, a scientific name. Examples include alder, beech, birch, blueberry, Forestiera, hickory, maple, oak, Potentilla, spruce.

The **General Habitat Feature** can be one of several types: (1) topographic feature, such as dune, levee, rapids, ridge, saddle, and slope (see Table 7, p. 39, for further examples); (2) distinctive physiognomic assemblages of plants combined with habitat and habit characters, such as bald, chapparal, pocosin, prairie, savannah, and tundra (this terminology frequently corresponds to the biome concepts); or (3) vegetational features resulting from the coherent massing together of plants, such as bed, float, and mat.

These entries are formulated as new Cover Classes are encountered; therefore, the hierarchical elements under Community Cover Class in Table 2 (pp. 38–39) are not exhaustive. Examples from the Community Types given above (p. 30) are Oak ridge, Pine–Hemlock flat, Reed marsh, Buckeye slope, Bulrush marsh, Hemlock slope, Oak slope, and Gum swamp.

The primary reason for this level is that most of these Cover Classes have been described and recognized in the literature over a long period of time and probably will be the level of the system most easily recognized by the layman. Considerable comparability occurs here with

A. W. Küchler's (1964) vegetation types in *Potential Natural Vegetation of the Conterminous United States*. Thus, once again, these entries may easily be integrated into the literature and be employed for communication purposes. In addition, the reasons for inclusion of the Community Class level (I.BB.) apply to this level as well.

Biotic Subsystem (I.AA.). The Biotic Subsystem (I.AA.) is based on **physiognomy**—the external appearance of vegetation (height, growth form, duration). Only the dominants or co-dominants (50% or 25% or more cover or importance values, respectively) of each layer are described. The primary purpose of the Biotic Subsystem is to describe further the appearance and characteristics of the Biotic System (I.A., see below) by utilizing appropriate modifiers included in various subject categories. Most of these modifiers have been used by a number of authors, in particular Mueller-Dombois and Ellenberg (1974) in their "Tentative Physiognomic-Ecological Classification of Plant Formations of the Earth" (Appendix B in *Aims and Methods of Vegetation Ecology*). Definitions for most terms utilized in the Biotic Subsystem (I.AA.) may be found in Appendix A of Mueller-Dombois and Ellenberg (1974). As the system presently stands, the elements under these categories are a complete list from which one selects an entry (see Table 2, pp. 37–38). We are examining physiognomy further, thus hoping to provide a more extensive and descriptive list of elements in the future.

Physiognomic characteristics are significant because they provide information on (1) the physical appearance of the plants; (2) the competitive and coexistent relationships between plants in a community; (3) the utilization of space; and (4) the response of plants and communities to environmental factors.

For the *Tsuga canadensis/Rhododendron maximum/ Mitchella repens* Community Type mentioned above, the Biotic Subsystem is Large, excurrent, evergreen trees/Tall, rhizomatous, evergreen shrubs/Small, reptant, evergreen forbs. One point of format in the Biotic Subsystem requires comment: when co-dominants are present and they have the same physiognomy, the physiognomic description is not repeated. For example, the Biotic Subsystem for the Community Types *Quercus alba–Q. rubra* or *Populus deltoides–Salix nigra–Ulmus americana* is Large, deliquescent, deciduous trees. On the other hand, because the physiognomy of the two species varies somewhat, the Biotic Subsystem entry for *Tsuga canadensis–Fagus grandifolia* reads Large, excurrent, evergreen trees–Large, deliquescent, deciduous trees.

Biotic System (I.A.). The Biotic System (I.A.) is the most inclusive rank in the scheme and contains two entries: Biotic System and Major Ecosystem. The **Biotic System** is defined by very broad, basic physiognomic types as acknowledged by various authors and is recognized on the basis of 50% or more cover or importance value of the top stratum. A fixed listing of System types from which to choose is delineated in the list of hierarchical elements (Table 2, p. 37). Once more, from the example Community Types above (p. 30), the Biotic Systems are Grass system, Broadleaf forest system, Grass system, Needleleaf forest system, Broadleaf forest system, and Broadleaf forest system.

The elements in the Biotic System basically correspond to the major plant habit types represented by the major biomes outlined in ecological literature (Radford et al., 1974; Whittaker, 1975). The various systems also represent successional stages (whether primary or secondary), for example, moss system → grass system → forb system → shrub system → needleleaf forest system → broadleaf forest system. Because the life form of the vegetation reflects the major features of climate and determines the structural nature of the habitat for animals, the biome (in this classification system the Biotic System) provides a sound basis for a natural ecological classification (Clements and Shelford, 1939; Odum, 1971).

Major Habitat is the second entry under Biotic System (I.A.) and includes three hierarchical elements from which to choose: terrestrial, wetland, and aquatic. These modifiers are to be placed before the Biotic System entry. Three examples are (1) Terrestrial needleleaf forest system for a *Tsuga canadensis/Rhododendron maximum* Community Type; (2) Wetland grass system for a *Phragmites communis* Community Type; and (3) Aquatic fern system for an *Azolla caroliniana* Community Type. These three major ecosystems are discussed extensively in Chapter 1, pages 8 to 12.

Master Species Presence List. After identifying and delimiting the above vegetational units, one should complete a master species presence list for each site in order to summarize the remainder of the biotic diversity, including threatened and endangered biota. Community Type (I.CC.) membership for each species is denoted on each study site master species presence list. When significant species microhabitat preferences are noted, these are also indicated.

On the species list the species are grouped in a taxonomic manner (according to Cronquist's orders for angiosperms, 1968; Foster and Gifford's orders for gymnosperms, ferns, and fern allies, 1974) by placing them into families, which are in turn placed into orders. Angiosperms, gymnosperms, ferns, fern allies, mosses, and lichens comprise the major divisions under which the orders and their respective families are arranged alphabetically. This matter is confronted here in order to defend the placement of species in a taxonomic scheme rather than in a purely alphabetical one. The family has served as the fundamental unit in classification and has persistently been used in identification manuals, other literature, and herbaria. We feel that this approach will provide a better basis for carrying out those intentions outlined in the fourth paragraph of the Prologue (p. xv).

Vegetation Summaries and Characterizations

1. Biologic Excerpts from Community Diversity Summaries

a. *Juncus subcaudatus/Rhynchospora capitellata/ Rhynchospora alba*: A. Wetland forb-grass system; AA. Medium, cespitose, deciduous forbs/Small, rhizomatous, deciduous graminoids/Very small, rhizomatous, deciduous graminoids. B. Rush fen zone; BB.. Juncales/ Cyperaceae/Cyperaceae. C. *Juncus subcaudatus*; CC. *Juncus subcaudatus/Rhynchospora capitellata/ Rhynchospora alba*.

b. *Quercus rubra* var. *borealis/Kalmia latifolia–Rhododendron catawbiense*: A. Terrestrial broadleaf forest system; AA. Stunted, deliquescent, deciduous trees/ Tall, rhizomatous, evergreen shrubs. B. Oak-heath flat; BB. Fagales/Ericales. C. *Quercus rubra* var. *borealis*; CC. *Quercus rubra* var. *borealis/Kalmia latifolia–Rhododendron catawbiense*.

c. *Rubus allegheniensis/Carex brunnescens*: A. Terrestrial shrub system; AA. Normal, rhizomatous, deciduous shrubs/Tall, cespitose, deciduous sedges. B. Blackberry bald; BB. Rosales/Cyperaceae. C. *Rubus allegheniensis*; CC. *Rubus allegheniensis/Carex brunnescens*.

d. *Acer rubrum*–Mixed bottomland hardwoods/ Mixed bottomland hardwoods/*Saururus cernuus*// Mixed lianas: A. Wetland broadleaf forest system; AA. Large, deliquescent, deciduous trees/Medium, deliquescent, deciduous trees/Tall, rhizomatous, rhizocarpic forbs//Large, tendril- and root-climbing, deciduous lianas. B. Maple–mixed hardwood swamp forest; BB. Sapindales–mixed hardwoods/Mixed hardwoods/ Piperales//Mixed lianas. C. *Acer rubrum*–mixed bottomland hardwoods; CC. *Acer rubrum*–mixed bottomland hardwoods/Mixed bottomland hardwoods/ *Saururus cernuus*//Mixed lianas.

e. *Spartina alterniflora*: A. Wetland grass system; AA. Tall, rhizomatous, rhizocarpic graminoids. B. Cord Grass marsh; BB. Poaceae. C. *Spartina alterniflora*; CC. *Spartina alterniflora*.

2. Vegetational Excerpts from Community Ecological Characterizations (Vegetationally)

a. *Juncus subcaudatus/Rhynchospora capitellata/ Rhynchospora alba*: Juncalean/Cyperaceous/Cyperaceous wetland forb-grass system with an open, upper herb layer of medium, cespitose, deciduous forbs, an open, middle herb layer of small, rhizomatous, deciduous graminoids, and a closed, lower herb layer of very small, rhizomatous, deciduous graminoids.

b. *Quercus rubra* var. *borealis/Kalmia latifolia–Rhododendron catawbiense*: Fagalean/Ericalean terrestrial broadleaf forest system with an open canopy of stunted, deliquescent, deciduous trees, an open subcanopy of transgressive, deliquescent, deciduous trees, a closed shrub layer of tall, rhizomatous, evergreen shrubs, and a sparse herb layer of medium to very tall, mixed forbs, ferns, and graminoids.

c. *Rubus allegheniensis/Carex brunnescens*: Rosalean/Cyperaceous terrestrial shrub system with a sparse canopy of scattered, transgressive, deliquescent, deciduous trees, a closed shrub layer of normal, rhizomatous, deciduous shrubs, and a closed herb layer of tall, cespitose, deciduous sedges.

d. *Acer rubrum*–mixed bottomland hardwoods/ Mixed bottomland hardwoods/*Saururus cernuus*// Mixed lianas: Sapindalean–mixed hardwoods/Mixed hardwoods/Piperalean//Mixed liana wetland broadleaf forest system with a closed canopy dominated by large, deliquescent, deciduous trees, a closed subcanopy of medium, deliquescent, deciduous trees, a sparse shrub layer of mixed, normal to tall, rhizomatous, deciduous shrubs, a sparse vine layer of normal, tendril-climbing, deciduous vines, a closed herb layer dominated by a tall, rhizomatous, rhizocarpic forb, and a closed liana layer of large, root- and tendril-climbing, deciduous lianas.

e. *Spartina alterniflora*: Poaceous wetland grass system with an open herb layer dominated by tall, rhizomatous, rhizocarpic graminoids.

Table 2. *Biologic Hierarchical Elements*

I.A. Biotic System

1. Biotic System

1.1	Algal system	1.8	Shrub system	1.15	Plankton system
1.2	Fungal system	1.9	Needleleaf forest system	1.16	Nekton system
1.3	Moss system	1.10	Broadleaf forest system	1.17	Neuston system
1.4	Lichen system	1.11	Unvegetated system	1.18	Periphyton system
1.5	Fern system	1.12	Epifauna system	1.19	Coral system
1.6	Forb system	1.13	Infauna system	1.20	Mollusc system
1.7	Grass system	1.14	Cave system	1.21	Echinoderm system

2. Major Habitats

2.1	Terrestrial	2.2	Wetland	2.3	Aquatic

I.AA. Biotic Subsystem

1. Size

Trees

1.1	dwarf trees	< 2 m
1.2	small trees	2- 5 m
1.3	medium trees	5-15 m
1.4	large trees	15-50 m
1.5	giant trees	>50 m

Shrubs

1.6	very low dwarf shrubs	< 3 cm
1.7	low dwarf shrubs	3- 10 cm
1.8	typical dwarf shrubs	10- 30 cm
1.9	tall dwarf shrubs	30-100 cm
1.10	normal shrubs	1- 2 m
1.11	tall shrubs	2- 5 m
1.12	giant shrubs	> 5 m

Herbs

1.13	very small herbs	< 3 cm
1.14	small herbs	3- 10 cm
1.15	medium herbs	10- 30 cm
1.16	tall herbs	30-100 cm
1.17	very tall herbs	1- 3 m
1.18	extremely tall herbs	> 3 m

2. Growth Form (Habit)

2.1 Trees

2.1.1	excurrent (H1)[1]	2.1.3	bottle (H3)	2.1.5	tall succulent (H5)
2.1.2	deliquescent (H2)	2.1.4	tuft (H4)		

2.2 Shrubs and Herbs

2.2.1	bulbous (H6)	2.2.5	reptant (H10)	2.2.8	scapose (H13)
2.2.2	caulescent (H7)	2.2.6	rhizomatous (H11)	2.2.9	stoloniferous (H14)
2.2.3	cespitose (H8)	2.2.7	root-budding (H12)	2.2.10	succulent (H15)
2.2.4	pulvinate (H9)				

[1]These symbols are to be used for growth forms (GF) in the analysis tables presented in the final reports for the study sites and natural areas.

Table 2. *(continued)*

2.3 Vines

2.3.1	root climbing (H16)	2.3.3	tendril climbing (H18)
2.3.2	spread climbing (H17)	2.3.4	winding climbing (H19)

2.4 Aquatics

2.4.1	free-floating thalloid (H20)	2.4.8	submergent scapose (H27)
2.4.2	free-floating leafy (H21)	2.4.9	submergent stoloniferous (H28)
2.4.3	free-floating stoloniferous (H22)	2.4.10	emergent caulescent (H29)
2.4.4	rooted-floating leaf (H23)	2.4.11	emergent cespitose (H30)
2.4.5	submergent caulescent (H24)	2.4.12	emergent rhizomatous (H31)
2.4.6	submergent cespitose (H25)	2.4.13	emergent scapose (H32)
2.4.7	submergent rhizomatous (H26)	2.4.14	emergent stoloniferous (H33)

3. Duration

Annual Herbs and Geophytes

3.1	rain-green	3.3	summer-green
3.2	spring-green	3.4	winter-green

Perennial Herbs, Shrubs, Trees

3.5	deciduous	3.7	evergreen
3.6	rhizocarpic		

I.B. Community Cover Class

1. Generic Class

Selected examples:

Alder	Bulrush	Hair Grass	Oat Grass
Amaranth	Cane	Hemlock	Pine
Amphianthus	Carex	Hemlock-Pine	Pine-Hemlock
Andromeda	Catalpa	Hickory	Pine-Oak
Arborvitae	Cattail	Holly	Pipewort
Arrowhead	Cedar	Horsetail	Pitcher Plant
Ash	Cherry	Huckleberry	Plantain
Aspen	Chestnut	Hudsonia	Pondweed
Azalea	Clubmoss	Iva	Poplar
Bald Rush	Cord Grass	Juniper	Primrose
Barley	Cotton Grass	Larch	Quillwort
Basswood	Cottonwood	Leatherwood	Redbay
Beak Rush	Cranberry	Litsea	Redbud
Beech	Cypress	Love Grass	Redwood
Beech-Birch	Dogtail	Magnolia	Rhododendron
Beech-Maple	Dogwood	Mangrove	Rye Grass
Birch	Douglas Fir	Manna Grass	Salt Grass
Birch-Cottonwood	Dropseed	Maple	Sandspur
Birch-Sycamore	Elderberry	Maple-Basswood	Sandwort
Blackberry	Elm	Muhly Grass	Sassafras
Bladderwort	Elm-Ash-Hackberry	Mulberry	Satureja
Blue Grass	Elm-Ash-Maple	Myrtle	Serviceberry
Blueberry	Fescue	Nemopanthus	Silverbell
Brome Grass	Fir	Nut Grass	Spikemoss
Buckeye	Gum	Oak	Spike-rush
Buckthorn	Hackberry	Oak-Hickory	Spruce

Table 2. *(continued)*

Spruce-Fir
Sumac
Sycamore
Tamarack
Three Awn Grass
Timothy
Umbrella Grass
Viburnum
Viguiera
Walnut
Water-lily
Water-meal
Water-milfoil
White-cedar
Wild Rye Grass
Willow
Willow-Cottonwood

2. General Habitat Features

Selected examples:

bald
barren
bay
bayhead
bed
bog
boulder field
brake
chapparal
cliff
dome
dune
fen
flat
flatrock
flatwoods
float
gap
grassland
hammock
head
levee
marsh
mat
outcrop
oxbow
peak
pocosin
prairie
rapids
ravine
ridge
saddle
savannah
scarp
scrub
seepage
slope
slough
strand
streambank
swamp
tundra
zone

I.BB. Community Class

1. Species Associates

Selected examples (also see pp. 68-69 for further examples):

Alder-Willow-Cottonwood
Ash-Basswood
Ash-Elm-Maple
Beech-Basswood-Buckeye
Birch-Beech/Viburnum alnifolium
Birch-Poplar
Birch/Ribes glandulosum/Circaea alpina
Birch-Spruce-Fir
Cherry-Maple
Cottonwood-Sycamore-Maple
Cypress-Cottonwood
Cypress-Tupelo
Hemlock-Birch
Hemlock-Spruce-Fir
Maple-Elm
Oak-Sweetgum
Pine-Birch
Pine-Tupelo
Sweetgum-Yellow Poplar
Tupelo-Magnolia-Maple

2. Taxonomic Order

Angiosperms

2.1 Alismatales
2.2 Arales
2.3 Arecales
2.4 Aristolochiales
2.5 Asterales
2.6 Batales
2.7 Bromeliales
2.8 Campanulales
2.9 Capparales
2.10 Caryophyllales
2.11 Casuarinales
2.12 Celastrales
2.13 Commelinales
2.14 Cornales
2.15 Cyclanthales
2.16 Cyperaceae
2.17 Diapensiales
2.18 Dilleniales
2.19 Dipsacales
2.20 Ebenales
2.21 Ericales
2.22 Eriocaulales
2.23 Eucommiales
2.24 Euphorbiales
2.25 Fagales
2.26 Gentianales
2.27 Geraniales
2.28 Haloragales
2.29 Hamamelidales
2.30 Hydrocharitales
2.31 Juglandales
2.32 Juncales
2.33 Lamiales
2.34 Lecythidales
2.35 Leitneriales
2.36 Liliales
2.37 Linales
2.38 Magnoliales
2.39 Malvales

Table 2. *(continued)*

2.40 Myricales	2.52 Polemoniales	2.64 Santalales
2.41 Myrtales	2.53 Polygalales	2.65 Sapindales
2.42 Najadales	2.54 Polygonales	2.66 Sarraceniales
2.43 Nymphaeales	2.55 Primulales	2.67 Scrophulariales
2.44 Orchidales	2.56 Proteales	2.68 Theales
2.45 Pandanales	2.57 Rafflesiales	2.69 Triuridales
2.46 Papaverales	2.58 Ranunculales	2.70 Trochodendrales
2.47 Piperales	2.59 Restionales	2.71 Typhales
2.48 Plantaginales	2.60 Rhamnales	2.72 Umbellales
2.49 Plumbaginales	2.61 Rosales	2.73 Urticales
2.50 Poaceae	2.62 Rubiales	2.74 Violales
2.51 Podostemales	2.63 Salicales	2.75 Zingiberales

Gymnosperms

2.76 Coniferales	2.78 Ginkgoales
2.77 Cycadales	2.79 Gnetales

Ferns and Fern Allies

2.80 Equisetales	2.84 Marattiales	2.87 Psilotales
2.81 Filicales	2.85 Marsileales	2.88 Salviniales
2.82 Isoetales	2.86 Ophioglossales	2.89 Selaginellales
2.83 Lycopodiales		

I.C. Community Cover Types

Selected examples:

(Taken from those Community Types listed in I.CC., below in table. For a more comprehensive list, see pages 253-63 in Radford, 1976. The selections are numbered simply to allow reference between I.C. and I.CC.)

1.-2.	*Abies fraseri*	19.	*Carex brunnescens-Rumex acetosella*
3.	*Acer saccharum-Fagus grandifolia*	20.	*Chamaecyparis thyoides*
4.	*Alnus crispa*	21.	*Cyrilla racemiflora-Vaccinium atrococcum*
5.	*Alnus rugosa*-mixed shrubs	22.	*Danthonia compressa*
6.	*Andropogon glomeratus*-mixed herbs	23.	*Danthonia compressa*-mixed sedges and forbs
7.-8.	*Andropogon scoparius*	24.	*Eleocharis equisetoides-Nuphar luteum*
9.	*Andropogon ternarius*-mixed graminoids	25.	*Eleocharis equisetoides-Rhynchospora inundata*
10.	*Angelica triquinata*	26.	*Eriophorum virginicum*
11.	*Azolla caroliniana*	27.-28.	*Fagus grandifolia*
12.	*Azolla caroliniana-Spirodela polyrrhiza*	29.	*Fagus grandifolia-Quercus alba*
13.	*Betula lenta*	30.	*Fagus grandifolia-Acer rubrum-Betula lutea*
14.-15.	*Betula lutea*	31.-32.	*Gaylussacia baccata*
16.	*Betula lutea-Fagus grandifolia*	33.	*Glyceria canadensis-Sparganium chlorocarpum*
17.	*Betula lutea-Tsuga canadensis*	34.	*Hottonia inflata*
18.	*Borrichia frutescens*	35.	*Hydrocotyle ranunculoides*

Table 2. *(continued)*

No.	Community
36.	*Iva imbricata-Myrica pensylvanica*
37.	*Juncus effusus*
38.	*Juncus roemerianus*
39.	*Juniperus virginiana*
40.-41.	*Kalmia latifolia*
42.	*Leersia virginica*
43.	*Leiophyllum buxifolium*
44.	*Lemna gibba*
45.	*Liquidambar styraciflua-Acer saccharum*
46.	*Liquidambar styraciflua-Nyssa sylvatica*
47.	*Liquidambar styraciflua*-mixed hardwoods
48.	*Liriodendron tulipifera*-cove hardwoods
49.	*Liriodendron tulipifera*-mixed hardwoods
50.	*Litsea aestivalis*
51.	*Mastichodendron foetidissimum*
52.	*Menziesia pilosa-Vaccinium angustifolium*
53.-54.	Mixed southern hardwoods
55.-56.	Mixed bottomland hardwoods
57.	Mixed upland hardwoods
58.	Mixed swamp hardwoods
59.	Mixed northern hardwoods
60.	Mixed heaths-*Cyrilla racemiflora*
61.	Mixed heaths
62.	Mixed marsh herbs
63.	Mixed oaks
64.	*Myrica cerifera*
65.	*Myriophyllum laxum*
66.	*Nuphar luteum*
67.	*Nymphaea odorata*
68.	*Nymphoides cordata*
69.	*Nyssa aquatica*
70.	*Nyssa biflora*
71.	*Panicum hemitomon*
72.	*Paronychia argyrocoma*
73.	*Persea borbonia-Acer rubrum*
74.	*Phragmites communis*
75.	*Picea rubens*
76.	*Picea rubens-Tsuga canadensis*
77.	*Pinus echinata*
78.	*Pinus elliottii*
79.-81.	*Pinus palustris*
82.-83.	*Pinus pungens*
84.-85.	*Pinus rigida*
86.-87.	*Pinus serotina*
88.	*Pinus strobus-Quercus prinus*
89.-92.	*Pinus taeda*
93.	*Pinus virginiana*
94.	*Pinus virginiana-Quercus prinus-Quercus marilandica*
95.	*Platanus occidentalis*-mixed hardwoods
96.	*Pontederia cordata*
97.-98.	*Quercus alba*
99.	*Quercus coccinea*
100.	*Quercus falcata-Carya ovalis*
101.	*Quercus laevis*
102.	*Quercus laurifolia*-mixed hardwoods
103.	*Quercus laurifolia*
104.-106.	*Quercus prinus*
107.	*Quercus prinus-Pinus virginiana*
108.	*Quercus rubra*
109.	*Quercus rubra-Liriodendron tulipifera*
110.	*Quercus rubra-Quercus alba*
111.	*Quercus rubra-Acer saccharum*
112.-118.	*Quercus rubra* var. *borealis*
119.	*Quercus virginiana*
120.	*Rhododendron catawbiense*
121.	*Rhododendron minus*
122.	*Rubus allegheniensis*
123.	*Sagittaria subulata-Myriophyllum exalbescens-Vallisneria americana*
124.	*Salicornia bigelovii*
125.	*Salicornia europaea*
126.	*Salicornia virginica*
127.	*Salix caroliniana*
128.	*Sarracenia flava*-mixed sedges
129.	*Scirpus americanus*
130.	*Scirpus cespitosus* var. *callosus*
131.	*Selaginella rupestris*
132.	*Selaginella tortipila*
133.	*Selaginella rupestris-Selaginella tortipila*
134.	*Solidago uliginosa-Carex* sp.
135.	*Sorbus melanocarpa*
136.	*Spartina alterniflora*
137.	*Spartina cynosuroides*
138.	*Spartina patens-Distichlis spicata*
139.	*Spartina patens*-mixed herbs
140.	*Spirodela polyrrhiza*
141.	*Spirodela polyrrhiza-Wolffiella floridana*
142.	*Spirodela polyrrhiza-Wolffia columbiana-Wolffiella floridana*
143.	*Taxodium ascendens*
144.	*Taxodium distichum*-mixed hardwoods
145.	*Thuja occidentalis*

Table 2. *(continued)*

146.	*Tsuga canadensis*	151.	*Vaccinium oxycoccos*
147.	*Tsuga caroliniana*	152.	*Vallisneria americana*
148.	*Typha domingensis*	153.	*Wolffiella floridana-Utricularia biflora*
149.	*Uniola paniculata*	154.	*Zostera marina*
150.	*Utricularia olivacea*	155.	*Zostera marina-Ruppia maritima*

I.CC. Community Types

Selected examples:

(Taken from reports done in the southeastern United States. The majority of these reports were completed personally by A. E. Radford or his Botany 235 classes at the University of North Carolina at Chapel Hill. The selections are numbered simply to allow reference between I.C. and I.CC.)

1. *Abies fraseri/Rhododendron catawbiense*
2. *Abies fraseri/Oxalis acetosella*
3. *Acer saccharum-Fagus grandifolia*/Mixed shrubs/Mixed herbs
4. *Alnus crispa/Rubus allegheniensis*
5. *Alnus rugosa*-mixed shrubs
6. *Andropogon glomeratus*-mixed herbs
7. *Andropogon scoparius*
8. *Andropogon scoparius/Hudsonia tomentosa*
9. *Andropogon ternarius*-mixed graminoids
10. *Angelica triquinata/Polytrichum* spp.
11. *Azolla caroliniana*
12. *Azolla caroliniana-Spirodela polyrrhiza*
13. *Betula lenta*/Mixed subcanopy/*Lindera benzoin*
14. *Betula lutea/Dennstaedtia punctilobula*
15. *Betula lutea-Rhododendron catawbiense*/Mixed heaths
16. *Betula lutea-Fagus grandifolia*/Mixed herbs
17. *Betula lutea-Tsuga canadensis*
18. *Borrichia frutescens/Spartina patens*
19. *Carex brunnescens-Rumex acetosella*
20. *Chamaecyparis thyoides*/Mixed evergreen shrubs
21. *Cyrilla racemiflora-Vaccinium atrococcum/Lyonia lucida*
22. *Danthonia compressa*
23. *Danthonia compressa*-mixed sedges and forbs
24. *Eleocharis equisetoides-Nuphar luteum*
25. *Eleocharis equisetoides-Rhynchospora inundata*
26. *Eriophorum virginicum/Sphagnum* spp.
27. *Fagus grandifolia*/Mixed vines
28. *Fagus grandifolia/Rhododendron catawbiense*
29. *Fagus grandifolia-Quercus alba*/Mixed shrubs/Mixed herbs
30. *Fagus grandifolia-Acer rubrum-Betula lutea*/Mixed ferns
31. *Gaylussacia baccata*
32. *Gaylussacia baccata/Gaultheria procumbens*
33. *Glyceria canadensis-Sparganium chlorocarpum*
34. *Hottonia inflata*
35. *Hydrocotyle ranunculoides/Spirodela polyrrhiza*
36. *Iva imbricata-Myrica pensylvanica*-mixed shrubs
37. *Juncus effusus*
38. *Juncus roemerianus*
39. *Juniperus virginiana/Bumelia lycioides*

Table 2. *(continued)*

40. *Kalmia latifolia/Potentilla tridentata*
41. *Kalmia latifolia/Epigaea repens/Polytrichum juniperinum*
42. *Leersia virginica*
43. *Leiophyllum buxifolium*
44. *Lemna gibba*
45. *Liquidambar styraciflua-Acer saccharum/Arundinaria gigantea*
46. *Liquidambar styraciflua-Nyssa sylvatica*
47. *Liquidambar styraciflua*-mixed bottomland hardwoods/*Ligustrum sinense*
48. *Liriodendron tulipifera*-mixed cove hardwoods/*Lindera benzoin*/Mixed herbs
49. *Liriodendron tulipifera*-mixed cove hardwoods/*Cornus florida/Arisaema triphyllum-Podophyllum peltatum*
50. *Litsea aestivalis/Lyonia lucida*
51. *Mastichodendron foetidissimum/Coccoloba diversifolia*
52. *Menziesia pilosa-Vaccinium angustifolium*
53. Mixed southern hardwoods
54. Mixed southern hardwoods/Mixed shrubs//*Lonicera japonica*
55. Mixed bottomland hardwoods/*Carpinus caroliniana*
56. Mixed bottomland hardwoods//*Rhus radicans*
57. Mixed upland hardwoods/*Oxydendrum arboreum-Cornus florida/Viburnum rafinesquianum*-mixed tall, deciduous shrubs
58. Mixed swamp hardwoods/Mixed shrubs/*Saururus cernuus*
59. Mixed northern hardwoods/*Acer saccharum-Fagus grandifolia* transgressives/ Mixed herbs
60. Mixed heaths-*Cyrilla racemiflora*
61. Mixed heaths/*Sphagnum* spp.
62. Mixed marsh herbs
63. Mixed oaks/Mixed small, deciduous trees
64. *Myrica cerifera/Andropogon glomeratus*
65. *Myriophyllum laxum*
66. *Nuphar luteum*
67. *Nymphaea odorata*
68. *Nymphoides cordata*
69. *Nyssa aquatica/Spirodela polyrrhiza///Tillandsia usneoides*
70. *Nyssa biflora*/Mixed shrubs
71. *Panicum hemitomon*
72. *Paronychia argyrocoma*
73. *Persea borbonia-Acer rubrum/Persea borbonia-Myrica cerifera/Myrica cerifera/Osmunda cinnamomea-Osmunda regalis* var. *spectabilis*
74. *Phragmites communis*
75. *Picea rubens/Viburnum cassinoides/Rubus hispidus/Polytrichum* spp.
76. *Picea rubens-Tsuga canadensis*/Mixed herbs
77. *Pinus echinata/Gaylussacia baccata-Vaccinium vacillans*
78. *Pinus elliottii/Sabal palmetto*
79. *Pinus palustris/Quercus laevis/Aristida stricta*
80. *Pinus palustris*/Mixed heaths
81. *Pinus palustris/Quercus laevis/Gaylussacia dumosa-Gaylussacia frondosa*
82. *Pinus pungens/Vaccinium vacillans-Gaylussacia baccata*
83. *Pinus pungens/Rhododendron catawbiense-Kalmia latifolia*
84. *Pinus rigida/Oxydendrum arboreum/Castanea pumila*
85. *Pinus rigida*/Mixed heaths
86. *Pinus serotina/Magnolia virginiana-Acer rubrum/Lyonia lucida/Sphagnum* spp.//*Smilax laurifolia*
87. *Pinus serotina/Cyrilla racemiflora-Zenobia pulverulenta/Sphagnum* spp.// *Smilax laurifolia*

Table 2. *(continued)*

88. *Pinus strobus-Quercus prinus/Kalmia latifolia-Rhododendron catawbiense*
89. *Pinus taeda/Persea borbonia//Rhus radicans*
90. *Pinus taeda/Quercus laurifolia*
91. *Pinus taeda/Gaylussacia frondosa-Vaccinium vacillans*
92. *Pinus taeda/Aralia spinosa/Mitchella repens*
93. *Pinus virginiana*/Mixed heaths
94. *Pinus virginiana-Quercus prinus-Quercus marilandica*
95. *Platanus occidentalis*-southern hardwoods/Mixed herbs
96. *Pontederia cordata*
97. *Quercus alba/Rhododendron catawbiense*
98. *Quercus alba/Rhododendron calendulaceum*/Mixed ferns
99. *Quercus coccinea*/Mixed heaths
100. *Quercus falcata-Carya ovalis*/Mixed subcanopy
101. *Quercus laevis*/Mixed heaths/*Cladonia* spp.
102. *Quercus laurifolia*-mixed hardwoods
103. *Quercus laurifolia/Ilex vomitoria*
104. *Quercus prinus/Kalmia latifolia*
105. *Quercus prinus*/Low heaths/Mixed herbs
106. *Quercus prinus/Rhododendron catawbiense*
107. *Quercus prinus-Pinus virginiana/Kalmia latifolia*
108. *Quercus rubra/Rhododendron catawbiense*
109. *Quercus rubra-Liriodendron tulipifera*/Mixed subcanopy/*Viburnum acerifolium*
110. *Quercus rubra-Quercus alba*
111. *Quercus rubra-Acer saccharum/Rubus* sp./*Polygonum cilinode*
112. *Quercus rubra* var. *borealis*/Mixed heaths
113. *Quercus rubra* var. *borealis*/Mixed tall herbs
114. *Quercus rubra* var. *borealis/Dennstaedtia punctilobula*
115. *Quercus rubra* var. *borealis/Carex pensylvanica*
116. *Quercus rubra* var. *borealis/Corylus cornuta*
117. *Quercus rubra* var. *borealis/Rhododendron maximum*
118. *Quercus rubra* var. *borealis/Rhododendron calendulaceum-Vaccinium constablaei*
119. *Quercus virginiana*/Mixed shrubs
120. *Rhododendron catawbiense*
121. *Rhododendron minus*
122. *Rubus allegheniensis/Carex brunnescens*
123. *Sagittaria subulata-Myriophyllum exalbescens-Vallisneria americana*
124. *Salicornia bigelovii*
125. *Salicornia europaea*
126. *Salicornia virginica*
127. *Salix caroliniana*
128. *Sarracenia flava*-mixed sedges
129. *Scirpus americanus*
130. *Scirpus cespitosus* var. *callosus*
131. *Selaginella rupestris*
132. *Selaginella tortipila*
133. *Selaginella rupestris-Selaginella tortipila*
134. *Solidago uliginosa-Carex* sp./*Sphagnum* spp.
135. *Sorbus melanocarpa/Deschampsia flexuosa/Polytrichum juniperinum*
136. *Spartina alterniflora/Salicornia europaea-Salicornia virginica*
137. *Spartina cynosuroides*
138. *Spartina patens-Distichlis spicata*
139. *Spartina patens*-mixed herbs

Table 2. *(continued)*

140. *Spirodela polyrrhiza*
141. *Spirodela polyrrhiza-Wolffiella floridana*
142. *Spirodela polyrrhiza-Wolffia columbiana-Wolffiella floridana*
143. *Taxodium ascendens*/Mixed herbs//*Rhus radicans*
144. *Taxodium distichum-mixed hardwoods*
145. *Thuja occidentalis/Mixed herbs*
146. *Tsuga canadensis/Rhododendron maximum*
147. *Tsuga canadensis/Rhododendron maximum/Mitchella repens*
148. *Typha domingensis*
149. *Uniola paniculata*
150. *Utricularia olivacea*
151. *Vaccinium oxycoccos/Eriophorum virginicum/Sphagnum* spp.
152. *Vallisneria americana*
153. *Wolffiella floridana-Utricularia biflora*
154. *Zostera marina*
155. *Zostera marina-Ruppia maritima*

II. Climate

Climate, particularly temperature and moisture conditions at ground level, exerts strong control over the vegetation of the earth and is probably the most significant factor in the differentiation of plant cover into major vegetational zones (Walter, 1973). According to Blumenstock and Thornthwaite (1941), three patterns dominate the earth: the pattern of climate, the pattern of vegetation, and the pattern of soils. They further state that when the three patterns are laid upon one another, their boundaries coincide to a remarkable degree because climate is the fundamental dynamic force shaping the other two. The interrelationships among these three forces are summarized in the following scheme by Walter (1973).

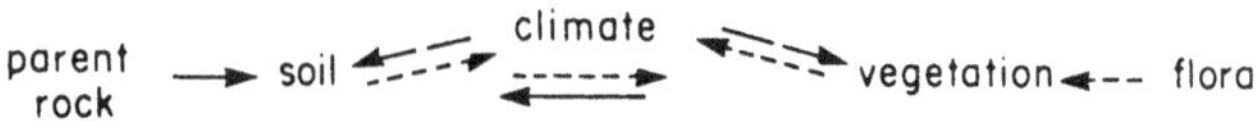

These authors are talking primarily about vegetation zones on a large scale, whereas we are concerned not only with these major vegetational zones (higher levels in the hierarchy) but also with zones much smaller in scale (lower levels in the hierarchy). These smaller vegetational zones are controlled by a combination of microclimatic and many other factors because regional climate would be, for all practical purposes, the same. Because a particular vegetation type could not exist in an area if climatic conditions were not favorable, it is clear why climate must be one of the seven major categories of concern in this system.

As in all our major thematic components, the climatic hierarchy is based upon a geographic, areal progression from a large regional area to an area occupied by a community and population(s)—hence, it is based on a concept of progressively smaller size. The first three levels discussed represent generalized climatic patterns for fairly large areas. Marked deviations frequently occur, and local climates, especially microclimates, may differ considerably from the more general pattern. Factors such as location of water bodies, wind patterns, mountain ranges, and degree of exposure influence these dissimilarities.

Climatic Hierarchies

Code	*Hierarchy*	*Inclusion*
II.A.	Climatic System	Climatic Regime
II.AA.	Climatic Subsystem	Climatic Subregime
II.B.	Climatic Class	Sectional Climate
II.BB.	Climatic Subclass	Local Climate
II.C.	Climatic Generitype	Natural Area Climatic Site Type
II.CC.	Climatic Type	Community Climatic Site Type

Climatic Regime (II.A.) and Climatic Subregime (II.AA.). The Climatic Regime (II.A.) and Climatic Subregime (II.AA.) consist of large, fairly well-defined, mapped, geographic areas that possess basically the same chief climatic characteristics based on temperature, precipitation, and season. These regimes and subregimes are those type regions and subtypes of Köppen (1931) and may be found in Goode and Espenshade (1950) or Trewartha (1954). They appear in Table 3 (p. 48).

An example of a Climatic Regime (II.A.) is **microthermal**, which is a snow-forest climate with at least one month 10°C to 18°C (50°F to 64.4°F) and at least four months under 1°C (33.8°F). **Boreal** is a Climatic Subregime (II.AA.) that falls under microthermal and has the warmest month under 22°C (71.6°F) and at least four months over 10°C (50°F).

Sectional Climate (II.B.). Sectional Climate (II.B.) includes data obtained from maps in *The National Atlas* (U.S. Department of the Interior, 1970): mean annual sunshine, average annual precipitation, freeze-free period, average mean maximum temperature of hottest month, average mean minimum temperature of coldest month, average mean precipitation of driest month, and average mean precipitation of wettest month. Descriptors denoting a range of values for each of these are utilized and are found in Table 3. Definitions for these descriptors are excerpted from *The National Atlas* (U.S. Department of the Interior, 1970). This information is on a somewhat more restricted scale than the large geographic regions of Climatic Regime (II.A.) and Climatic Subregime (II.AA.).

Sectional Climate (II.B.) information from a dune community on Ocracoke Island in the Coastal Plain of North Carolina serves as an example—cool and moist yearly, with moderately hot and moderately wet summers, moderately warm and moderately dry winters, and a very long freeze-free period. Descriptors are used and not the actual data.

Local Climate (II.BB.). Local Climate (II.BB.) is more confined in area than Sectional Climate (II.B.). Information is gathered from climatic supplements (U.S. Department of Commerce, 1965) available for each state. The reference point is the weather station closest to the natural area. The data acquired and descriptors used are the same as those listed above for Sectional Climate (Table 3, pp. 48–49).

Local Climate (II.BB.) for the Ocracoke example used above is warm and moist yearly, with moderately hot and moderately wet summers, warm and moderately dry winters, and a very long freeze-free period. As can be seen from this example, the Local Climate (II.BB.) and Sectional Climate (II.B.) often will be very similar.

Natural Area Climatic Site Type (II.C.) and Community Climatic Site Type (II.CC.). The Natural Area Climatic Site Type (II.C.) and the Community Climatic Site Type (II.CC.) are categories in which one subjectively compares the climate of the natural area to the Local Climate (II.BB.) and that of the community site to the Natural Area Climatic Site Type (II.C.). Is the climate of a specific site cooler than, warmer than, or similar to the local or natural area climate? Likewise, is it drier than, wetter than, or similar to the local or natural area climate?

Thus far, data have been procured from literature sources, but when the lowest levels are reached, field observations and data provide the major sources of information. Table 3 (p. 49) provides hierarchical elements for temperature and precipitation relations. The ideal method would be to measure the criteria listed under Sectional Climate (II.B.) at the natural area and/or community site, but this would require sophisticated instrumentation over at least a year-long period.

The element entry for II.C. (Natural Area Climatic Site Type) for the Ocracoke example above is similar to local climate. Likewise, the element entry for II.CC. (Community Climatic Site Type) is similar to natural area climate.

At these levels it is important to note microhabitat climatic differences or prominent minor climatic features, such as thermal belts, frost pockets, and air drainage patterns. These features pertinent to the natural area and/or the community should be recorded and included in the report discussion. For example, is there a wetter, cooler seepage area in a cold air drainage in the general site analyzed?

Population Climatic Site Type (II.D.). This is discussed on page 90 in the last section of this chapter.

Climatic Summaries and Characterizations

1. Climatic Excerpts from Community Diversity Summaries

a. *Juncus subcaudatus/Rhynchospora capitellata/ Rhynchospora alba*: A. Mesothermal; AA. Warm temperate. B. Moderately cold and moist yearly, with warm and moderately wet summers, moderately cold and moderately wet winters, and a short freeze-free period; BB. Moderately cold and moist yearly, with moderately warm and moderately wet summers and moderately cold and moderately wet winters, and a short to very short freeze-free period. C. Cooler and wetter than local climate; CC. Similar to natural area climate.

b. *Quercus rubra* var. *borealis/Kalmia latifolia–Rhododendron catawbiense*: A. Mesothermal; AA. Warm temperate. B. Cool and moist yearly, warm and moderately wet summers, moderately cold and moderately dry winters, with a short freeze-free period; BB. Local climate same as regional climate. C. Natural area climate cooler and ranges from wetter to drier than local climate; CC. Community climate cooler and drier than natural area climate.

c. *Rubus allegheniensis/Carex brunnescens*: A. Microthermal; AA. Boreal. B. Moderately cold and moist yearly, with warm and moderately wet summers, moderately cold and moderately wet winters, with a short freeze-free period; BB. Cool and moist yearly, with warm and moderately wet summers, cold and moderately wet winters, with a short freeze-free period. C. Cooler and wetter than local climate, with high elevation fogs and mists; CC. Similar to natural area.

d. *Acer rubrum*–mixed bottomland hardwoods/ Mixed bottomland hardwoods/*Saururus cernuus*// Mixed lianas: A. Mesothermal; AA. Warm temperate. B. Cool and moderately moist yearly, with moderately hot and moderately wet summers, moderately warm and moderately dry winters, and a long freeze-free period; BB. Similar to sectional climate. C. Cooler and wetter than local climate; CC. Similar to natural area climate.

e. *Spartina alterniflora*: A. Mesothermal; AA. Warm temperate. B. Cool and moist yearly, moderately hot and moderately wet summers, moderately warm and moderately dry winters, with a very long freeze-free period; BB. Warm and moist yearly, moderately hot and moderately wet summers, warm and moderately dry winters, with a very long freeze-free period. C. Similar to local climate; CC. Similar to natural area climate.

2. Climatic Excerpts from Community Ecological Characterizations (Climatically)

a. *Juncus subcaudatus/Rhynchospora capitellata/ Rhynchospora alba*: Warm temperate, mesothermal climate, cooler and wetter than the local moderately cold and moist yearly climate, which has moderately warm and moderately wet summers and moderately cold and moderately wet winters, and a short to very short freeze-free period.

b. *Quercus rubra* var. *borealis/Kalmia latifolia–Rhododendron catawbiense*: Warm temperate, mesothermal climate, cooler and drier than the local and regional yearly cool and moist climate, which has warm and moderately wet summers, moderately cold and moderately dry winters, and a short freeze-free period.

c. *Rubus allegheniensis/Carex brunnescens*: Boreal, microthermal climate, cooler and wetter than the regional climate, which is moderately cold and moist yearly and which has warm and moderately wet summers, moderately cold and moderately wet winters, and a short freeze-free period.

d. *Acer rubrum*–mixed bottomland hardwoods/ Mixed bottomland hardwoods/*Saururus cernuus*// Mixed lianas: Warm temperate, mesothermal climate, cooler and wetter than the local and sectional cool and moderately moist yearly climate, which has moderately hot and moderately wet summers, moderately warm and moderately dry winters, and a long freeze-free period.

e. *Spartina alterniflora*: Warm temperate, mesothermal climate, similar to the local yearly warm and moist climate, which has moderately hot and moderately wet summers, warm and moderately dry winters, and a very long freeze-free period.

Table 3. *Climatic Hierarchical Elements*

II.A. Climatic Regime

1. Tropical (A)[1]
2. Arid (B)
3. Mesothermal (C)
4. Microthermal (D)
5. Cryothermic (E)

II.AA. Climatic Subregime

Tropical
1. Tropical monsoon (Am)[1]
2. Tropical wet (Af)
3. Tropical savanna (Aw)

Arid
4. Desert (BWs)
5. Desert (BWw)
6. Steppe (BSs)
7. Steppe (BSw)

Mesothermal
8. Warm temperate (Cf)
9. Mediterranean (Cs)

Microthermal
10. Cool temperate (Da)
11. Boreal (Db)
12. Taiga (Dc)

Cryothermic
13. Tundra (Et)
14. Glacial (Ef)

II.B. Sectional Climate

1. Mean Annual Sunshine (hours)

1.1	extremely hot	>4,000
1.2	very hot	3,600-4,000
1.3	moderately hot	3,200-3,600
1.4	warm	2,800-3,200
1.5	cool	2,400-2,800
1.6	moderately cold	2,000-2,400
1.7	cold	1,600-2,000
1.8	very cold	1,200-1,600
1.9	extremely cold	<1,200

2. Average Annual Precipitation (inches)

2.1	extremely wet	>96
2.2	very wet	80-96
2.3	moderately wet	64-80
2.4	moist	48-64
2.5	moderately moist	32-48
2.6	moderately dry	24-32
2.7	dry	16-24
2.8	very dry	8-16
2.9	extremely dry	< 8

3. Freeze-Free Period (days)

(Precipitation to be calculated when data are available.)

3.1	freeze-free all year	365
3.2	extremely long	>300
3.3	very long	240-300
3.4	long	210-240
3.5	average	150-210
3.6	short	120-150
3.7	very short	60-120
3.8	extremely short	0- 60
3.9	permanently frozen	< 0

[1]Symbols used in Goode's School Atlas (Goode and Espenshade, 1950).

Table 3. *(continued)*

4. Average Mean Maximum Temperature of Hottest Month (°F)

4.1	extremely hot	>100
4.2	very hot	90–100
4.3	moderately hot	80– 90
4.4	warm	70– 80
4.5	moderately warm	60– 70
4.6	moderately cold	50– 60
4.7	cold	40– 50
4.8	very cold	30– 40
4.9	extremely cold	< 30

5. Average Mean Maximum Temperature of Coldest Month (°F)

5.1	extremely hot	>70
5.2	very hot	60–70
5.3	moderately hot	50–60
5.4	warm	40–50
5.5	moderately warm	30–40
5.6	moderately cold	20–30
5.7	cold	10–20
5.8	very cold	0–10
5.9	extremely cold	< 0

6. Average Mean Precipitation of Driest Month (inches)

6.1	extremely wet	>16
6.2	very wet	8–16
6.3	moderately wet	4– 8
6.4	moderately dry	2– 4
6.5	very dry	1– 2
6.6	extremely dry	< 1

7. Average Mean Precipitation of Wettest Month (inches)

7.1	extremely wet	>16
7.2	very wet	8–16
7.3	moderately wet	4– 8
7.4	moderately dry	2– 4
7.5	very dry	1– 2
7.6	extremely dry	< 1

II.BB. Local Climate

Same characters as Sectional Climate (II.B.).

II.C. Natural Area Climatic Site Type

1. Temperature Relations

1.1 Warmer than local climate
1.2 Cooler than local climate
1.3 Similar to local climate

2. Precipitation Relations

2.1 Wetter than local climate
2.2 Drier than local climate
2.3 Similar to local climate

II.CC. Community Climatic Site Type

1. Temperature Relations

1.1 Warmer than natural area climate
1.2 Cooler than natural area climate
1.3 Similar to natural area climate

2. Precipitation Relations

2.1 Wetter than natural area climate
2.2 Drier than natural area climate
2.3 Similar to natural area climate

III. Soils

The Soil Survey Staff gives several different definitions for the word soil. The simplest definition describes **soil** as the natural medium for the growth of land plants, whether or not it has discernible horizons. The Soil Survey Staff (1975, p. 1) more precisely defines **soil** as

> the collection of natural bodies on the earth's surface, in places modified or even made by man of earthy materials, containing living matter and supporting or capable of supporting plants out-of-doors. Its upper limit is air or shallow water. At its margins it grades to deep water or to barren areas of rock or ice. Its lower limit to the not-soil beneath is perhaps the most difficult to define. Soil includes the horizons near the surface that differ from the underlying rock material as a result of interactions, through time, of climate, living organisms, parent materials, and relief. In the few places where it contains thin, cemented horizons that are impermeable to roots, soil is as deep as the deepest horizon. Most commonly soil grades at its lower margin to hard rock or to earth materials virtually devoid of roots, animals, or marks of other biologic activity. The lower limit of soil, therefore, is normally the lower limit of biologic activity, which generally coincides with the common rooting depth of native perennial plants. Yet in defining mapping units for detailed soil surveys, lower layers that influence the movement and content of water and air in the soil of the root zone must also be considered.

As with most aspects of the natural world, soils can be mapped at several levels of differentiation. United States soil maps illustrate large geographic regions predominantly covered by a particular soil order, the highest and most inclusive level of soil classification. Regionally similar soils over extensive stretches of land surface develop under similar environmental conditions, conditions produced by regional similarities in climate, vegetation, source rock, topography, and length of time that a region has existed without major geologic alterations of the land surface.

When looking at similarly constructed regional vegetation, climatic, geologic, hydrologic, and physiographic maps, one can see a great deal of uniformity in the size, shape, and location of the regions for each of these major themes. These regions have similar boundaries because large-scale correspondences in characters differ significantly from the major controlling characters in neighboring regions. These defining characters are intricately interrelated, and all depend on the others for their existence. Thus, certain vegetation patterns, climates, pedologies, geologies, hydrologies, and topographies frequently are found together.

Regional similarities in arrangement and association of various biotic and abiotic factors can be differentiated into smaller, even more similar geographic levels. Regardless of criteria used for mapping, each of the major regions contains numerous variations within its large-scale features and also includes some features common to neighboring regions. These differences result in a wide degree of variability within a region but not a variation so random or uncontrolled that it cannot be recognized, identified, and classified.

As with most aspects of nature, soil is a continuum and covers the earth in all places, except where it is interrupted by deep water, bare rock, or ice. This continuum exhibits numerous changes in its internal characters, changes brought about by modifications of factors that have direct or indirect influence on the soil's development (Soil Survey Staff, 1975). These influential factors include five major categories—topography, parent material, climate, vegetation, and time—that contain innumerable subdivisions. When one considers the infinite combinations of these subdivisions, one can see how soils have developed into the complex system of small-scale variations now recognized. In the United States over 10,500 series, the lowest level of soil classification, are recognized, and this number is likely to increase with more extensive, more comprehensive field surveys.

The Soil Survey Staff (1975) recognizes that soils vary from extensive, regional trends down to the smallest observable and mappable units (pedons). Numerous features affect local, small-scale differences—variations in deposits left by running water, burrowing animals, taprooted plants, fallen trees, and plants that collect different elements (Soil Survey Staff, 1975). A major control on local soil diversity is the variation in composition, grain size, fracture patterns, bedding, and other features common to the underlying lithologies. Changes in these factors can produce many small-scale differences, even in the space of a mere foot or yard.

Soil scientists recognize the importance of small-scale variations by establishing the pedon as the official unit of sampling soils. The Soil Survey Staff (1975, p. 5) states:

> A **pedon** [boldfacing added by the editor] is a three-dimensional body of soil that has lateral dimensions large enough to include representative variations in the shape and relation of horizons and in composition

of the soil. Its area ranges from 1 to 10 m², depending on the nature of the variability in the soil. Where the cycle of variations is less than 2 m long and all horizons are continuous and of nearly uniform thickness, the pedon has an area of approximately 1 m². Where horizons or other properties are intermittent or cyclic and recur at linear intervals of 2 to 7 m, the pedon includes one-half of the cycle. If horizons are cyclic but recur at intervals greater than 7 m, the pedon reverts to an area of approximately 1 m² and more than one soil is usually represented in each cycle.

Although controlled by and dependent on many other aspects of nature for its development and maintenance, soil plays a very important part in natural systems and significantly influences plant and animal diversity and distribution. The soil in which plants are rooted acts as a stabilizing medium for plants to anchor themselves and serves as a reservoir from which plants draw needed nutrients and water. Soil properties of nutrient and water acquirement and retention and soil depth vary and greatly control the distribution of plant life. Plant life in turn exerts great control over animal distribution. These small-scale variations on the series, pedon, or polypedon levels and minor soil changes are most important to recognize and identify when determining factors that control community and species distributions. Sometimes these variations can be so great that small pockets of soil belonging to completely different orders (the highest level of soil classification) may be found in a community dominated by another order.

The Soil Survey Staff (1975) developed the soil classification system adopted for use in this ecological diversity classification scheme. The system is described in detail in Agriculture Handbook No. 436, *Soil Taxonomy: A Basic System of Soil Classification for Making and Interpreting Soil Surveys* (Soil Survey Staff, 1975). Russian soil scientists formulated the concept upon which the soil classification is based.

> Soils were conceived as "independent natural bodies, each with a unique morphology resulting from a unique combination of climate, living matter, earthy parent materials, relief, and age of land form. The morphology of each soil, as expressed by a vertical section through the differing horizons, reflected the combined effects of the particular set of genetic factors responsible for its development." [Soil Survey Staff, 1975, p. 1]

This classification is divided into six hierarchical levels, arranged from the highest and most inclusive taxon to the lowest and most differentiated: order, suborder, great group, subgroup, family, and series. Each level contains a set of divisions defined at approximately the same scale of generalization or abstraction, and each level contains all soils. *Soil Taxonomy* (Soil Survey Staff, 1975) describes and explains this classification in detail. It would be fruitless to try to summarize such a monumental scheme in a few short pages; anyone needing additional information concerning this aspect of the Ecological Diversity Classification System should refer to *Soil Taxonomy*.

The soil classification system is truly hierarchical, such that once a level is identified, all higher levels for a particular soil are also known. It is important to determine and to record the higher levels of each soil series. Once a soil series is ascertained, the more general levels may be determined by consulting *Soil Series of the United States, Puerto Rico, and the Virgin Islands: Their Taxonomic Classification* (Soil Survey Staff, 1972). Each name from order through subgroup (III.A.–III.BB.) consists of a set of modifying prefixes attached to a root term based on the name of the corresponding soil order, with a separate modifying term added at the subgroup level. Each of these modifiers has a specific meaning and identifies a distinguishing character of the soil being described. As an example, the subgroup (III.BB.) Typic Quartzipsamment can be divided into the following modifying term, modifying prefixes, and root term: Typic/Quartzi/psamm/ent, meaning it is of the Entisol order (-ent), the Psamment suborder (psamm-), the Quartzipsamment great group (quartzi-), and the Typic Quartzipsamment subgroup (Typic).

The soil hierarchies section of this discussion contains only a minimum of explanation. Table 4 (p. 55) includes a complete listing of soil orders and suborders (III.A. and III.AA.) and selected examples for the other hierarchical levels. For an excellent soils text, see Buol et al. (1980).

Soil Hierarchies

Code	*Hierarchy*	*Inclusion*
III.A.	Soil System	Soil Order
III.AA.	Soil Subsystem	Soil Suborder
III.B.	Soil Class	Soil Great Group
III.BB.	Soil Subclass	Soil Subgroup
III.C.	Soil Generitype	Soil Family
III.CC.	Soil Type	Soil Series

Soil Order (III.A.). The order represents the greatest degree of soil generalization and consists of 10 hierarchical elements, each with an identifying name ending in *-sol*: Entisol, Vertisol, Inceptisol, Aridisol, Mollisol, Spodosol, Alfisol, Ultisol, Oxisol, and Histosol. Soil orders are "differentiated by the presence or absence of diagnostic horizons or features that are marks in the soil of differences in the degree and kind of the dominant sets of soil-forming processes that have gone on" (Soil Survey Staff, 1975). **Soil horizons** are soil layers that lie approximately parallel to the land surface and differ from adjacent genetically related layers in physical, chemical, and biological properties or characteristics, such as color, structure, texture, consistency, kinds and numbers of organisms, and degree of acidity or alkalinity.

A set of characters used as the distinguishing criteria for a particular soil order may not be consistent and parallel with characters used in the differentiation of other orders. Steila (1976) gives the following examples. Specific compositions differentiate some orders, such as **Histosols**, which are composed of organic debris. A mixture of compositions and textures characterizes some orders, such as **Vertisols**, which are clayey soils that exhibit deep, wide cracks at some time during the year. Certain diagnostic horizons distinguish other orders, such as **Spodosols**, which have a spodic or a placic horizon, or **Mollisols**, which have a mollic epipedon or its equivalent.

Each order characteristically occurs in certain environments and contains specific types of horizon development and other features associated with those environments. In an environmental continuum soil orders rank from immature Entisols to mature Ultisols. **Entisols** are "soils that have little or no evidence of development of pedogenic horizons" (Soil Survey Staff, 1975). The absence of these pedogenic horizons may be caused by several factors related to time, topography, or parent material. Many Entisols have not had time to develop horizons. Others exist on steep, actively eroding slopes or on floodplains or glacial outwash plains that receive new deposits of alluvium at frequent intervals, thus preventing horizon development (Soil Survey Staff, 1975). Some Entisols consist of quartz or other minerals that do not alter and form horizons (Soil Survey Staff, 1975). In midlatitude climates, the **Ultisols** are "more thoroughly weathered and have experienced greater mineral alteration than any other soil" (Steila, 1976). As their name implies, these soils are the ultimate, or most mature, soils. If the majority of soils in most environments remain undisturbed and subsequently stabilize and develop, they will weather eventually into an Ultisol.

Soil Suborder (III.AA.). The suborders are order subdivisions based on chemical and/or physical properties that indicate drainage conditions or that possess genetic differences caused by climate and vegetation (Steila, 1976). Forty-seven suborders are currently recognized in the United States. These are listed in the soil hierarchical elements (Table 4, p. 55).

Suborders within a given order are usually readily discernible. For example, Entisols are divided into five suborders, which Steila (1976) defines in the following manner. **Aquents** exhibit evidence of wetness, which plays an important role in soil development. **Arents** "lack horizons because they have been deeply mixed by plowing, spading, or moving by man." **Fluvents** "form in recent water-deposited sediments, primarily in flood plains, fans, and deltas of rivers and small streams, but not in back swamps where drainage is poor." **Orthents** develop on recent erosional surfaces. **Psamments** form in "poorly graded sands of shifting or stabilized sand dunes, cover sands, or parent materials sorted in an earlier geologic cycle."

Soil Great Group (III.B.). The suborders are broken down into great groups on the basis of the kind and array of diagnostic horizons. Thus far 185 great groups are known to occur in the United States. Soils at this level are given the same name if they have the following properties in common (Soil Survey Staff, 1975): (1) close similarities in kind, arrangement, and degree of expression of horizons; (2) close similarities in soil moisture and temperature regimes; and (3) similarities in base saturation. The Soil Survey Staff (1975, p. 77) further explains the great group concept:

> At as high a categoric level as possible, it is desirable to consider all the horizons and their nature collectively as well as the moisture and temperature regimes. The moisture and temperature regimes are causes of properties, and they also are properties of the whole soil rather than of specific horizons. At the levels of the order and suborder, only a few of the most important horizons could be considered because there are few taxa in those categories. At the great group level, therefore, we try to consider the whole soil, the assemblage of horizons, and the most significant properties of the whole soil, selected on the basis of the numbers and importance of accessory properties. Although the definition of a great group may involve only a few differentiae, the accessory properties are many times that number. In a few taxa, soil moisture and soil temperature regimes are not defined but are accessory properties to some that are defined.

The following represent the great groups of the Aquent suborder (taken from Steila, 1976). **Cryaquents** are cold, wet soils of high mountains, tundra, or cold coastal marshes. **Fluvaquents** are wet soils of floodplains and deltas of mid and low latitudes. **Haplaquents** are wet Aquents in upland depressions where fresh sediments do not accumulate. **Hydraquents** are clayey soils of tidal marshes that are permanently saturated with water. **Psammaquents** have sandy textures and gray or mottled gray colors. **Sulfaquents** have sulfidic materials within 20 inches of the surface. **Tropaquents** are permanently warm and wet soils in depressions of intertropical regions.

Soil Subgroup (III.BB.). Great groups are further divided into subgroups, of which 970 presently are known in the United States.

> Through the categories of order, suborder, and great group, emphasis has been placed on marks or causes of sets of processes that appear to dominate the course or degree of soil development. In addition to these dominant marks, many soils have properties that, although apparently subordinate, are still marks of important sets of processes. Some of these appear to be marks of sets of processes that are dominant in some other great group, suborder, or order, but in a particular soil they only modify the marks of other processes. . . . Other properties are marks of sets of processes that are not used as criteria of any taxon at a higher level. [Soil Survey Staff, 1975, p. 79]

Thus three major types of subgroups are acknowledged (Soil Survey Staff, 1975):

> The central concept of the great group.—This is not necessarily the most extensive subgroup.
>
> The integrades or transitional forms to other orders, suborders, or great groups.—The properties may be the result of sets of processes that cause one kind of soil to develop from or toward another kind of soil, or otherwise to have intermediate properties between those of two or three great groups.
>
> Extragrades.—These subgroups have some properties that are not representative of the great group but that do not indicate transitions to any other known kind of soil.

The division of the great group Cryaquent into the two subgroups Typic Cryaquents and Andaqueptic Cryaquents supplies an example of the subgroup concept. Brief definitions (Soil Survey Staff, 1975) of these two subgroups suffice for this discussion. A **Typic Cryaquent** has a cover of water-loving trees, mostly black spruce, with a ground cover of *Sphagnum* spp. and sedges, lies on a broad outwash plain in southeastern Alaska, and exhibits colors typical of a moist soil. An **Andaqueptic Cryaquent** differs from the Typic Cryaquent in that it has a layer in the upper 75 cm that is 18 cm or more thick and rich in pyroclastics. The clays in this layer do not disperse well and have a high pH-dependent charge. The rare Andaqueptic Cryaquents occur only on wet floodplains in Alaska and in the high mountains of the northwestern states.

Soil Family (III.C.). In the family category, "the intent has been to group the soils within a subgroup having similar physical and chemical properties that affect their responses to management and manipulation for use" (Soil Survey Staff, 1975). About 4,500 families currently are discerned in the United States. Families are defined primarily to provide groupings of soils with restricted ranges in (1) particle-size distribution in horizons of major biologic activity below plow zone; (2) mineralogy of the same horizons that are considered in naming particle-size classes; (3) temperature regime; (4) thickness of the soil penetrable by roots; and (5) a few other properties that are used in defining some families to produce the needed homogeneity.

The technical family name is descriptive. Unlike the names for the preceding four levels, a family name includes the subgroup and a set of adjectives that describe the particle-size class, the mineralogy, and the temperature regime. A few families also incorporate descriptive terms for depth of soil, consistency, moisture equivalent, and other properties (Soil Survey Staff, 1975).

An example of a family is "fine, montmorillonitic, thermic Typic Hapludalf," which is an Alfisol of the Typic Hapludalf subgroup. This family encompasses all soils of the given subgroup that has 35% through 59% clay in its fine-earth fraction (fine), that has more than half of this clay being montmorillonite and nontronite by weight or has more montmorillonite than any other one clay mineral (montmorillonitic), and that has a mean annual soil temperature ranging from 15° to 22°C (59° to 72°F) at a depth of 50 cm or at a lithic or paralithic contact, whichever is shallower, with the difference between mean summer and mean winter soil temperature being greater than 5°C (thermic).

Soil Series (III.CC.). The series is the lowest hierarchical level. Approximately 10,555 series have been discovered in the United States. "The differentiae used for series are mostly the same as those used for classes in other categories, but the range permitted in one or more properties is less than is permitted in a family or in some other higher category" (Soil Survey Staff, 1975, p. 80). The process for determining soil series is discussed in Part IV on pages 389 to 390.

The Soil Survey Staff (1975) feels that the function of the series is pragmatic and that the differences within a family that are important to soil use should be considered in classifying soil series. These differences embody particle size, texture, mineralogy, amount of organic matter, structure, and other characters that are not family differentiae. The Soil Survey Staff (1975, p. 80) makes two kinds of distinctions between series: "First, the distinctions between families and between classes of all higher categories are also distinctions between series; a series cannot range across the limits between two families or between two classes of any higher category. Second, distinctions between similar series within a family are restrictions in one or more but not necessarily between all of the ranges in properties of the family."

Nomenclature of soil series is discussed by the Soil Survey Staff (1975, p. 84). Names of series as a rule are abstract place names, usually taken from a place near the one where the series was first recognized. It may be the name of a town, a county, or some local feature, although some series in sparsely settled regions have coined names. Most of the series names have been carried over from earlier classifications, some having been in use since 1900. The carryover of series names from earlier soil classifications helps maintain a link with the older soil maps and other publications, thereby enabling these older data to be easily updated and related to the modern classification.

Population Soil Site Type (III.D.). This is discussed on page 90 in the last section of this chapter.

Soil Summaries and Characterizations

1. Soils Excerpts from Community Diversity Summaries

a. *Juncus subcaudatus/Rhynchospora capitellata/Rhynchospora alba*: A. Histosol; AA. Saprist. B. Borosaprist; BB. Lithic Borosaprist. C. Not determined; CC. Not determined.

b. *Quercus rubra* var. *borealis/Kalmia latifolia–Rhododendron catawbiense*: A. Ultisol; AA. Udult. B. Hapludult; BB. Humic Hapludult. C. Mixed, mesic, acidic, skeletal, friable loam; CC. Clifton or Porters.

c. *Rubus allegheniensis/Carex brunnescens*: A. Inceptisol; AA. Umbrept. B. Haplumbrept; BB. Typic Haplumbrept. C. Coarse-loamy, mixed, mesic, acidic; CC. Burton stony loam.

d. *Acer rubrum*–mixed bottomland hardwoods/Mixed bottomland hardwoods/*Saururus cernuus*//Mixed lianas: A. Ultisol; AA. Udult. B. Hapludult; BB. Aquic Hapludult. C. Fine-loamy, mixed, thermic; CC. Altavista.

e. *Spartina alterniflora*: A. Entisol; AA. Aquent. B. Psammaquent; BB. Typic Psammaquent. C. Mixed, thermic; CC. Carteret.

2. Soils Excerpts from Community Ecological Characterizations (Pedologically)

a. *Juncus subcaudatus/Rhynchospora capitellata/Rhynchospora alba*: Lithic Borosaprist soil.

b. *Quercus rubra* var. *borealis/Kalmia latifolia– Rhododendron catawbiense*: Mixed, mesic, acidic, skeletal, friable Humic Hapludult, Clifton or Porters loam soil.

c. *Rubus allegheniensis/Carex brunnescens*: Coarse-loamy, mixed, mesic, acidic Typic Haplumbrept, Burton stony loam soil.

d. *Acer rubrum*–mixed bottomland hardwoods/Mixed bottomland hardwoods/*Saururus cernuus*//Mixed lianas: Fine-loamy, mixed, thermic Aquic Hapludult, Altavista soil.

e. *Spartina alterniflora*: Mixed, thermic Typic Psammaquent, Carteret soil.

Table 4. *Soil Hierarchical Elements*

III.A. Soil Order

1. Alfisols
2. Aridisols
3. Entisols
4. Histosols
5. Inceptisols
6. Mollisols
7. Oxisols
8. Spodosols
9. Ultisols
10. Vertisols

III.AA. Soil Suborder

Alfisols
1. Aqualfs
2. Boralfs
3. Udalfs
4. Ustalfs
5. Xeralfs

Aridisols
6. Argids
7. Orthids

Entisols
8. Aquents
9. Arents
10. Fluvents
11. Orthents
12. Psamments

Histosols
13. Fibrists
14. Folists
15. Hemists
16. Saprists

Inceptisols
17. Andepts
18. Aquepts
19. Ochrepts
20. Plaggepts
21. Tropepts
22. Umbrepts

Mollisols
23. Albolls
24. Aquolls
25. Borolls
26. Rendolls
27. Udolls
28. Ustolls
29. Xerolls

Oxisols
30. Aquoxs
31. Humoxs
32. Orthoxs
33. Ferroxs
34. Ustoxs

Spodosols
35. Aquods
36. Ferrods
37. Humods
38. Orthods

Ultisols
39. Aquults
40. Humults
41. Udults
42. Ustults
43. Xerults

Vertisols
44. Torrerts
45. Uderts
46. Usterts
47. Xererts

III.B. Soil Great Group

Selected examples:

Haplumbrepts   Cryandepts   Argiustolls

III.BB. Soil Subgroup

Selected examples:

Aquic Xerofluvents   Xerollic Haplargids   Vertic Haplaquepts

III.C. Soil Family

Selected examples:

Fine-loamy, siliceous, thermic Humic Hapludults
Cindery, frigid Typic Vitrandepts
Euic, hyperthermic Lithic Medihemists

III.CC. Soil Series

Selected examples:

Cobb   Foxhome   Waverly
Fontana   Owego   Wedowee

IV. Geology

The earth, a multibillion-year-old sphere of complexly interacting geologic processes and products, is composed of countless rock compositions, lithologic associations, and geologic structures. To understand the evolution, distribution and diversity of life, one must consider all aspects of geology, ranging from such comprehensive topics as the origin of the earth and plate tectonics to such small-scale, seemingly minor features as local outcrop patterns and rock structure. Every facet of geology has probably exercised some control in the origin, evolution, and migration of the world's flora and fauna.

The classification scheme proposed in this publication is intended for employment in natural area work and consequently is designed for observation and analysis of the biosphere on community and population levels. Furthermore, because the geologic portion is necessarily constructed to aid in recognizing and understanding geologic features that control community and population distributions, it is limited to the petrologic, petrographic, lithologic, and structural processes and products controlling these distributions. Thus, for the purposes of this treatment, **geology** involves the study of the local origin, chemical and mineral composition, classification, outcrop pattern, and structure of rocks. **Structure** refers to physical features of rocks, such as fractures, foliation, bedding, jointing, particle size, mineral alignment, porosity, and permeability; to the general disposition, attitude, arrangement, and relative position of these features within a rock; and to alteration of these features by deformational processes, such as folding and faulting (Gary et al., 1972). Structural features may or may not be related directly to chemical composition.

We attempt to arrange the major geologic processes and products important for natural area work into a system that is at least partially hierarchical, that is, one in which an entry at any level of the system permits the completion of all higher levels within the system. All levels of this geologic classification provide information that aids in the understanding of how rocks or weathered products of rocks affect the life inhabiting them. As will be seen, the Community Geologic Site Type (IV.CC.) and Population Geologic Site Type (IV.D.) describe the geologic features that have direct, immediate control over a particular community and population, respectively. To comprehend better the presence of these community- and population-level features and the control they exert over species distribution, it is important to understand several rock characteristics: origin, method of formation, chemical and mineral composition, internal structure, and deformational history. Levels IV.A. through IV.C. provide this information. In fact, all geologic variations within a study site should be recognized, especially if they regulate differences in biotic systems or in other abiotic systems, such as pedology, hydrology, or topography. In most cases, a competent field geologist with some prior knowledge of the regional geology will be able to complete most, if not all, of the information requested from his personal observations in a study site or from consideration of preexisting available geologic data and/or hand samples gathered by other qualified field scientists.

In summation, the major goal of geologic analyses is to increase our understanding of geologic diversity and of biotic-abiotic relationships from a geologic perspective.

Geologic Hierarchies

Code	*Hierarchy*	*Inclusion*
IV.A.	Geologic System	Rock System
IV.AA.	Geologic Subsystem	Rock Subsystem
IV.B.	Geologic Class	Rock-Sediment Chemistry
IV.BB.	Geologic Subclass	Rock-Sediment Occurrence
IV.C.	Geologic Generitype	Natural Area Geologic Site Type
IV.CC.	Geologic Type	Community Geologic Site Type

Rock System (IV.A.). The Rock System consists of three major classes of rocks recognized on the basis of origin: igneous, sedimentary, and metamorphic. Geologists agree that these three systems include essentially all known rocks. Categorizing the rocks at this level provides the basic framework for adding more detailed information at subsequent levels.

A detailed discussion of these three systems may be found in any introductory or advanced petrology book. **Igneous** rocks are those "solidified from molten or partly molten material, i.e., from a magma" (Gary et al., 1972). **Sedimentary** rocks are those "resulting from the consolidation of loose sediment that has accumulated in layers; consisting of mechanically formed fragments of older rock transported from its source and deposited in water, or from air or ice, or formed by precipitation from solution, or consisting of the remains or secretions of plants and animals" (Gary et al., 1972). **Metamorphic** rocks include "any rock derived from pre-existing rocks by mineralogical, chemical, and structural changes, essentially in the solid state, in response to marked changes in temperature, pressure, shearing stress, and chemical environment at depth in the Earth's crust, i.e., below the zones of weathering and cementation" (Gary et al., 1972).

Rock Subsystem (IV.AA.). Rock Subsystems are subdivisions of the Rock Systems and display attributes unique to each subsystem. These subsystems may overlap in various chemical or physical properties, but they differ significantly in their mode of origin and hence in certain characteristic features.

The igneous system contains two subsystems: plutonic and volcanic rocks. **Plutonic** rocks form at considerable depth by crystallization of magma or by chemical alteration and are characteristically medium to coarse grained (Gary et al., 1972). **Volcanic** rocks result from volcanic action at or near the earth's surface by explosive ejection or lava extrusion and are generally finely crystalline or glassy (Gary et al., 1972).

The sedimentary system includes three subsystems: chemical, clastic, and organic. **Chemical sedimentary** rocks form either from direct water precipitates or from inorganic remains of life forms, such as calcium carbonate or siliceous shells (Gary et al., 1972). **Clastic sedimentary** rocks form principally from broken fragments of preexisting, mechanically transported and deposited rocks (Gary et al., 1972). **Organic sedimentary** rocks form principally from organic remains and include such substances as muck, peat, and the various coals (Gary et al., 1972). The nature of sediment deposition and subsequent consolidation into sedimentary rock causes frequent interbedding of the sedimentary subsystems.

The metamorphic system encompasses three subsystems based on the primary igneous or sedimentary origin of the rock: metaplutonic, metavolcanic, and metasedimentary. A **metaplutonic** rock is derived from a plutonic igneous rock, a **metavolcanic** rock from a volcanic igneous rock, and a **metasedimentary** rock from a sedimentary rock. As with sedimentary rocks, the metamorphic subsystems sometimes are associated, especially metavolcanic and metasedimentary rocks. If such a mixture is recognizable, it should be recorded. The process of metamorphism commonly destroys identifying characters of the preexisting rock (especially in the higher grades) and thus makes it impossible to identify readily the rock's origin. In this case, the origin should be listed as undetermined.

Recognition of these subsystems provides a geologist with pertinent information thereby allowing inferences of rock characteristics, such as particle size, composition, maximum or minimum rock body size, laterally adjacent variation in rock type, and occurrence of various types of planar features, such as fractures. This information aids in the interpretation of many features listed in the lower levels.

Rock-Sediment Chemistry (IV.B.). Rock-Sediment Chemistry identifies the major chemical components of the rock type underlying the community or population being studied. Rock chemistry, identified both elementally and mineralogically, is important because it controls many abiotic and biotic features of a community or population. It can affect weathering rates, soil composition, elements and nutrients released from the rock, and ground and surface water chemistry. Rock chemistry can also influence such larger-scale geologic features as the mode of implacement of plutonic rocks, the mode of ejection of volcanic rocks, the degree and/or type of metamorphism, the method or site of deposition of sediments, and even the manner in which a rock deforms when subjected to stress and/or strain.

Rock chemistry is employed differently for each rock system. Igneous rocks are classified chemically in four manners: (1) differences in and/or the abundance of certain minerals (Bayly, 1968); (2) differences in and/or the abundance of chemical components of the minerals (Bayly, 1968); (3) absolute or relative abundance of quartz, potash feldspar, sodium plagioclase, calcium plagioclase, Fe Mg silicates, and feldspathoidal minerals; and (4) the calculated change in absolute or relative abundance of the different oxides within the rock. Nevertheless, the resulting classifications generally agree and divide most igneous rocks into four classes based on weight percentage of SiO_2: acidic (over 66%), intermediate (52%–66%), basic (45%–52%), and ultrabasic (under 45%) (Bayly, 1968). Many geologists also employ another igneous rock classification that proves useful: felsic, mafic, and ultramafic. **Felsic** is a mnemonic adjective derived from *fe*ldspar + *l*enad (feldspathoid) + *si*lica + *c* and is applied to an igneous rock having light-colored minerals in its mode (Gary et al., 1972). **Mafic** is a mnemonic adjective derived from *ma*gnesium + *f*erric + *ic* and is applied to an igneous rock composed chiefly of one or more ferromagnesian, dark-colored minerals in its mode (Gary et al., 1972). An **ultramafic** rock has a low silica content (less than that of a basic rock).

Problems exist with these two igneous rock classifications because some geologists use the terms synonymously (for example, felsic for acidic, mafic for basic, ultramafic for ultrabasic), whereas others utilize each term as defined here. For instance, Gary et al. (1972) state that ultramafic is not necessarily synonymous with ultrabasic, for example, some ultramafic rocks have a high SiO_2 content and are not ultrabasic, and some ultrabasic rocks (SiO_2 = 43.2%) are not necessarily ultramafic. Both igneous classifications serve as hierarchical elements in this Ecological Diversity Classification System.

The chemical classification of sedimentary rocks is based on the presence of certain minerals, such as calcite, aragonite, dolomite, quartz, phosphate minerals, salts, sulfur-bearing minerals, iron-bearing minerals, and organics (Blatt et al., 1972). Unlike an igneous rock, which can belong to only one chemical class, a sedimentary rock can belong to several. As examples, a calcareous limestone can also contain a significant amount of dolomite and thus also be dolomitic, and a quartz arenite, predominantly quartz and therefore siliceous, can contain calcite cement or abundant fossils and thus also be calcareous or can contain hematite and thus also be ferruginous. If more than one important chemical component occurs in a rock, especially if the different components exert some control over the biota, each of the chemical classes should be recognized and recorded.

All metamorphic rocks ultimately derive from preexisting igneous or sedimentary rocks and therefore follow the chemical classifications of these two rock systems. The chemical classification that most appropriately describes the chemistry of the metamorphic rock being analyzed should be used.

In conjunction with this general information, the mineral or chemical either quantitatively or qualitatively determined to be dominant in the study site should be indicated at this higher level (IV.B.). The ideal procedure would be to determine rock chemistry by quantitatively analyzing the rock both chemically and mineralogically and to list the results under the Community Geologic Site Type (IV.CC.). Table 5 (p. 63) includes a list of terms frequently used.

Rock-Sediment Occurrence (IV.BB.). Rock-Sediment Occurrence deals with the specific origin of a particular rock body: was it magmatically emplaced, sedimentarily deposited, or metamorphosed? Although Rock System (IV.A.) and Rock Subsystem (IV.AA.) provide information regarding the origin of a particular rock, they do not concern specifics. Information relating to detailed methods of rock formation can provide a better understanding of numerous controlling factors within the study site. For instance, knowing the exact origin of a rock can supply a geologist with clues or data as to how the exposed surface area of a particular rock type may change, compositionally and texturally, over the terrain.

Plutonic igneous rocks are described according to their **mode of emplacement**—the process of intrusion of magma into preexisting rock (country rock). The terminology is based primarily on whether the plutonic rock body is **concordant**—the contacts are parallel to the bedding or foliation of the country rock (Trowbridge et al., 1962)—or **discordant**—the contacts are not parallel and hence cut across the bedding or foliation of the country rock (Trowbridge et al., 1962). Terminology depends secondarily upon size and shape of the emplacement. Volcanic igneous rocks are classified according to the manner of ejection and the resultant rock body size and shape.

Sedimentary rocks are identified according to their depositional environment. Because sedimentary rocks form at the earth's surface and are therefore relatively easily recognizable and distinguishable, many different depositional environments compose the hierarchical elements under this category. In contrast, most plutonic igneous and metamorphic rocks form at great depths within the earth and are therefore less well understood, as is reflected by the simpler origin terminology.

Metamorphic rocks are differentiated according to the particular type of metamorphism to which they were subjected. The metamorphic rock terminology in the geologic hierarchical elements (Table 5, p. 64) contains selected examples, some of which refer to the chemical changes a rock has undergone, others to the heat and pressure source, and still others to the extent of metamorphism that has occurred in a region. Possibly several terms referring to these three attributes can be applied correctly to a single rock.

Selected examples of terms used to describe rock origin are listed in the geologic hierarchical elements (Table 5, p. 63). Additional terminology may be found in advanced geology books that deal with specific lithologies. The most useful information concerning the origin of particular rock is published in articles specifically treating the rock body under investigation.

Natural Area Geologic Site Type (IV.C.) and Community Geologic Site Type (IV.CC.). The Natural Area and Community Geologic Site Types comprise a set of geologic attributes for the entire study site and each community within the site, respectively: (1) rock-sediment type underlying the community or study site; (2) deformational history of the rock-sediment type; (3) percentage of the area that consists of exposed rock; (4) the manner in which the rock is exposed; (5) the degree to which the exposed rock is weathered; and (6) rock-sediment fabric, which includes a number of structural and textural features. At these two levels geologic features that directly or indirectly influence what lives in the study site or community are circumscribed. Some of these geologic characteristics remain constant throughout a study site or community, hence requiring only a single hierarchical element entry. The entries can be identical for both the natural area and community levels, although frequently a feature exhibits a range of variables within a study site or community. In this case, either continuous variation is evident and only the end members of the range should be listed (for example, highly to slightly weathered rock) or discontinuous variation is evident and the entire spectrum of element entries should be listed (for example, shale, sandstone, and limestone). Throughout the following discussion definitions for terms are excerpted from Gary et al. (1972) unless otherwise indicated.

The **rock-sediment name** designates the rock or sediment type and is the communication symbol and reference base for information storage, retrieval, and use. This name constitutes the most essential feature to be determined, because it often indicates important characteristics: (1) pH (limestone—basic, quartz arenite—acidic); (2) texture (shale—clay, sandstone—sand); (3) composition (granite—feldspar, quartz, and micas; silica-cemented quartz arenite—high percentage SiO_2; fossiliferous, glauconitic, calcite-cemented quartz arenite—most clasts are quartz, cement is calcite, and rock contains some fossils and glauconite); (4) weathering resistance (in the eastern United States, quartz sandstone—highly resistant, shale—very easily weathered); and many others. The rock-sediment name also contributes to completion of the system's higher levels.

The minimum entry for the rock-sediment name should be gross lithology, such as granite, diorite, and gabbro for igneous rocks; conglomerate, sandstone, siltstone, shale, and limestone for sedimentary rocks; and gneiss, schist, phyllite, and slate for metamorphic rocks. See geologic hierarchical elements (Table 5) for additional sample entries. Although in some cases the gross lithologies imply general mineral compositions, most are based on rock textures.

The majority of rock types can be differentiated into more specific lithologies according to local changes in mineral composition (for example, hornblende gneiss), cementing agent (for example, calcite-cemented quartz arenite), particle size (for example, graywacke), and other characteristics used in rock classification. For example, sandstones (arenites) are classified primarily according to percentages of three major components: (1) quartz and chert, (2) unstable rock fragments (clays, micas, shale, slate, phyllite, and schist), and (3) feldspar, granite, and feldspathic gneiss. Three rock types exemplify this classification scheme: **quartz arenite**—contains 90%–100% quartz; **arkose**—contains 25% or more feldspars; **litharenite**—contains 25% or more unstable rock fragments, three-fourths of which are not feldspars. An additional aspect considered in their classification is the percentage of matrix (in a rock in which certain grains are larger than the others, the grains of smaller size compose the **matrix**). For example, an arenite (0%–15% matrix) becomes a graywacke (15%–50% matrix) with the addition of matrix. Each of these sandstones (quartz arenite, arkose, litharenite, graywacke) can be differentiated even further by stating specific mineral compositions, cementing agents, grain sizes, accessory minerals, maturity, and so forth. Several examples follow: (1) submature, medium- to coarse-grained, fossiliferous, glauconitic, calcite-cemented quartz arenite; (2) mature, medium- to very coarse-grained, silica-cemented quartz arenite; (3) submature, fine- to very coarse-grained, silica-cemented arkose; (4) immature, coarse to very coarse-grained, very fine to fine pebbly arkose. A number of selected examples of more specific rock-sediment names are presented in the geologic hierarchical elements (Table 5, p. 65).

With regard to communities and species, any minor changes in rock type composition should be noted, especially if these changes appear to affect community and species distributions. Apparently minor differences in rock composition commonly occur and often aid in the creation of microhabitats. For example, small pockets of calcite-cemented quartz arenite in a rock body predominantly cemented with quartz produce pockets of soil with a higher pH and a higher calcium content. In turn, these possibly can affect the occurrence and distribution of plant species possessing varying reaction and/or nutrient affinities. Likewise, small interbeds of shale in an area dominated by sandstone form pockets of clayey soil. Similarly, interbeds of sandstone on a slope underlain predominantly by shale often protrude as small ledges. If these same sandstone beds dip out of the slope at the proper angle, they also can serve as aquifers for ground water and act as outlets for small seepages or springs.

One must exercise caution when determining community or study site rock type(s). Rock very frequently outcrops in a nonrandom manner within a community or study site; if a rock of similar composition, texture, and structure underlies an entire community or study site, the rock would be expected to be exposed in a random manner. This often happens where scattered boulders of similar size, shape, composition, and distribution cover the terrain. Isolated, well-defined ledges, cliffs, or concentrated piles or rows of loose rock might not be representative of the entire community or study site. On the contrary, these nonrandom exposures suggest differential weathering, in which the more resistant rock becomes exposed while the less resistant rock erodes into soil. These variations are important to note if they account for microhabitats, but the identification of the overall habitat lithology is crucial.

Determination of the major habitat lithology for a community or study site can be accomplished by consolidating several bits of evidence. Local stream exposures and man-made road cuts often aid in this determination. Careful examination of subsurface rock fragments procured while digging soil pits should be undertaken to see if these fragments are of similar composition to exposed lithologies. Also important to note is the possibility that the soil could have derived from the parent material present as outcrops. Taken together, these clues may indicate whether or not the exposed rock is representative of the entire community or study site. If the investigator is not sufficiently trained or qualified to identify correctly the rock-sediment type, a hand sample(s) should be collected and presented to a professional geologist for examination.

Deformation is a general term for the process of folding, faulting, shearing, compression, or extension of rocks as a result of various earth forces. The overall deformational history of the area can be determined by consulting geologic maps and reports, by observing the rock outcrops, and by measuring strikes and dips.

Folding and faulting represent the two most common types of deformations and serve here as examples. **Folding** is the curving or bending of a planar structure (see definition below), such as rock strata, bedding planes, foliation, or cleavage, by deformation; the resultant structure constitutes a **fold**. **Faulting** is the process of fracturing and displacement that produces a **fault**, a surface or zone of rock fracture along which there has been displacement. For further selected examples see the geologic hierarchical elements (Table 5, p. 65).

Deformational history furnishes a physiographic and topographic picture of the area and information regarding overall rock outcrop patterns—how the rocks are exposed and therefore eroded. For an example on a very broad scale, the Valley and Ridge Province consists of a series of folds and hence erodes into elongate ridges rather than rounded hills. On the other hand, the Cumberland Plateau does erode into rounded hills because it has horizontal beds of rock that have not undergone folding. These same erosion and topographic patterns resulting from deformational history apply also to smaller-scale features, such as a ridge or a hill that a community might occupy.

Exposed rock outcrops at the earth's surface. Gary et al. (1972) define **exposed** as applying to a continuous area in which a rock formation or geologic structure is visible, either naturally or artificially, and is unobscured by soil, vegetation, water, or works of man. A chart listing modifiers and respective cover values for percentage of the study site or the community that consists of exposed rock appears in Table 5 (p. 65).

The manner in which the rock is exposed presents another feature requiring mention in these site types. In what way do the rocks lie or outcrop in the area? Does the exposed rock exist as scattered boulders, broken fragments found in the soil, ledges, cliffs, talus materials, stream exposures, or road cuts? In general, how were the rocks deposited in the area or how were they exposed? Are the rocks **in place** materials (in the situation in which they were originally formed or deposited) or **float** materials (not in their original situation)? In addition, appropriate characteristics of certain rock features should be noted, such as size and spacing of ledges. The fact that no exposed rock occurs in the area holds just as much significance as the above positive statements; therefore, this absence should be recorded as a hierarchical element. Notice that the hierarchical elements in Table 5 (p. 65) serve as selected examples.

The amount and the manner of rock exposure often establish microhabitats within an area and control species population distributions as well as entire communities. The need for caution in determination of a community or study site rock type was discussed in the preceding subsection.

Weathering encompasses the destructive process or group of processes constituting that part of erosion whereby earthy and rocky materials on exposure to atmospheric agents at or near the earth's surface are changed in character (color, texture, composition, firmness, or form), with little or no transport of the loosened or altered materials. Specifically, weathering is defined as the physical disintegration and chemical decomposition of rock. These two processes produce an in situ mantle of waste and prepare sediments for transportation. Some geologists include biologic changes and the corrosive action of wind, water, and ice.

The degree of weathering can be measured in a number of ways: (1) qualitatively—by thickness of the weathering rind on exposed rock surfaces, by the ease with which a weathered rock breaks in comparison with a fresh sample of the same lithology, and by estimation of the degree of alteration of the weatherable minerals because of their superficial appearance, for example, how completely the feldspars have been altered to clay; and (2) quantitatively—by several methods, one of which is the weathering-potential index, which measures the degree of susceptibility to weathering of a rock or mineral by computation from a chemical analysis (Reiche, 1943). In this Ecological Diversity Classification System a subjective estimation using modifiers ranging from fresh (unweathered) to completely decomposed suffices (see Table 5, p. 66, for a complete list of modifiers).

Knowledge of the degree of weathering of different rock types provides information within a given area on the rate of erosion, the rate of soil genesis, and the manner in which new rock is exposed. In addition, it can aid in predicting the relative age of the site and in identifying unnatural disturbances. Weathering rates also supply information concerning release of nutrients from the rock.

Rock-sediment **fabric** incorporates the notion of function or behavior (the correlative physical properties) as well as of form (arrangement of structural and textural components) and is the sum of all the structural and textural features of a rock. Many geologic factors control moisture availability and directly or indirectly influence the rate of water and dissolved nutrient movement into and out of the system. For this reason, the following fabric attributes are considered important in this system: structure, strike and dip of these structural features, porosity, permeability, degree of consolidation, and particle size. These attributes are defined below.

Structure is the general disposition, attitude, arrangement, or relative position of the rock masses of a region or an area. Structural features result from such rock movement processes as faulting, folding (both defined above under deformation), and igneous intrusion (see IV.BB.) or from original deposition processes. The term **fracture**—a general term for any break in a rock, whether or not it causes displacement, caused by mechanical failure by stress—exemplifies a structural feature. More specifically, faults (defined previously), **cracks** (partial or incomplete fractures), and **joints** (surfaces of actual or potential fractures or partings in a rock, without displacement) are types of fractures. **Foliation** represents a somewhat different type of structural feature in that it usually results during metamorphism from heat and pressure, which cause parallel aligning or banding of minerals or mineral concentrations. More generally, it refers to a planar arrangement of textural or structural features in a rock.

Very commonly structural features are **planar features** that lie in planes, usually implying more or less parallel planes, such as those of bedding. **Bed** refers to a layer of sedimentary rock that is distinguishable from the layers above and below by virtue of some discontinuity in rock type, internal structure, or texture (Blatt et al., 1972). References containing extensive discussions of structural features include Lahee (1961), Compton (1962), Billings (1972), and Pettijohn (1975). It would be helpful to consult some of these when determining hierarchical element entries in this category.

Strike is defined as the direction or trend that a structural surface, such as a bedding or fault plane, takes as it intersects the horizontal. It is measured by finding the compass direction of a horizontal line on the surface of the feature (Compton, 1962). **Dip** is the angle that a structural feature makes with the horizontal, measured perpendicular to the strike. Because strike and dip of planar structures (beds, fractures, joints, foliation, and others) directly influence rate of water and nutrient movement into and out of the system, it becomes important to note whether or not these structures dip into or out of the community.

Porosity and permeability concern the way in which constituent rock or sediment parts fit or pack together. Rock, soil, or other materials containing interstices exhibit these two properties. **Porosity** is commonly expressed as a percentage of the bulk volume of material occupied by interstices, whether isolated or connected. **Permeability** refers to the capacity of a porous rock, sediment, or soil for transmitting a fluid without impairment of the structure of the medium. Both these properties can be measured quantitatively by sophisticated techniques; however, in this system a subjective description of the rock serves the purpose, for example, highly porous, nonporous, slightly permeable, moderately permeable. A list of hierarchical elements from which to choose appears in Table 5 (p. 66).

Consolidation concerns any process whereby loosely aggregated, soft, or liquid earth materials become firm and coherent rock and includes such processes as **solidification**—consolidation of magma to form an igneous rock, **lithification**—consolidation of loose sediments to form a sedimentary rock, **welding**—consolidation by pressure either from the weight of superincumbent material or earth movement, and **cementation**—the process of precipitation of a binding material around grains or minerals in rocks. Once again, as in the previous categories, a subjective estimation of the degree of consolidation will be made by using the modifiers in the geologic hierarchical elements (Table 5, p. 66). If the specific type of consolidation is unknown, use the set of modifiers formed with the past participle "consolidated."

Particle size pertains to the general dimensions (such as average diameter or volume) of the particles in a sediment or rock, or of the grains of a particular mineral that make up a sediment or rock, based on the premise that the particles are spheres or that the measurements made can be expressed as diameters of equivalent spheres. The size classes generally used are given in the geologic hierarchical elements (Table 5, p. 67). Frequently a geologist can simply estimate particle sizes unaided or with a hand lens. The inexperienced should have available a slide composed of actual sets of grains corresponding to the smaller size classes.

Population Geologic Site Type (IV.D.). This is discussed on page 90 in the last section of this chapter.

Geologic Summaries and Characterizations

1. Geologic Excerpts from Community Diversity Summaries

a. *Juncus subcaudatus/Rhynchospora capitellata/ Rhynchospora alba*: A. Metamorphic; AA. Mixed metaigneous and metasedimentary. B. Intermediate to basic; BB. Regionally metamorphosed. C. Folded hornblende gneiss, with major foliation striking N65°E and dipping 35°W, and major fractures striking N22°W and dipping 73°E; CC. Hornblende gneiss, exposed as scattered, moderately weathered outcrops and transported cobbles and gravel.

b. *Quercus rubra* var. *borealis/Kalmia latifolia–Rhododendron catawbiense*: A. Metamorphic; AA. Undetermined. B. Intermediate to basic; BB. Regionally metamorphosed. C. Folded and fractured, variously exposed gneiss, with major foliation planes striking N65°E and dipping 35°W, and major fracture planes striking N22°W and dipping 73°E; CC. Hornblende gneiss, exposed as relatively fresh, extensive, ground-level flats and cobble- to boulder-sized float covering about 50% of the community.

c. *Rubus allegheniensis/Carex brunnescens*: A. Metamorphic; AA. Mixed metasedimentary and metaigneous. B. Intermediate to basic; BB. Regionally metamorphosed. C. Strongly folded gneisses and schists; CC. Hornblende gneiss, exposed as scattered rock flats on the ridge crest and ledges on the upper slopes.

d. *Acer rubrum*–mixed bottomland hardwoods/ Mixed bottomland hardwoods/*Saururus cernuus*// Mixed lianas: A. Sedimentary; AA. Clastic. B. Siliceous; BB. Floodplain stream deposit. C. Mixed fine clastics and organics; CC. Unbedded, unaligned, low permeability, unconsolidated silt.

e. *Spartina alterniflora*: A. Sedimentary; AA. Mixed clastic and organic. B. Siliceous and sulfaceous; BB. Estuarine tidal marsh deposit. C. Mixture of eolian sands and water-deposited fine clastics and organics; CC. Porous, unconsolidated, water-saturated, silty muck.

2. Geologic Excerpts from Community Ecological Characterizations (Geologically)

a. *Juncus subcaudatus/Rhynchospora capitellata/ Rhynchospora alba*: Intermediate to basic, regionally metamorphosed, mixed metaigneous and metasedimentary, folded hornblende gneiss exposed as scattered, moderately weathered outcrops and transported cobbles and gravel, with major foliation striking N65°E and dipping 35°W, and major fractures striking N22°W and dipping 73°E.

b. *Quercus rubra* var. *borealis/Kalmia latifolia–Rhododendron catawbiense*: Regionally metamorphosed, folded and fractured intermediate to basic hornblende gneiss, with major foliation planes striking N65°E and dipping 35°W, and major fracture planes striking N22°W and dipping 73°E, exposed as relatively fresh, extensive, ground-level flats and cobble- to boulder-sized float covering about 50% of the community.

c. *Rubus allegheniensis/Carex brunnescens*: Strongly folded, regionally metamorphosed, intermediate to basic, hornblende gneiss, exposed as scattered rock flats on the ridge crest and occasional ledges on the upper slopes.

d. *Acer rubrum*–mixed bottomland hardwoods/ Mixed bottomland hardwoods/*Saururus cernuus*// Mixed lianas: Clastic, siliceous, sedimentary, floodplain stream deposit, consisting of unbedded, unaligned, low permeability, unconsolidated silt.

e. *Spartina alterniflora*: A mixed clastic and organic, sedimentary, estuarine tidal marsh deposit, composed of siliceous and sulfaceous, porous, unconsolidated, water-saturated, silty muck.

Table 5. *Geologic Hierarchical Elements*

IV.A. Rock System

1. Igneous	2. Sedimentary	3. Metamorphic

IV.AA. Rock Subsystem

Igneous	Sedimentary	Metamorphic
1. Plutonic	3. Chemical	7. Metaplutonic
2. Volcanic	4. Clastic	8. Metavolcanic
	5. Organic	9. Metasedimentary
	6. Mixed (specify)	10. Mixed (specify)
	11. Undetermined	

IV.B. Rock-Sediment Chemistry

Selected examples:

Igneous and Metamorphic	Sedimentary and Metamorphic	
acidic	calcareous	salinaceous
intermediate	dolomitic	sulfaceous
basic	ferruginous	siliceous
ultrabasic	organic (carbonaceous)	mixed (specify)
felsic	phosphatic	others (specify)
mafic		
ultramafic		
others (specify)		

IV.BB. Rock-Sediment Occurrence

Selected examples:

Igneous-plutonic	discordant bodies	Igneous-volcanic
concordant bodies	discordant bodies	cinder cone
laccolith	batholith	composite cone
lopolith	cone dike	lava flow
placolith	dike	plateau basalt
sill	ring dike	pyroclastic air fall
	stock	pyroclastic flow
	general term	shield cone
	pluton	volcanic neck
		volcanic plug

Sedimentary		
alluvial fan	glacial till	oxbow lake
beach	glacioeolian	point bar
braided stream	glaciofluvial	precipitate
channel	glaciolacustrine	residual
deep ocean basin	lagoonal	road fill
delta	landslide	saline lake
eolian dunes	levee	slump
estuary	marine	swamp
evaporite	mine dump	talus
floodplain	organic	turbidite
fresh water lake		

Table 5. *(continued)*

Metamorphic

contact
dynamic
dynamothermal
isochemical
metasomatic
prograde
regional
retrograde
thermal
undetermined

IV.C. Natural Area Geologic Site Type

1. Rock-sediment type

Selected examples of gross lithologies:

Igneous

adamellite
andesite
anorthosite
ash deposit
basalt
dacite
diorite
dunite
felsite
gabbro
granite
granodiorite
lapillistone
latite
monzonite
obsidian
pegmatite
peridotite
perlite
phonolite
pitchstone
pumice
rhyodacite
rhyolite
scoria
syenite
trachyandesite
trachyte
traprock
tuff
vitrophyre
volcanic agglomerate
volcanic breccia
welded tuff

Sedimentary

arkose
bog iron
breccia
chert
clay
clay shale
claystone
coal
conglomerate
diatomite
feldspathic litharenite
geyserite
gravel
graywacke
guano
impact breccia
intraformational conglomerate
iron formation
lignite
limestone
- mudstone
- wackestone
- packstone
- grainstone
- boundstone
- crystalline

litharenite
lithic arkose
mud
mud shale
mudstone
oil shale
orthoconglomerate
paraconglomerate
peat
quartz arenite
radiolarite
residual breccia
sand
sandstone
silt
silt shale
siltstone
spiculite
solution breccia
subarkose
sublitharenite

Metamorphic

amphibolite
argillite
cataclasite
gneiss
granofels
hornfels
marble
metabasalt
metarhyolite
mylonite
phyllite
quartzite
scarn
schist
slate

Table 5. *(continued)*

Selected examples of more specific lithologies:

Sedimentary

kaolinitic clay shale
montmorillonitic clay shale
illitic clay shale
muscovite phyllitharenite
biotite phyllitharenite
calcite-cemented, limestone pebble paraconglomerate
hematite-cemented, granite pebble orthoconglomerate
calcite-cemented quartz arenite
quartz-cemented quartz arenite
phosphatic biomicrite
dolomitic biomicrudite

Igneous

quartz granite
aplite
quartz granodiorite
rhyolitic tuff
quartz monzonite

Metamorphic

hornblende gneiss
quartz-biotite gneiss
muscovite-biotite gneiss
garnet-bearing biotite gneiss
calcitic marble
dolomitic marble

2. Deformational history of rock-sediment type

Selected examples:

Folded

anticlinally folded
arched
chevron folded
fan folded
folded
homoclinally folded
isoclinally folded
monoclinally folded
overturned fold
recumbently folded
synclinally folded
undeformed

Faulted

dip-slip fault
fault
normal fault
reverse fault
strike-slip fault
thrust fault
transform fault
transverse fault
undeformed

3. Percentage of area that consists of exposed rock

		Cover value
3.1	very sparse	1 = 1-5%
3.2	sparse	2 = 5-12.5%
3.3	scattered	3 = 12.5-25%
3.4	dense	4 = 25-50%
3.5	very dense	5 = 50-100%

4. Manner in which the rock is exposed

Selected examples:

scattered boulders (in place or float)
broken fragments found only in the soil
ledges (give size and spacing)
cliff (give size)
talus material (give size of fragments)
no rocks present
stream exposures
artificial road cuts

Table 5. *(continued)*

5. Degree to which the exposed rock is weathered

5.1 fresh (unweathered)
5.2 slightly weathered
5.3 moderately weathered
5.4 highly weathered
5.5 completely decomposed

6. Rock-sediment fabric

6.1 Structure (plus thickness of or distance between features)

Selected examples:

bed
crack
fault
foliation
fracture
joint
schistosity
zone

6.2 Strike and dip of features in 6.1 (plus whether or not structures dip into or out of the area)

Selected example:

Foliation strikes N70^{o}W and dips 40°W into the slope.

6.3 Porosity

6.3.1 nonporous
6.3.2 slightly porous
6.3.3 moderately porous
6.3.4 highly porous
6.3.5 extremely porous

6.4 Permeability

6.4.1 impermeable
6.4.2 slightly permeable
6.4.3 moderately permeable
6.4.4 highly permeable
6.4.5 extremely permeable

6.5 Degree of consolidation

6.5.1 highly consolidated
6.5.2 moderately consolidated
6.5.3 loosely consolidated
6.5.4 unconsolidated

6.5.5 solidified

6.5.6 highly lithified
6.5.7 moderately lithified
6.5.8 loosely lithified
6.5.9 nonlithified

6.5.10 highly welded
6.5.11 moderately welded
6.5.12 loosely welded
6.5.13 nonwelded

6.5.14 highly cemented
6.5.15 moderately cemented
6.5.16 loosely cemented
6.5.17 uncemented

Table 5. *(continued)*

6.6	Particle size		
6.6.1	clay		<1/256 mm
6.6.2	silt		1/256–1/16 mm
6.6.3	sand		1/16–2 mm
	6.6.4	very fine sand	1/16–1/8 mm
	6.6.5	fine sand	1/8–1/4 mm
	6.6.6	medium sand	1/4–1/2 mm
	6.6.7	coarse sand	1/2–1 mm
	6.6.8	very coarse sand	1–2 mm
6.6.9	granule		2–4 mm
6.6.10	pebble		4–8 mm
6.6.11	cobble		8–256 mm
6.6.12	boulder		>256 mm
6.6.13	mixed (specify)		

IV.CC. Community Geologic Site Type

Same characters as Natural Area Geologic Site Type (IV.C.).

V. Hydrology

Hydrology is defined as "the science that deals with continental water (both liquid and solid), its properties, circulation, and distribution, on and under the earth's surface and in the atmosphere, from the moment of its precipitation until it is returned to the atmosphere through evapotranspiration or is discharged into the ocean; in recent years it has been expanded to include environmental and economic aspects" (Gary et al., 1972). Study of the properties of ocean water is not usually included in hydrology but instead in the independent science of oceanography or, more properly, oceanology. For simplicity's sake, in this particular classification system some characteristics of ocean water are included under hydrology; however, we are not concerned here with oceanic deep-water habitats.

We consider hydrology to be extremely important, not only in "wetland" situations but also in terrestrial systems because a gradual continuum exists between wet and dry environments (see the discussion in Section III in Chapter 1). Because all life depends upon water for existence, hydrology is related intricately to biotic systems. Hydrologic factors strongly influence distributions of species and communities.

Hydrologic Hierarchies

Code	*Hierarchy*	*Inclusion*
V.A.	Hydrologic System	Hydrologic System
V.AA.	Hydrologic Subsystem	Hydrologic Subsystem
V.B.	Hydrologic Class	Water Chemistry
V.BB.	Hydrologic Subclass	Water Regime
V.C.	Hydrologic Generitype	Natural Area Hydrologic Site Type
V.CC.	Hydrologic Type	Community Hydrologic Site Type

Hydrologic System (V.A.). The Hydrologic System (V.A.) refers to "a complex of wetland and deep-water habitats that share the influence of one or more dominant hydrologic, geomorphologic, chemical, or biological factors" (Cowardin et al., 1977). To this concept we have added the terrestrial habitat because hydrologic features play a significant role in this situation also. Six fixed categories constitute the Hydrologic System (V.A.): terrestrial, marine, estuarine, lacustrine, riverine, and palustrine. The last five are taken directly from Cowardin et al. (1977) and are defined within their report, whereas the terrestrial system has been added by us and is defined in Chapter 1 (p. 12). The concept of these major systems and their characteristics appears throughout the literature. Disagreements exist, however, as to precisely which attributes bound each system in space (pragmatically speaking).

In general, the boundaries and definitions proposed by Cowardin et al. (1977) are adopted for this diversity classification system. We disagree on a major point, however: we feel that if a Community Type is influenced by one of these particular systems, then the Community Type should fall under that system. For example, a *Phragmites communis* Community Type along a river edge would be placed under palustrine system according to Cowardin et al. (1977). We feel that it should be placed under riverine system because the river is the major influencing environment. Likewise, an *Eleocharis equisetoides* Community Type along a lake edge would be palustrine according to Cowardin et al. (1977), whereas it would be lacustrine according to our system. Another example would be a floodplain still strongly influenced by a river. We would call this riverine; Cowardin et al. (1977) would call this palustrine.

Hydrologic Subsystem (V.AA.). The Hydrologic Subsystems (V.AA.) are subdivisions of the above systems because within a given system the habitats can be grouped or differentiated based on dominant ecological factors, such as tidal influence, water duration, gradient, and velocity. At this level hierarchical elements are chosen from an established list (Table 6, p. 71). An explanation of the following elements may be found in Cowardin et al. (1977): marine—subtidal, intertidal; estuarine—subtidal, intertidal; lacustrine—limnetic, littoral. The supratidal subsystem under marine and estuarine and the subsystems under riverine, palustrine, and terrestrial are proposed by us and are discussed here.

The marine system includes three subsystems: supratidal, intertidal, and subtidal. The **supratidal zone**, the splash zone, occurs above the extreme high water mark. The moisture received from the ocean spray sometimes results in a wetland ecosystem.

The riverine system is divided into four subsystems: tidal, perennial, intermittent, and ephemeral. In the **tidal** subsystem the gradient is low and water velocity fluctuates under tidal influence (Cowardin et al., 1977). Leopold and Miller (1956) define the other three subsystems. A **perennial** stream carries some flow at all times. An **intermittent** stream is one in which, at low flow, dry reaches alternate with flowing ones along the stream length. **Ephemeral** streams carry water only during storms and are generally smaller but much more numerous than perennial ones.

The palustrine system contains two subsystems: aqueous and interaqueous. **Aqueous** includes that part of the palustrine system in which the substrate is continuously submerged. **Interaqueous** includes that part of the palustrine system in which the substrate is variously exposed and flooded by palustrine waters.

The terrestrial system contains five subsystems: wet, mesic, dry-mesic, dry-xeric, and very dry-xeric. The wet subsystem is a concept proposed by us, and the other four subsystems are adapted from the U.S.D.A. *Soil Survey Manual* (Soil Survey Staff, 1951).

A **wet** subsystem is one in which the soil remains saturated with water a substantial length of time but retains oxygenating conditions because of continual flow of water, thus allowing good aeration in the root zone of plants. Conditions do not exist that would support typical aquatic or wetland vegetation or animal life. Examples are seepage areas populated by *Impatiens pallida* (Jewel-weed), *Chrysosplenium americanum* (Golden Saxifrage), or *Desmognathus fuscus fuscus* (Northern Dusky Salamander) and cave entrances populated by ferns, liverworts, and mosses.

A **mesic** subsystem is one in which the soils are moderately well drained. Water drains from the soil somewhat slowly, so that the profile is wet for a small but appreciable part of the year. Moderately well-drained soils commonly have a medium-high water table, a slowly permeable layer within or immediately beneath the solum, additions of water through seepage, or a combination of these conditions. The soils have uniform colors in the A and upper B horizons, with mottling in the lower B and C horizons.

A **dry-mesic** subsystem is one in which the soils are well drained. Water drains from the soil readily but not rapidly. Well-drained soils commonly are medium in texture, although soils of other textures also may be well drained. The soils are free of gray mottlings. The A horizons may be gray, brown, or red, and the B horizons red, yellow, yellowish-red, or brown. After rain the soils commonly retain optimum amounts of moisture for plant growth.

A **dry-xeric** subsystem is one in which the soils are somewhat excessively drained. Water drains from the soil rapidly. Many of these soils have low degrees of horizon differentiation and frequently are sandy and very porous.

A **very dry-xeric** subsystem is one in which the soils are excessively drained. Water drains from the soil very rapidly. The soils commonly are very sandy, very gravelly, stony, steep, shallow, or some combination of these conditions. Excessively drained soils usually are yellow, light gray, brown, or reddish in color and free of mottling throughout the profile. This category also includes rock outcrops with rapid runoff.

Water Chemistry (V.B.). Water Chemistry pertains to two environmental parameters: (1) the presence of salts in inland habitats (saline) or in marine or estuarine habitats (haline), and (2) the hydrogen ion concentration (pH) in fresh-water habitats. As stated in Cowardin et al. (1977), the accurate characterization of water chemistry is difficult, both because of problems in measurement and because values vary with season, weather, time of day, and other factors. Water chemistry is obtained for a terrestrial system by testing a soil solution. Only three elements (haline, saline, and fresh) compose this hierarchical level.

Water chemistry is considered a very important aspect of this classification system because minute changes will determine what organisms are present where, for example, *Zostera marina* is restricted to haline waters. It is placed at this level in Hydrology to make it equivalent to the B. level in Geology (IV.B. Rock-Sediment Chemistry).

Water Regime (V.BB.). Water Regime has as its basis the duration and timing of surface inundation or of ground water fluctuation. The hierarchical elements in this level are a complete list from which one selects an entry (Table 6, p. 71). The terms under tidal and nontidal are excerpted from Cowardin et al. (1977) and are clarified in their publication. The selections under terrestrial are formulated by us and are delimited below.

1. **Permanently saturated**—soil is saturated but oxygenated at all times of the year in all years.
2. **Sporadically exposed**—soil is saturated but oxygenated throughout the year except in years of extreme drought.
3. **Semipermanently saturated**—soil is saturated but oxygenated throughout the growing season in most years.
4. **Seasonally saturated**—soil is saturated but oxygenated for extended periods early in the growing season but is dry by the end of the season in most years.
5. **Temporarily saturated**—soil is saturated but oxygenated for brief periods during the growing season but is usually dry most of the season.
6. **Intermittently saturated**—soil is dry or moist but is oxygenated when saturated for variable periods without detectable seasonal periodicity. Weeks, months, or even years may intervene between periods of saturation. The dominant plant communities may change as soil moisture changes.
7. **Permanently exposed**—soil is dry at all times of the year in all years.

For an excellent example of how important flooding duration can be in the control of species or community distributions, see the chart on page 234 from the Swift Creek report. For further elaboration, refer to the discussion from pages 231 to 248.

Natural Area Hydrologic Site Type (V.C.) and Community Hydrologic Site Type (V.CC.). The Natural Area Hydrologic Site Type and the Community Hydrologic Site Type list controlling hydrologic parameters, the first for the entire natural area (therefore, it may represent a range of characters, such as haline and fresh) and the second for each community analyzed within the natural area. The Natural Area and Community Hydrologic Site Types can be exactly the same, for example, if the natural area contains only one community. The more significant parameters are circumscribed below and are outlined in Table 6 (pp. 72–73).

Water chemistry at this level is used with the same meaning as above in V.B., although more specific water chemistry modifiers are employed. The modifiers of haline and saline systems are selected from a list of terms defined by Cowardin et al. (1977).

Coastal Modifiers	*Inland Modifiers*	*Salinity (0/00)*
hyperhaline	hypersaline	>40
euhaline	eusaline	30–40
mixohaline	mixosaline	0.5–30
polyhaline	polysaline	18–30
mesohaline	mesosaline	5–18
oligohaline	oligosaline	0.5–5
fresh	fresh	<0.5

Fresh water modifiers and their respective pH classes are as follows:

pH Class	*pH*
Very strongly acidic	<4.5
Strongly acidic	4.5–5.4
Acidic	5.5–6.4
Circumneutral	6.5–7.4
Basic	7.5–8.4
Strongly basic	8.5–9.0
Very strongly basic	>9.0

Drainage concerns surface discharge of water from an area by stream flow and sheet flow and the removal of excess water from soil by downward flow (Gary et al., 1972). We are interested mainly in the latter process here because it is important in aquatic, wetland, and terrestrial situations. For example, even though an area may be regularly flooded, it could be either excessively drained (as in *Salicornia bigelovii* salt flats) or very poorly drained (as in the *Spartina alterniflora* zone of a salt marsh). Descriptions of drainage classes can be found in the U.S.D.A. *Soil Survey Manual* (Soil Survey Staff, 1951).

Water source—the place from which the water in a site originates—represents another significant parameter in these site types. For example, is the water supplied by fresh rain, fog, ground water, seepage, or a combination of these? For further selected examples, see Table 6, page 72.

The substrate should be noted for aquatic communities that do not overlie soil, in concordance with the Soil Survey Staff's definition (1975) given in Section III of this chapter. Two general terms apply to this category—mineral and organic. Within each of these respective groups belong various substrate types, such as cherty, gravelly, shaly, and clayey in mineral and peaty and mucky peat in organic. A more extensive list of these substrate types appears in Table 6 (p. 72). For definitions and extensive discussion consult Appendix I in *Soil Taxonomy* (Soil Survey Staff, 1975).

Water depth or depth to water table should be measured where feasible. The importance of this parameter is obvious, particularly with regard to aquatic habitats. For example, the sequence of vegetation types in a lake from deep to progressively shallower water is as follows: plankton and floating vascular plants → submergents → floating-leaved anchored hydrophytes → emergent anchored hydrophytes.

Color and turbidity give hue and other optical qualities to water and determine light transmission in natural waters, consequently "regulating" biological processes within the water. Both provide some qualitative indication of productivity and indicate the general chemical nature of water.

Reid (1961) defines two types of color—true color (specific color) and apparent color. **True color** is derived from substances in solution or from materials in colloidal state. **Apparent color** is the result of interplay of light on suspended particulate materials together with such factors as bottom or sky reflection. The determination of color often is based on the platinum-cobalt scale of the United States Geological Survey (see Welch, 1948, for methods). The following list provides some examples of colors and their causes.

Lakes
dark green—blue-green algae
yellow or yellowish brown—diatoms
red—zooplankton, especially microcrustaceans, or ferric hydroxide
green or yellow-brown—humus
dark brown—peat
greenish—calcium carbonate
yellow-green—sulfur

Streams
clear—lack a true plankton
light to dark amber or brown—dissolved plant substances, such as tannin
black—acids
greenish tint—algae

Water turbidity is the term describing the degree of opaqueness produced by suspended particulate matter (Reid, 1961). Reid further explains that the concentration of the materials, if sufficiently high, determines the transparency of the water by limiting the light transmission within it. Note that turbidity is not a uniform parameter, for example, it may vary seasonally. Measurements of turbidity are based on readings from such instruments as the U.S. Geological Survey turbidity rod, the Jackson Turbidimeter, and the Hellige Turbidimeter. For details concerning these methods, see Reid (1961) or Welch (1948).

Various other factors—such as water temperature, water velocity, light penetration, volume of water moving past a given point per given time, turnover rate, length of time ice cover remains, type and amount of dissolved solids and gases and colloidal substances—can be determined if practical. If an intensive study is undertaken, many other community-controlling factors may be discovered and should be listed.

Population Hydrologic Site Type (V.D.). This is discussed on page 90 in the last section of this chapter.

Hydrologic Summaries and Characterizations

1. Hydrologic Excerpts from Community Diversity Summaries

a. *Juncus subcaudatus/Rhynchospora capitellata/ Rhynchospora alba*: A. Palustrine; AA. Interaqueous. B. Fresh; BB. Intermittently exposed. C. Variously drained, acid to circumneutral waters, with drainage predominantly to the northeast; CC. Very poorly drained, wetted by fresh rains, springs, and seeps.

b. *Quercus rubra* var. *borealis/Kalmia latifolia–Rhododendron catawbiense*: A. Terrestrial; AA. Dry-mesic. B. Fresh; BB. Permanently exposed. C. Variously drained; CC. Well drained, wetted by fresh rains and high elevation fogs.

c. *Rubus allegheniensis/Carex brunnescens*: A. Terrestrial; AA. Dry-mesic. B. Fresh water; BB. Permanently exposed. C. Well drained; CC. Wetted by fresh rains, fogs, and downslope drainage.

d. *Acer rubrum*–mixed bottomland hardwoods/ Mixed bottomland hardwoods/*Saururus cernuus*// Mixed lianas: A. Palustrine; AA. Interaqueous. B. Fresh; BB. Intermittently flooded. C. Variously drained and wetted; CC. Poorly drained, wetted by fresh rains, occasional flooding, and a high water table.

e. *Spartina alterniflora*: A. Estuarine; AA. Intertidal. B. Haline; BB. Regularly flooded. C. Mixed estuarine and terrestrial systems ranging from regularly flooded to permanently exposed; CC. Poorly drained euhaline system, wetted by haline tidal waters and fresh rains.

2. Hydrologic Excerpts from Community Ecological Characterizations (Hydrologically)

a. *Juncus subcaudatus*/*Rhynchospora capitellata*/ *Rhynchospora alba*: Very poorly drained, intermittently exposed, fresh, interaqueous palustrine system, with drainage predominantly to the northeast, and wetted by fresh rains, springs, and seeps.

b. *Quercus rubra* var. *borealis*/*Kalmia latifolia*–*Rhododendron catawbiense*: Well-drained, permanently exposed, dry-mesic, terrestrial system, wetted by fresh rains and high elevation fogs.

c. *Rubus allegheniensis*/*Carex brunnescens*: Well-drained, permanently exposed, dry-mesic, terrestrial system, wetted by fresh rains, high elevation fogs, and downslope drainage.

d. *Acer rubrum*–mixed bottomland hardwoods/ Mixed bottomland hardwoods/*Saururus cernuus*// Mixed lianas: Poorly drained, intermittently flooded, interaqueous palustrine system, wetted by fresh rains, occasional flooding, and a high water table.

e. *Spartina alterniflora*: Poorly drained, regularly flooded, intertidal, euhaline estuarine system wetted by haline tidal waters and fresh rains.

Table 6. *Hydrologic Hierarchical Elements*

V.A. Hydrologic System

1. Marine
2. Estuarine
3. Riverine
4. Lacustrine
5. Palustrine
6. Terrestrial

V.AA. Hydrologic Subsystem

Marine
1. Subtidal
2. Intertidal
3. Supratidal

Estuarine
4. Subtidal
5. Intertidal
6. Supratidal

Riverine
7. Tidal
8. Lower perennial
9. Upper perennial
10. Intermittent
11. Ephemeral

Lacustrine
12. Limnetic
13. Littoral

Palustrine
14. Aqueous
15. Interaqueous

Terrestrial
16. Wet
17. Mesic
18. Dry-mesic
19. Dry-xeric
20. Very dry-xeric

V.B. Water Chemistry

1. Haline
2. Saline
3. Fresh

V.BB. Water Regime

Tidal
1. Subtidal
2. Irregularly exposed
3. Regularly flooded
4. Irregularly flooded

Nontidal
5. Permanently flooded
6. Intermittently exposed
7. Semipermanently flooded
8. Seasonally flooded
9. Saturated
10. Temporarily flooded
11. Intermittently flooded
12. Artificially flooded

Terrestrial
13. Permanently saturated
14. Sporadically exposed
15. Semipermanently saturated
16. Seasonally saturated
17. Temporarily saturated
18. Intermittently saturated
19. Permanently exposed

Table 6. *(continued)*

V.C. Natural Area Hydrologic Site Type

1. Water chemistry

Haline	Saline	Fresh
1.1 hyperhaline	1.7 hypersaline	1.13 very strongly acidic
1.2 euhaline	1.8 eusaline	1.14 strongly acidic
1.3 mixohaline	1.9 mixosaline	1.15 acidic
1.4 polyhaline	1.10 polysaline	1.16 circumneutral
1.5 mesohaline	1.11 mesosaline	1.17 basic
1.6 oligohaline	1.12 oligosaline	1.18 strongly basic
		1.19 very strongly basic

2. Drainage

2.1 excessively drained	2.5 somewhat poorly drained
2.2 somewhat excessively drained	2.6 poorly drained
2.3 well drained	2.7 very poorly drained
2.4 moderately well drained	

3. Water source

3.1 fresh rain	3.4 ground water
3.2 acidic rain	3.5 flood water
3.3 seepage	3.6 fog

4. Substrate

4.1 mineral	
4.2 bedrock	4.14 sandy
4.3 channery	4.15 loamy sand
4.4 cherty	4.16 sandy loam
4.5 cobbly	4.17 loamy
4.6 flaggy	4.18 silt loamy
4.7 gravelly	4.19 silty
4.8 marly	4.20 clay loam
4.9 rocky	4.21 sandy clay
4.10 shaly	4.22 silty clay
4.11 shelly	4.23 clay
4.12 slaty	
4.13 stony	
4.24 organic	
4.25 muck	4.27 mucky peat
4.26 peat	4.28 peaty muck

5. Water depth or depth to water table

6. Color

6.1 true color	6.2 apparent color

7. Turbidity

8. Water temperature

9. Water velocity

10. Light penetration

11. Volume of water moving past a given point per given time

Table 6. *(continued)*

12. Turnover rate

13. Length of time ice cover remains

14. Type and amount of dissolved solids and gases and colloidal substances

V.CC. Community Hydrologic Site Type

Same characters as Natural Area Hydrologic Site Type (V.C.).

VI. Topography

The earth's surface exhibits a seemingly endless variety of topographic features, ranging from the highest mountains and the deepest ocean trenches to the tiniest anthills and shallowest runoff gullies in an old field, from the most extensive continent and the largest ocean to the most diminutive island and the smallest water puddle. Primary causative agents that develop the present-day landscape are animals and plants, gravity, ice, igneous activity, man, meteorite impacts, tectonic deformation, water, and wind. Man is considered separately because of his ability to alter the face of the earth to a much greater degree than any other extant or extinct species. These agents act through numerous processes, the most important being erosion and deposition, to shape the surface of the earth through time.

The configuration of the land surface and the shape, size, and composition of water bodies constitute very important environmental components, not only in and of themselves but also because they control many other environmental factors. On a global scale, the distribution of continents and oceans can affect features such as worldwide wind current patterns, temperature patterns, ocean currents, and moisture availability. On a continental and oceanic scale, the existence of mountain ranges and/or proximity to large water bodies can influence regional climatic patterns; for example, deserts occur on the leeward side of mountain ranges and warm, humid climates occur along coastlines. On a local scale, individual mountains, hills, streams, lakes, and the like, together with their sizes, shapes, distributions, and associations, largely determine small-scale climatic variations. On an even lesser scale, small-scale topographic features, such as the relative site position on a slope, the degree of habitat shelter, the rockiness or unevenness of a site, the habitat aspect, and the angle of the slope on which the habitat is located, exert environmental control over community and species distributions. Knowledge of topographic controlling factors at every level of consideration is essential to understanding these distributions, that is, what lives where, and why. All aspects of topography should be recorded and, where possible, related to communities, populations, life forms, and species distribution patterns.

An area in which all topographic, geologic, hydrologic, pedologic, and climatic factors remain uniform throughout would support one vegetational community. Any significant change in species frequency and/or density could indicate some modification of controlling ecologic factors. Because of the intricate interdependence of many ecologic factors, a variation in one can produce alterations in others and frequently can cause an accompanying change in topography. Conversely, a shift in topographic patterns can result in different ecologic factors. Often the first and most easily observed factors affecting species numbers and kinds appear in the topography of the area.

The various agents and processes that form, alter, and destroy landforms and water bodies have functioned in the past, are active today, and will continue to work in the future. Different agents and processes frequently interact to develop similar landforms, water bodies, and, ultimately, habitats. Therefore, species having similar or overlapping habitat requirements combine and develop into distinct communities distributed over repeated expressions of their required habitat, often in widely separated regions. Because the landscape continually changes as locations with specific environmental conditions are destroyed through time, new, similar habitats are also formed. Thus, if a species with appropriate habitat requirements invades these new, available environments, it can perpetuate its own kind. Likewise, if a sufficient number of species from a particular community can become established in these new, available habitats, entire communities can perpetuate themselves. Recognition of these topographic habitat factors aids in the location of repeated examples of unusual Community Types and populations of threatened and endangered species.

As discussed above, the earth's surface results from different processes acting in many different combinations on innumerable rock types and geologic structures for varying lengths of time. With such an infinite number of variables, developing a classification scheme that incorporates and arranges all controlling factors and the subsequent results of the interaction of these factors into a truly hierarchical system is a difficult task. A hierarchically arranged topographic classification system is developed here on the basis of increasing size and inclusiveness of topographic features. In relation to natural area inventory, the smallest (lowest level) topographic feature of direct concern is that particular feature on which the community or population in question actually is located. Depending on the size of the community or population as well as that of the topographic feature itself, the lowest level of topographic importance in one situation may be further subdivided in another. As an example, at one locality an entire, uniform, north-facing slope of a mountain may contain only one distinctive community, evenly distributed over the entire slope—therefore, the lowest level of topography is the mountain slope. Elsewhere a north-facing mountain slope with minor topographic variations may contain a number of different communities, one in drainages, another on interdrainage divides, a third on boulder fields, and still another on steep, rocky portions of the slope—in this case, the lowest level of topography is the smaller-scale features, while the entire mountain slope belongs in a higher, more inclusive leve!.

As outlined above, the problem of similar topographic features fitting into different levels of the hierarchy, dependent upon the size of the community, study site, or topographic feature itself, renders it impossible to compile lists of continually more inclusive features to be used exclusively at a single hierarchical level. Therefore, to avoid confusion, one must realize that in actuality a "sliding scale," so to speak, is employed in levels VI.B. through VI.D. of the scheme presented—the appropriate entry at one or more of these levels can differ according to the situation encountered. The base (lowest) level (VI.CC. or VI.D.) of the scale is always the most minute, yet still recognizable, topographic feature that influences species or community distributions. As indicated previously, identical terms describing features at successively higher levels can vary in their level of placement from situation to situation, dependent upon the size of the vegetational community or population and relative to the topographic variation encompassed. Consequently, each population, community, and study site must be treated separately by working from the lowest recognizable level of topography through the more inclusive features at successively higher levels.

Topographic Hierarchies

Code	*Hierarchy*	*Inclusion*
VI.A.	Topographic System	Topographic System
VI.AA.	Topographic Subsystem	Topographic Subsystem
VI.B	Topographic Class	Landscape
VI.BB.	Topographic Subclass	Landform
VI.C.	Topographic Generitype	Natural Area Topographic Site Type
VI.CC.	Topographic Type	Community Topographic Site Type

Topographic System (VI.A.). The Topographic System broadly describes the regional topography for the area in which a study site is located. Small-scale features that influence community and species distributions are important, but larger, more regional controls are equally important. Nearly identical small-scale features in two different topographic regions normally contain different floras and faunas. For example, the biota of a river valley located on a plain probably will differ considerably from that of a river valley located in a mountainous area.

Terminology follows Hammond (1964), who divides the landscape of the United States into five systems: plains, tablelands, plains with hills or mountains, open hills and mountains, and hills and mountains. Each study site should be placed in one of these five topographic systems. These systems are based on changes in slope, local relief, and profile type. **Slope** refers to the amount of land consisting of gently sloping terrain.

*Slope**

A. greater than 80% of terrain is gently sloping
B. 50%–80% of terrain is gently sloping
C. 20%–50% of terrain is gently sloping
D. less than 20% of terrain is gently sloping

*Numbering and lettering systems used by Hammond (1964) and U.S. Department of the Interior (1970).

Local relief refers to the change in elevation between the highest and lowest points in the region.

*Local Relief**

1. 0–100 feet
2. 100–300 feet
3. 300–500 feet
4. 500–1,000 feet
5. 1,000–3,000 feet
6. greater than 3,000 feet

Profile type refers to the amount of gentle slope present as lowland or as upland.

*Profile Type**

a. greater than 75% of gentle slope is lowland
b. 50%–75% of gentle slope is lowland
c. 50%–75% of gentle slope is upland
d. greater than 75% of gentle slope is upland

A more detailed explanation of these terms can be found in Hammond (1964) or in *The National Atlas* (U.S. Department of the Interior, 1970). At this level of classification it is best to determine the proper system by consulting either of these sources. It would be difficult to determine the regional landscape pattern in the field without the aid of extensive field surveys to determine large-scale slope, relief, and profile patterns.

Topographic Subsystem (VI.AA.). Topographic Subsystems are subdivisions of the five topographic systems and consist of 21 subsystems according to increasing similarity in slope, local relief, and profile type. All of these are listed in the topographic hierarchical elements (Table 7) and delimited in Hammond (1964). These subsystems present greater detail in the description of the surrounding terrain. As an example, plains (Topographic System—VI.AA.) include flat plains, smooth plains, irregular plains with slight relief, and irregular plains. In **flat plains**, greater than 80% of their surface occurs as gently sloping terrain and the local relief is between zero and one hundred feet. **Smooth plains** possess the same amount of gently sloping terrain as flat plains, but the local relief has increased to between one hundred and three hundred feet. On the other hand, **irregular plains with slight relief** have a lesser amount (50%–80%) of gently sloping terrain, although their local relief remains between zero and one hundred feet, the same as flat plains. **Irregular plains** also contain between 50% and 80% gently sloping terrain, but the local relief ranges between one hundred and three hundred feet.

These subsystems provide the investigator with an idea of the overall configuration of the landscape, including the amount of irregularity in the terrain. Knowledge concerning the regional landscape pattern also supplies an investigator with insight into additional information, such as the manner in which winds move across the terrain, the degree to which specific slopes are sheltered from adverse climatic conditions, the types of surface runoff that might be present, and many other important factors that directly or indirectly affect plant and animal life.

Landscape (VI.B.). Landscape includes two categories: (1) Landscape-Forming Agents and Processes, and (2) Landscape Type. As suggested by the title, the first category recognizes those agents and processes that act in forming the landscape. Generally the overall development of a large-scale feature (such as a mountain, river valley, or island) can be attributed to a single major agent and process. For example, an individual mountain can exist primarily as the result of volcanic eruptions, a river valley as the result of water erosion, or an island as the result of aggregation of coral reefs. A list of the important agents and generalized processes by which these agents work is included in the topographic hierarchical elements (Table 7, p. 79). This list is not all-inclusive: many small-scale, more specialized processes further influence landscape development. Where these can be identified, they should be incorporated into the analysis.

Landscape Type, the second category, identifies the local accumulation of landforms that creates the distinctive landscape on or in which the study site occurs. The size scale of this landscape depends on several factors, most important being the geographical extent of similar landforms, the complexity of the landscape, and the size of the study site. When the study site is located in or on a geographically extensive topographic feature, such as on a large mountain or in a large river valley, the individual topographic feature becomes the Landscape Type.

When the Landscape Type does not indicate a particular origin, it becomes necessary to add formative terms to the type. Examples of these agent-process topographic terms include ice-scoured basin, fault valley, water solution lake, stream-eroded cliff, and glacial outwash plain. Many Landscape Types not only physically describe a feature but also imply a particular origin without the need of a modifier. Examples of such terms include esker, beaver pond, pothole, playa, drumlin, and levee.

The list of selected Landscape Types provided in the

---

*Numbering and lettering systems used by Hammond (1964) and U.S. Department of the Interior (1970).

topographic hierarchical elements (Table 7, p. 80) is not complete but simply serves to illustrate the different types of features that possibly could be Landscape Types. For additional terminology, consult Lahee (1961) or one of the many introductory geomorphology books, such as Hinds (1943), Fairbridge (1968), Thornbury (1969), Garner (1974), or Twidale (1976), or books dealing with specific geomorphic terrains, such as Sweeting (1973) on karst landforms, Cooke and Warren (1973) on deserts, Gregory and Walling (1973) on drainage basins, or Goldthwait (1975) on glacial deposits.

Landform (VI.BB.). The Landform represents that particular, individual topographic feature or smaller portion of a single extensive feature on or in which the study site exists. Again, as with Landscape (VI.B.), the size and complexity of this feature depends on the extent of the study site. Occasionally a study site occupies only part of a single mountain or of a floodplain. At other times the study site can include a large number of topographic features that form a topographic complex, such as a cluster of several hills and valleys or a series of islands along a coastline. In summary, the landform name is contingent on the landform size and its relative location in the landscape. Whatever the case, the single feature or grouping of features needs to be listed. A list of selected examples appears in the topographic hierarchical elements (Table 7, p. 81). Additional terms applicable at this level include many that are provided in the lists at other topographic levels. For additional terminology, consult the references cited above in the discussion concerning Landscape (VI.B.).

In addition to denoting Landform type, the agent and process that shaped the landform should be noted if they differ from those at the Landscape level (VI.B.). Within a landscape whose overall origin can be attributed to a single, major agent and process, many minor agents and processes typically also are at work. Mountainous terrain can originate primarily from downcutting action of streams; however, an individual mountain within this terrain can contain a rocky peak formed by frost heaving, a cliff structured by local faulting, talus slopes built by gravity deposition at the base of the cliff, and steep artificial cuts made by man while building roads. The major agent and process that produced each of these successively smaller topographic features should be designated. The list of formative agents and processes provided at the Landscape level (VI.B.) in the topographic hierarchical elements (Table 7, p. 79) can be used, or more specialized terminology may become necessary.

Natural Area Topographic Site Type (VI.C.) and Community Topographic Site Type (VI.CC.). As do the Geologic and Hydrologic Natural Area and Community Site Types, the Natural Area and Community Topographic Site Types involve a set of characteristics: landform type, shelter, aspect, slope angle, profile, surface pattern, and position. These topographic levels function in two ways: (1) to delimit the portion of the Landform (VI.BB.) in or on which the study site or community occurs, and (2) to characterize topographically the study site or community.

Once again, as in geology and hydrology, the Natural Area and Community Site Types can have identical hierarchical element entries, for example, if only one community occupies the study site. In addition, each characteristic contained in a site type description can be composed of (1) a number of individual entries (for example, landforms in an area might include a beach, dune field, and blowout within a dune field); (2) a continuous range of variables (for example, the slope angle might vary from strongly sloping to very steep); or (3) a discontinuous range of variables (for example, the study site might contain north- and south-facing slopes).

The topographic characterization is requisite to comprehending fully the influence that topography can exert over communities and species. Variations within and between these topographic characters often control the dominant species within a community. Moreover, these small-scale differences produce microhabitats within the community. The larger the area a microhabitat encompasses, the more abundant a given species that occupies that microhabitat. The topographic site type features (items 1–7 under VI.C. and VI.CC., pp. 81–82) are defined below.

The **natural area** or **study site landform** occasionally encompasses the entire landform, in which case one can repeat the same descriptive term used for Landform (VI.BB.) or simply state that the natural area or study site covers the entire landform. If the natural area is situated on a distinct portion of the landform, this particular feature should be described. Again, if this smaller feature has a more specific or different origin from the larger feature, the method of formation should also be included.

The **community landform** represents the particular topographic feature that has direct, immediate control over the community. Once more, depending on the community size, this topographic feature may be an entire landform or a small part of a larger landform. Whatever the case, this feature should be described with adequate topographic terminology. As in the higher levels of topography, if this community-level feature has an origin different from the higher-level features, it should be noted.

With regard to natural area and community landform types, as with the previous two levels (VI.B. and VI.BB.), the lists of landform types (Table 7, pp. 81 and 82) are not comprehensive but rather contain only a few selected examples. A larger selection of terms can be found in the references cited previously in the discussion concerning Landscape (VI.B.).

Shelter (or exposure) describes the nature and degree of openness of a slope or place to wind, sunlight, weather, oceanic influences, and the like (Gary et al., 1972). Shelter strongly influences vegetation distribution. For example, in the Great Smoky Mountains a gradient exists from xeric, open ridges and peaks inhabited by xeric pines to deeply sheltered coves and canyons inhabited by mesic cove hardwoods (Whittaker, 1956). See the topographic hierarchical elements (Table 7, p. 81) for a complete list of entries. A subjective determination of the degree of shelter is based on observation in the study site and inspection of topographic maps.

Aspect gives the direction toward which a slope faces with respect to the compass or to the sun's rays (Gary et al., 1972). A community located on a north-facing portion of a slope whose general aspect is northeast possibly can differ from another community on a north-facing portion of a slope whose general aspect is northwest. A compass reading provides the hierarchical element entry (for a comprehensive list of entries, see Table 7, p. 81). For directions concerning use of a compass, consult Lahee (1961), Compton (1962), or an instruction pamphlet that generally accompanies the instrument.

Slope angle indicates the difference in angle degrees between the inclined surface of the earth and the horizontal plane. It can furnish a clue to various factors affecting plant occurrence and distribution, such as the stability of the slope if composed of unconsolidated sediments, the exposure of the slope, and the amount of downslope drainage of water and nutrients. A clinometer is used to obtain the slope angle reading, which then is associated with a modifier. Modifiers and their corresponding degree values (modeled after Soil Survey Staff, 1951) appear in Table 7 (p. 82). References such as Lahee (1961), Compton (1962), and the pamphlet accompanying the instrument give instructions for using the clinometer.

Profile indicates the configuration or slope of the surface of the ground as it would appear if intersected by a vertical plane perpendicular to the horizontal, from the top of the site downward. As in several other topographic characteristics, profile can provide information regarding drainage patterns and downslope movement of soils or unconsolidated sediments. A **concave slope** exemplifies the concept as the lower part of a hillside surface, tending to become concave below the constant slope, having an angle that decreases continuously downslope as the hillside extends to the valley floor or other local base level. A slope can exhibit an overall concave profile but also contain many small, localized slope variations—some areas of the slope are concave, others convex, and still others constant. A community on a local convexity of an overall concave slope usually differs from one on a local convexity of a constant slope. Site observations and topographic maps aid in the designation of a profile type from the complete list of hierarchical elements in Table 7 (p. 82). For definitions of these terms, refer to Gary et al. (1972).

Surface patterns generally consist of small-scale topographic features that represent minor variations in relief or configuration of the land. Surface patterns contribute greatly to microhabitat formation and hence affect Community Type and species occurrences and distribution. An example of a surface pattern is **polygons**, which form horizontal patterned ground whose mesh is dominantly polygonal, as tetragonal, pentagonal, or hexagonal (Gary et al., 1972). The topographic hierarchical elements (Table 7, p. 82) contain selected examples. The element entries are based on careful observation of the study site and of the distribution patterns of Community Types and species.

Position generally is defined as the place occupied by a point on the surface of the earth. More specifically, we utilize the term to place relatively a study site or community on a particular portion of an entire slope—upper slope, midslope, lower slope, or entire slope. Use of position as a category applies only when dealing with sites or communities on a slope; otherwise the element entry is "not applicable." As is the case with several previous topographic characteristics, position can indicate the degree of exposure of a particular point and provide information about the quantity of nutrients and water that point can receive from downslope drainage. For instance, generally the higher the slope position, the greater the exposure to sunlight, wind, and other climatic factors. Once again, Whittaker's (1956) article provides an example—a cove hardwood community on a lower slope becomes a northern hardwood community at a higher elevation on the same slope.

Population Topographic Site Type (VI.D.). This is discussed on page 90 in the last section of this chapter.

Topographic Summaries and Characterizations

1. Topographic Excerpts from Community Diversity Summaries

a. *Juncus subcaudatus/Rhynchospora capitellata/Rhynchospora alba*: A. Hills and mountains; AA. Low and high mountains. B. Water-eroded mountain; BB. Plateau on the upper southeast-facing slope. C. Entire mountain; CC. Sheltered, small, disturbed area on a gently sloping flat in the middle of the upper portion of a northeast-southwest-oriented fen.

b. *Quercus rubra* var. *borealis/Kalmia latifolia–Rhododendron catawbiense*: A. Hills and mountains; AA. High mountains. B. Water-eroded mountain; BB. Northeast-southwest-running ridge. C. Entire mountain; CC. Partly sheltered, nearly level, irregular, and rocky outer portion of a plateau on the upper southeast-facing slope of a northeast-southwest-running ridge.

c. *Rubus allegheniensis/Carex brunnescens*: A. Hills and mountains; AA. High mountains. B. Water-eroded mountain; BB. Upper slopes and crest. C. Entire upper slopes and crest; CC. Open, south- to north-facing, rocky and irregular, sloping to strongly sloping convex upper slopes and flat to gently sloping crest of an east-west-running ridge.

d. *Acer rubrum*–mixed bottomland hardwoods/Mixed bottomland hardwoods/*Saururus cernuus*//Mixed lianas: A. Plains; AA. Flat plain. B. Broad, east-west-running, shallow stream valley; BB. Water-deposited and organically accumulated floodplain. C. Entire floodplain and adjoining valley slopes; CC. Open, nearly level, slightly irregular, alluvial flat.

e. *Spartina alterniflora*: A. Plains; AA. Flat plain. B. Water-deposited barrier island; BB. Water-deposited and organically accumulated estuarine tidal marsh. C. Mixture of tidal flats and active and relict dunes; CC. Open, variously exposed, smooth, nearly level mud flats in the lower tidal marsh on the northwest side of the barrier island.

2. Topographic Excerpts from Community Ecological Characterizations (Topographically)

a. *Juncus subcaudatus/Rhynchospora capitellata/Rhynchospora alba*: Sheltered, small, disturbed area on a gently sloping flat in the middle of the upper portion of a northeast-southwest-oriented fen, located on a plateau on the upper southeast-facing slope of a water-eroded mountain in a region of low and high mountains.

b. *Quercus rubra* var. *borealis/Kalmia latifolia–Rhododendron catawbiense*: Partly sheltered, nearly level, irregular, and rocky outer portion of a plateau on the upper southeast-facing slope of a northeast-southwest-running ridge on a water-eroded mountain in a region of high mountains.

c. *Rubus allegheniensis/Carex brunnescens*: Open, south- to north-facing, rocky and irregular, sloping to strongly sloping convex upper slopes and flat to gently sloping crest of an east-west-running ridge on the upper slopes and crest of a water-eroded mountain in a region of high mountains.

d. *Acer rubrum*–mixed bottomland hardwoods/Mixed bottomland hardwoods/*Saururus cernuus*//Mixed lianas: Open, nearly level, slightly irregular, alluvial flat on a water-deposited and organically accumulated floodplain in a broad, east-west-running portion of a shallow stream valley situated on a flat plain.

e. *Spartina alterniflora*: Open, variously exposed, smooth, nearly level mud flats in the lower tidal marsh on the northwest side of a water-deposited barrier island located on the seaward edge of a flat plain.

Table 7. *Topographic Hierarchical Elements*

VI.A. Topographic System

1. Plains
2. Tablelands
3. Plains with hills or mountains
4. Open hills and mountains
5. Hills and mountains

VI.AA. Topographic Subsystem

Plains

1. Flat plains (A1)[1]
2. Smooth plains (A2)
3. Irregular plains with slight relief (B1)
4. Irregular plains (B2)

Tablelands

5. Tablelands with moderate relief (B3cd)
6. Tablelands with considerable relief (B4cd)
7. Tablelands with high relief (B5cd)
8. Tablelands with very high relief (B6cd)

Plains with Hills or Mountains

9. Plains with hills (AB3ab)
10. Plains with high hills (B4ab)
11. Plains with low mountains (B5ab)
12. Plains with high mountains (B6ab)

Open Hills and Mountains

13. Open low hills (C2)
14. Open hills (C3)
15. Open high hills (C4)
16. Open low mountains (C5)
17. Open high mountains (C6)

Hills and Mountains

18. Hills (D3)
19. High hills (D4)
20. Low mountains (D5)
21. High mountains (D6)

VI.B. Landscape

1. Landscape-forming agents and processes

Selected examples:

Animals and Plants
- erosion
- deposition

Gravity
- erosion
- deposition

Ice
- erosion
- deposition

Igneous Activity
- vent eruption
- fissure eruption
- caldera formation
- ash fall
- ash flow
- others (specify)

Man
- erosion
- deposition

Meteorite Impact
- erosion
- deposition

Tectonic Deformation
- folding
- faulting
- subsidence
- uplift

Water
- erosion
- deposition
- solution
- precipitation
- organic accumulation
- weathering

Wind
- erosion
- deposition

Others (specify)

---

[1]Symbols used by Hammond (1964) and U.S. Department of the Interior (1970). Refer to text above under Topographic System (VI.A.) for meanings of symbols, which pertain to slope, local relief, and profile type.

Table 7. *(continued)*

2. Landscape types

Selected examples:

<u>Basins</u>

barrier basin	lacustrine basin	tectonic basin
caldera	landslide basin	volcanic basin
estuarine basin	marine basin	weathered basin
fault basin	meteorite crater	wind-scoured basin
ice-scoured basin		

<u>Estuaries</u>

bar estuary	fjord	lagoon
drowned estuary	glacial estuary	tectonic estuary

<u>Hills-Ridges-Mountains</u>

ash cone	lava dome	scoria mound
cinder cone	lava mound	tower karst
divide ridge	lava shield	turret mountain
dome	mesa	volcanic cone
fault ridge	pinnacle	volcanic ridge
fold ridge	plug dome	volcanic spine
hogback	plug neck	water-eroded hill
lava cone	scarp ridge	water-eroded mountain
lava disk	scoria cone	water-eroded ridge

<u>Islands</u>

atoll	estuarine island	riverine island
barrier island	lacustrine island	stack
depositional island	marine island	tectonic island
erosional island	reef-built island	volcanic island

<u>Marine</u>

near shore marine
open ocean

<u>Plains-Flats</u>

alluvial fan	frontal apron	peneplain
alluvial flat	glacial ice field	till plain
alluvial plain	lava plain	veneered limestone plain
coastal plain	marine plain	volcanic ash plain
deltaic plain	outwash plain	

<u>Valleys</u>

canyon	glacial valley	hanging valley
collapse doline	graben	karst valley
corrosion valley	gorge	solution valley
fault valley	gully	stream valley
faultline valley		

VI.BB. Landform

1. Landform-forming agents and processes

See VI.B.1. above. Should be noted only if it differs from agent and process in VI.B.1.

Table 7. *(continued)*

2. Landform types

Selected examples:

arete	hum	mountain peak
baraboo	inselberg	mountain ridge
basin floor	island beach	mountain slope
basin slope	island dune field	pinnacle
basin rim	island tidal flat	quarry
cirque	kame	ridge crest
coulee lake	kettlehole	ridge gap
drumlin	lake beach	stack
esker	lake bottom	stream channel
estuarine beach	lake flats	stream shore
estuarine subtidal zone	lake levee	talus cone
estuarine tidal flats	lake shore	talus slope
gap	lake terrace	terrace flat
hill crest	marine shore	terrace slope
hill slope	monadnock	valley floor
hoodoo	mountain crest	valley slope

VI.C. Natural Area Topographic Site Type

1. Natural area landforms

Selected examples:

lower floodplain
upper floodplain

oxbow lake
cirque lake

boulder field

cave entrance

fault cliff on a mountain slope
karst cliff on a valley slope
sea cliff on ocean shore

beach
blowout within a dune field
foredune of dune field
entire dune field
upper tidal flat
lower tidal flat
entire tidal flat

2. Natural area shelter

2.1 open
2.2 partly sheltered
2.3 sheltered
2.4 deeply sheltered

3. Natural area aspect

3.1 north
3.2 north-northeast
3.3 northeast
3.4 east-northeast
3.5 east
3.6 east-southeast
3.7 southeast
3.8 south-southeast
3.9 south
3.10 south-southwest
3.11 southwest
3.12 west-southwest
3.13 west
3.14 west-northwest
3.15 northwest
3.16 north-northwest

Table 7. *(continued)*

4. Natural area slope angle

4.1	nearly level	0– 2°	4.4	strongly sloping	10–15°
4.2	gently sloping	2– 6°	4.5	moderately steep	15–25°
4.3	sloping	6–10°	4.6	steep	25–45°
			4.7	very steep	>45°

5. Natural area profile

5.1	constant	5.5	concavoconvex	5.8	convexoconstant
5.2	concave	5.6	convexoconcave	5.9	constantconcave
5.3	convex	5.7	concavoconstant	5.10	constantconvex
5.4	flat				

6. Natural area surface patterns

Selected examples:

blocky	imbricated	reticulately grooved	sandy
bouldery	irregular	reticulately ridged	shaly
crevices	jointed	ridged	shingled
flats and mounds	muddy	ridged and grooved	silty
flats and pans	pans	rills	smooth
gravelly	pillars	ripples	swell and swale
grooved	pitted	rocky	wind throw depressions
gullied	polygons	rolling	undulating
hummocky	reticulate	rubbley	

7. Natural area position

7.1	upper slope	7.3	lower slope
7.2	midslope	7.4	entire slope
		7.5	not applicable

VI.CC. Community Topographic Site Type

1. Community landforms

Selected examples:

shore of cirque lake	littoral zone of rift lake
beaver pond	tidal creeks in upper tidal flat
dune depression pool	pan in upper tidal flat
entire oxbow lake	cove on middle slope
slope of old meander scar	entire boulder field
slope of solution depression	drainages on boulder field
shore of rift lake	interdrainage divides on boulder field

2. Community shelter: See VI.C.2. for list of elements.

3. Community aspect: See VI.C.3. for list of elements.

4. Community slope angle: See VI.C.4. for list of elements.

5. Community profile: See VI.C.5. for list of elements.

6. Community surface patterns: See VI.C.6. for list of elements.

7. Community position: See VI.C.7. for list of elements.

VII. Physiography

Physiography, generally considered a part of physical geography, concentrates on the physical description of landforms (Gary et al., 1972). In this physiographic classification scheme, the earth's surface is divided into identifiable, recognizable units—region, province, section, and landform. Each of these units undergoes a unique developmental history resulting from thousands and often millions of years of interaction among several factors, the most important being climate, vegetation, and underlying lithologies. The variation in rate of erosion of different lithologies and structures determines the position, size, and shape of positive and negative features on a local scale and the positions and associations of entire landscapes on a larger, regional scale. This interaction produces the present landscape patterns that characterize the physiographic units.

The physiographic setting of any study site should be described. Present-day physiography frequently supplies information that allows the history of an area to be deciphered (see Fenneman, 1931, 1938). The physiographic location of a study site provides a working framework for an investigator and also gives a great deal of control in comparative studies. Because each individual physiographic unit at each level of classification was formed by a unique combination of processes over different lengths of time, recognition of these units helps in understanding present-day community and population diversity as well as community and species origin, evolution, and migration within and between physiographic areas of similar and/or different histories.

Geology ultimately controls physiographic development of the earth's surface, whereas climate governs the expression of geology topographically. Consequently, geology and climate should function as the basis of any major physiographic classification scheme. This classification system divides the earth into regions, provinces, sections, and individual landforms of varying sizes, with each progressively smaller unit based on increasing similarity in lithologies, geologic structures, and climate. The lowest level has the most unified geomorphic history. Each **physiographic unit** at any level consists of a pattern of relief features that differs significantly from that of adjacent areas (Gary et al., 1972).

The physiography of only the conterminous United States has been incorporated into this system. Other areas of the world have not been considered. The three highest levels (Region [VII.A.], Province [VII.AA.], and Section [VII.B.]) and respective terminology are based on physiographic units established by Fenneman (1931, 1938) and therefore are taken directly from these two publications. The terminology for the three lowest levels (Local Landform [VII.BB.], Natural Area Physiographic Site Type [VII.C.], and Community Physiographic Site Type [VII.CC.]) consists of proper names from official maps or publications. If no proper name is applicable to a particular landform or portion of a landform, descriptive geomorphic terminology, as presented in the topography section above, can be used to describe a particular feature.

Physiographic Hierarchies

Code	*Hierarchy*	*Inclusion*
VII.A.	Physiographic System	Physiographic Region
VII.AA.	Physiographic Subsystem	Physiographic Province
VII.B.	Physiographic Class	Physiographic Section
VII.BB.	Physiographic Subclass	Local Landform
VII.C.	Physiographic Generitype	Natural Area Physiographic Site Type
VII.CC.	Physiographic Type	Community Physiographic Site Type

Physiographic Region (VII.A.). A Physiographic Region outlines a very large, physiographically similar, geographic area of the earth. Rock of similar age (for example, all Paleozoic, all Precambrian, or all Mesozoic and Cenozoic) typically underlies these regions and usually has structural control over the present topography. Examples of such regions include: (1) **Appalachian Highlands**—encompasses all the exposed, folded, Precambrian and Paleozoic, eastern North American rocks, the deformational history of which is related to plate collisions during openings and closings of the Atlantic Ocean; (2) **Atlantic Plain**—lies to the east of the Appalachian Highlands and contains essentially undeformed, seaward-dipping, unconsolidated to highly cemented beds of Mesozoic and Cenozoic sediments; and (3) **Interior Plains**—lies to the west of the Appalachian Highlands and rests on relatively undeformed, flat-lying beds of Paleozoic sedimentary rocks.

The terminology for this level follows Fenneman (1931, 1938). The entries listed in the physiographic hierarchical elements (Table 8, p. 87) cover all the regions of the conterminous United States.

Physiographic Province (VII.AA.). Physiographic Provinces consist of smaller geographic units that are subdivisions of Physiographic Regions. These provinces display greater similarity in topography than do the regions. In turn, more closely related geologic, erosive, depositional, and climatic factors control their topography. Provinces also share some large-scale, controlling factors. The provinces of the Atlantic Plain and of the Appalachian Highlands regions serve as examples at this level. The development of the Atlantic Plain region of the eastern United States directly related to and depended on the rising and falling of sea level during the Mesozoic and Cenozoic eras. This region is subdivided into the **Continental Shelf Province**—that portion of the Plain that is presently submersed by marine waters—and the **Coastal Plain Province**—that portion that is presently emersed. The Appalachian Highlands region is subdivided into seven provinces based on topographic changes controlled by underlying structural features and lithologies. Fenneman (1931, 1938) similarly subdivides all other regions.

The terminology for this level follows Fenneman (1931, 1938). The selections listed in the physiographic hierarchical elements (Table 8, p. 87) comprise all the provinces of the conterminous United States.

Physiographic Section (VII.B.). Physiographic Sections are subdivisions of Physiographic Provinces, just as the provinces are subdivisions of the Physiographic Regions. In most cases these subdivisions are recognized as smaller areas of increasingly similar topography. Geography rather than topography sometimes is the basis for subdividing some provinces, such as the Blue Ridge Province, which contains the Northern and Southern Sections. Other provinces include sections of geologic origin drastically different from that of the greater portion of the province, such as the Piedmont Province, which is divided into the Upland and Lowland Sections. The **Uplands** comprise areas underlain by typical Piedmont lithologies—Precambrian and Paleozoic metamorphics and volcanics and later Paleozoic plutonics. The **Lowlands**, however, encompass areas underlain by Triassic clastic sediments deposited in large graben-type basins that formed during early rifting stages of the modern Atlantic Ocean.

The terminology for this level follows Fenneman (1931, 1938). The selections listed in the physiographic hierarchical elements (Table 8, pp. 87–88) include all the sections of the conterminous United States.

Local Landform (VII.BB.). The Local Landform is the proper name (if officially designated and listed on a topographic map or in some other official publication) of the specific topographic feature on which or in which the study site is situated. This includes names of (1) mountains, such as Roan Mountain, Bluff Mountain; (2) hills, such as Iron Mine Hill; (3) islands, such as Ocracoke Island, Topsail Island; and (4) any other distinct topographic features. For wetland or aquatic environments, the major river drainage system (or other water body) in which or along which the study site is found should be recorded. This includes names of (1) riverine systems, such as Dan River drainage basin, Ohio River drainage basin; (2) lacustrine systems, such as Lake Michigan, Great Lake; (3) estuarine systems, such as Pamlico Sound, Albemarle Sound; (4) marine systems, such as Atlantic Ocean, Pacific Ocean; and (5) palustrine systems, such as Ditch Pond, Muskrat Pond.

If no proper name exists for a given feature, a depictive term should be used to describe physically the particular landform or water body. These terms may be analogous to the terms used for the Topographic Landform (VI.BB.) as discussed on page 147. No comprehensive list of hierarchical elements can be composed for the Local Landform; thus the entries in the physiographic hierarchical elements (Table 8, p. 89) represent selected examples. Data should be obtained from published maps or other official sources. Wetland and aquatic information is important to studies concerning aquatic plants and animals that depend on water for dispersal and migration.

Natural Area Physiographic Site Type (VII.C.). The Natural Area Physiographic Site Type deals with the exact location of the study site. If the study area encompasses the entire local landform or water body, then it can be recorded as such. As examples, all of Bluff Mountain in Ashe County, North Carolina, is being listed as a natural area, and all of Great Lake in Craven County, North Carolina, is being recommended as a natural area.

If the study site does not include the entire local landform or water body, then the particular portion of the local landform or water body which the study site occupies should be entered as the Natural Area Physiographic Site Type. Once more, if a proper name is available, it should be used, such as The Plains and the First Hammock Hills on Ocracoke Island, or Nags Head Woods on Bodie Island. Again, if a proper name is not available, then a descriptive term should be employed, such as "the upper slopes and crest of Roan Mountain" or "the hills surrounding the upper reaches of Bolin Creek." For wetland and aquatic situations, that particular portion of the major drainage system in which the study site is found should be named, such as "the Swift Creek portion of the Tar River drainage basin" or "the Bolin Creek portion of the New Hope River–Haw River–Cape Fear River drainage basin."

As in the Local Landform (VII.BB.), because no comprehensive lists of hierarchical elements can be composed for the Natural Area Physiographic Site Type, the entries in the physiographic hierarchical elements (Table 8) represent selected examples. Data should be obtained from published maps or other official sources.

Community Physiographic Site Type (VII.CC.). The Community Physiographic Site Type presents a physiographic description of the particular topographic feature on which or in which the community being analyzed is located and also includes the age and name of the underlying geologic formation. A community usually occupies only one distinctive topographic feature. A change in topography is usually the result of a change, or results in a change, in other factors that control vegetation (soils, drainage, lithology, and the like).

Occasionally a study site contains only one recognizable community. In such a case the entry at this level of classification is equivalent to the entry at the natural area level. On the contrary, most study sites include community complexes so that one must delimit each community's position within the study site. Unless the community covers an entire single landform or water body, it is unlikely that a proper name will be completely applicable. Most often a descriptive term will be necessary at this level of refinement. Examples of Community Physiographic Site Types from Roan Mountain Natural Area include "the north slope of Round Bald," "the crest of Grassy Ridge," and "the north-facing cove on Roan High Knob." For the Great Lake Natural Area, community sites include "the terrace along the south shore" and "the levee on the east shore."

Once again, no comprehensive list of hierarchical elements can be composed for the Community Physiographic Site Type, and the entries in the physiographic hierarchical elements (Table 8) represent selected examples. Data should be obtained from published maps, other official reports, or on-site observations.

The geologic formation and age should be determined and recorded because a single formation commonly underlies a community and plays a major role in controlling the shape of the topographic feature being described. Official formation names and ages should be taken from official geologic maps or other publications. If no formal name has been proposed, or if the necessary publications are not available, the rock type (the same as that described in the Community Geologic Site Type [IV.CC.]) should be recorded. If more than one formation or rock type underlies the community, especially if a difference in formation or rock type controls species distributions, then each formation or rock type should be listed. Included in the physiographic hierarchical elements (Table 8) are selected examples of particular geologic formations and their ages.

The amount of detailed stratigraphic work done in a particular area determines how accurately a particular lithology can be dated and named. In some areas one can determine merely the age, such as Cambrian or Ordovician, and sometimes only the era, such as Paleozoic or Mesozoic. In other areas accurate ages can be given. The same applies to naming the formation. Some areas may contain thousands of feet of rocks containing numerous lithologies listed under one name, whereas other areas worked on in great detail contain sequences of rock divided into numerous small formal units, each with its own name. For identification of the latter units, local geologic reports should be consulted.

Population Physiographic Site Type (VII.D.). This is discussed on page 90 in the last section of this chapter.

Geographic Location. In addition to characterizing the physiographic setting of a study site, one should locate the area geographically. So that the site can be found, it is important to give directions and distances along roads or other features, such as trails or streams, from a permanent landmark to the study site. The exact latitude and longitude of a site should be determined in order to locate a site readily on topographic maps because a permanent reference mark can be destroyed, through time, by man or nature. For large sites, in which a range of readings would be needed to incorporate the entire area, a single reading is sufficient. This reading is based on a centralized point in the site. Elevation or water depth should also be recorded because it can help locate a given site and can prove to be an environmentally significant factor useful in comparative studies. This geographic information is part of the introduction to each natural area report (see format in Table 9, p. 97, and reports in Part II). Generally it also is incorporated into a report discussion.

Physiographic Summaries and Characterizations

1. Physiographic Excerpts from Community Diversity Summaries

a. *Juncus subcaudatus/Rhynchospora capitellata/Rhynchospora alba*: A. Appalachian Highlands; AA. Blue Ridge Province. B. Southern Section; BB. Bluff Mountain. C. Bluff Mountain; CC. Fen, underlain by Precambrian hornblende gneiss, located on a plateau on the upper, southeast-facing slope.

b. *Quercus rubra* var. *borealis/Kalmia latifolia–Rhododendron catawbiense*: A. Appalachian Highlands; AA. Blue Ridge Province. B. Southern Section; BB. Bluff Mountain. C. Northeast-southwest-running ridge; CC. Plateau, underlain by Precambrian Roan Gneiss.

c. *Rubus allegheniensis/Carex brunnescens*: A. Appalachian Highlands; AA. Blue Ridge Province. B. Southern Section; BB. Roan Mountain. C. Upper slopes and crest; CC. Grassy Ridge, underlain by Precambrian Roan Gneiss.

d. *Acer rubrum*–mixed bottomland hardwoods/ Mixed bottomland hardwoods/*Saururus cernuus*// Mixed lianas: A. Atlantic Plain; AA. Coastal Plain. B. Embayed Section; BB. Tar River drainage system. C. Swift Creek floodplain; CC. Alluvial flat, underlain by Recent alluvium.

e. *Spartina alterniflora*: A. Atlantic Plain; AA. Coastal Plain. B. Embayed Section; BB. Ocracoke Island. C. The Plains tidal marsh; CC. Lower tidal flats, underlain by Recent fine clastics and organics.

2. Physiographic Excerpts from Community Ecological Characterizations (Temporally and Spatially)

a. *Juncus subcaudatus/Rhynchospora capitellata/ Rhynchospora alba*: Seral stage of a lithosere, located in a fen, underlain by Precambrian hornblende gneiss, on a plateau on the upper, southeast-facing slope of Bluff Mountain in the Southern Section of the Blue Ridge Province of the Appalachian Highlands.

b. *Quercus rubra* var. *borealis/Kalmia latifolia–Rhododendron catawbiense*: A topoedaphic climax community of a lithosere, on a plateau on a northeast-southwest-running ridge, underlain by Precambrian Roan Gneiss, on Bluff Mountain in the Southern Section of the Blue Ridge Province of the Appalachian Highlands.

c. *Rubus allegheniensis/Carex brunnescens*: A seral stage of a lithosere on Grassy Ridge, underlain by Precambrian Roan Gneiss, on the upper slopes and crest of Roan Mountain in the Southern Section of the Blue Ridge Province of the Appalachian Highlands.

d. *Acer rubrum*–mixed bottomland hardwoods/Mixed bottomland hardwoods/*Saururus cernuus*//Mixed lianas: A climax community of a pelopsammosere on an alluvial flat, underlain by Recent alluvium, on the Swift Creek floodplain in the Tar River drainage system in the Embayed Section of the Coastal Plain Province of the Atlantic Plain.

e. *Spartina alterniflora*: A topoedaphic climax of a hydrosere on the Plains lower tidal flat, underlain by Recent estuarine tidal flat sediments, on Ocracoke Island in the Embayed Section of the Coastal Plain Province of the Atlantic Plain Region.

Table 8. *Physiographic Hierarchical Elements*

VII.A. Physiographic Region

1. Laurentian Upland (A)[1]
2. Atlantic Plain (B)
3. Appalachian Highlands (C)
4. Interior Plains (D)
5. Interior Highlands (E)
6. Rocky Mountain System (F)
7. Intermontane Plateaus (G)
8. Pacific Mountain System (H)

VII.AA. Physiographic Province

Laurentian Upland
1. Superior Upland (A1)[1]

Atlantic Plain
2. Continental Shelf (B2)
3. Coastal Plain (B3)

Appalachian Highlands
4. Piedmont Province (C4)
5. Blue Ridge Province (C5)
6. Valley and Ridge Province (C6)
7. St. Lawrence Valley (C7)
8. Appalachian Plateaus (C8)
9. New England Province (C9)
10. Adirondack Province (C10)

Interior Plains
11. Interior Low Plateaus (D11)
12. Central Lowland (D12)
13. Great Plains Province (D13)

Interior Highlands
14. Ozark Plateaus (E14)
15. Ouachita Province (E15)

Rocky Mountain System
16. Southern Rocky Mountains (F16)
17. Wyoming Basin (F17)
18. Middle Rocky Mountains (F18)
19. Northern Rocky Mountains (F19)

Intermontane Plateaus
20. Columbia Plateaus (G20)
21. Colorado Plateaus (G21)
22. Basin and Range Province (G22)

Pacific Mountain System
23. Cascade-Sierra Mountains (H23)
24. Pacific Border Province (H24)
25. Lower California Province (H25)

VII.B. Physiographic Section

Superior Upland
1. no sections differentiated (1)[1]

Continental Shelf
2. no sections differentiated (2)

Coastal Plain
3. Embayed Section (3a)
4. Sea Island Section (3b)
5. Floridian Section (3c)
6. East Gulf Coastal Plain (3d)
7. Mississippi Alluvial Plain (3e)
8. West Gulf Coastal Plain (3f)

Piedmont Province
9. Piedmont Uplands (4a)
10. Piedmont Lowlands (4b)

Blue Ridge Province
11. Northern Section (5a)
12. Southern Section (5b)

Valley and Ridge Province
13. Tennessee Section (6a)
14. Middle Section (6b)
15. Hudson Section (6c)

[1]Symbols used by Fenneman (1931, 1938).

Table 8. *(continued)*

St. Lawrence Valley
16. Champlain Section (7a)
17. Northern Section (7b)

Appalachian Plateaus
18. Mohawk Section (8a)
19. Catskill Section (8b)
20. Southern New York Section (8c)
21. Allegheny Mountain Section (8d)
22. Kanawha Section (8e)
23. Cumberland Plateau Section (8f)
24. Cumberland Mountain Section (8g)

New England Province
25. Seaboard Lowland Section (9a)
26. New England Upland Section (9b)
27. White Mountain Section (9c)
28. Green Mountain Section (9d)
29. Taconic Section (9e)

Adirondack Province
30. no sections differentiated (10)

Interior Low Plateaus
31. Highland Rim Section (11a)
32. Lexington Plain (11b)
33. Nashville Basin (11c)
34. possible western section (not delimited) (11d)

Central Lowland
35. Eastern Lake Section (12a)
36. Western Lake Section (12b)
37. Wisconsin Driftless Section (12c)
38. Till Plains (12d)
39. Dissected Till Plains (12e)
40. Osage Plains (12f)

Great Plains Province
41. Missouri Plateau, glaciated (13a)
42. Missouri Plateau, unglaciated (13b)
43. Black Hills (13c)
44. High Plains (13d)
45. Plains Border (13e)
46. Colorado Piedmont (13f)
47. Raton Section (13g)
48. Pecos Valley (13h)
49. Edwards Plateau (13i)
50. Central Texas Section (13j)

Ozark Plateaus
51. Springfield-Salem Plateaus (14a)
52. Boston "Mountains" (14b)

Ouachita Province
53. Arkansas Valley (15a)
54. Ouachita Mountains (15b)

Southern Rocky Mountains
55. no sections differentiated (16)

Wyoming Basin
56. no sections differentiated (17)

Middle Rocky Mountains
57. no sections differentiated (18)

Northern Rocky Mountains
58. no sections differentiated (19)

Columbia Plateaus
59. Walla Walla Plateau (20a)
60. Blue Mountain Section (20b)
61. Canyon Lands (20c)
62. Snake River Plain (20d)
63. Harney Section (20e)

Colorado Plateaus
64. High Plateau of Utah (21a)
65. Uinta Basin (21b)
66. Canyon Lands (21c)
67. Navajo Section (21d)
68. Grand Canyon Section (21e)
69. Datil Section (21f)

Basin and Ridge Province
70. Great Basin (22a)
71. Sonoran Desert (22b)
72. Salton Trough (22c)
73. Mexican Highland (22d)
74. Sacramento Section (22e)

Cascade-Sierra Mountains
75. Northern Cascade Mountains (23a)
76. Middle Cascade Mountains (23b)
77. Southern Cascade Mountains (23c)
78. Sierra Nevada (23d)

Pacific Border Province
79. Puget Trough (24a)
80. Olympic Mountain (24b)
81. Oregon Coast Range (24c)
82. Klamath Mountains (24d)
83. California Trough (24e)
84. California Coast Ranges (24f)
85. Los Angeles Ranges (24g)

Lower California Province
86. no sections differentiated (25)

Table 8. *(continued)*

VII.BB. Local Landform

1. Terrestrial systems and self-contained wetland systems

Selected examples:

Roan Mountain	Ocracoke Island
Grandfather Mountain	Patsy Pond

2. Wetland systems that are parts of larger systems

Selected examples:

Dan River drainage basin	Croatan National Forest lake complex
Ohio River drainage basin	Rio Grande drainage basin

VII.C. Natural Area Physiographic Site Type

Selected examples:

Roan Mountain	Ocracoke Island	Croatan National Lake Complex
Round Bald	The Plains	Lake Ellis
Grassy Ridge	First Hammock Hills	Great Lake
Roan High Knob		
Roan High Bluff		

VII.CC. Community Physiographic Site Type

1. Position of community on natural area physiographic feature

Selected examples:

Round Bald
north slope
crest
south slope

Roan High Knob
north-running ridge crest
north-facing cove

The Plains
tidal flats
foredune

Lake Ellis
hummocks in center of lake
south shore

Great Lake
north shore
levee on east shore

2. Age and name of underlying geologic formation

Selected examples:

Precambrian Roan Gneiss	Lower Jurassic Navajo Sandstone
Upper Ordovician Fairview Formation	Cretaceous Black Creek Formation
Lower Silurian Tuscarora Sandstone	Middle Eocene Castle Hayne Limestone
Pennsylvanian Atoka Formation	Pleistocene Wisconsin glacial till

If the name of the formation is not known, list the rock type (see Table 5, IV.C., p. 64).

Population Level Inventory

Hierarchical level D. (Population Level Inventory) in each theme (I.–VII.) deals with site characteristics of populations of plants. Generally the data for Population Level Inventory will be gathered only for species of special concern, such as endangered, threatened, rare, disjunct, or endemic species. Each species of concern at a site will be treated separately. The following is a list of categories at the Population Inventory Level (D.) in the Ecological Diversity Classification System:

I.D. Biotic Associates
II.D. Population Climatic Site Type
III.D. Population Soil Site Type
IV.D. Population Geologic Site Type
V.D. Population Hydrologic Site Type
VI.D. Population Topographic Site Type
VII.D. Population Physiographic Site Type

Biotic Associates (I.D.) is self-explanatory (see definition, Chapter I, Section IV). In categories II.D. through VII.D. the information collected is the same as that for the type level (CC.) in the seven themes. The results may be the same as the Community Site Types (I.CC.–VII.CC.), for example, if a species grew throughout the study site without any noticeable microhabitat preference (such as, *Panax quinquefolium* or *Cladrastis lutea*). On the contrary, a species within a community may occupy a very specialized microhabitat feature which then must be characterized; for example, *Chrysosplenium iowense* grows in ice cave openings within a forest (Weber, 1978, unpublished data).

Microhabitat preferences for these species of special concern always must be noted and characterized. Other species characters to be recorded include:

- Species Legal Status
- Successional Stage
- Population Inventory
  - Population Numerical Relations
    - Population Size
    - Abundance
    - Density
    - Frequency
  - Distributional Relations
  - Areal Relations–Cover

These are explained briefly below. Detailed discussion and tables can be found in Chapter 4.

The **Species Legal Status** is the present legal condition or standing of a particular species throughout its range of occurrence. Is it endangered, threatened, or of special concern locally, statewide, nationally, or internationally? Is the status a proposed listing (candidate) or does it actually have legal standing (protected)? Presently, few species have true legal status. This information may be obtained from various local and state lists (for example, *Endangered and Threatened Plants and Animals of North Carolina*, edited by Cooper et al., 1977), national lists (for example, *Endangered and Threatened Plants of the United States*, by Ayensu and DeFilipps, 1978; *The Biota of North America*, Part 1, *Vascular Plants*, Volume 1, *Rare Plants*, by Kartesz and Kartesz, 1977), and international lists (for example, "International Trade in Endangered Species of Wild Fauna and Flora," by Fish and Wildlife Service, 1977).

The **Successional Stage** is the state of habitat development. Is the habitat in a pioneer (beginning or early successional) stage, a seral (transient or middle successional) stage, or a climax (stable or late successional) stage? Massey and Whitson (1978) state that this information may be used with other findings to prioritize our studies and searches by identifying those species that occur in relatively stable habitats or areas and those that are likely to disappear because of normal habitat, vegetation, and community dynamics.

Population Inventory (based upon Massey and Whitson, 1978) is an assessment of various characters of a population and includes numerical, distributional, and areal relations. The amount of data collected depends primarily upon available time; hence, the investigator must use personal judgment in working with these characters. A summary chart of these three population relationships and their respective subdivisions appears below.

- POPULATION INVENTORY
  - POPULATION NUMERICAL RELATIONS
    - POPULATION SIZE
      - definition of unit-selected examples:
        - individual
        - tuft
        - cushion
        - mat
        - colony
      - number of units
      - approximate number of subunits per unit–selected examples:
        - stem

culm
ABUNDANCE
very rare
rare
infrequent
abundant
very abundant
DENSITY
very sparse
sparse
scattered
dense
very dense
FREQUENCY
Examples of how frequencies might be grouped into classes:
1–20%
21–40%
41–60%
61–80%
81–100%
DISTRIBUTIONAL RELATIONS
random individuals
clumped random
uniform
AREAL RELATIONS (COVER)
Example of cover classes (taken from Radford et al., 1974):

R	one specimen in area, few if any nearby
+	Cover less than 1%
Cov. 1	Cover 1–5%
Cov. 2	Cover 5–12.5%
Cov. 3	Cover 12.5–25%
Cov. 4	Cover 25–50%
Cov. 5	Cover 50–100%

The following categories comprise Population Numerical Relations: Population Size, Abundance, Density, and Frequency. **Population Size** contains (1) definition of a unit (individual, tuft, mat, and others); (2) the number of these units; and (3) the approximate number of subunits per unit (for example, stems, culms). **Abundance** is an estimation of plentifulness of individuals of a species within a region (Radford et al., 1974) and ranges from very rare to infrequent to very abundant (see above for a list of terms). **Density** is an actual count and determination of the number of individuals on a unit area basis or the average number of individuals per area sampled (Radford et al., 1974). If time does not permit actual density to be determined, a scale ranging from very sparse to scattered to very dense might be employed (see chart above for a list of terms). **Frequency** is the percentage of sample plots in which a species occurs.

Distributional Relations describe the dispersion of a species within an area. The following terms are employed (definitions taken from Radford et al., 1974): (1) **random individuals**—dispersed irregularly without a definite pattern; (2) **clumped random**—aggregates of individuals dispersed irregularly without definite pattern; and (3) **uniform**—individuals dispersed in a regular and definite pattern.

Areal Relations, in this particular system cover, refers to the amount of area covered by a species within a region or sample plot. This may be done by estimating cover. A table of cover classes given in the chart above serves as a selected example.

The meaning of **Disturbance Type(s) and Agent(s)** should be self-evident. Selected examples include:

browsing—deer
grazing—cows
disease—fungus
infestation—insect
predation—insect
parasitism—insect
bulldozing—man
chemical spraying—man
washing out—rain
timbering—man

Population Vulnerability or Threat includes these questions: (1) How susceptible is the population to injury? (2) Is the threat potential or actual? and (3) Is the population threatened or endangered at the population or habitat level? Fire, urban expansion, flooding, and plans to drain a savannah exemplify such threats.

Other recorded characteristics might indicate how successful the population is: Size Classes, Phenological Classes, Reproductive Classes, and Population Vigor. **Size Classes** refers to the number of individuals of certain sizes within an area of population (for example, estimated percentages of seedlings, saplings, and mature individuals in a population). **Phenological Classes** pertains to the periodic biological phenomena that are occurring in the population—flowering, fruiting, vegetative phases (again, percentages of population might be recorded). **Reproductive Classes** deals with the presence of sexual or asexual forms of the plants (as previously, percentages might be estimated). **Population Vigor** is a description of the appearance of the plants based on their capacity for natural growth and survival—are the plants yellow and weak or robust and doing well? Once again, percentages might be determined.

The relationship between the Population Level Inventory (D.) of our Ecological Diversity Classification System and Massey and Whitson's (1978) Species Information System (Chapter 4) is apparent. Obtaining the information and data in Level D of our system is only very preliminary work to the understanding of our species of special concern. We hope that this Population Level Inventory will begin the attainment of the following goals:

1. To supply information for a fundamental background for working out analytical procedures and experimental design for advanced research on aspects of species biology.
2. To assist in wise decision making on conservation and preservation issues.
3. To better our understanding of species-habitat relationships at the population level.
4. To aid in identification of significant species.
5. To provide Species Ecological Characterizations for better understanding of species habitat preferences.
6. To help in preventing the inadvertent destruction of populations of species of special concern.

Summary Statement

Essentially all classification systems represent man-made devices for coping with complexity. **Natural diversity classification** is the process of grouping together like biotic and abiotic elements of diversity by recognizing character correlations and by subsequently placing groups together into larger groups, forming hierarchies of ranks and categories in each of the distinct components. Hopefully, the effort produces a **classification system** (that is, a group of interacting, interrelated, or interdependent elements forming a collective entity) that provides an efficient and effective means of communication, as well as an orderly organization for information storage, retrieval, and use.

Natural diversity embodies all of the variety and attendant phenomena in the biosphere. **Ecological diversity** includes the variety of organisms in relation to their various environments or habitats. Because organisms have abiotic as well as biotic environments and the emphasis of this system is biological, natural and ecological diversity have been used interchangeably in this book. Geologic diversity, per se, is natural diversity, but when used in relation to the biological component, it is considered ecological—thus the dual use of diversity types in this classification.

In this system the diversity in nature has been classified primarily as components (I. Biology–VII. Physiography) and secondarily into hierarchical levels (A. System–D. Population Level). Each level of each component represents a varietal or diversity class, for example, I.A. Biologic System, IV.C. Natural Area Geologic Site Type, and so on (see the chart below). Each varietal class is composed of two or more elements of diversity (for example, the number of elements in class IV.A. is three). The distinctive elements in each class are circumscribed varietal entities (taxa or objects) delimited from each other within the class and desirably, but not actually, from all other elements in all of the varietal classes.

Our basic classification includes six components (I.–VI.) and six hierarchical levels (A.–CC.), which embrace 36 varietal classes. A seventh component, physiography, is used regularly as a regional approach, and a seventh hierarchical level, population, has been made available as an option for studies of species of special concern. Thus 49 classes are recognized in the system with the seven components and levels. The maximum number of elements officially determined for a class to date is 10,555 in III.CC.

This classification system with its varietal classes gives command of diversity knowledge, makes efficient and effective use of information, and leads directly to the acquisition of more data, information, and knowledge of ecological and/or natural diversity. Use, application, and experience will provide the key to the success and revision of the classification presented in this chapter.

	I	II	III	IV	V	VI	VII
A	1 I.A 1.1–2.3[1]	2 II.A 1–5	3 III.A 1–10	4 IV.A 1–3	5 V.A 1–6	6 VI.A 1–5	7 VII.A 1–8
AA	8 I.AA 1.1–3.7	9 II.AA 1–14	10 III.AA 1–47	11 IV.AA 1–11	12 V.AA 1–20	13 VI.AA 1–21	14 VII.AA 1–25
B	15 I.B 1–;2–	16 II.B 1.1–7.6	17 III.B 1–185	18 IV.B 1–	19 V.B 1–3	20 VI.B 1–;2–	21 VII.B 1–86
BB	22 I.BB 1.1–2.89	23 II.BB 1.1–7.6	24 III.BB 1–970	25 IV.BB 1–	26 V.BB 1–19	27 VI.BB 1–;2–	28 VII.BB 1–;2–
C	29 I.C 1–	30 II.C 1.1–2.3	31 III.C 1–4,500	32 IV.C 1–;6.6.13	33 V.C 1.1–12–	34 VI.C 1–;7.5	35 VII.C 1–
CC	36 I.CC 1–	37 II.CC 1.1–2.3	38 III.CC 1–10,555	39 IV.CC 1–;6.6.13	40 V.CC 1.1–12–	41 VI.CC 1–;7.5	42 VII.CC 1–;2–
D	43 I.D 1–	44 II.D 1.1–2.3	45 III.D 1–10,555	46 IV.D 1–;6.6.13	47 V.D 1.1–12–	48 VI.D 1–;7.5	49 VII.D 1–;2–

1. See Tables 1–8.

Classes of Ecological Diversity with Number of Elements Per Class

3

Inventory of Our Natural Heritage

Inventory of the natural diversity in our heritage can be made by many types of individuals—observant farmers, hikers, students, teachers, weekend outdoor enthusiasts. Everyone has a role to play in preserving our natural heritage. The public can provide information on the known locations of elements of ecological diversity of potential significance to local and state heritage personnel. Academic institutions are repositories for much valuable natural diversity information that can be retrieved by the interested public for heritage purposes. Citizens with the proper background can make immeasurable contributions to conservation by searching for significant elements of diversity in unknown regions or by discovering desirable elements within previously surveyed areas. The lay and professional public alike possess much diversity experience that needs to be tapped for the preservation of our heritage.

An inventory of this heritage consists of the collecting of data and information on the diversity within natural areas, which can be accomplished efficiently and effectively by layman and scientist alike. Inventory requires a classification of ecological diversity for the identification of the elements of diversity within our counties, parishes, states, and physiographic regions. Inventory data and information on diversity for a natural area must be accumulated, analyzed, and synthesized for the determination of significance.

I. Types of Inventory

Inventory is hierarchical: public, information, reconnaissance, basic, and heritage, according to this system. **Public inventory** includes the collection of diversity data and information by interested citizens. **Information inventory** is the accumulation of diversity information from the literature, museums, and knowledgeable scientists and naturalists trained in the components of diversity. **Reconnaissance inventory** is the field substantiation of reported significant elements of diversity within an area. **Basic inventory** includes reconnaissance inventory with additional detailed, on-site information on diversity. **Heritage inventory** consists of an ongoing collection of diversity information resulting from advanced field studies of many types. **Species population status inventory** is a special type designed primarily for endangered, threatened, and rare species study (see Chapter 4, page 118, and Table 11, page 101).

A. Public Inventory

Most hunters, fishermen, bird watchers, and outdoor service personnel from many agencies know the location(s) of some rare plant or animal, of a unique community, possibly of a virgin stand of trees, or of a large population of a significant species. Professional heritage workers and conservationists should make a systematic search for knowledgeable foresters, wildlife personnel, park rangers, soil conservation service workers, extension service agents, garden club enthusiasts, local biologists, planners, and amateur naturalists who possess significant heritage information on that diversity. This information should be utilized routinely for public inventory of natural diversity. The public should be encouraged to report significant elements of diversity to local conservation leaders, to regional heritage coordinators, or directly to state heritage offices.

B. Information Inventory

Most heritage personnel systematically search the literature as well as herbarium and museum specimens for documented sources of information on endangered and threatened species, unique communities, unusual abiotic features, and other phenomena. Heritage workers normally interview professional biologists, geologists, and naturalists to obtain desired information. Successful heritage programs incorporate a classification of natural diversity as a guide to information inventory (see Chapter 1, Section II).

The gathering of these documented data and information on the diversity within an area is a prerequisite for an on-site reconnaissance or basic inventory. This level of inventory can be done adequately by individuals trained at the bachelor or master's level in field biology, environmental sciences, or natural history programs who know how to use the library, herbarium, and museum as resources.

C. Reconnaissance Inventory

Normally a reconnaissance of a natural area is made after the information inventory, although sometimes it is made after public reports have been obtained. Heritage workers and trained conservationists make a site reconnaissance to verify the occurrences of significant elements of diversity that have been reported in the literature or by individuals.

Field work is expensive and time consuming. Effective reconnaissance requires careful planning and systematic study before the site visit. Minimal homework for a reconnaissance includes study of topographic, geologic, and soils maps. Land cover, land use, and vegetation maps are used whenever available. Procuring and studying reports or references on ecological, biological, geological, and other studies from the heritage program files and institutional libraries should be standard procedure. Efficient reconnaissance teams will study predictive models or make their own for the area to be visited. Invariably, other significant elements will be found when substantiating the presence of one or more endangered or threatened, unique, relict, disjunct, or endemic elements. While reconnoitering, astute heritage workers will be searching for excellent representations of typical, threatened, or unique communities and populations not known previously within the state or region.

The reconnaissance team should make careful notes of the following: (1) exact locality and physiographic data; (2) ownership and administration of the area; (3) threat to the area and vulnerability of the site; (4) potential management and protection problems; (5) the presence of endangered or threatened species—population size, density, and condition; (6) the age and size of a virgin or an excellent stand of a species; (7) the uniqueness of a community with an indication of the cause of uniqueness; (8) the unique abiotic features—type and condition; and (9) archeological and other site values.

The reconnaissance team should make a well-documented written summary for items 1–9 for the site visit. Recommendations as to future action, based on documented elements of significance, should be made to the heritage staff. See Tables B1 and B2 in Part IV.

Normally a reconnaissance inventory for a significant site should be a two-stage procedure, the first resulting in a field data report and the second in a significant site report (see Part II, Reports F and G and Report E, respectively). Each report should follow a standard format (see Table 12 for significance criteria). Potentially significant natural areas should be inventoried by professional conservationists or field scientists. For cost-effectiveness several significant sites within a limited area could be reconnoitered within one day by a professional team of two individuals trained in different component disciplines.

D. Basic Inventory

Basic inventory is the on-site procurement of data and information on diversity features within a natural area according to a standardized classification system. This type of inventory is applied to sites thought to be potentially significant on the basis of information inventory and site reconnaissance. On-site inventory is usually made after deliberation and evaluation by two or more groups, such as heritage staff and advisory body and/or higher administration. The basic inventory should be made by professionally trained naturalists, field biologists, geologists, and pedologists. This type of inventory is economically feasible only for selected communities, habitats, and populations of ecological significance within a large natural area. Normally this work will be a team effort resulting in a comprehensive report on the natural diversity within the area. Fundamental inclusions within the report will be indications of the significance of the site, management recommendations, protection suggestions, and detailed documentation for diversity elements of importance. A basic on-site inventory for a natural area provides accumulated, analyzed, and synthesized documented information and data that can be used as a basis for decision on acquisition, as a basis for placement on a state or national landmark register, as a basis for the development of a master plan for that area, and as a basis for sound management and protection recommendations (see Reports A–D in Part II).

Natural heritage staffs should have one or more employees trained in basic inventory work primarily for supervisory purposes, contract writing, and report evaluation. At least one staff member should be a trained field scientist thoroughly knowledgeable in field inventory and natural areas analysis procedures. We are convinced that, to be as cost-effective as possible, basic inventory should be done by trained professionals under contract. Basic inventory expenses are standard costs for any heritage program.

E. Heritage Inventory

Heritage inventory becomes an ongoing process after the natural area has been acquired and management and protection have begun. The economics of basic inventory do not permit a complete cataloging of natural elements. Microorganisms, soil animals, liverworts, lichens, and details of soil, geologic, and topographic features will be inventory products resulting from future research in our conserved natural areas. Heritage inventories and studies will be made by advanced students and professionals in biology, climatology, pedology, hydrology, geology, and topography.

II. Inventory Report

The basic inventory report includes a summary statement that is an abstract of the report; a discussion section that places the report in perspective from diversity, geographic, and significance standpoints; community data that represent the detailed documentation for the natural area; and a master species presence list that represents the biotic foundation for conservation (see Table 9, p. 97, for natural area report format and Part II for natural area reports).

A. Natural Area Summary

The initial summary statement includes four sections: geography and physiography, land administration and use, priorities and recommendations, data sources and documentation (see Table 9, p. 97, and reports in Part II for format).

Locality data include the political boundaries, distance and direction from an established, easily recognized map point, and latitude and longitude with map reference for the center of the natural area. Physiographic data are arranged hierarchically to give regional perspective. Maximum and minimum elevation (and depths for aquatic sites) are given as well as the approximate size of the area.

Legal ownership of property should be determined before inventory proceeds at any level. Permission should be obtained for natural area study from the owner or the administrator of the property. Past use of the land should be established for an understanding of the present condition of the area. Land classification information is needed in order to know the potential future use of the property. Commercially zoned property faces a greater loss of integrity sooner than land classified as an area of environmental concern. Some dangers to natural area integrity are burning, development, mechanical destruction, and trash dumping. Threats to integrity and obvious avenues of vulnerability should be indicated in the report. Publicity sensitivity is an important item in property acquisition and land valuation considerations. In general, the less said publicly the better for acquiring property for heritage and conservation purposes.

In the section on significance and recommendations, the comparative value of a natural area should be determined according to evaluation criteria (see evaluation suggestions in Table 13, p. 104). Significant elements of diversity are listed to indicate the reasons for the priority rating, and management recommendations and protection suggestions are included. Documentation for the elements of significance should be provided and the bases of management recommendations and suggestions for protection presented in detail in the discussion section of the inventory report.

Data sources for the documentation section pertain to general data for the whole site. General documentation and scientific references for the entire natural area are included here; specific references cited in the discussion are listed in the bibliography at the end of that section, for each community at the end of the community diversity report, and so forth. If soils determinations or rock identifications have been made or authenticated for all

Table 9. *Natural Area Report Organization* (See examples in Part II for format.)

A. Natural Area Summary (See Part IV, Table A1.)

Physiographic-geographic data
Land administration and use
Dangers to integrity
Significance and priority rating
Management recommendation
General documentation

B. Discussion

1. Site Description

2. Perspective

   Biotic features
   Abiotic features
   Biotic-abiotic relationships
   Spatial-temporal dynamics
   Local relationships
   Regional perspective
   Pertinent literature survey

3. Significance and Management (See Tables 12-15.)

4. Bibliography

5. Topographic Map and Visual Aids

C. Diversity Summaries

1. Natural Area Diversity Summary (See Part IV, Table A2.)

2. Community Diversity Summary (See Table 10.)

3. Population and Microhabitat Summary (See Table 11.)

D. Master Species Presence List

community reports within the area, then the individual responsible should be listed in this section. The underlying principle for documentation is to include the information at the most appropriate place within the report without undue repetition of any single general reference.

B. Discussion

This section is included to give a site description and perspective on the diversity within the natural area, to aid in understanding the significance of diversity features, and to make management recommendations and protection suggestions based on this perspective and significance.

1. Site Description

The site description should convey a visual impression of the natural area. The major habitats, including the primary vegetational, topographic, geologic, and hydrologic features, must be described. Large animal populations and communities should be mentioned. System through class level elements of the components should be emphasized. The description of the natural area must be brief and concise.

2. Perspective

The plant and animal Community Cover Types and Community Types should be described. A purview of the pedology, geology, topography, hydrology, and climate must be included for perspective on the distribution and composition of the biotic assemblages. The significant abiotic features should be enumerated and the biotic-abiotic relationships stressed. In general, biotic diversity should be related to specific abiotic component elements whenever possible.

The types of seres within the area should be listed and the stage of succession indicated for each Community Type. Biotic, geologic, topographic, hydrologic, and soils dynamics within the site should be discussed for perspective on present and future communities. Pioneer, early successional, ecotonal communities and/or conditions within the natural area should be described when not analyzed. Most management problems are directly related to the biotic and abiotic dynamics of the natural area and its surroundings.

The communities within the site should be related to adjacent communities outside the natural area, particularly in relation to changes in elevation, moisture, slope aspect and degree, rock type, soil type, and other factors. Available literature on similar diversity within a wide area should be used for geographic perspective. A review of pertinent literature should be made for the significant biotic and abiotic elements within the area for regional and national perspective. Fire and disturbance history should be included whenever possible. (See Part III, Section H.)

3. Significance and Management

All endangered and threatened, relict, disjunct, and endemic species within the area should be listed by type. Unique, disjunct, relict, and endemic communities should be recorded. Unusual abiotic features, excellent stands or populations, fine examples of landscape, soil, and rock types should be noted. Unusual relationships between, or excellent examples of, organisms and habitat should be included in this section. Every significant element of diversity must be documented as to collection, photographic evidence, and/or literature citation.

The bases for recommendations on management and protection should be explained. Considerations of the significant elements of diversity should provide the foundation for the recommendations and suggestions. (See Section III of this chapter for discussion of significance and priority criteria and Part III, Sections F and G.)

4. Bibliography

All literature cited in the discussion must be included in the references at the end of the discussion. (See p. 383 and the bibliography at the end of the book for citation format.)

5. Topographic Map and Visual Aids

Standard visual aids used in the inventory report are maps, profiles, and photographs. The minimal map inclusion is a topographic map, preferably with a 1:24,000 scale, showing location of the natural area with clearly and sharply defined boundaries. A map legend inset must include name of quadrangle, scale, and date of publication. Topographic maps should show an easily recognized landmark for ready location of the site in order to eliminate the frustration of trying to locate it in the middle of a mass of contour lines and unnamed topographic features. (See maps for natural area reports in Part II.)

Detailed soil and geologic maps are desirable. Profiles showing the distribution of communities in relation to relief and elevation or depth are particularly helpful in gaining perspective on biotic diversity within the area. Profiles showing distribution of biotic communities in relation to mapped soils and/or geologic formations are most pertinent. All maps should have a legend indicating name and type of map, scale, publisher, and publication date. Photographs of communities and distinctive topographic geologic features provide helpful visual documentation.

C. Diversity Summaries

1. Natural Area Diversity Summary

The Natural Area Diversity Summary is a collection of all entries at each hierarchical level for each component in the site. This section is included primarily for data banking and manual file purposes. This summary, however, provides an easily accessible listing of abiotic features, such as all of the soil series, rock types, landforms, and climatic regimes, as well as biotic features. An investigator wishing to know the Community Cover Types and other features found in the natural area can find them easily in this coded summary. (See Table 9 on p. 97, Table A2 on p. 396, and discussion on p. 383 for format and reports in Part II for examples.)

2. Community Diversity Summary

In a natural area the elements of diversity are determined for each hierarchical level of all seven components in the basic inventory of a Community Type. Height, dbh, and age usually are determined for the largest representative of each canopy and subcanopy species within the Community Type. Size classes are ascertained for the woody species. An analysis, using standard ecological procedures, is made for each layer in the community. Analytical summaries are presented in table form for each layer of vegetation. Species presence is noted for those species that are not part of the analysis. References and general documentation pertinent to that Community Type are made. The Community Ecological Characterization includes a physiognomic description of the vegetation, a standard habitat summary for the abiotic features of the Community Type, notes on type of succession and stage of development of the community, and a general physiographic summary for the community.

References pertinent to the Community Type must be cited. Sampling procedure needs to be indicated. Any determiner or authenticator, other than the author(s), of component elements should be identified. (See Tables 9 and 10, p. 97 and p. 100, for format and procedure; also see the natural area reports in Part II.)

3. Population and Microhabitat Summary

Microhabitat notes are made for selected species when appropriate. Endangered and threatened species are listed. Population inventory status reports usually are made on any endangered and threatened species within the community. Endangered or threatened species population size, density, and condition are determined; threat to and vulnerability of the population are recorded; and distinctive microhabitat features are described. (See population level section at end of Chapter 2 and species biology treatment in Chapter 4; for format, see Table A18, pp. 411–412.)

D. Master Species Presence List

The key element in our natural heritage is the species. One of our primary conservation goals is to acquire, manage, and protect as many natural areas as necessary to give us the diversity of habitat required for the preservation and perpetuation of our native species. An alphabetized species list for each major group within a natural area is a convenient way to catalog and collate species diversity for local areas, states, regions, and the nation. The availability of species presence lists by natural area should eliminate the continued collecting of specimens by succeeding students and classes. One documentation for a species from a natural area should be adequate. Additions to the flora and fauna lists of an area can be made easily when the Master Species Presence List is available.

Most inventory reports will be based on vascular plant species, which are standard inclusions in any list unless absent in the area inventoried. Mycetozoa, crustose li-

Table 10. *Community Diversity Summary Format*

NATURAL AREA
County, State; Locality

Biotic System
Biotic Subsystem

Community Cover Class
Community Class

COMMUNITY TYPE
Community Cover Type

CLIMATE: A. Climatic System; AA. Climatic Subsystem. B. Climatic Class; BB. Climatic Subclass. C. Climatic Generitype; CC. Climatic Type.
SOILS: A. Soil System; AA. Soil Subsystem. B. Soil Class; BB. Soil Subclass. C. Soil Generitype; CC. Soil Type.
Description of soil horizons.
GEOLOGY: A. Geological System; AA. Geological Subsystem. B. Geological Class; BB. Geological Subclass. C. Geological Generitype; CC. Geological Type.
HYDROLOGY: A. Hydrologic System; AA. Hydrologic Subsystem. B. Hydrologic Class; BB. Hydrologic Subclass. C. Hydrologic Generitype; CC. Hydrologic Type.
TOPOGRAPHY: A. Topographic System; AA. Topographic Subsystem. B. Topographic Class; BB. Topographic Subclass. C. Topographic Generitype; CC. Topographic Type.
PHYSIOGRAPHY: A. Physiographic System; AA. Physiographic Subsystem. B. Physiographic Class; BB. Physiographic Subclass. C. Physiographic Generitype; CC. Physiographic Type.

CANOPY (by species): Height, dbh, age.
CANOPY DOMINANT(S): Species.
CANOPY PHYSIOGNOMY: Size, growth form, and duration of canopy dominants.

CANOPY ANALYSIS

Number of points: _____ d: _____ Number of individuals/acre: _____
CANOPY SPECIES PRESENT, BUT NOT IN ANALYSIS: Species list

SUBCANOPY (by species): Height, dbh, age.
SUBCANOPY DOMINANT(S): Species.
SUBCANOPY PHYSIOGNOMY: Size, growth form, and duration of subcanopy dominants.

SUBCANOPY ANALYSIS

Number of points: _____ d: _____ Number of individuals/acre: _____
SUBCANOPY SPECIES PRESENT, BUT NOT IN ANALYSIS: Species list.

SHRUB LAYER DOMINANT(S): Species.
SHRUB LAYER PHYSIOGNOMY: Size, growth form, and duration of shrub dominants.

SHRUB ANALYSIS

Number of relevés: _____ Relevé size: _____
SHRUB SPECIES PRESENT, BUT NOT IN ANALYSIS: Species list.

HERB LAYER DOMINANT(S): Species.
HERB LAYER PHYSIOGNOMY: Size, growth form, and duration of herb dominants.

HERB ANALYSIS

Number of relevés: _____ Relevé size: _____
HERB SPECIES PRESENT, BUT NOT IN ANALYSIS: Species list.

COMMUNITY REFERENCES:
COMMUNITY DOCUMENTATION:

COMMUNITY ECOLOGICAL CHARACTERIZATION: Vegetationally: I.BB., I.A., I.AA. plus description of other vegetational strata. Climatically: II.AA., II.A., II.B., II.BB., II.CC. Pedologically: III.C., III.CC. Geologically: IV.A.-IV.CC. in readable, sensible manner. Hydrologically: V.A.-V.CC. in readable, sensible manner. Topographically: VI.A.-VI.CC. in readable, sensible manner. Temporally and Spatially: Succession stage, sere type, VII.A.-VII.CC. in readable, sensible manner.

Table 11. *Species Population Summary Format*

NATURAL AREA
County, State; Locality

SPECIES
Species Legal Status (es)

BIOLOGY: I.D. Biotic Associates
CLIMATE: II.D. Population Climatic Site Type
SOILS: III.D. Population Soil Site Type
GEOLOGY: IV.D. Population Geologic Site Type
HYDROLOGY: V.D. Population Hydrologic Site Type
TOPOGRAPHY: VI.D. Population Topographic Site Type
PHYSIOGRAPHY: VII.D. Population Physiographic Site Type

SUCCESSIONAL STAGE:

POPULATION INVENTORY
POPULATION SIZE:
ABUNDANCE:
DENSITY:
FREQUENCY:
DISTRIBUTIONAL RELATIONS:
AREAL RELATIONS:

DISTURBANCE TYPE(S) AND AGENT(S):

POPULATION VULNERABILITY:

SIZE CLASSES:

PHENOLOGICAL CLASSES:

REPRODUCTIVE CLASSES:

POPULATION VIGOR:

MICROHABITAT FEATURE:

SPECIES REFERENCE(S):

SPECIES DOCUMENTATION:

SPECIES ECOLOGICAL CHARACTERIZATION: I.AA., I.BB., I.CC., I.A. The remainder of the characters are the same as in the Community Ecological Characterization, with I.D.-VII.D. being included in each respective theme. Normally only endangered and/or threatened species would be characterized in this format because each species within a community is characterized in the Community Characterization.

chens, soil algae, protozoa, and the like represent groups of organisms that usually will be added to the master species list by heritage inventory workers.

The Master Species Presence List should be documented as to depository for voucher specimens, for example, the herbarium for plants, the museum for animals and fossils. The collector and his collection number should be included for each species listed. The reference base for nomenclature should be indicated for each major group. (See p. 35 for format and p. 394 for notes; also see reports in Part II for examples of Master Species Presence Lists.)

III. Significance and Priority in our Natural Heritage

The preservation of natural diversity is the maintenance of genetic and ecological diversity. Preservation decisions on management and protection must be based on studies of carefully selected criteria that represent the significant genetic and ecological features of natural areas. The establishment of priorities for the acquisition of natural areas must be based on documented evaluations of the important features within the sites inventoried.

A. Criteria for Determination of Significance

Criteria used in the determination of significance of natural features should be comprehensive enough to ensure preservation of most of our natural diversity. These criteria should be sufficiently adaptable to evaluation to permit introduction of some objectivity in the establishment of priorities for natural areas within a state, a region, or the nation.

The selected criteria in Table 12, page 103, are classified as follows:

A. Endangered and threatened species, communities, and habitats—the imperilment or extinction jeopardy of each.

B. Biotic and abiotic diversity—kinds or types of elements for each hierarchical level of each component.

C. Natural features condition—physical properties of the elements for each hierarchical level of each component.

D. Distribution features—geographic ranges of biotic and abiotic elements.

E. Humanistic features—human values and interests for the abiotic components, communities, and/or natural areas.

F. Productivity—total amount produced in natural area or community; yield per unit area.

These criteria have been used in part, but not as a whole, in numerous studies for the determination of significance (see Table 14). Many other criteria of equal diversity significance could have been listed, and some of those listed might be combined under certain conditions. Experience reveals a definite need for an organized set of criteria for the determination of significance for abiotic components, communities, and natural areas. Without a set of selected criteria, it is too easy to forget some elements of significance and to overemphasize others. We must preserve the best of our ecological diversity.

B. Evaluation of Criteria

An effort should be made to produce an evaluation scheme for the significance of the criteria even though there never will be full agreement with any system developed. The suggestions in Table 13, page 104, are presented as guidelines, not as absolutes. In the author's scheme, for instance, an endangered endemic, an irreplaceable element, is given the highest rating of 10; a good second growth community, 4; and so forth. In general, the reasons, basis, or comparative basis for the significance values for selected criteria for any natural area should be explained in the discussion section of the site report. The following comments on rated criteria, arranged according to Tables 12 and 13, are included as aids to greater comparability in value judgments and better documentation for significance values. Diversity elements rated average or below usually are not mentioned in significance summaries.

(A) *Geum radiatum*, an endangered species according to Ayensu and DeFilipps (1978), is given a value of 10 because it is a highly localized endemic known from very few localities and is considered an irreplaceable element deserving the greatest protection possible. *Panax quinquefolium*, a threatened species according to Ayensu and DeFilipps (1978), is given a value of 6 because it is fairly widespread in its distribution but exploited for commercial purposes.

Table 12. *Criteria Used in the Determination of Significance*

A. Endangered or Threatened Species and Communities
1. Presence of endangered (rare) species
2. Presence of threatened (rare) species
3. Presence of endangered community type
4. Presence of threatened community type
5. Presence of endangered habitat

B. General Biotic and Abiotic Diversity
6. Presence of characteristic species of a habitat type (rock type, soil series, water chemistry type, topographic site type)
7. Presence of characteristic community types/community cover types
8. Presence of pioneer, transient, and climax communities for a sere
9. Presence of climax zonation
10. Presence of communities with unique species composition
11. Presence of habitats with unusual species relationships
12. Presence of communities or habitats with discordant species or community composition
13. Presence of five or more soil orders
14. Presence of five or more landscape-forming processes and products
15. Presence of three or more material-forming processes and products
16. Presence of three or more major water chemistry types

C. Natural Features Condition
17. Example of stand or community
18. Example of population of rare species
19. Example of geological feature(s)
20. Example of soil series
21. Example of topographic feature(s)
22. Example of hydrological feature(s)

D. Distribution Features
23. Presence of relict species and/or community
24. Presence of endemic species and/or community
25. Presence of disjunct species and/or community
26. Presence of transition communities and zones
27. Presence of excellent continua
28. Presence of discrete communities
29. Presence of relict or disjunct soil
30. Presence of unique geological feature
31. Presence of unique topographic feature
32. Presence of unique hydrological feature

E. Humanistic Features
33. Obvious aesthetic value
34. Obvious scenic value
35. Scientific research value
36. Historical value

F. Productivity
37. Example of biomass
38. Example of animal cover
39. Example of food productivity
40. Example of breeding grounds or territory

Table 13. *Evaluation of Criteria*

INFORMATION SYSTEM I	INFORMATION SOURCES I
A. ENDANGERED AND THREATENED SPECIES (total: ___)	
Endangered endemic 10 Endangered throughout 9 Endangered disjunct 8 Threatened endemic 7 Threatened throughout 6 Threatened disjunct 5 Endangered peripheral 4 Threatened peripheral 3 Infrequent endemic 2 Infrequent peripheral 1	International lists Smithsonian list State lists Field observation, determination, and authentication
B. BIOTIC AND ABIOTIC DIVERSITY (total biotic: ___; total abiotic: ___)	
Biotic Systems Pedologic Cover Classes Geologic Cover Types Hydrologic Species Topographic Excellent, 5; Good, 4; Average 3; Mediocre, 2; Poor, 1.	Vegetation maps, studies, and reports Pedologic maps, studies, and reports Hydrologic maps, studies, and reports Topographic maps, studies, and reports Field observation, determination, and authentication
C. NATURAL FEATURES CONDITION (total: ___)	
Communities Hydrology Pedology Topography Geology Virgin or excellent, 5; Good, 4; Average, 3; Mediocre, 2; Poor, 1.	Pertinent reports and studies Field observation and determination
D. NATURAL FEATURES DISTRIBUTION (total: ___)	
Community Hydrology Pedology Topography Geology Endemic, 5; Unique, 4; Infrequent, 3; Common, 2; Very Common, 1.	Vegetation maps, studies, and reports Pedologic maps, studies, and reports Geologic maps, studies, and reports Hydrologic maps, studies, and reports Topographic maps, studies, and reports Hydrographic maps, studies, and reports Field observation and determination
E. HUMANISTIC FEATURES (total: ___)	
Aesthetic value Scenic value Scientific value Historical value Excellent, 5; Good, 4; Average, 3; Mediocre, 2; Poor, 1.	Field experience and reports Scientific reports Historical reports Land use reports
F. PRODUCTIVITY (total: ___)	
Biomass Cover Food Breeding territory Excellent, 5; Good, 4; Average, 3; Mediocre, 2; Poor, 1.	Wildlife reports Economic reports NATURAL AREA EVALUATION Total: _______

(B) Heggies Rock in Georgia is rated excellent (5) for the number of species characteristic of granitic flatrocks. McVaugh (1943) lists 26 plant species as being characteristic of these flatrocks, with our experience indicating 6 to 10 species as average (3), 11 to 16 as good (4), and more than 16, excellent (5). Nineteen characteristic species have been reported from the rock. The Roanoke River Bluffs Natural Area on the Coastal Plain of North Carolina is rated good (4) for soil order diversity. Again, our experience indicates that the orders Entisol, Inceptisol, and Ultisol are typical (3) for slopes and mature stream bottoms. A site with a fourth order is rated good (4), and one with more than four in that area, excellent (5).

(C) The Canadian Hemlock stand in the Joyce Kilmer Memorial Forest is considered virgin (5) because of age and size according to the Spector *Handbook of Biological Data* (1956). Trees in that region in the 24"–30" dbh range with ages ranging up to 250 years should be rated good (4) on an extrapolation basis from the *Handbook*. Orbicular diorite on the Cooleemee Plantation is a "spectacular rock with geometric texture and contrasting color," according to Butler et al. (1975). We would certainly apply an excellent (5) rating for such a statement. It is unknown from any other locality (Butler et al., 1975) and thus would also have a distribution value of 5.

(D) Heggies Rock in Georgia has 12 of the 19 endemics reported from granitic flatrocks (McVaugh, 1943). Three to five endemic species per rock seems to be average (3); six to nine, good (4); and more than ten, excellent (5) (our rating system). Heggies Rock is rated excellent (5) for its endemic species diversity and distribution. The White Cedar/Fetterbush/Evergreen Blueberry (sp. nov.) community is known only from the Scouter Creek drainage in the sandhills of South Carolina, hence, the community and the new species of Evergreen Blueberry are very localized endemics (5). The White Cedar stand on a seepage slope, rather than in the typical bog, is also in a highly unique habitat (5).

(E) The Hutcheson Memorial Forest in New Jersey has been rated from a scientific research standpoint as excellent (5). According to Radford and Martin (1975), "The stand has been intensively studied over the past twenty-five years and has yielded valuable information on, among other topics, the role of fire in pre-settlement forests and the effect of windthrow and drought on a mature forest. Through 1971 sixteen papers had been published that pertained to the forest."

(F) The Roanoke River Bluffs Natural Area has been rated by Lynch (personal communication, 1979) as excellent (5) breeding territory, indicated by nests found for 54 species of birds in a tract of approximately 10 acres.

In general, an effort should be made to acquire any site with a top-evaluated element among the 40 criteria listed. Any natural area with a single top-rated criterion, for example, with an excellent virgin stand of White Oak, with a majority of the characteristic species on a granitic flatrock, with an outstanding example of a hoodoo (topographic feature), or with an excellent fossil locality, should have first priority for acquisition, protection, and management. If funds are limited at a particular time, a natural area with two or more top-rated features would have preference over a site with one significant feature. Effort should continue, however, for preservation of the site with the one excellent or outstanding diversity characteristic.

C. Establishment of Priorities

The establishment of priorities for natural areas must be based on significance values determined for accepted criteria. Endangered, unique, and irreplaceable natural areas or ecosystems, based on excellent or outstanding rating values, should be given a high degree of national significance and first priority (see Tables 13, 14, and 15, pp. 104, 106, and 108). Sites with outstanding examples of communities, soils, geologic features, hydrologic elements, landforms or water bodies, and other features also should rate number one in national significance and priority. Natural areas with threatened species and/or communities or excellent examples of diversity and/or equally excellent distribution, humanistic, and productive features would appear to be nationally significant with second priority. If the various criteria rate good to excellent, the natural area should be considered of regional or state significance. Average biotic and abiotic features would appear to be locally significant. Natural areas with an established or apparent high degree of national significance automatically would be of top priority on a regional (interpreted as physiographic region) or state basis.

D. Management and Protection

Past land use and present land classification and use will dictate protection priorities or suggestions. Management recommendations should be made after a basic inventory has been completed. Advisory committees and planners at various levels should have input before final management decisions are made or implemented. Management and protection decisions must be made in relation to inventory and the desired goals for the natural area.

Table 14. *First Priority Sites, Piedmont Region, Eastern United States, Based on Several Criteria*[1]

Site	Type
I. GREATEST NUMBER OF ENDEMIC, RARE, CHARACTERISTIC, DISJUNCT SPECIES FOR EACH ROCK TYPE	
Heggie's Rock (GA)	Granitic flatrock
Stevens Creek (SC)	Syenite
Nottingham Barrens (PA)	Serpentine
Flint River Wilderness (GA)	Quartzite
Soldiers Delight (MD)	Serpentine
York County (SC)	Gabbro (Enon soil)
II. GREATEST COMMUNITY AND SPECIES DIVERSITY FOR EACH ROCK TYPE	
Heggie's Rock (GA)	Granitic flatrock
Soldiers Delight (MD)	Serpentine
Flint River Wilderness Area (GA)	Quartzite
Nockamixon Rocks (PA)	Shale cliff
James River (VA)	Marble
Joe-Little Joe Mountain (NC)	Granitic monadnock
Rocky River, Stanly Co. (NC)	Slate-bedded argillite
III. BEST COMMUNITIES BASED ON DIVERSITY, ENDEMIC, AND RARE SPECIES IN RELATION TO LANDFORM	
A. Upland slopes	
Stevens Creek (SC)	Syenite (north slope)
Panther Creek (GA)	Marble (sheltered slopes)
Pumpkinvine Creek (GA)	Slate-bedded argillite (north slope)
Octoraro Creek Region (PA)	Serpentine (north slope)
Flint River Wilderness (GA)	Quartzite (all slopes)
B. Flatland	
York County (SC)	Gabbro depression
C. Bottomland	
The Great Swamp (NJ)	Old Pleistocene lake bed sediments
Alcovy River (GA)	Mature river floodplain sediments
Accotink Creek (VA)	Creek clay loam
D. Rock outcrops	
Heggie's Rock (GA)	Granitic flatrock
Soldiers Delight (MD)	Serpentine flatrock
Stone Mt. (GA)	Granitic dome
IV. MOST REPRESENTATIVE TYPICAL COMMUNITIES OR STANDS	
Heggie's Rock (GA)	Pioneer and early successional, granite
Soldiers Delight (MD)	Pioneer and early successional, serpentine
John De La Howe (SC)	Oak-pine climax
Brandywine/Woodlawn (DE)	Yellow poplar and beech stands
Sweet Briar College (VA)	White oak
Hawlings River (MD)	Chestnut oak

[1] A. E. Radford and D. L. Martin. 1975. Potential Ecological Natural Landmarks, Piedmont Region, Eastern United States. Privately published for National Park Service.

Table 14. *(continued)*

V. MOST UNIQUE COMMUNITY COMPOSITION AND DIVERSITY	
Stevens Creek (SC)	Mixed tree--shrub--herb
Panther Creek (GA)	"Cove hardwoods" mixed tree, shrub, herb, vine
Nottingham Barrens (PA)	Pitch pine--dwarf chinkapin oak
Tye River (VA)	Hemlock--beech--mixed herb
Octoraro Creek Red oak (PA)	Red oak--benzoin--mixed herb

VI. BEST DISJUNCT STANDS	
Rocky River, Chatham Co. (NC)	White pine
Otter Creek (PA)	Hemlock

VII. BEST SITES FOR GREAT ABUNDANCE OF UNCOMMON NATIVE SPECIES Endemic to Piedmont* Very rare throughout range' Rare disjuncts"	
Heggie's Rock (GA)	Amphianthus pusillus*'; Cyperus granitophilus*'; Forestiera ligustrina; Viguiera porteri*; Oenothera fruticosa var. subglobosa*
Panther Creek (GA)	Athyrium pycnocarpon; Collinsonia verticillata*; Panax quinquefolium; Schisandra glabra'
Stevens Creek (SC)	Ribes echinellum' (one of two known populations); Trillium discolor*
Nottingham Barrens (PA)	Quercus prinoides; Cerastium arvense var. villosissimum*
John De La Howe (SC)	Obolaria virginica; Dirca palustris
Brandywine/Woodlawn (DE)	Viburnum acerifolium; Symplocarpus foetidus
Bogg's Rock (SC)	Aster avitus*"
Joe-Little Joe Mt. (NC)	Pellaea wrightiana"
York County (SC)	Camassia scilloides"
Flint River Wilderness (GA)	Hymenocallis occidentalis
Octoraro Creek Red Oak (PA)	Myosoton aquaticum
Long Green Creek (MD)	Smilacina racemosa

VIII. BEST SPECIALTY SITES	
Stone Mt. (GA)	Type specimen locality
Hutcheson Woods (NJ)	Scientific research area

IX. COMBINATION SITES BASED ON CRITERIA I-VIII	
1. Heggie's Rock (GA)	2. Stevens Creek (SC)
3. Soldiers Delight (MD)	4. Panther Creek (GA)
5. Nottingham Barrens (PA)	

X. OUTSTANDING COMBINATION SITES BASED ON COMMUNITY AND SPECIES DIVERSITY; ENDEMIC, RARE, AND DISJUNCT SPECIES; AND EDAPHIC, TOPOGRAPHIC, AND GEOLOGIC FEATURES
1. Forty Acre Rock-Flat Creek Dike (SC)
2. Flint River Water Gap Wilderness Area (GA)
3. Susquehanna River Area, Otter Creek, etc. (PA)

Table 15. *Priority Rating Systems*[1]

SIGNIFICANCE PRIORITY

1. High degree of national significance, recommended without reservation
2. Appears to be nationally significant
3. Information lacking for a confident recommendation, but may prove to be nationally significant upon further investigation
4. Not recommended, not of national significance
5. Appears to be regionally significant
6. Appears to be of state significance
7. Appears to be locally significant
8. Appears to be of no significance regionally, for the state, or locally

PROTECTION PRIORITY

A. Site is in serious impending jeopardy
B. Site is in some jeopardy
C. Site is in no apparent jeopardy
D. Relative jeopardy of site is unknown
E. Site is protected
F. Site is protected and managed

MANAGEMENT PRIORITY
(Recommendations)

I. Natural area should remain as wilderness
II. Natural area should be partly managed, partly preserved as wilderness
III. Natural area should be managed for endangered and/or threatened species preservation
IV. Natural area should be managed for unique community preservation
V. Natural area should be managed for scientific research
VI. Natural area should be managed for multiple use
VII. Natural area should be managed as education resource
VIII. Natural area should be managed as wildlife resource
IX. Natural area should be managed as economic resource
X. Natural area should be managed for recreation
XI. Natural area should not be managed

[1]Significance and protection priorities based primarily on those established for the Landmarks Program of the National Park Service.

E. Summary

Carefully selected standardized criteria should be used in the determination of significance. The ascertainment of national, regional, and state significant natural areas should have a common basic inventory foundation to ensure comprehensive comparability and consistency in evaluation. The resulting priority classification of natural areas certainly will be more meaningful and significant from a preservation and natural heritage standpoint.

IV. Local, State, and Regional Inventory Procedures

The procedures presented herein represent major efforts in a time frame for the inventory of our natural heritage. These operations are included as suggestions for an orderly sequence for a comprehensive inventory of the ecological diversity within a physiographic unit and/or climatic zone. These suggestions are equally pertinent to state heritage and national land-managing agency inventories, drainage basin surveys, regional floristic and faunistic studies, and area habitat summaries. These operations are included also as suggestions for the most efficient ways of inventorying the maximum ecological diversity within an area for the least expenditure of time, effort, and money. All procedures, however, might occur simultaneously or some before others, with common sense dictating changes in sequence. For instance, a site with an endangered species must be acquired and/or protected immediately. In the case of a planned development, a landform with a lithology that is unknown from the standpoint of communities over that combination of abiotic features must be inventoried directly for potentially significant biotic elements of diversity. Priority based on the intrinsic natural heritage value of the site coupled with threat and availability are the major factors that determine action on natural area acquisition and/or protection.

Initially, basic decisions must be made on the ecological diversity classification to be used and the physiographic or geographic area to be covered. The first phase of inventory work is essentially a compilation of known data and information on significant elements of diversity gathered from institutional repositories and individuals. The second phase is largely an inventory of the unknown or unreported elements within the region, which, when completed, represents new knowledge of diversity. Both phases of inventory are absolute requirements for the preservation of the ecological diversity in our natural heritage. This procedural synopsis should provide some perspective for inventory programs of different heritage organizations and national land-managing agencies.

Basic Decisions

1. Adopt, adapt, or develop an ecological diversity classification as a basis for the state or regional inventory. (See Chapters 1 and 2.)

2. Determine the physiographic unit(s) and/or climatic zones to be used as the geographic basis for inventory. (See Chapter 1, Section II; Chapter 2, Sections VII and II.)

First Phase

3. List reported and predicted potentially significant elements of diversity for the physiographic unit(s) and/or climatic zones. (See Chapter 3, Sections III and IB.) Suggested classes of elements are listed below; also see Chapter 1, Section VIII.

A
Threatened species
Endangered species
Endemic species
Disjunct species
Exploited species
Peripheral rare species

B
Breeding territories
Nesting sites
Spawning areas
Rookeries
Migratory routes
Feeding grounds
Cover areas

C
Virgin communities
Unique communities
Endemic communities
Disjunct communities
Rare communities

D
Rock outcrops
Formation exposures
Rare mineral sites
Fossil localities
Topographic features
Soil series
Geomorphic processes
Geologic processes

E
Critical habitats
Type localities
Scientifically valuable areas
Historically significant areas
Potential wildland areas
Potential wilderness areas
Potential scenic areas

4. Initiate species general information studies on threatened and endangered species or others in 3A above. (See Chapter 4.)

5. Plot distribution of significant elements of diversity from 3 and 4 on topographic maps. (See Chapter 3, Section IC.)

6. Reconnoiter sites for significant elements of diversity. Prepare reconnaissance reports and determine significance of sites. (See Part IV; Chapter 3, Section III.)

7. Make an on-site basic inventory for reconnoitered sites of potential significance. Prepare report. Evaluate criteria for determination of significance. Establish priorities. (See Chapter 3, Sections II and III, and Part IV.)

8. Institute acquisition and/or protection proceedings for sites deemed worthy of nomination to the Registry of Natural Areas. (See Part III C; Chapter 3, Section III.)

9. Periodically summarize elements of diversity protected (preferably by class, 3A–E above), indicating the degree of protection, type of management established or planned, and agency or organization administering site. (See Part III D.)

Second Phase

10. Systematically search for significant unprotected elements of diversity by regular updating of public and information inventory. (See Chapter 3, Sections IA and IB.)

11. Prepare population status inventories for threatened and endangered species. (See Chapter 4.)

12. Determine critical habitat for threatened and endangered species and communities. (See Chapter 4.)

13. List the community cover classes, landforms and water bodies, rock-sediment chemistry types, water chemistry types, soil great groups reported, mapped, and/or predicted for the physiographic unit and/or climatic zone. This work can be done most effectively by mountain range, plateau, and drainage basin within the physiographic unit. (See Tables 3–7 at the B. and BB. levels and consult appropriate references in the bibliography.)

14. Relate each rock-sediment type, each soil great group, and each water chemistry type to each mapped topographic feature. (See Chapter 2, Sections III–VI.)

15. Using satellite land cover maps and data, aerial photos, and vegetation maps, select the best reconnaissance sites for community cover classes on each rock-sediment type and topographic feature, each soil great group and topographic feature, and each water chemistry type and topographic feature. (See Chapter 2, Sections III–VI.)

16. Make preliminary reconnaissance by air or boat, as appropriate, for selection of the best reconnaissance sites and then make a ground reconnaissance noting the community cover types for each abiotic combination. With CC. level modifiers in mind, particularly slope characteristics, note community types present. Be particularly aware of pioneer and transient communities as well as climax ones. Prepare reconnaissance reports and determine significance of sites. (See Part IV; Chapter 3, Section III.)

17. Make an on-site basic inventory for reconnoitered sites of potential significance. Prepare report. Evaluate criteria for determination of significance. Establish priorities. (See Chapter 3, Sections II, III, and Part IV.)

18. Institute acquisition and/or protection proceedings for sites deemed worthy of nomination to the Registry of Natural Areas. (See Part IIIC; Chapter 3, Section III.)

19. Systematically continue the search for significant unprotected elements of diversity by regular updating of public and information inventory and the development of predictive models for the discovery of elements of diversity unknown to the physiographic unit(s) and/or climatic zone or new to science. (See Chapter 2.)

20. Periodically summarize all elements of diversity protected (by class), indicating the degree of protection, type of management established or planned, and agency or organization administering site. Analyze summaries for ecologic and genetic diversity preserved and for effort yet required to adequately protect our natural heritage. (See Chapter 3, Section III; Part IIID.)

4

An Endangered Plant Information Program:

Current Taxonomic, Distribution, Population, Habitat, and Threat Status

The Endangered Species Act of 1973 provided the impetus and the mechanism for government agencies, the botanical community, and conservation-preservation organizations to initiate serious and effective actions to counteract alteration and destruction of habitats and the extinction of plant species. These activities have involved, for the most part, the compiling of state and national lists of threatened species and the acquisition of related general information (Massey and Whitson, 1980). It appears that the availability and reliability of information needed for preserving plant species and for making serious decisions on conservation and preservation issues is still less than adequate. During the past three years, we have devoted our efforts toward developing, testing, and evaluating a philosophy and a program to assist in securing a better understanding of the status and basic biology of special plant species. This program is composed of a set of priority questions and a series of specific information-documentation systems.

Although designed to handle issues specifically related to endangered and threatened species, these priority questions and information-documentation systems can be applied directly to sound basic research programs. Because the concepts and approaches of the fields of systematics and ecology are incorporated into the program, the objectives, the methods employed, and the information sought are appropriate for the study of populations, population systems, and species. The program, as presented, is neither complete nor exhaustive but is suggestive and should generate many additional questions and investigations as well as stimulate the production of a number of significant products for the various fields of basic and applied science.

The rapid destruction of habitats and the extirpation of plant populations and species clearly indicate the necessity for maximum cooperation and information exchange between all concerned persons, agencies, and institutions. The recognition of special species as elements of natural diversity, the legal requirement of critical habitat designation for the listing of endangered and threatened species, the high cost of field work, and the pressing need for natural area and habitat protection and management also emphasize the urgency for maximum cooperation and information exchange. We hope that the portion of the program presented here will enhance understanding of special species, stimulate information flow, reduce the duplication of effort, and, above all, promote efficient identification of the elements of our natural diversity. This program should lead to the acquisition of information and data that will produce justification for conservation of special species or their declassification as species of special concern.

The basic goals of our approach are (1) to retrieve pertinent information on a species from herbaria, literature, and related sources; (2) to provide the procedures for determining the current status of each population; (3) to promote the compilation of documented status reports on special plant species to be evaluated (Henifin et al., 1980); and (4) to initiate studies on the basic biology of those species threatened with extinction. Our program consists of four major units: I. Species General Information; II. Species Population, Habitat, and Threat Inventory Information; III. Species Biology Information; and IV. Environmental Factor Information (Table 16). Only the first and second units are treated in detail in this chapter. An overview of the third and a circumscription of the fourth are included as general information on the entire system. The objectives, principles, and elements of the first two units are discussed below.

Table 16. *Overview of Species Information Program*

UNIT I. SPECIES GENERAL INFORMATION

- Species Taxonomic Status
- Species Phenology
- Species Legal Status
- Historical Distribution
- Land Status and Ownership
- Habitat Preference
- Habitat Development Status

UNIT II. SPECIES POPULATION, HABITAT, AND THREAT INVENTORY INFORMATION

- Locality Reconnaissance
- Authentication of Species
- Precise Population Location
- Land Inventory
- Population Inventory
- Habitat Inventory
- Threat Inventory
- Author Information

UNIT III. SPECIES BIOLOGY INFORMATION

- Reproduction Status
- Dispersion Status
- Establishment Status
- Maintenance Status

UNIT IV. ENVIRONMENTAL FACTOR INFORMATION

- Influence on Reproduction Status
- Influence on Dispersion Status
- Influence on Establishment Status
- Influence on Maintenance Status

I. Objectives, Principles, and Elements of the Program

A. Objectives of the Units

We have adopted the following set of objectives to alleviate many of the current difficulties encountered in the acquisition, verification, and documentation of special species information. The specific objectives of the units are to:

1. Provide a basic information system to guide information acquisition and documentation on plant taxa under consideration for protection and preservation.
2. Provide direction for specific information searches and retrieval in herbaria, libraries, and the field.
3. Provide documented information for preliminary evaluation of species status.
4. Provide a means of determining and comparing historical and current distribution of a taxon.
5. Provide a preliminary basis for population monitoring and threat evaluation.
6. Provide detailed and documented information for efficient initiation of more detailed investigations on a species or specific population.
7. Provide assistance in preventing the inadvertent destruction of populations through lack of information.

B. Guiding Principles of the Units

The principles used to guide the development of the units of our program are included to assist the reader in understanding, using, and evaluating the units.

1. A uniform information and documentation system is desirable in order to acquire comparable information on populations and species and to promote inter- or multiple disciplinary and agency activities.
2. Species General Information is necessary for the establishment of priorities in species protection and preservation and for the promotion of efficient inventories and species studies.
3. Species population inventories should be based on a survey of all historical distribution sites.
4. Current population localities should be precisely delimited and recorded.
5. Species population and threat status inventories should include provisions for habitat, disturbance, and population analyses, as well as population monitoring.
6. A species status evaluation should be based upon an accumulation of individual population status inventories.
7. A species status evaluation is necessary and should be available for all legal listing and delisting decisions and activities.
8. Sound species biology studies should be founded upon both general information and population status inventories.
9. Comprehensive status reports, habitat management programs, and long-term preservation often will require information from both species biology and environmental factor studies.

C. Elements of the Units

The two information units—Species General Information and Species Population, Habitat, and Threat Inventory Information—are comprised of three basic elements: (1) Question-Procession, (2) Information-Answer, and (3) Documentation. According to this conceptual logic, an investigator studying a species or population would ask or use a series of questions or methods; seek or enumerate a series of potential answers or information items; and provide standardized documentation of the results and methods used.

II. Species Information Program

A. Species General Information Unit

This unit is designed to guide the capture or retrieval of information from herbaria, literature, diaries, maps, statutes, and personal communications or observations. This basic information is necessary to select species for further consideration or investigation and to initiate Species Population, Habitat, and Threat Inventories.

Basic questions (Question-Procession) addressed by this unit are:

1. What is the taxonomic status of the species?
2. How may the taxon be recognized?

3. What is the legal status of the species? Historical distribution?

4. What is the ownership, designation, protection, and use status of the land for each population site?

5. What are the habitat preferences of the species and the developmental or successional states of the habitat? The Information-Answer system is presented in Table 17. Each item or response in the information system must be documented with an appropriate Citation Source, Code, and Number as presented in the Documentation System (see Section 2, below).

1. Instructions for Conducting a Species General Information Search

The brief instructions that follow are given to assist a researcher in the compilation of Species General Information according to the system presented in Table 17. The directions are coded in the same Item Code sequence as the Information System.

Item Code

1. *Species Taxonomic Status*. Each investigator should review the taxonomic status (position, rank) and nomenclature of the taxon in question. Issues relating to the correct name (1.1), synonymy (1.3), location and examination of type specimen (1.4), and original description (1.2.1) should be studied. A detailed technical description (1.2.2) including the basic attributes listed in the information system should be prepared. If good or adequate descriptions exist in manuals or floras, these may be used if documented. In any event, the number of specimens and/or types examined plus literature references used in the preparation of descriptions should be indicated.

2. *Species General Characteristics*. Plant habit, inhabitant type, nutrition type, and life span should be noted and documented.

2.5–2.7 *Species Phenology*. The three major phenophases should be documented carefully to assist in the planning of field work for Population Status Inventories as well as species biology studies and population monitoring. Although one may report a summary of a particular phenophase (for example, fruiting, June–September), it is desirable to document each month in the range by a reference to a herbarium specimen, other voucher, or observation. This will associate a specific phenophase and time with a particular locality or site.

3. *Species Legal Status*. To complete this section, one should review the various statutes or listings for protection of a species throughout the area of occurrence. Particular care should be taken to discriminate between the numerous proposed listings and those that indeed have legal status under local, state, or national law. Presently, few species have legal status. When using item 3.3, indicate category and status, for example, "Threatened-Candidate, North Carolina" or "Protected, Georgia."

4. *Historical Distribution*. Historical distribution refers to the past known documented distribution of a taxon within biogeographic, physiographic, geographic, and specific regions or areas. Only field investigations can provide the present distribution. Historical distribution will involve a survey of herbaria and literature. Because time and resources will be limited, it is imperative that the herbaria, specimens, and literature be carefully documented. The Federal Information Processing Standards Publication (FIPS 6–2) or *Worldwide Geographic Location Codes* published by General Services Administration, Office of Finance (1976) (based on FIPS), should be used to number states and counties. The systematic use of these standard codes will greatly assist information accumulation for use by federal, state, and private agencies.

4.5.1 *Locality Name or Number*. Each locality should be based on specimen label, literature report, or personal observation or communication and should be documented in the appropriate section of the Documentation System. Numbering of localities will be artificial but should be consistent with Item Number on the documentation sheet. Locality name should be noted carefully, for example, "top of Roan Mountain" should be cited as "Roan Mountain," whereas "4.5 miles ESE of Pittsboro" should not be cited as "Pittsboro." In the latter case, a reference number from the Documentation System would better serve our purposes than listing "Pittsboro." Precise locality information—state, county or parish, collector, date of collection, herbarium, and so forth—should appear in the Herbarium or Specimen Documentation.

5. *Ownership Status of Land*. Ownership status of lands from which specimens have been collected should be noted. Many herbarium labels indicate national parks, state parks, national forests, and so forth. Ownership should be designated only if specified on labels or in the literature.

6. *Land Status*. This section is designed to indicate specific land designation or use of a historical distribution or collection site. In some cases a species population may be protected already. This information will be useful in the population status inventory to direct field investigators to note the use as reported and whether it appears that the use is compatible with population survival.

7. *Habitat Preference*. This section provides a general

Table 17. *Species General Information System* (after Whitson and Massey, in press)

1. SPECIES TAXONOMIC STATUS
    1. Species name
    2. Description
        1. Original
        2. Recent or revised
    3. Synonomy
    4. Type specimen(s) location
    5. Family
    6. Common names
2. SPECIES GENERAL CHARACTERISTICS
    1. Plant habit
        1. Herb
        2. Shrub
        3. Tree
        4. Vine
    2. Inhabitant type
        1. Aquatic
        2. Semiaquatic
        3. Terrestrial
        4. Epiphytic
            1. Host
    3. Nutrition type
        1. Autophytic
        2. Parasitic
            1. Host
        3. Saprophytic
        4. Hemiparasitic
            1. Host
    4. Life span
        1. Annual
        2. Biennial
        3. Perennial
    5. Vegetative phenophase
        1. January
        2. February
        3. March
        4. April
        5. May
        6. June
        7. July
        8. August
        9. September
        10. October
        11. November
        12. December
    6. Flowering phenophase
        1. January
        2. February
        3. March
        4. April
        5. May
        6. June
        7. July
        8. August
        9. September
        10. October
        11. November
        12. December
    7. Fruiting or sporulating phenophase
        1. January
        2. February
        3. March
        4. April
        5. May
        6. June
        7. July
        8. August
        9. September
        10. October
        11. November
        12. December
3. SPECIES LEGAL STATUS
    1. Threatened
        1. International
        2. National
        3. State
    2. Endangered
        1. International
        2. National
        3. State
    3. Other
        1. International
        2. National
        3. State
        4. Other
4. HISTORICAL DISTRIBUTION
    1. Biogeographic regions
        1. Nearctic
        2. Palearctic
        3. Neotropical
        4. Ethiopian
        5. Oriental
        6. Australian
        7. Antarctic
    2. Physiographic regions
        1. Appalachian Highlands
        2. Coastal Plain
            1. Atlantic
            2. Gulf
            3. Floridian
        3. Interior Highlands
        4. Interior Plains
        5. Laurentian Upland
        6. Other

Table 17. *(continued)*

4.
    3. Country(ies)
    4. State(s) and county(ies)[1]
    5. Locality(ies)
        1. Location name or number
        2. Latitude
        3. Longitude

5. OWNERSHIP STATUS OF LAND
    1. National
    2. State
    3. County
    4. Municipal
    5. Private
    6. Uncertain

6. LAND STATUS
    1. Designation
        1. Research natural area
        2. National forest
        3. State forest
        4. Experimental forest
        5. National natural landmark
        6. State natural area
        7. State park
        8. Commercial forest
        9. Farm
        10. National Park
        11. Uncertain
        12. Other
    2. Protection
        1. Protected
        2. Unprotected
        3. Uncertain
    3. Apparent use
        1. Commercial
        2. Recreation
        3. Other

7. HABITAT PREFERENCE
    1. Habitat types
        1. Bog
        2. Swamp
        3. Marsh
        4. Grassland
        5. Savannah
        6. Chaparral--shrub
        7. Desert
        8. Tundra
        9. Ice
        10. Forest
            1. Gymnosperm
            2. Angiosperm
            3. Mixed
        11. Other
    2. Topographic types
        1. Mountains, hills, and ridges
        2. Scarps, bluffs, cliffs, and escarpments
        3. Benches and terraces
        4. Basins
        5. Lakes, bays, ponds, and pools
        6. Valleys, gorges, and channels
        7. Plains and flats
        8. Beaches
        9. Other
    3. Substrate types
        1. Water
            1. Fresh
            2. Saline
        2. Soil
            1. General type
            2. Texture
        3. Rock
            1. Origin type
                1. Igneous
                2. Sedimentary
                3. Metamorphic
                4. Uncertain
            2. Specific type
                1. Limestone
                2. Sandstone
                3. Shale
                4. Granite
                5. Other
        4. Humus or organic layers
    4. Disturbance types
        1. Biotic
            1. Roads, paths, or right-of-ways
            2. Cultivated lands
            3. Waste and spoil areas
            4. Grazed or browsed areas
            5. Mowed
            6. Forest management
                1. Clear cutting
                2. Selective cutting
            7. Wildlife areas subject to burrowing, rooting, and mounding
            8. Abandoned
            9. Chemically treated (weed or pest control)
            10. Disease and pests
                1. Types
                    1. Pine bark beetle

Table 17. *(continued)*

```
7.   4. 1.10. 1. 2. Dutch elm disease        8.  HABITAT DEVELOPMENT STATE
                   3. Other                       1. Early successional or pioneer
              11. Uncertain                           state
              12. Other                           2. Middle successional or transient
           2. Abiotic                                 state
              1. Fire                             3. Late successional or climax
              2. Flood                                state
              3. Erosion or slides                4. Uncertain
              4. Other
```

[1]States and counties numbered according to Federal Information Processing Standards Publication (FIPS Pub.) 6.2.

assessment of the habitat and should incorporate some general information on and insight into types of habitats, areas, sites, and topographic and substrate types to be investigated in both a literature search and population status inventory work in the field.

8. *Habitat Development State*. This category represents an attempt to place each taxon in a successional phase(s) or state(s). This information may be used with other findings to establish priorities for our studies and searches by identifying those species that occur in relatively stable habitats (climax) or areas and those that are likely to disappear because of normal habitat, vegetation, and community dynamics.

2. Documentation System

The Documentation System is based upon the concept of providing each information response with a specific citation. We have found that the preparation of a separate documentation sheet for each citation source (literature, specimens, maps, statutes, observations, and so forth) with consecutively numbered entries is quite effective. Following this procedure for each species provides efficient, open-ended, and up-to-date information documentation. Suggested formats for major citation categories and entries are given below.

Literature, Statutes, and Maps—Citation Code L

Code and number. Author. Date. Title. Journal (with pages) or publisher (with city).

L1. Mackenzie, K. K. 1910. Notes on Carex VI. Bull. Torrey Bot. Club 37:231–50.

L2. U.S. Geological Survey. 1942. North Carolina. Montreat quadrangle. 1:24,000. Topographic map.

L3. Hooker, J. D., and B. D. Jackson et al. 1893–95. Index Kewensis . . . Oxford.

L4. Kartesz, J. T., and R. Kartesz. 1977. The biota of North America. Part 1, Vascular plants. Volume 1, Rare plants. B.O.N.A.C. Pittsburgh.

Specimen Citations—Citation Code H

Code and number. Herbarium acronym and herbarium accession number. Collector and collector's number. Locality, date of collection. State, county.

H1. NCU 450678. Whetstone 9887. Beech-maple forest, Pittsboro State Park, 16 miles SE of Pittsboro on County 1728, 4 June 1974. North Carolina, Chatham Co.

Observations—Citation Code O

Code and number. Observer. Date. Observation. Institution or agency, address.

O1. Moore, J. K. 1978. Seeds harvested in June 1978 and planted in peat-pearlite; germinated in 9 days. Germination 96%. North Carolina Botanical Garden, Chapel Hill, N.C. 27514.

O2. Morgan, S. 1977. Roan Mountain population on Cloudland Trail, Mitchell County, N.C. began flowering on 5 June 1977. Only two flowers were fully expanded on this day. Department of Botany, University of North Carolina–Chapel Hill, N.C. 27514.

---

3. An Example of a Completed Species General Information Unit

Information on a southern Appalachian endemic is presented here as a specific example of the documentation procedure for the Species General Information unit (Table 17, above). The Documentation System identifies each item of information by an Item Code Number. Each applicable item response is documented with an appropriate Citation Code and Number. Both codes provide for the convenient addition of information categories and citation sources, as well as an up-to-the-moment summary for inspection or duplication.

The information for *Solidago spithamaea* was compiled under a cooperative agreement (Contract No. 18–606) between the Southeastern Forest Experimental Station, U.S. Forest Service, and The Highlands Biological Station, Inc. This information and that for 48 other southern Appalachian vascular plants in North Carolina represent a portion of a continuing effort on the endangered species and natural areas programs sponsored by The Highlands Biological Station. Specific item responses for *Solidago* are presented in Table 18, whereas citation source documentation from literature, statutes, and maps, specimen citations, and observations are presented in Tables 19 and 20.

B. Species Population, Habitat, and Threat Inventory Information Unit

This unit of the program is designed to direct an on-site inventory of a population and its habitat. The on-site investigation initially can refute, verify, or add to the data in the Species General Information Unit. The significant contribution of the on-site inventory is the establishment of qualitative and quantitative baseline data for both a population and its habitat. This information can be of significance in making current population comparisons so that a species status can be developed for legal and other considerations. Likewise, the baseline information can permit population and habitat monitoring through time. These kinds of comparisons may contribute to an identification and evaluation of threats to both population and habitat integrity, provide management and protection implications for a species and its habitat, and suggest specific basic research problems.

Basic questions (Question-Procession) addressed by the unit include:

1. Where are the members of the taxon?
2. How many population units exist at a particular site?
3. What threats to the habitat and population integrity can be documented based on an analysis of evident disturbances?
4. What is the type or nature of the habitat?
5. What is the general condition of the population?

Table 21 includes the Information-Answer elements of the unit.

1. Instructions for Conducting a Species Population, Habitat, and Threat Inventory

Our current concept is that the inventory information will be acquired and maintained as a complete documentation report. To facilitate this approach a set of worksheets is provided (pp. 420–428) to assist in recording the desired information. Because one function of the inventory is to update Species General Information, a form also is provided to accomplish this task. Suggested instructions for each section of the inventory are indexed to the appropriate information system item code.

Table 18. *General Information Documentation for* Solidago spithamaea[1]

ITEM CODE	DOCUMENTATION RESPONSE	CITATION CODE AND NUMBER
1.1.	*Solidago spithamaea* M. A. Curtis	L1
1.2.1.	"*S. spithamaea* M. A. Curtis. Stem villous-pubescent, leafy; leaves oval or oblong-lanceolate, ciliate, nearly glabrous, sharply serrate above the middle, or the uppermost entire; the lowest and radical spatulate-oblong, tapering into winged petioles; heads (middle-sized) disposed in a compound glomerate corymb; peduncles and pedicels villous; scales of the involucre somewhat equal, lanceolate, ciliate; rays 6 or 7, short; achenia pubescent. — Gray! in Sill. Jour. 42. p. 42."[2]	L1
1.2.2.	"Plants somewhat mephitic. Stems 1-4 dm tall from a short, stout rhizome or branched caudex, rough-puberulent or shortly spreading hirsute, or glabrate below. Leaves basally disposed, glabrous or nearly so, sharply serrate, the largest ones with elliptic to ovate or subrhombic blade mostly 5-10 x 1.5-4 cm. Infloresecense densely corymbiform, up to 10 (15) cm wide; involucre 5-6 mm high, with firm, green-tipped, rather narrow bracts; rays *ca.* 8 (13), 2-3.5 mm long; disk hairy or eventually subglabrate."	L2
1.3.	*Aster spithamaeus*?	L3
1.4.	?[3]	
1.5.	Asteraceae (Compositae)	L1
1.6.	Blue-Ridge Goldenrod	L5
	Skunk Goldenrod	L2
2.1.1.	Herb	L2
2.2.3.	Terrestrial	L1
2.3.1.	Autophytic	
2.4.3.	Perennial	L2
2.5.7.	July	H10
2.5.8.	August	H3
2.5.9.	September	H8, H12
2.5.10.	October	H7
2.6.7.	July	H10
2.6.8.	August	H3, H4, H5, H6, H14
2.7.7.	July	H10, H11
2.7.9.	September	H8, H12
2.7.10.	October	H7
3.3.3.	North Carolina--Candidate Endangered	L5
	Tennessee--Candidate Threatened[4]	L7
4.1.1.	Nearctic	H1
4.2.1.	Appalachian Highlands	H1
4.3.	United States	H1
4.4.13.	Georgia?[5]	H1?
4.4.37.	North Carolina	H1?, H2
	Avery (011)	H2?, H3?, H4, H5
	Mitchell (121)	H6, H7, H8, H9, H10 H11, H12, H13, H14
4.4.47.	Tennessee	L2

Table 18. *(continued)*

ITEM CODE	DOCUMENTATION RESPONSE	CITATION CODE AND NUMBER
4.5.1.	North Carolina	
	-Grandfather Mountain	H2
	Avery-Grandfather Mountain	H4
	Mitchell-Roan High Bluff	H6
	Roan Mountain	H8
4.5.2.	36°05'N	L11
4.5.3.	81°52'W-82°10'W	L11
5.1.	National	L12
5.5.	Private	L12
6.1.2.	Pisgah National Forest	L12
6.2.1.	Protected	L12
6.3.1.	Private-Commercial Recreation Area	L12
7.1.4.	Grassland	L8
7.1.6.	Shrub (heath)	H5
7.2.1.	Mountains, hills, and ridges	H1
7.2.2.	Scarps, bluffs, cliffs, and escarpments	H5
7.3.2.1.	Rocky soil	H5
7.3.3.1.3.	Metamorphic	L8
7.3.3.2.5.	Gneiss	L8
7.4.2.1.	Fires	L9
8.1.	Early successional or pioneer state	L9
8.2.	Middle successional or transient state	H5

[1]Compiled by J. R. Massey, R. David Whetstone, and T. A. Atkinson, Herbarium, Department of Botany, University of North Carolina, Chapel Hill, N.C., under cooperative agreements between The Highlands Biological Station and the U.S. Forest Service (Agreements 18-606 and 18-668).

[2]Gray in Sill. Jour. 42 is a *nomen nudum*.

[3]Holmgren and Keuken (L4) cite the following collections as primary holders of Curtis's specimens: Brown University Herbarium (BRU), Providence, Rhode Island; and the University of Maine Herbarium (MAINE), Orono, Maine.

[4]Although listed by Kartesz and Kartesz (L7) this species was eliminated from the Tennessee list of plants because "no Tennessee specimens seen" (L6). A. Cronquist (L2) apparently has seen specimens from Tennessee and reports "at upper altitudes of BR (Blue Ridge) of NC and Tenn, notably on Roan Mt. and Grandfather Mt."

[5]This report for Georgia is certainly suspect and is included here only in the pursuit of "thoroughness." Buckley (H1) lists the locality as "in montibus Carolinae et Georgiae." We have seen neither reports nor collections of this species from Georgia.

Table 19. *Citation Sources for* Solidago spithamaea

DOCUMENTATION: LITERATURE, STATUTES, AND MAPS (L)

L1. Torrey, J., and A. Gray. 1838-40. A flora of North America. Wiley & Putnam, New York.

L2. Cronquist, A. n.d. Asteraceae. *In* A. E. Radford, ed. Vascular flora of the southeastern United States. Manuscript.

L3. Kuntze, O. 1891. Revisio Generum Plantarum . . . Pars 1. Arthur Felix, Leipzig.

L4. Holmgren, P. K., and W. Keuken. 1974. Index herbariorum. Part 1. The herbaria of the world. 6th ed. Oosthoek, Scheltema & Holkema, Utrecht, Netherlands.

L5. Committee on Vascular Plants. 1977. Vascular plants. Reprinted from J. E. Cooper, S. S. Robinson, and J. B. Funderburg, eds. Endangered and threatened plants and animals of North Carolina. Bookstore, University of North Carolina, Charlotte.

L6. Committee for Tennessee Rare Plants. n.d. The rare vascular plants of Tennessee. B. E. Wofford, Rapporteur for the Committee, Department of Botany, University of Tennessee, Knoxville, Tenn., 37916. Manuscript.

L7. Kartesz, J. T., and R. Kartesz. 1977. The biota of North America. Part 1, Vascular plants. Vol. 1, Rare plants. B.O.N.A.C., Pittsburgh.

L8. Otte, L. 1977. Roan Mountain bald communities. Class report for Botany 235 (A. E. Radford, professor). Department of Botany, University of North Carolina, Chapel Hill.

L9. Brown, D. M. 1941. Vegetation of Roan Mountain: A phytosociological and successional study. Ecol. Monogr. 11:61-97.

L10. Gray, A. 1842. Notes of a botanical excursion to the mountains of North Carolina . . . Amer. J. Sci. 42:1-49. Published also in London J. Bot. (1842+). Vol. 1.

L11. North Carolina State Highway Commission. 1970. Municipal, state primary, and interstate highway systems. (Map.) Raleigh?

L12. Pittillo, J. D. 1976. Potential natural landmarks of the Southern Blue Ridge Portion of the Appalachian Ranges Natural Region. Department of Biology, Western Carolina University, Cullowhee, North Carolina.

Table 20. *Documentation-Specimen Citations for* Solidago spithamaea

DOCUMENTATION: SPECIMEN CITATIONS (H)

H1. GA 75328. Buckley, *s.n.* North Carolina-Georgia?

H2. GA 66706. Sargent 7360. 18 September 1955. North Carolina, Avery Co.

H3. NCSC 10760. Wells and Shunk, *s.n.* 7 August 1929. North Carolina, Avery Co.

H4. NCU 305367. Smyth 1217. 18 August 1967. North Carolina, Avery Co.

H5. NCU 213738. Ramseur 3986. 6 August 1957. North Carolina, Avery Co.

H6. NCU 213737. Ramseur 1216. 6 August 1956. North Carolina, Mitchell Co.

H7. TENN. Shanks, Norris, and Clebsch 28018. 3 October 1960. North Carolina, Mitchell Co.

H8. TENN. Berhend, Pearman, and Odenwelder 72-191. 7 September 1972. North Carolina, Mitchell Co.

H9. NCU 79225. No collector. 11 August ____. North Carolina, Mitchell Co.?

H10. 79226. Ashe, *s.n.* ___July 1893. North Carolina, Mitchell Co.

H11. NCU 164087. Gray and Carey, *s.n.* ___July 1841. North Carolina, Mitchell Co. [Replicate GH!]

H12. NCU 304308. Canby, *s.n.* ___September 1876. North Carolina, Mitchell Co.

H13. NCSC 37549. (Curtis?, *s.n.*) n.d. North Carolina, Mitchell Co.

H14. TENN. Morton 3860. 5 August 1969. North Carolina, Mitchell Co.

H15. VPI 16000. Smyth 3351. 9 August 1973. North Carolina, Avery Co.

H16. GH. Steele 141. 6 September 1915. North Carolina, Avery Co.

H17. GH. Steele 79. 1 September 1915. North Carolina, Avery Co.

H18. GH. No collector. October 1879. Cultivated, Botanical Garden, Harvard, probably *ex* North Carolina, Avery Co.

H19. GH. Gray and Sullivant, *s.n.* September 1843. North Carolina, Avery Co.

H20. GH. Smith, *s.n.* 10 September 1884. North Carolina, Mitchell Co.

H21. GH. Randolf and Randolf, *s.n.* 5 August 1922. North Carolina, Mitchell Co.

H22. GH. Hunnewell 12993. 1 August 1933. North Carolina, Avery Co.

H23. GH. Curtis (?), *s.n.* 1845. North Carolina, Avery Co.

H24. GH. Merriam, *s.n.* 11 September 1892. North Carolina, Mitchell Co.

Table 21. *Species Population, Habitat, and Threat Inventory Information*

1. LOCALITY RECONNAISSANCE
    1. Initial sources of reconnaissance locality information
    2. Date(s) of reconnaissance
    3. Location of reconnaissance area
        1. State
        2. County/Parish
        3. Distances and directions from landmarks
        4. Locality name(s)
        5. Reconnaissance area map with distances, directions, and landmarks indicated; indicate area searched on a U.S. Geological Survey Quadrangle map (include quadrangle name).
    4. Approximate duration of search (hours)

2. AUTHENTICATION OF SPECIES
    1. Specimen(s) used
    2. Photograph(s) and illustration(s) used
    3. Expert determination(s)
    4. Description(s) used

3. PRECISE POPULATION LOCATION
    1. State
    2. County/Parish
    3. Distances and directions from landmarks
    4. Locality name(s) and/or number(s)
    5. Latitude/Longitude
    6. Range/Township/Section
    7. Location map with distances, directions, and landmarks indicated (preferably U.S. Geological Survey Quadrangle map; include quadrangle name)

4. LAND INVENTORY
    1. Owner
    2. Contact
    3. Designation
    4. Current use or management activities
    5. Past use or management (if evident)
        1. Recent past (1-10 years)
        2. Historic past (>10 years)
    6. Comments:

5. POPULATION INVENTORY
    1. Provide a scaled population map denoting boundary lines, distinct landmarks or features that can link it to the U.S. Geological Survey Quadrangle map.
    2. Population size
        1. Definition of units (for example, individual, colony, mat)
        2. Number of units
        3. Approximate size of unit (cover range, cm, m)
        4. Approximate number of subunits per unit (for example, stems, culm)

Table 21. *(continued)*

```
5.  3.  Unit distribution
        1.  Uniform within boundary line; density if possible
        2.  Clustered within boundary line; locate clusters on diagram
    4.  Unit demography
        1.  Reproduction evidence or origin of units
            1.  Asexual forms (for example, bulbs, bulbils, rhizomes, stolons);
                proportion of population
            2.  Sexual forms (for example, seedling, sapling); proportion of
                population
        2.  Size or age class evidence
            1.  Definition of classes
            2.  Class percentages in the population
            3.  Actual count of units in the classes
    5.  Unit phenology
        1.  Phenophase classes
            1.  Vegetative only
            2.  Flowers or inflorescences
            3.  Fruits or infructescences
            4.  Seeds (mature)
        2.  Class percentages in the population
        3.  Actual count of units in the classes
    6.  Unit biotic interrelationships
        1.  Evidence comments on the following:
            1.  Browsing, grazing, sucking
            2.  Diseases
            3.  Infestations
            4.  Predation (especially of flowers, fruits, and seeds)
            5.  Flower visitors
            6.  Parasites
            7.  Other
    7.  Unit vigor (indicate if not uniform in the population; specify proportion
        percentage)
        1.  Evidence comments on the following:
            1.  Robustness
            2.  Vigor
            3.  Other
    8.  Other comments:

6.  HABITAT INVENTORY
    1.  Population habitat.  The habitat or area within the boundary lines or
        confines of the population units.  A response of habitat factor uniformity
        is imparted by a yes and a single choice response, whereas nonuniformity is
        designated by no and two or more choice responses.
        1.  Vegetation and biotic associates
            1.  Vegetation (yes; no)
                1.  Community type
            2.  Biotic associates
                1.  Species list with stratum category designation
                2.  Stratum categories
                    1.  Canopy (C)
                    2.  Subcanopy (c)
```

Table 21. *(continued)*

```
6.  1.  1.  2.  2.  3.  Shrub (S)
                    4.  Herb (H)
                    5.  Cryptogams (K)
        2.  Climate (yes; no)
            1.  Population climatic site type
        3.  Soils (yes; no)
            1.  Population soil site type
        4.  Geology (yes; no)
            1.  Population geologic site type
        5.  Hydrology (yes; no)
            1.  Population hydrologic site type
        6.  Topography (yes; no)
            1.  Population topographic site type
        7.  Physiography (yes; no)
            1.  Population physiographic site type
        8.  Disturbance evidence
            1.  Vegetation
                1.  Removed (complete to near-complete removal from habitat)
                    1.  Quality
                        1.  Stratum (canopy, subcanopy, shrub, herb, cryptogam)
                        2.  Species
                    2.  Quantity
                        1.  Total
                        2.  Partial
                    3.  Activity type (apparent, for example, clear-cutting,
                        timbering, mowing, browsing, grazing, routing, rooting,
                        bulldozing, chaining, chemical spraying, burning,
                        planting, rotting, washing-out)
                    4.  Agent (apparent, for example, man, livestock, wildlife,
                        infestation, disease, fire, wind, rainfall, icefall,
                        flooding)
                    5.  Apparent objectives
                        1.  Selective
                        2.  Nonselective
                    6.  Comment:  If a comment replaces the above five categories,
                        each should be explicitly cited if discernible.
                2.  Uprooted (material primarily present)
                    1.  Quality
                        1.  Stratum
                        2.  Species
                    2.  Quantity
                        1.  Total
                        2.  Partial
                    3.  Activity type
                    4.  Agent
                    5.  Apparent objectives
                    6.  Comment:
                3.  Cut, broken, or trampled (material may or may not be present)
                    1.  Quality
                        1.  Stratum
                        2.  Species
```

Table 21. *(continued)*

6\. 1. 8. 1. 3. 2. Quantity

            1. Total
            2. Partial
        3. Activity type
        4. Agent
        5. Apparent objectives
        6. Comment:
    4. Dead or dying (apparent death signs)
        1. Quality
            1. Stratum
            2. Species
        2. Quantity
            1. Total
            2. Partial
        3. Activity type
        4. Agent
        5. Apparent objectives
        6. Comment:
    5. Species introduction and invasion
        1. Quality
            1. Stratum
            2. Species
        2. Quantity
            1. Total
            2. Partial
        3. Activity type
        4. Agent
        5. Apparent objectives
        6. Comment:
2. Litter or duff
    1. Removed
        1. Quantity
            1. Total
            2. Partial
        2. Activity type (apparent, for example, bulldozing, plowing, digging, rooting, routing, timbering, burning, planting, washing-out)
        3. Agent (apparent, for example, livestock, wildlife, fire, rainfall, flooding)
        4. Apparent objectives
            1. Selective
            2. Nonselective
        5. Comment:
    2. Upheaved or overturned
        1. Quantity
            1. Total
            2. Partial
        2. Activity type
        3. Agent
        4. Apparent objectives
        5. Comment:

Table 21. *(continued)*

```
6.  1.  8.  2.  3.  Trampled or crushed
                    1.  Quantity
                        1.  Total
                        2.  Partial
                    2.  Activity type
                    3.  Agent
                    4.  Apparent objectives
                    5.  Comment:
                4.  Added litter or duff
                    1.  Quantity
                        1.  Total
                        2.  Partial
                    2.  Activity type
                    3.  Agent
                    4.  Apparent objectives
                    5.  Comment:
            3.  Soil (topsoil) as substrate
                1.  Removed
                    1.  Quality
                        1.  Horizons
                    2.  Quantity
                        1.  Total
                        2.  Partial
                    3.  Activity type (apparent, for example, bulldozing, scraping,
                        plowing, rooting, routing, frost heaving, eroding,
                        treading, trampling, driving upon, depositing, windblowing,
                        flooding)
                    4.  Agent (apparent, for example, man, livestock, wildlife,
                        rainfall, wind, frost, flood, vehicles)
                    5.  Apparent objectives
                        1.  Selective
                        2.  Nonselective
                    6.  Comment:
                2.  Upheaved or overturned
                    1.  Quality
                        1.  Horizons
                    2.  Quantity
                        1.  Total
                        2.  Partial
                    3.  Activity type
                    4.  Agent
                    5.  Apparent objectives
                    6.  Comment:
                3.  Compacted or trampled
                    1.  Quality
                        1.  Horizons affected
                    2.  Quantity
                        1.  Total
                        2.  Partial
```

Table 21. *(continued)*

```
6.  1.  8.  3.  3.  3.  Activity type
                    4.  Agent
                    5.  Apparent objectives
                    6.  Comment:
                4.  Overburdened (as by sedimentation or deposition)
                    1.  Quality
                        1.  Source of overburden
                        2.  Depth of overburden
                        3.  Kind of overburden
                    2.  Quantity
                        1.  Total
                        2.  Partial
                    3.  Activity type
                    4.  Agent
                    5.  Apparent objectives
                    6.  Comment:
                5.  Exchanged or refilled
                    1.  Quality
                        1.  Source of fill
                        2.  Depth of fill
                        3.  Kind of fill
                    2.  Quantity
                        1.  Total
                        2.  Partial
                    3.  Activity type
                    4.  Agent
                    5.  Apparent objectives
                    6.  Comment:
            4.  Water as substrate, not agent
                1.  Removed (as in diverted and drained)
                    1.  Quality
                        1.  Frequency
                        2.  Duration
                    2.  Quantity
                        1.  Total
                        2.  Partial
                    3.  Activity type (apparent, for example, draining, diverting,
                        evaporating, pumping, irrigating, precipitating, polluting,
                        salting, fertilizing, heating, eroding, change in pH)
                    4.  Agent (apparent, for example, man, livestock, wildlife,
                        algae, higher plants, sewage, geologic faults, earth
                        slide, rainfall, sheet erosion, runoff, ice melt, flooding,
                        wind)
                    5.  Apparent objectives
                        1.  Selective
                        2.  Nonselective
                    6.  Comment:
                2.  Increased (as in irrigation and impoundment)
                    1.  Quality
                        1.  Source of increase
                        2.  Depth of increase
```

Table 21. *(continued)*

```
6.  1.  8.  4.  2.  1.  3.  Quality of increase
                        4.  Frequency
                        5.  Duration
                    2.  Quantity
                        1.  Total
                        2.  Partial
                    3.  Activity type
                    4.  Agent
                    5.  Apparent objectives
                    6.  Comment:
                3.  Enriched (as in pollution)
                    1.  Quality
                        1.  Source of enrichment
                        2.  Kind of enrichment
                        3.  Frequency
                        4.  Duration
                    2.  Quantity
                        1.  Total
                        2.  Partial
                    3.  Activity type
                    4.  Agent
                    5.  Apparent objectives
                    6.  Comment:
            5.  Other (specify disturbance and explain)
        2.  Extrapopulation habitat.  The habitat or area surrounding the
            population habitat to at least the distance of 10 m.  A response of
            uniformity (yes)-nonuniformity (no) is again important.
            1.  Vegetation (yes; no)
                1.  Community type
            2.  Climate (yes; no)
                1.  Community climatic site type
            3.  Soils (yes; no)
                1.  Community soil site type
            4.  Geology (yes; no)
                1.  Community geologic site type
            5.  Hydrology (yes; no)
                1.  Community hydrologic site type
            6.  Topography (yes; no)
                1.  Community topographic site type
            7.  Physiography (yes; no)
                1.  Community physiographic site type
            8.  Disturbance evidence (see 6.1.8)

7.  THREAT INVENTORY
    Using the accumulated information from the disturbance inventory and population
    inventory, a reasonable list of actual pressures upon the population and
    habitat should be evident.  It should be clear that some disturbances may
    enhance the population; thus, only negative pressures should be given threat
    inventory status.
```

Table 21. *(continued)*

7\. 1. Actual threats to the following:
   - 1. Population integrity
     - 1. Removal (taking)
       - 1. Activity
       - 2. Agent
       - 3. Apparent objectives
       - 4. Comment:
     - 2. Alteration, modification
       - 1. Activity
       - 2. Agent
       - 3. Apparent objectives
       - 4. Comment:
   - 2. Habitat integrity
     - 1. Destruction
       - 1. Activity
       - 2. Agent
       - 3. Apparent objectives
       - 4. Comment:
     - 2. Degradation, alteration, modification
       - 1. Activity
       - 2. Agent
       - 3. Apparent objectives
       - 4. Comment:

   2\. Potential threats to the following:
   - 1. Population integrity
     - 1. Activity
     - 2. Agent
     - 3. Apparent objectives
     - 4. Comment:
   - 2. Habitat integrity
     - 1. Activity
     - 2. Agent
     - 3. Apparent objectives
     - 4. Comment:

8\. AUTHOR INFORMATION
   - 1. Name(s)
   - 2. Address
   - 3. Date of site visit(s)
   - 4. Voucher deposition (specimen, photo, notes)
   - 5. Permanent plots established
   - 6. Monitoring plans
   - 7. Signature
   - 8. Date

Item Code

1. The **locality reconnaissance** item is the manner by which an investigator can document a historical locality search that may prove unsuccessful. This information may assist others who desire to search the putative locality further. Should the search in locating the population be successful, the item establishes a current verification of the locality.

2. The **authentication of species** item documents the investigator's procedure for verifying the species at the site. Voucher specimens should be collected only if existing collections are inadequate and if the population is large. Frequently, former specimens and current photographs can provide adequate evidence for authentication.

3. The **precise population location** item enables the investigator to assemble and present adequately detailed information that will help a subsequent investigator to locate the population with relative ease.

4. The **land inventory** item outlines the status of the land occupied by the population.

5. The **population inventory** item details a minimal set of information an investigator should acquire to document the population adequately. The investigator, depending upon available time and resources, may secure additional or more detailed information. The inventory should be considered to be an initial step in a monitoring program and should be conducted accordingly. Permanent plots, transects, or sampling points should be marked with unobtrusive markers and clearly designated on the accompanying maps. Appropriate landmarks or features also should be included to facilitate future boundary, point, or triangulation determinations.

5.1 The investigator should provide a detailed map or diagram of the population or population boundary, subpopulations or subpopulation boundaries, and individuals or units to as specific a level as is practical. Consideration and care should be given to supply the level of detail that can furnish evidence of population change through time. For example, new subpopulations or individuals should be identifiable in resampling efforts.

5.2 The investigator should seek to choose a unit that may be recognized readily or defined by a subsequent investigator and possibly may provide some biological insight into population dynamics.

5.3 The investigator should provide by map or diagram the relative distribution of units within the population boundary.

5.4. The investigator should seek to give evidence of unit origin by either sexual or asexual reproduction and the proportion of each in the population. Also a small series of age or maturation classes (possibly combine with phenophases, 5.5.) and their proportion in the population should be established to provide some insight into both origin and maturation success.

5.5. The investigator should define a series of phenophase classes and state the proportion of each class in the population. These data may be of great value to more specific studies that may be conducted in the future.

5.6. The investigator should comment on forms of biotic interrelationships that may have been observed during the inventory. The prevailing weather conditions should also be included.

5.7. The investigator should describe the evident vigor of the population units by class, area sector, or other meaningful manner. A relative definition of vigor classes is of value as are the proportions in the population.

6. The **habitat inventory** is a detailed analysis of the population and extrapopulation habitats. The **population habitat** (6.1.) is defined as the area within the boundary lines or confines of the outermost population units. The **extrapopulation habitat** (6.2.) is that area outside or surrounding the population habitat to at least a distance of 10 m. The investigator should provide as much of the information requested as possible. Many sections of the Habitat Inventory are based upon the preceding contents of this volume developed by Radford et al. Most of the information choices may be secured from the various classification systems or themes presented and their selection assisted by the suggested methods and procedures. In many instances it would behoove the population investigator to secure the services of an appropriate specialist; but with study, assistance, and experience, the desired detail can be achieved.

6.1.1. *Vegetation and biotic associates*. The investigator should indicate whether the vegetation within the population habitat is or is not uniform. An answer of yes suggests that only one community type is present, whereas an answer of no indicates that at least two types are present.

To determine the types, the vegetation is sampled first and then classified by a detailed analysis of the vegetation strata. The strata presently recognized include canopy, subcanopy, shrub, herb, and cryptogam. The general methods for vegetation sampling are discussed on page 27, and their nomenclature-classification on page 29. Because of time constraints we have used an estimate of cover as employed and detailed in the relevé method, even for the tree stratum. For most population habitats we have utilized either the entire area or the following subsample plot sizes: canopy and subcanopy, 10 m × 10 m; shrub, 5 m × 5 m; herb and cryptogam, 1 m × 1 m. Total species cover was estimated and scored for one or more plots and used to determine nomenclature. Strata that possess either a single species with 50% or greater cover or two species that each possess 25% or greater cover may qualify for nomenclatural recognition. This is accomplished by listing the appropriate species in sequence from canopy through cryptogam strata. Many examples are detailed in Chapter 2 under biologic hierarchies. The investigator is encouraged to employ as detailed sampling techniques as feasible to secure the greatest quantitative representation of the vegetation. The nomenclature must be founded upon quantitative data with definable criteria.

The investigator is encouraged also to list as many of the biotic associates as possible that occur in both habitats. A report form is provided in the set of worksheets. Collection numbers also can be recorded on the form. The lists provide a valuable source of corroborative information for between population habitat comparisons as well as current measures of relative diversity.

6.1.2. *Climate*. The Population Climatic Site Type is equivalent to the Community Climatic Site Type and is discussed on page 46. The separate elements are listed on page 49. We have sought to contrast the temperature and precipitation relations of the population habitat with the extrapopulation habitat rather than with the natural area climate. We also have sought to incorporate a moisture effectiveness concept that includes soil, vegetation cover, aspect, slope, exposure, wind, and other related factors that seemingly could influence the population of concern. The relative states include more or less effective than or similarly effective as the extrapopulation habitat.

6.1.3. *Soils*. The investigator should indicate whether the soil within the population habitat is or is not uniform. An answer of yes requires only one response, whereas an answer of no requires at least two responses. The investigator can find a discussion of the soil family and series on pages 53–54. Additional information can be secured by consulting the Soil Survey Staff (1975) reference. Soil depth (range in cm), pH, and texture qualities can be obtained from probe samples. Several field kits are available to determine pH. We have resorted generally to a relative texture evaluation, although an analytical evaluation is recommended. The investigator is advised to check the Soil Conservation Service or state soil laboratory facilities and services.

6.1.4. *Geology*. The investigator should indicate whether the geology within the population habitat is or is not uniform. An answer of yes requires only one response, whereas an answer of no requires at least two responses. The several characteristics included in the Population Geologic Site Type are cross-referenced under the Natural Area Geologic Site Type on page 58 and the response choices are presented on pages 64–67. The population investigator may desire to collect a rock sample and secure additional professional assistance to gain greater detail about the Rock System. Each sample should be appropriately labeled.

6.1.5. *Hydrology*. The investigator should indicate whether the hydrology within the population habitat is or is not uniform. An answer of yes requires only one response, whereas an answer of no requires at least two responses. The several characteristics included in the Population Hydrologic Site Type are cross-referenced under the Natural Area Hydrologic Site Type on page 69 and the response choices are presented on pages 72–73.

6.1.6. *Topography*. The investigator should indicate whether the topography within the population habitat is or is not uniform. An answer of yes requires only one response, whereas an answer of no requires at least two responses. The several characteristics of the Population Topographic Site Type are cross-referenced under the Natural Area Topographic Site Type on page 76 and the response choices are presented on pages 81–82.

6.1.7. *Physiography*. The investigator should indicate whether the population habitat is or is not uniform. An answer of yes requires only one response, whereas an answer of no requires at least two responses. The characteristics of the Population Physiographic Site Type are cross-referenced under the Community Physiographic Site Type on page 85 and selected response choice examples are presented on page 89. The investigator will have incorporated most of this information into the locality name requested by the locality reconnaissance section of the inventory.

6.1.8. *Disturbance evidence*. In this section of the habitat analysis, the investigator is assisted in discerning the major kinds of disturbance evidence in both the population and extrapopulation habitats. The purpose of disturbance analysis is to identify current changes by addition, reduction, or alteration of the vegetation, litter, soil, or water of either habitat. The repeating paradigm under each is directed toward (1) the quality or kind of change; (2) the area, extent, quantity, or degree of change; (3) the type of disturbance activity seemingly in motion; (4) the agent or cause of the disturbance; and (5) the apparent objective (if any) of the agent. By objective we mean whether the agent has seemingly exhibited a systematic intent of disturbance activity. Frequently, many biotic agents will exhibit a form of selective disturbance (for example, deer browsing, rodent grazing, insect predation, human timbering). Abiotic agents will not exhibit selective disturbance (for example, forest fire, rock slide, stream flooding, soil heaving). The investigator is encouraged to analyze the categories systematically as to their potential effects and evaluate them in light of the paradigm while at the site. These evidences, when adequately recorded, can provide a significant addition to monitoring efforts.

6.2. *Extrapopulation habitat*. This section is designed to secure habitat information for the extrapopulation habitat comparable to that acquired for the population habitat. To complete this section the investigator is directed to item code 6.1.1. in the instructions for assistance. All worksheets have sought to incorporate the information requested for both habitats onto the same form for comparative purposes.

7. *Threat Inventory*. The investigator is encouraged to use the accumulated information from the disturbance inventory and population inventory to develop a reasonable list of actual and potential threats to the population and both habitats. We have discovered that some disturbances may enhance the population; thus, only negative pressures should be given threat inventory status.

8. *Author Information*. The investigator or author of the forms is requested to provide the information to assist potential verification or clarification of included data. The author may wish to include other forms of documentation.

2. Documentation System

Documentation for the Population, Habitat, and Threat Status Information responses may be made by reference to either a field observation list or field data or worksheet. Copies of worksheets may be appended to status reports (documentation responses) thus providing immediate documentation for summaries and conclusions as well as other data that have been collected. Documentation for historical distribution, species authentication, and locality information may be referenced to the Species General Information Documentation. Voucher specimens should be collected only with extreme discretion. We suggest that existing herbarium specimens be annotated to note the existence or condition of a population and that new vouchers not be collected unless the population represents a new station or existing specimens are inadequate. The relative ease with which specimens may be borrowed from herbaria and the willingness of many curators to have specimens annotated should make us question our motives for collecting rare or special species. Preservation, not extirpation, of populations and species should be the goal of scientists and natural heritage workers.

3. An Example of a Population, Habitat, and Threat Inventory Information Status Report

A summary or population status report for a single population of *Solidago spithamaea* is presented here to illustrate the information and documentation procedures for Population, Habitat, and Threat Inventory. The numbers for the various entries refer to the item code numbers in Table 21. The information for the *Solidago spithamaea* population on Roan Mountain, Mitchell County, North Carolina, was compiled under a cooperative agreement between the U.S. Fish and Wildlife Service and the Highlands Biological Station of the University of North Carolina, Highlands, North Carolina. Complete documentation for this report has been deposited in the Herbarium of the University of North Carolina. Sample field data forms used to compile this population information are included in Part IV of this book. Permanent plots were not established, but the population map or diagram could be used for this purpose. No monitoring plans have been established. This report was compiled in September 1979.

SPECIES POPULATION, HABITAT, AND THREAT INVENTORY
STATUS REPORT FOR *SOLIDAGO SPITHAMAEA* ON ROAN MOUNTAIN,
NORTH CAROLINA

1. LOCALITY RECONNAISSANCE
    1. Herbarium specimens from NCU, GH, TENN, and NCSC (see General Species Information Documentation in Table 20 for specimen citations, H6–H14) were used as locality sources.
    2. A reconnaissance of the area was made on 16 July 1979.
    3. Location—North Carolina, Mitchell County, Roan High Bluff.

2. AUTHENTICATION OF SPECIES
    The field investigators studied herbarium specimens prior to conducting

the field work. Plants in the field were also compared with illustrations prepared by S. Sizemore from herbarium specimens and with descriptions in the *Manual of the Vascular Flora of the Carolinas*. This is a distinctive species when in flower. At least one other species of *Solidago* occurs within the *S. spithamaea* population boundary. These other plants are more robust and also flower later than *S. spithamaea*.

3. PRECISE POPULATION LOCATION
   The population occurs on Roan High Bluff in Pisgah National Forest, Mitchell County, North Carolina, approximately 0.5 mile west of west end of Vista Circle on Roan High Bluff Trail, 36° 05′ 35″ N and 82° 08′ 45″ W. See USGS Quad Map Bakersville, NC–TENN (1:24,000, 7.5 min) below.

4. LAND INVENTORY
   The population area is owned by the U.S. Forest Service and is designated as Pisgah National Forest. Currently the area is used as a recreation and scenic area. Cutting of timber on slopes below the population occurred more than 10 years ago.

5. POPULATION INVENTORY
   1. Population Map. (See Figure 2.)
      Note the three (I, II, III) sampling (subpopulation) stations.

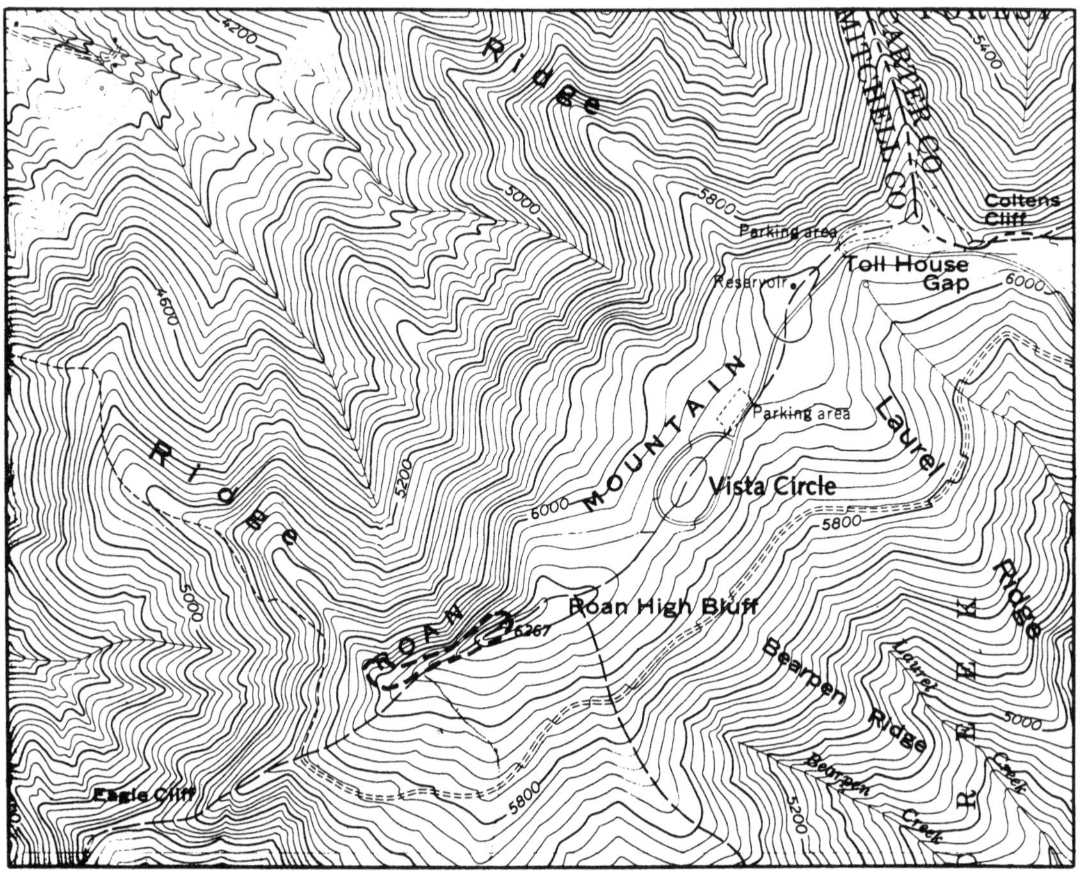

Map 1. *Population location of* Solidago spithamaea.

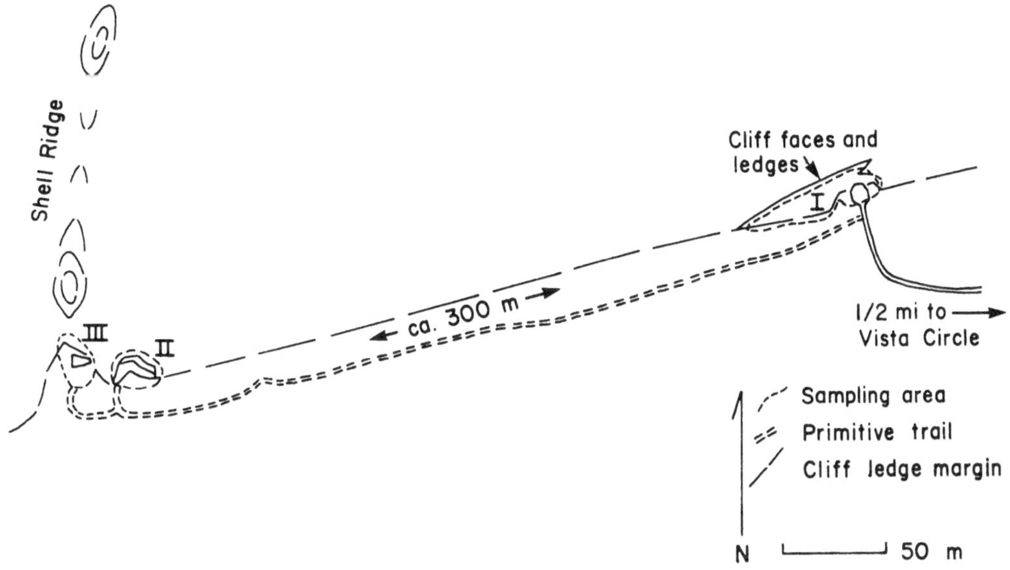

Figure 2. *Population map of* Solidago spithamaea.

---

2. Population Size.
   Population units recognized were small or large clumps of rosettes ranging from 2–3 cm to 50 cm in diameter. A total of 233 units were counted within the three sample plots. No subunits were identified.
3. Unit Distribution.
   The density of the population is approximately 0.23–0.30 clumps/meter^2 within the subpopulation habitats. The area of the three subpopulations is 500 m^2, 200 m^2, 150 m^2 respectively. The units are clustered within the subpopulation boundaries, as well as the general area on Roan High Bluff.
4. Unit Demography.
   The sexual or asexual origin of the population units counted could not be determined. Four size and maturation classes were recognized:
   - A 0–25 cm in diameter, vegetative at time of inventory
   - B 26–50 cm in diameter, vegetative
   - C 0–25 cm in diameter, flowering or with one or more inflorescences
   - D 26–50 cm in diameter, flowering

   Class percentages and actual number of units per class:

Class	*Percent*	*Total number units counted*	*Subpopulation distribution*
A	48	113	50 I, 36 II, 27 III
B	5	11	4 I, 2 II, 5 III
C	23	54	42 I, 11 II, 1 III
D	24	55	37 I, 6 II, 12 III

5. Unit Phenology.
   Flowering or vegetative units were used in class designation; see 4 above for phenology.
6. Unit Biotic Interrelationships.
   No biotic interrelationships observed; cool and cloudy.

7. Unit Vigor.
Most clumps and rosettes were vigorous. The units with inflorescences rarely had more than one or two open flowers. Some units recorded as vegetative might well produce some flowers before the end of the growing season.
8. Other Comments.
The boundary of this population of *Solidago spithamaea* overlaps that of two other proposed endangered, threatened, or rare species—*Geum radiatum* (Rosaceae) and *Carex misera* (Cyperaceae).

6. HABITAT INVENTORY
1. Population Habitat.
1. Vegetation and Biotic Associates.
The vegetation of the population habitat is uniform—a *Solidago spithamaea* population within a *Carex misera*–mixed herb community type. Neither *Geum radiatum* nor *Solidago spithamaea* has sufficient cover value for inclusion in the community type name.
Biotic Associates with stratum designations. The following species were observed in the population habitat. No specimen collections were made, and many species were still vegetative at the time of this inventory; therefore some species determinations could not be made.

Biotic Associates	*Stratum*	*Population Habitat*	*Extrapopulation Habitat*
Abies fraseri	C	+	+
Rhododendron catawbiense	S	+	+
Alnus crispa	S	+	+
Sorbus americana	S	+	+
Leiophyllum buxifolium	S	+	+
Carex misera	H	+	–
Solidago spithamaea	H	+	–
Geum radiatum	H	+	–
Aster acuminatus	H	+	+
Solidago spp.	H	+	+
Saxifraga michauxii	H	+	+
Heuchera villosa	H	+	+
Chelone sp.	H	+	+
Poa sp.	H	+	+
Houstonia montana	H	+	+
Impatiens pallida	H	+	+
Thalictrum clavatum	H	+	–
Lycopodium selago	H	+	+
Athyrium asplenioides	H	+	–
Carex spp.	H	+	+
Clintonia sp.	H	+	+

2. Climate.
The climate is uniform within the population area. It is a boreal microthermal climate and is cooler and drier than local and sectional climate. This general area has warm and moderately wet summers and moderately cold and dry winters and a short freeze-free period.
3. Soils.
The humus and clay loam soils vary in depth from 2 to 40 centimeters,

have a pH of 4, and occur as small patches and pockets covering approximately 20% of the area.

4. Geology.
   The population is on a cliff face or bluff which is composed of quartz diorite, an igneous rock, and is exposed over approximately 80% of the area.
6. Topography.
   The topography of the area is not uniform. *Solidago spithamaea* is on both constant and convex ledges, with cracks and crevices, which range from nearly level to moderately steep, approximately O°–25° angles. These ledges are located on the upper slopes of a very steep (approximately 90-degree angle) cliff face or bluff, which is open with a northwest aspect.
7. Physiography.
   The population occurs at an elevation of approximately 6,267 feet on the northwest face of Roan High Bluff. This is on Roan Mountain in the Southern Section of the Blue Ridge Province of the Appalachian Highlands.
8. Disturbance Evidence.
   1. Vegetation.
      The only evidence of disturbance is limited trampling of vegetation including *Solidago spithamaea* and its other herbaceous associates. The trampling by hikers and other visitors to this recreation area is limited and nonselective. At present the disturbance is not severe, but should be monitored.
   2. Soil.
      The surface of the A horizon of the soil is trampled and compacted by visitors. The amount of compaction seems limited and nonselective. No immediate erosion problems exist, but the site should be monitored.

2. Extrapopulation Habitat.
   The extrapopulation habitat is similar to the population habitat with respect to climate, soils, geology, hydrology, topography, and physiography. Away from the cliff, the community type is an *Abies fraseri* community. Most of the biotic associates of the population habitat are present, but not as numerous. See p. 136 for list. Evidence of windthrow and windshearing is present in this community. Disturbance evidence other than the effects of wind and ice is the same as found within the population area—very limited trampling and compacting of vegetation and soil.

7. THREAT INVENTORY
   1. Actual Threats.
      The threats to population integrity are best described as alteration and modification through trampling by visitor activities. The activities are, however, nonselective and the effects apparently minimal at this time. Threats to habitat integrity are the same as those to population and are also minimal at this time.
   2. Potential Threats.
      The most likely threats to this population and its habitat are increased visitor traffic, rock climbing, trail construction, and other activities associated with tourism.

8. AUTHOR INFORMATION
The field data and status report were compiled by J. R. Massey, University of North Carolina, Chapel Hill; Paul D. Whitson, University of Northern Iowa, Cedar Falls; and T. A. Atkinson, University of North Carolina, Chapel Hill. Site visit was conducted 16 July 1979. Field data and notes were deposited in the Herbarium, Department of Botany, UNC–Chapel Hill and Office of Endangered Species, U.S. Fish and Wildlife Service, Washington, D.C.

---

4. An Example of a Preliminary Species Status Summary

A Preliminary Species Status Summary based on the population status inventory for all populations of *Solidago spithamaea* is presented below. This summary is taken from a report submitted to the U.S. Fish and Wildlife Service under cooperative agreement No. 14–16004–78–108 between the service and the Highlands Biological Station and is used here with permission. Documentation for this summary is included in the individual population inventory status reports.

SPECIES STATUS SUMMARY FOR *SOLIDAGO SPITHAMAEA*, A SOUTHERN APPALACHIAN ENDEMIC

1. LOCALITIES VISITED AND LAND STATUS
The two *Solidago spithamaea* historical localities visited during the project were:

No.	*Locality*	*Date*	*Species*	*Ownership*
1	NC, Mitchell Co., Roan High Bluff	16 July	Present	Pisgah Natl. Forest
2	NC, Avery Co., Grandfather Mtn.	22 July	Present	Grandfather Mtn., Inc.

To the best of our knowledge no other localities have been cited for the species.

2. POPULATION INFORMATION
The population unit was defined as clumps of rosettes with or without inflorescences. Only four classes were recognized: clump diameters to 25 cm, and from 25 cm to 50 cm; either vegetative or reproductive (inflorescence).

Population units, classes, and extent.
The following table provides a summary of *Solidago spithamaea* at the two visited localities.

Locality	*Sample*	*No. Units*	*Classes*	*Phenophases*	*Sample/Population Area (m^2)*
1	Partial	233	A–D	V, FL	850
2	Partial	249	A, B	V, FL	243

Both localities were so inaccessible that a total count was not practical. The samples probably represent less than half of the individual populations. The number of units for Grandfather Mountain does not include the few individuals sighted at the MacRae Peak and Attic Window sites. Visibility from inclement weather and inaccessible habitat positions made enumeration of these small populations (12 units each) unreliable.

Population unit distribution, vigor, and biotic relations.
The units were rather clumped in distribution within the habitat. The distribution pattern was predominantly a function of suitable soils interspersed between rock outcrops and boulders. All units were seemingly vigorous. No biotic interrelations were observed; no units were observed to be in flower, only bud.

3. HABITAT INFORMATION
The following is a summary of habitat information for *Solidago spithamaea* at the two localities inventoried.

Community types
Population habitat—*Carex misera*–mixed herbs, Mixed herbs.
Extrapopulation habitat—*Abies fraseri*, *Abies fraseri*/Mixed shrubs, Mixed herbs.

Biotic associates

	Population Habitat					*Extrapopulation Habitat*	
Species	*C*	*c*	*S*	*H*	*K*	*Present*	*Absent*
Abies fraseri	+					+	
Sorbus americana	+					+	
Leiophyllum buxifolium			+			+	
Rhododendron catawbiense			+			+	
Aster acuminatus				+		+	
Carex spp.				+		+	
Heuchera villosa				+		+	
Saxifraga michauxii				+		+	
Solidago spp.				+		+	

Climate
The temperature of both the population and extrapopulation habitats was cooler than local. The moisture relations of both habitats were drier than local.

Soil
The clay loam to humus soils of both habitats were from 0 cm to 40 cm in depth and possessed pH values of 4.0.

Geology
The exposure of the population habitat ranged from 20% to 90%. Both exposures were primarily due to cliff faces and mountain peaks. The igneous and metasedimentary rock systems provided substrates of quartz diorite, metagraywacke, metaconglomerates, and metaarkoses rich in feldspar and chlorite.

Hydrology
Both habitats were intermittently saturated, but excessively drained to moderately poorly drained.

Topography
The general landform at the localities was mountain peaks. Both habitats were open to partly sheltered, approximately 0°–45° slopes on NW exposures. The upper slope positions with constant to convex profiles had surface features such as cracks, crevices, and small ledges.

Physiography
The altitudes recorded at the localities were 5,280, 5,600–5,800, and 6,267 feet.

Disturbance
There was no major evidence of disturbance to the populations of *Solidago* or the associated vegetation. Only limited visitor trampling and concurrent soil compaction were observed.

4. SPECIES THREAT EVALUATION
Although there was only limited current evidence of disturbance trampling upon *Solidago spithamaea* and its associated vegetation, the intense use and potential impact by increased use and development pose serious levels of disturbance to the limited populations and their habitats. Current use levels and development appear compatible; however, should additional development and use occur at either locality, the habitats and population could be seriously threatened. Fortunately, many of the population units are rather inaccessible. Unfortunately, only the smaller population is on land with potential legal protection.

5. SPECIES MANAGEMENT IMPLICATIONS
Management implications at both populations currently indicate compatible habitat use. The small populations and habitats suggest that management considerations should be directed toward controlling development and use of the immediate habitat.

6. SPECIES RECOMMENDATIONS
On the basis of the rarity of the species (two populations), smallness of the populations and their habitat, unprotected status of one population (private land), and apparent need to control use and development within the habitat, we recommend that *Solidago spithamaea* be given consideration for being listed as threatened. Although the current use and designation of the Roan High Bluff population appears to be compatible with the species, the U.S. Forest Service should be encouraged to assign the area to a land status that will protect the species and supportive mountain peak habitat. Appropriate information should be shared with the private landholder. Fortunately, the landholder has demonstrated a concern for natural land value, use, and management.

C. Species Biology Information Unit: An Overview

It is our firm belief that in many cases Species General Information and Population, Habitat, and Threat Inventory information will not be sufficient to make sound protection-management decisions. The various aspects of the biology of a species and its habitat relationships will require more detailed consideration. Protection for threatened species will, in many cases, require varying degrees of habitat manipulation (management). This will promote one or a few species at a site while inhibiting others. The Endangered Species Act of 1973, in our view, clearly states that such manipulations are required for the protection of designated species.

The manipulations necessary for the adequate management of a species within an area must be founded upon a sound understanding of the biology of a species and its habitat relations. To acquire this knowledge one must adopt a holistic view of the species by studying individuals, populations, and population systems utilizing the structures, processes, and habitat relations of each major life cycle phase within a particular time reference. This approach is called Species Biology, which is the subject of this unit.

As in units I and II, we have developed questions, information, and documentation formats to direct information acquisition. The major directive ideas of the unit are incorporated into a hierarchical matrix of questions for each of the phases of a generalized life cycle (Table 22). Detailed information systems for each of the major phases have been developed to assist information acquisition (Whitson and Massey, in press). A summary of the major information categories is presented in Table 23.

Major environmental factors affecting phase success or failure should be indicated by comparable and comparative life cycle information gained from studies of several populations. Careful manipulation of particular factors and concurrent monitoring of a species' response should suggest or allow evaluation of remedial efforts and in turn foster long-term preservation. A model research proposal to study two closely related species illustrating the value and use of the Species Biology approach is presented in Part IV, Section F.

Table 22. *Question Matrix for Major Life Phases*

REPRODUCTION	DISPERSION	ESTABLISHMENT	MAINTENANCE
Is reproduction occurring?	Are propagules present?	Are new individuals present?	Is there a range of classes?
What types of reproduction are occurring?	What types of viable propagules are present?	What are the origins of the new individuals?	What are the origins of the classes?
What breeding systems are operative?	What dispersal systems are operative?	What establishment processes are operative?	What are the percentages of each class in the population?
What pollination systems are operative?	What are the dispersal units and/or agents?	What are the spatial relations of establishment processes?	What are the spatial relations of the classes?
What is the reproductive capacity or status of the population?	What is the dispersal effectiveness of the population?	What are the establishment effectivenesses based on origin?	What is the survivorship of each class progressing to the next class?

Table 23. *Summary of the Species Biology Information Unit*

REPRODUCTION	ESTABLISHMENT
REPRODUCTIVE SYSTEM TYPES	NEW INDIVIDUAL ORIGIN
Sexual	Sexual
Asexual	Asexual
BREEDING SYSTEM TYPES	PREESTABLISHMENT PROCESSES
Selfing	Kind by origin
Outcrossing	Distribution
POLLINATION SYSTEM TYPES	ESTABLISHMENT PROCESSES
Pollination types	Origin census states
Vectors	
REPRODUCTIVE CAPACITY	ESTABLISHMENT DISTRIBUTION
Census by origin	Origin census states
Population census	Population census states
REPRODUCTIVE EFFECTIVENESS	ESTABLISHMENT EFFECTIVENESS
Actual	Population percentages
Potential	Origin percentages
DISPERSION	**MAINTENANCE**
DIASPORE SYSTEM TYPE	POPULATION CLASSES
Diaspore origin	Specific
Diaspore type	Relative
DISPERSAL SYSTEM TYPE	CLASS ORIGINS
Release mechanism	Sexual
Transport-vector	Asexual
DISPERSAL STATUS	POPULATION COMPOSITION AND DISTRIBUTION
Dispersal unit type	Sociability of individuals
Dispersal census	Distribution census
DISPERSAL EFFECTIVENESS	MAINTENANCE EFFECTIVENESS
Actual	Class survivorship
Expected	Vitality/vigor of individuals

D. Environmental Factor Analysis

To complete a logical and hierarchical progression of activities and studies, Unit IV of our program is designed to specifically test and identify environmental factors that appear to control or limit the success of each life cycle phase. These studies should be based on information from Species Biology studies. In some cases these factor studies will be academic and in others clearly needed for management and preservation. The field and laboratory experiments necessary for testing and identifying limiting factors generally will require expertise and techniques from such disciplines as plant physiology, physiological ecology, genetics, and others.

Summary

1. The use of a life cycle model and a four-unit program promotes a holistic view of a species as an integral part of our natural heritage.
2. The hierarchical units provide a systematic approach to the problems of species preservation, promote intermediate decision-making, and facilitate the establishment of preservation management priorities.
3. Units I and II provide systems for the acquisition of information that enables us to assess the current status of populations, to identify threats to populations of special concern, to initiate monitoring, and to establish priorities for populations and species in urgent need of protection.
4. Units III and IV provide a sound basis for management and protection and also stimulate basic research even when detailed factor analysis and systematic relationships are not required for immediate preservation efforts.
5. The integrated results of Units I–IV provide superior information for use in developing options and in making specific commitments and decisions for species and habitat preservation and management.
6. This program should stimulate multiple disciplinary investigations, reduce redundancy, maximize cost efficiency, and provide the best information that the scientific community can produce for resolving problems associated with species of special concern in our natural heritage.

Part II

Natural Area Reports

(Basic Inventory Reports, A–D; Significant Site Reconnaissance Report, E; Field Data Reconnaissance Reports, F–G)

A. Iron Mine Hill
B. Swift Creek
C. Carolina Beach
D. Ocracoke Island
E. Roanoke River Bluff
F. South Hyco Creek
G. North Mayo River

The natural area reports are included to illustrate the application of the ecological diversity classification system to the inventory and analysis of a wide variety of sites. The diversity presented covers terrestrial, wetland, and aquatic ecosystems; successional and climax communities; and endangered, threatened, endemic, disjunct, and restricted species. The communities range from upland forests and savannahs to salt marshes, grass dunes, and aquatic beds; the soils, from Entisols through Inceptisols to Alfisols, Spodosols, and Ultisols; the rocks, volcanic and plutonic to sedimentary and from basic to acid; and the waters, from estuarine salt to fresh. The natural areas encompass the regionally to locally significant. Even though the reports are based on natural areas in North Carolina, the diversity described exhibits the potential for the use of the system in the remainder of the nation. Each report retains the style of the individual contributor. Reports also differ because they were written during various stages of the development of the ecological diversity classification system.

A

Iron Mine Hill Natural Area*

Orange County, North Carolina; ca. 2.5 miles northwest of Coker Hall,
UNC, Chapel Hill, on NC 86, ca. 0.16 mile W on SSR 1780,
ca. 0.6 mile N on SSR 1843; along Bolin Creek and the Southern Railroad
35°56'08"N, 79°04'38"W; Chapel Hill, N.C., quad. 7.5 min., 1946
Appalachian Highlands; Piedmont Province; Piedmont Uplands
400-510 ft; 122-156 m
ca. 3.0 square miles

Ownership: University of North Carolina at Chapel Hill and private.

Administration: University of North Carolina at Chapel Hill and private.

Land use: University-owned land is used as a buffer zone for the Horace Williams Airport. Private land is unused, farmed, or residential.

Land classification: Unknown.

Danger to integrity: Expansion of the Horace Williams Airport could jeopardize the best hardwood stands in the natural area; private development could jeopardize much of the natural area.

Publicity sensitivity: None presently known.

Significance and protection priority: The hardwood stands in the natural area are definitely of regional significance. The area is in some jeopardy from development.

Reasons for priority rating:

Biotic Diversity: Endangered or threatened species and communities: No threatened or endangered species occur within the complex. One could say, however, that the entire complex is an endangered entity because much of the Piedmont Province has been or is being cut over for a variety of reasons, and many areas have been cut several times. A mature, undisturbed complex of climax hardwood communities is becoming increasingly rare in the North Carolina Piedmont.

Botanical Diversity: The climax hardwood communities contain 34 species of angiosperms and gymnosperms in the canopy and subcanopy. This represents almost 70% of the common canopy species found in the Piedmont of North Carolina (Table A3, p. 186). These species segregate into numerous

---

*Report originally compiled by Lee J. Otte, Department of Geology, UNC at Chapel Hill. Report edited by Deborah K. Strady Otte, Laurie S. Radford, and Julie Smith, Department of Botany, UNC at Chapel Hill.

Community Types composed of different combinations and densities of trees, shrubs, and herbs. These segregations result from differences in moisture, nutrients, and local climate, which are controlled by differences in topography, soils, and underlying lithologies. The natural area is relatively small (3.0 square miles) but includes land varying from recently cleared fields to mature climax hardwood forests. Natural pine lots are present, ranging from recently invaded fields to pine stands containing trees well over 100 years old.

Abiotic Diversity: The proximity of the natural area to the Durham Triassic Basin (see discussion) has resulted in the development of a terrain that changes from highly dissected to relatively flat upland over a very short distance. The area is situated geologically in the Carolina Slate Belt. Within the area the lithologies range from acidic to ultramafic plutonic igneous complexes, from fine-grained, water-deposited tuffs to volcanic breccias, from metamorphosed slates to chlorite schists to a small iron deposit that once was excavated commercially. The above-mentioned topographic and lithologic diversity causes the development of a wide variety of soils. The area contains four soil orders and 16 soil series, representing 77% of all the soil series presently recognized in Orange County.

Management recommendations: The numerous uses to which the land is now being subjected necessitate a variety of management recommendations.

1. The mature hardwood communities in the complex should be left undisturbed to preserve a part of the hardwood diversity of the Piedmont Province.

2. With its complex changes in species distributions and densities, this area offers a great opportunity to investigate the microhabitat requirements of many Piedmont trees and shrubs. The area should become a research center for this line of investigation.

3. The area is also excellent for conducting successional studies over a variety of soils, lithologies, and landforms. The boundaries of the natural area were drawn intentionally to include roads and actively farmed land. To follow all stages of succession, one also must see how the land was disturbed originally and how this disturbance has injured or aided original recolonization. This type of natural area provides an excellent opportunity to conduct controlled experiments with the factors, both natural and man-made, that affect plant and animal distribution and succession.

Data sources: None.

General scientific references: None.

General documentation and authentication: Data for this investigation were collected in 1977, 1978, and 1979 by Lee J. Otte of the Geology Department of the University of North Carolina at Chapel Hill and by Deborah K. Strady Otte and Alan Weakley of the Botany Department of the University of North Carolina at Chapel Hill. Specimens were deposited in the NCU Herbarium by Lee J. Otte and Deborah K. Strady Otte in 1978.

Discussion

The Iron Mine Hill Natural Area, covering approximately 3.0 square miles, lies immediately west of the Horace Williams Airport, one mile north of Chapel Hill, Orange County, North Carolina (Map A1, p. 150). The entire area, contained within the Bolin Creek drainage of the Cape Fear River system, ranges in topography from deeply cut terrain to relatively flat and gently rolling Piedmont uplands. Geologically, the site is underlain by acidic to ultramafic plutonic rocks and acidic to intermediate metavolcanic and metasedimentary rocks.

It is unusual for a tract of land so close to a populated area (Chapel Hill and Carrboro) still to support a large number of relatively undisturbed hardwood communities. The fact that much of the land is owned and managed by the University of North Carolina at Chapel Hill probably explains why some of the area has escaped cutting for so long. The adjacent Horace Williams Airport is university owned, however, and expansion of this facility could jeopardize the existence of this natural area (see conclusions, p. 189). Portions of the university owned forest immediately north of the airport have already been clear-cut within the past 10 years. Although the remainder of the natural area is owned privately, not all of it has been converted to farm land, probably because of the extreme unevenness and rockiness of the terrain (steep slopes facing Bolin Creek and its tributaries, rocky hill tops). It is unclear why this privately owned part has not been cut for timber.

This area contains portions of a hardwood community complex that has remained essentially undisturbed for several hundred years and is probably a segment of the Piedmont climax hardwood forest (Fig. A1, p. 152). Various combinations of *Quercus*, *Carya*, and *Acer* species, *Liriodendron tulipifera*, *Liquidambar styraciflua*, *Fagus grandifolia*, and other miscellaneous hardwoods form the canopy of the hardwood stands. The subcanopy, shrub, and herb layers are equally diverse and varied. Detailed vegetational analyses were conducted in five communities as well as canopy analyses on an additional six communities (Table A2, p. 169, and community diversity summaries). These 11 communities, consisting of different combinations and concentrations of 34 species of trees, are controlled by local variations in topography, underlying lithology, soils, and moisture and nutrient availability. The development and present distribution of the communities are interrelated with all of these factors.

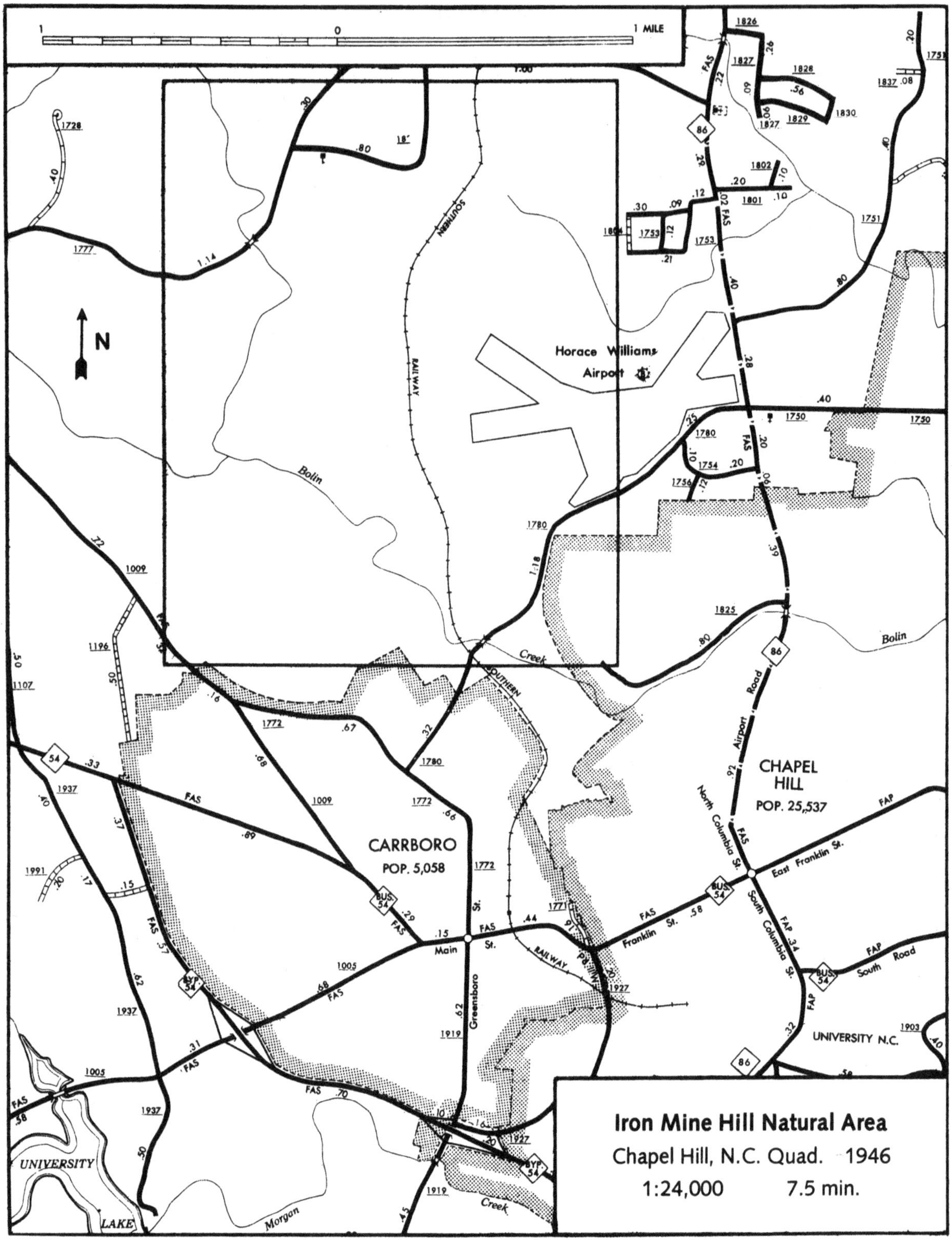

Map A1. *Location of Iron Mine Hill Natural Area, Orange County, North Carolina.*

The large size and old age of the canopy trees (see community diversity summaries) in the 11 analyzed communities indicate that much of the vegetation has developed into a mature, climax forest. All the communities except the "*Quercus stellata*-mixed upland hickories-oaks" community have closed canopies (this exception will be discussed later, p. 178). These facts plus the virtual absence of cut tree stumps indicate that selective cutting has not taken place at all or at least not for a very long time. "Weedy" trees, such as *Robinia pseudo-acacia* and others, are absent from most of the canopy and from most lower vegetation levels. Other "weedy" plants, such as *Rhus radicans* and *Lonicera japonica*, are present in very limited numbers, although they are common in the adjacent pine lots.

Much of the area has been cut over at different times and now ranges from open fields to young pine lots to forests composed of mixtures of *Pinus taeda*, *P. echinata*, and *P. virginiana* (Fig. A1, p. 152). Even though pines now dominate large portions of the area, essentially no pines exist as seedlings or transgressives in the hardwood communities. Mature pines are present within the communities (1) in a narrow zone of mixing where a pine and a hardwood community meet and (2) scattered throughout the forest as occasional large specimens of canopy height. Some of these mature pines probably became established in the hardwood communities when the canopy was opened by the death of a large hardwood tree (see Mixed upland oak community discussion, p. 175). Others may be remnants of the preclimax mixed hardwood-pine forest. These large, scattered individual pines appear to be a minor, natural component of this type of climax forest.

In the hardwood forests along Bolin Creek and its tributaries, a variety of small yet distinct communities can be found. Although some of these overlap, they nevertheless remain distinct enough to permit recognition of definite associations between Community Types and specific combinations of major controlling ecological factors. In the case of the Iron Mine Hill Natural Area, as in most plant communities, these major controlling factors consist of different combinations of moisture and nutrient availability plus protection from various adverse climatic agents. Variation in these factors is influenced by differences in topography, hydrology, soils, and underlying lithologies. It is therefore important to understand in a broad sense how each factor influences vegetation so that one then can study individual communities, such as at Iron Mine Hill, to determine major ecological controls.

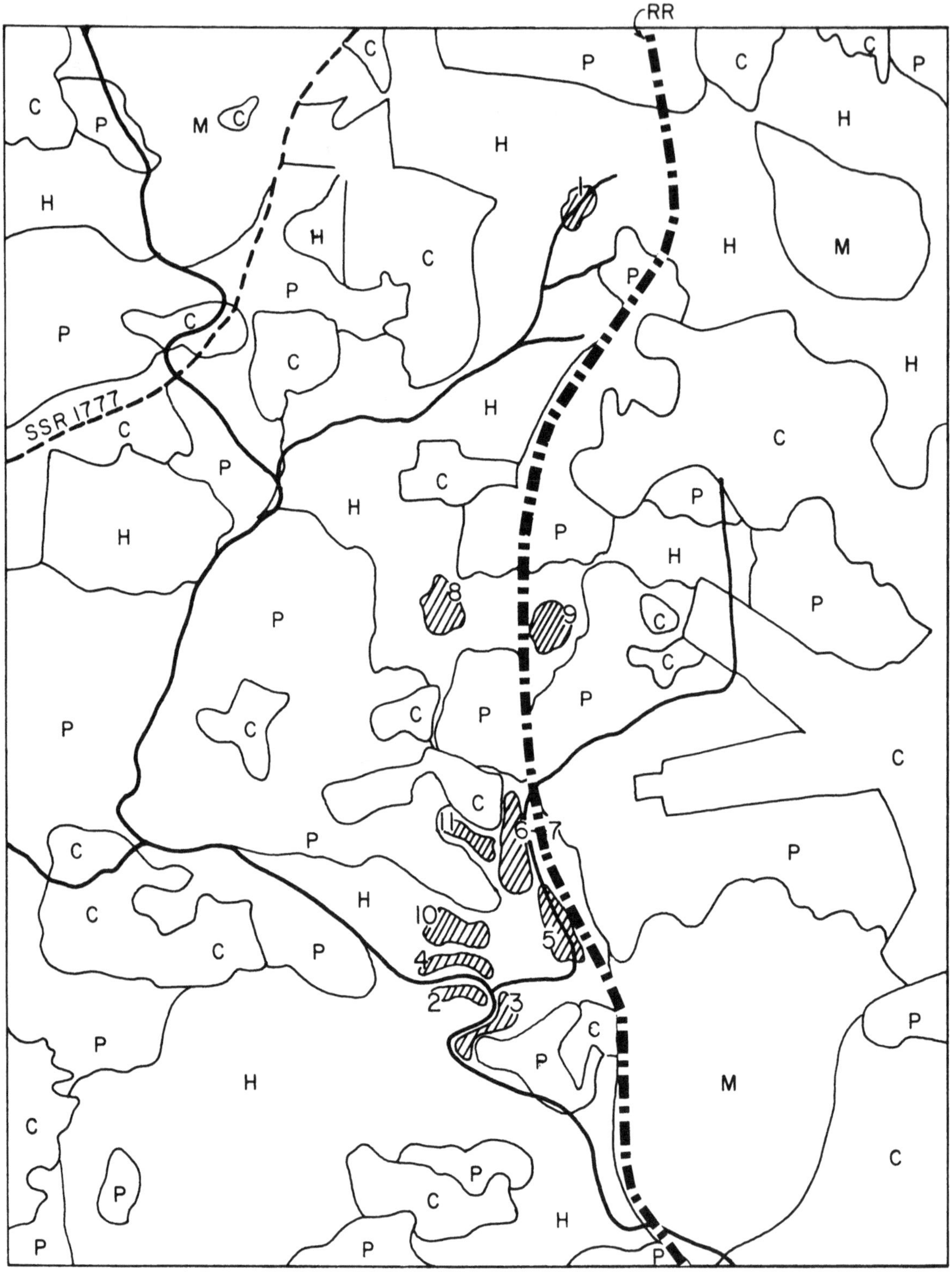

C–Cleared, P–Pine, H–Hardwood, M–Mixed pine and hardwood

Figure A1. *Major vegetational patterns based on aerial photography*. The shaded, numbered areas represent the Community Types analyzed. (See Table 2 for community canopy compositions and Community Diversity Summaries for detailed vegetational analyses.)

The remainder of this discussion will deal with three major topics: (1) the abiotic diversity of the Iron Mine Hill Natural Area (origin, distribution, and diversity); (2) the biotic diversity of the area (origin, distribution, and diversity); and (3) the diversity of hardwood canopy species in the North Carolina Piedmont. The discussion is based on personal observations of Iron Mine Hill and on a search of the literature pertinent to the area.

I. ABIOTIC DIVERSITY OF THE IRON MINE HILL NATURAL AREA

A. Geologic Diversity

The eastern Piedmont of North Carolina consists of four major lithologic associations which form large northeast-southwest-striking belts (Stuckey, 1965). The first and oldest (Precambrian) of these four contains a mixture of gneisses and schists, dominated by mica-rich mineral assemblages, and also includes mafic, amphibole-rich rocks. The second group, the Carolina Slate Belt, contains a series of Late Precambrian to Early Paleozoic, felsic to mafic, volcanic sediments and flows interbedded with fine-grained slates and argillites. The third group consists of a series of plutonic complexes of Middle to Late Paleozoic age intruded into the first two groups. These intrusives are predominantly acidic or intermediate, but can range to ultramafic in composition. The fourth major rock group consists of the unconsolidated to highly cemented clastic sediments that fill the Triassic-aged grabens, formed during the early rifting stage of the modern Atlantic Ocean. These Triassic sediments contain various mixtures of clays, silts, sands, and gravels.

Each of the above four lithologic belts contains numerous, minor to major, local variations in lithologies, textures, structures, and associated susceptibilities to erosion. This geologic variation accounts for the wide variety of topographic features and soils found in the eastern part of the North Carolina Piedmont and plays a very important role in shaping the Iron Mine Hill Natural Area.

The Iron Mine Hill Natural Area contains portions of the Carolina Slate Belt and the igneous intrusive complexes of group three (Fig. A2, p. 154). The volcanic rocks of the Carolina Slate Belt include acidic to basic tuffs and breccias as well as rhyolites. The sedimentary rocks of the Slate Belt predominantly are water-deposited clays and silts and reworked volcanic ash. The

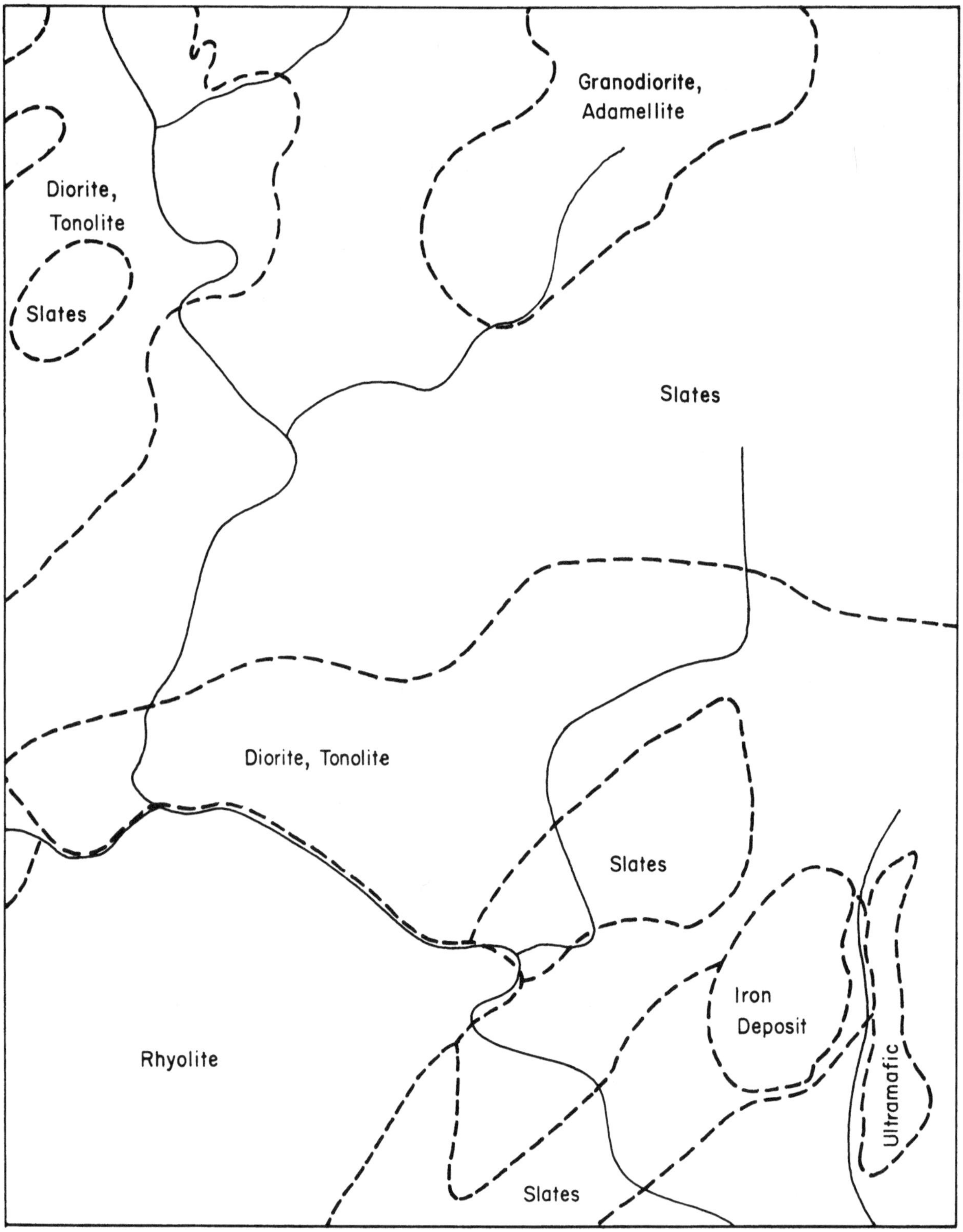

Figure A2. *Lithologic geology*. Based on Mann et al. (1965).

plutonic igneous rocks range from granites (acidic) to gabbros (basic) to isolated occurrences of ultramafic rocks. Recent alluvium contains various mixtures of clays, silts, and sands. Different metamorphic grades, ranging to chlorite schists (Mann et al., 1965), are found as contact metamorphic zones in the volcanic and sedimentary rocks, in association with the intruded plutonic rocks. The area itself is named after a hill in the southeast portion of the natural area, on which a small deposit of iron ore once was mined (Mann et al., 1965). The control this varied lithology has over the development of topographic features and of soils within the natural area will be discussed in detail in the appropriate sections.

B. Topographic Diversity

The eastern Piedmont of North Carolina, although not exhibiting a great deal of relief (absolute change in elevation between high and low points), shows a fairly wide range of topographic variation in its landscape. The Triassic Basins of this physiographic province, notably the Durham and Sanford Basins, are mostly low lying and relatively flat as compared with the rest of the Piedmont. The remainder of the province contains a variety of topographic features, ranging from isolated monadnocks of highly resistant rock and relatively deep valleys cut by the major rivers and their larger tributaries to essentially flat uplands, gently rolling hills, and broad, gently sloping interstream divides.

The history of the development of this landscape can be traced back through the Cenozoic Era. In the first half of this era the Piedmont was eroded to a broad peneplain over which the larger rivers meandered (Stuckey, 1965). With renewed uplift of the entire region near the beginning of the Miocene (Stuckey, 1965), the larger streams began to cut down to rejuvenated base levels without drastically altering their courses. These rivers then and now flow mostly from the northwest to the southeast and cut across the northeast-southwest-striking lithologic belts (Stuckey, 1965). The smaller streams, because of their lesser erosive powers, were controlled more readily by the underlying lithologies and adjusted their courses to flow over weaker rocks, which could be eroded more easily than the physically and chemically stronger rocks. These adjustments and downcuttings provided the beginnings of the present-day topographic expression of the Piedmont Province.

As the major rivers flow through the Piedmont, the river valleys and their tributary valleys change topographically along their lengths. A pattern emerges: the larger streams form narrow valleys when cutting through hard, crystalline, igneous and metamorphic rock and wider, more open valleys when cutting through more easily eroded, low-grade metamorphics and fine-grained sedimentary rocks. The smaller tributaries are found in valleys that are often relatively deep and narrow where they feed into the valley of a major river. They become shallower and broader upstream where they have not as yet cut down to local base level. These topographic changes in valley form can take place over a distance of several hundred miles for the major rivers and their larger tributaries and over shorter distances for the smaller tributaries.

Lesser streams that flow into the Piedmont Triassic Basins differ from streams of equivalent size elsewhere in the Piedmont because they usually become entrenched deeply where they enter the basins and undergo all or most of the topographic changes that the larger streams exhibit along their much longer lengths. These changes occur over a much shorter distance, often less than 10 or 20 miles. Bolin Creek, which flows through the Iron Mine Hill Natural Area, is less than 10 miles long from its headwaters to where it feeds into the Durham Basin. The creek ranges from a deeply incised stream to one flowing through gently rolling hills and valleys and typical, flat, Piedmont uplands.

Several examples of lithologic control over the topographic shaping of the Iron Mine Hill Natural Area deal with ease of erosion of the underlying rocks. The mid-Cenozoic uplift of the Piedmont, aided by the drastic lowering of base level during the Pleistocene glacial episodes, allowed the structurally weak Triassic Basins to be eroded into topographic lowlands. The smaller streams flowing into these lowlands from the adjacent crystalline portion of the Piedmont have kept pace with the lowering of their local base level (the Triassic Basins), but most of their energy is expended on downcutting at the edge of the lowland, with very little sidecutting or backcutting into the crystalline rocks surrounding the lowlands. This downcutting is most evident where the streams feed directly from plutonic igneous rock into the lowland. Here they form deep, narrow valleys as opposed to equally deep but wider valleys where the streams feed from the more easily eroded slates and volcanic rocks into the lowlands. This results in a series of small streams whose valleys are narrow and often as deep as those of the larger Piedmont rivers. Unlike the larger rivers, however, these smaller streams lose their depth and narrowness very rapidly upstream away from the lowlands.

Bolin Creek flows first across a mixture of volcanics, slates, and small intrusive bodies and then cuts across the Chapel Hill Granite just before entering the Durham Triassic Basin. This granite has prevented the creek from cutting a broad, shallow valley where it feeds into the basin. The valley through the granite is nearly 200 feet deep and not over half a mile wide. Two miles upstream from where the creek flows into the basin in the natural area, the valley is 150 feet deep but still relatively narrow. Near the headwaters of the stream, six linear miles from where the stream flows into the basin, the valley is more open and gentle and only about 80 feet deep. As a result of this decreased downcutting as one moves away from the Triassic Basins and away from the larger rivers, the Piedmont terrain loses its localized, dissected appearance and becomes characteristically gently rolling and relatively flat, with only occasional, isolated monadnocks projecting above the surrounding countryside.

This landscape pattern can be observed by looking at U.S. Geological Survey topographic maps, which include a basin edge or a large river valley. As an example, on the Chapel Hill quadrangle (7.5 minutes), on which the Iron Mine Hill Natural Area is located, the contact between the western edge of the flat Triassic Basin and the higher but equally flat Piedmont Uplands is marked by a north-south line of locally rugged topography.

Another example of lithologic control of topography in the Iron Mine Hill Natural Area concerns the directional orientation of the positive and negative landforms (Fig. A3, p. 158). The hills and valleys in the area are aligned mostly in a northeast-southwest or in a north to northwest-south to southeast direction. General trends of fracture patterns, measured in both the plutonic igneous rocks and altered slates and volcanics, reveal a series of patterns that correspond with the two major weathering trends. One set of fractures trends from N25°E to N40°E and a second from N5°W to N20°W. This similarity in fracture pattern and landscape development suggests that the structural deformation of the rocks within the natural area, outwardly expressed by the fracture systems within the rocks, influences the positioning of the smaller streams and the shape and alignment of the hills.

For its geographic size, the Iron Mine Hill Natural Area contains a wide variety of landforms as compared with much of the eastern Piedmont. This variation, in association with changes in underlying lithologies, provides many different topoenvironments. These environments, in turn, affect water drainage, soil formation, nutrient supply, protection from adverse climatic factors, and,

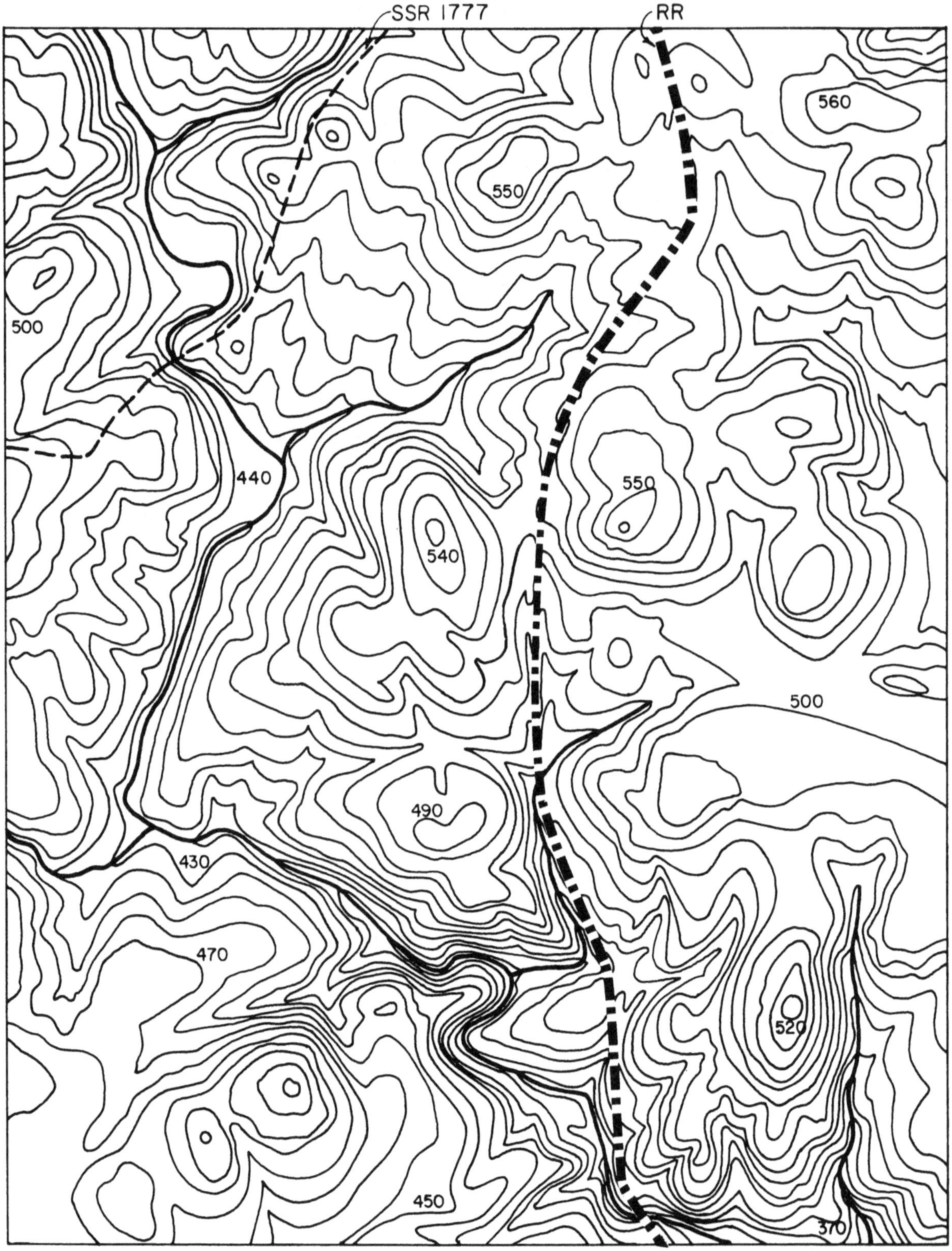

Figure A3. *Topographic map of Iron Mine Hill Natural Area*. Enlarged from the Chapel Hill, N.C., 7.5 minute topographic quadrangle (1946). Contour interval: 10 ft Scale: 1 cm = 370 m or 1214 ft

finally, the biota. The overall small areal extent of each topoenvironment within the natural area causes small yet distinct communities to lie close to each other. Occasionally these communities display some degree of overlap, but usually they are separate entities whose major controlling factors are distinct and observable.

C. Soil Diversity

"Taxonomic classification of soils on the series level is based on similarities and differences in numerous criteria important in understanding the physical make-up of the material. These criteria include color, texture, structure, reaction, consistency, and mineral and chemical composition" (Dunn, 1977). Climate, plant and animal life, parent material, topography, and time produce differences in the above characteristics (Dunn, 1977). Climate and plant and animal life are active forces in soil formation, whereas parent material, topography, and time are passive contributors. Although the relative importance of each of these factors differs from one place to another, all interrelate with each other and all contribute to the formation of most soils.

Climate affects the type of weathering of the underlying rocks, influences the alteration of the different minerals in the rock, and controls translocation and leaching of elements and compounds within and from the rock and overlying soil. Furthermore, the erosive and depositional forces of climate can greatly affect soils by either physically removing or bringing in soil-forming components. Plants and animals, both micro- and macroscopic organisms, contribute by opening the soil to water and air and by returning certain materials to the soil through waste products and decay after death. Organic acids, too, aid in chemical weathering. The parent material (underlying sediment or rock) is the direct source of the chemical compounds and mineral constituents found in most soils. Topography controls surface drainage and movement of water through the soils, thus affecting soil development and depth and the distribution of vegetation. Time is one of the most important factors in soil development because it affects soil maturity. Generally, the older the soil the better the development of horizons and the more mature the soil. In a given area, however, soils with the same initial time of development can mature at different rates because of differing degrees of influence by the other controlling factors.

The wide variety of soil series found in the Iron Mine Hill Natural Area arises from variation in two major factors: (1) the underlying parent material, and (2) the local topography, which in turn is controlled partially by the composition, texture, and structure of the bedrock. Other major soil-controlling factors (climate, plants and animals, and the length of time to which the land has been exposed to all the other factors) remain essentially the same over the entire natural area.

A total of 22 soil series occurs in Orange County; the Iron Mine Hill Natural Area contains 16 of these (Table A1, p. 161, and Fig. A4, p. 162) representing four different soil orders--Entisols, Inceptisols, Alfisols, and Ultisols. Entisols, here represented by the Congaree fine sandy loam, are found on isolated, narrow floodplains parallel to the larger streams and are composed of alluvial debris. Inceptisols, represented by Chewacla loams and Goldston slaty silt loams, are found on lower slopes and valley floors. They appear as isolated occurrences over slaty rock. Alfisols, located over intermediate to basic rocks (diorite, gabbro, hornblende schist), occur from the floodplains and valley floors up to the ridge crests. They are represented here by Enon loam, Iredell gravelly loam, Orange silt loam, and Wilkes gravelly loam. Ultisols, which predominantly develop over acidic rocks (including granites, aplites, slates, volcanic tuffs, and rhyolites), occur on all topographic features, each feature usually supporting a particular series. These Ultisols include Appling sandy loam, Cecil fine sandy loam, Georgeville silt loam, Helena sandy loam, Herndon silt loam, Hiwassee clay loam, Lignum silt loam, Tatum silt loam, and Wedowee sandy loam.

If the variations in lithology found in this area were associated with a more subdued topography, the overlying soils probably would be less well differentiated and the individual soil series more widespread. This can be seen in the flatter, more topographically subdued portions of Orange County just to the west of the Iron Mine Hill Natural Area (see Dunn, 1977). The more varied topography in the southeastern section of the county where Iron Mine Hill is located results from interaction of the localized occurrences of highly crystalline plutonic igneous rock in an area of chemically and physically weaker, low-grade metamorphic rocks. This causes differential erosion, with the highly crystalline rocks forming the higher hills and the metamorphic rocks forming the flatter, more subdued uplands. The local downcutting by Bolin Creek and its larger tributaries highly dissects the terrain, which grades rapidly upstream into the flatter uplands typical of this portion of the Piedmont. A wide variety of soils develops in this local area because of varied topography underlain by diverse lithologies--relatively deep valleys, narrow valley floors, and steep valley slopes grading into broader,

Table A1. *Soil diversity summary presenting characteristics of soil series in the Iron Mine Hill Natural Area* (after Dunn, 1977) for soil distribution

Series	Order	Parent Material	Slope Degree	Topographic Position
1 Congaree (CP)	Entisol	loamy alluvium	0–2	narrow bands parallel to streams or floodplains or at the base of slopes
2 Chewacla (CH)	Inceptisol	recent alluvium	0–2	long flat areas parallel to major streams
3 Goldston (GI)	Inceptisol	fine-grained, felsic slates	6–45	narrow interstream ridges and sides of ridges between intermittent and perennial streams
4 Enon (EN)	Alfisol	diorite, gabbro, hornblende schist	2–12	tops and sides of ridges between intermittent and perennial streams in the uplands
5 Iredell (IR)	Alfisol	diorite, gabbro, hornblende schist	1–4	on broad ridges in the uplands
6 Orange (OR)	Alfisol	fine, quartz monzonite, hornblende schist, dacite, diorites	0–3	on broad ridges in the uplands
7 Wilkes (WX)	Alfisol	diorite, hornblende schist, or mixtures of acidic and basic rocks	8–45	narrow side slopes adjacent to major drainageways
8 Appling (AP)	Ultisol	acidic igneous and metamorphic rocks	2–10	on broad ridges and narrow side slopes that are crossed by intermittent drainageways in the uplands
9 Cecil (CF)	Ultisol	acidic igneous and metamorphic rocks	2–6	on broad ridges and narrow side slopes in the uplands
10 Georgeville (GE)	Ultisol	phyllites and Carolina slates	2–10	on broad ridges and narrow side slopes in the uplands
11 Helena (HE)	Ultisol	mixed aplitic granites or granite gneiss cut by dikes of gabbro and diorite	2–8	on broad ridges in the uplands
12 Herndon (HR)	Ultisol	phyllites and slates	2–10	on broad ridges and narrow side slopes in the uplands
13 Hiwassee (HW)	Ultisol	old alluvium or residuum of basic or mixed basic and acidic crystalline rocks	2–6	on broad ridges and narrow side slopes in the uplands
14 Lignum (LG)	Ultisol	slates, fine schists	0–3	interstream divides and around the head of drainageways
15 Tatum (TA)	Ultisol	sericite schist	8–25	on side slopes in the uplands
16 Wedowee (WM)	Ultisol	acidic crystalline rocks	8–25	on side slopes adjacent to major drainageways

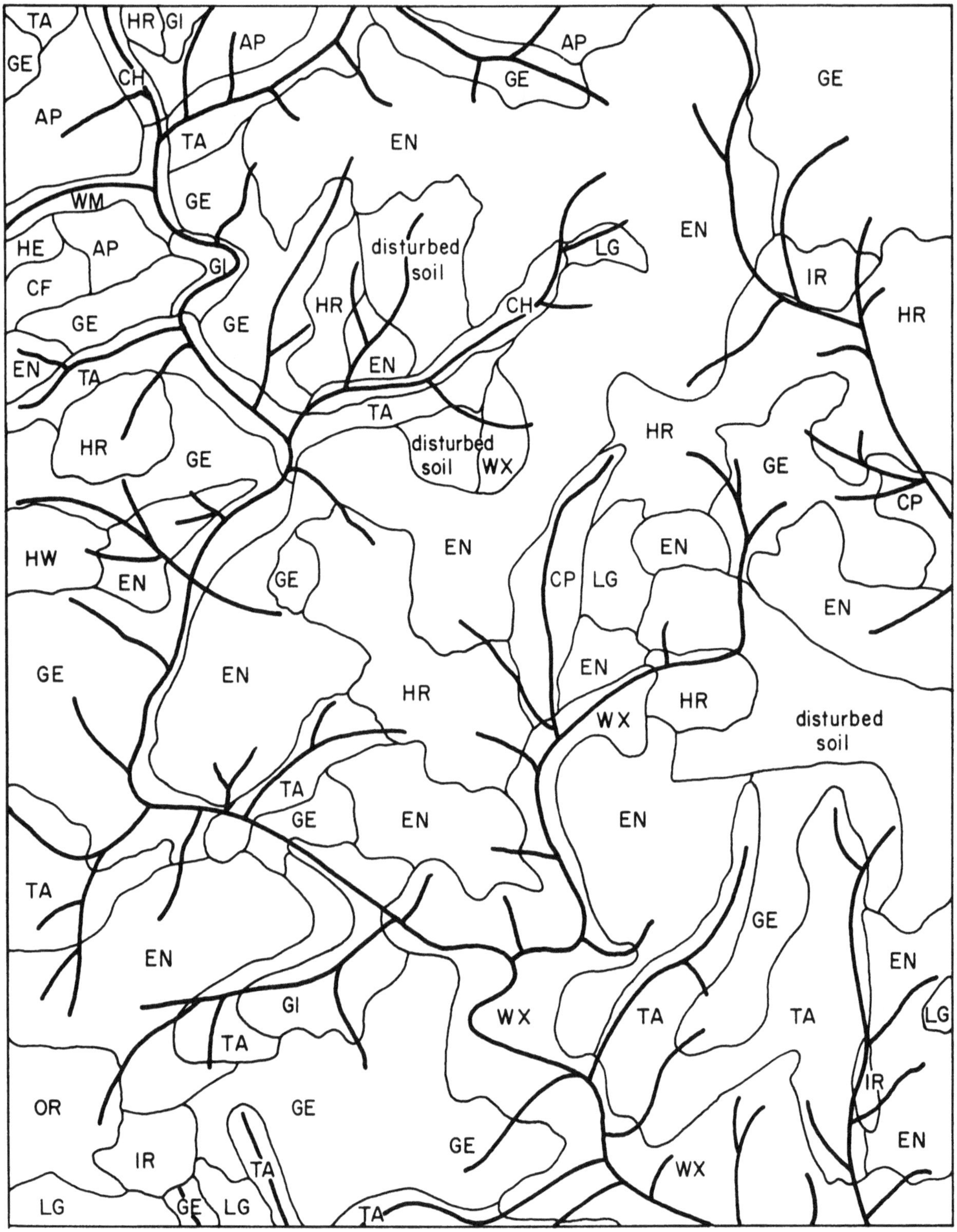

Figure A4. *Soil diversity—distribution of soil series.* Modified from Dunn (1977). (See Table A1 for symbols and descriptions of the individual series.)

shallower valleys with low, rounded ridges and isolated higher hills of more resistant rock.

Soils derived under similar controlling factors usually develop into the same or very similar soil series. A number of situations cause soil series differentiation. Soils derived from chemically and/or physically different lithologies in topographically identical environments often belong to different soil series and frequently even to different soil orders. Different lithologies under similar topographic and climatic conditions produce, upon weathering, different amounts of clays, silts, and sands. These grain sizes and resulting textures yield different degrees of porosity and permeability, which produce and control total nutrient availability and the rate at which the nutrients leach out of or come into the soil. Soils derived from chemically and physically similar lithologies but found in different topographic environments can or cannot develop into different soils. This depends on how drastically the topography changes and how that change affects other controlling factors.

An example of the last situation follows. A hill completely underlain by one rock type very often is covered by one soil series on the hill top and uppermost slopes, a different series on the middle and lower slopes, and yet a different one on the lowest slopes and adjacent stream valley floor. These changes result from differences in drainage that cause more intense leaching on the hill top and upper slopes and greater accumulation of sediments, moisture, and nutrients on the lower slopes. Gentler slopes often have a different soil series than steeper slopes over the same lithology because of more intense erosion on the steeper slopes, resulting in less well-developed soil horizons. Similar slopes receiving different amounts of moisture (such as rainfall, runoff, drainage), which affect erosion rates, weathering rates, vegetation, and nutrient availability, often have different soil series.

The associations of factors that can change soil types seem almost endless. Because so many possible combinations of controlling factors exist, different combinations can produce the same soils if other factors compensate for changes brought on by one or more controlling factors. Such is the case with several of the soil series in the Iron Mine Hill Natural Area. Here, soils developing from similar lithologies but under different topographic conditions result in the same series. As topography changes from the lower reaches of Bolin Creek to its upper reaches and smaller tributaries, a distinctive soil pattern emerges. Several soil series, most notably the Tatum and Georgeville, occur on the upper ridge slopes and ridge crests in the more highly dissected terrain adjacent to the larger streams. In the gentler, less dissected terrain associated with the upstream portion of Bolin Creek and its tributaries, these same soils occur adjacent to the streams and on the lower slopes. Other series replace the Tatum and Georgeville series on the adjacent upper slopes and ridge crests. This pattern suggests that the major soil-forming factors necessary to develop a particular soil series can occupy different topographic, and possibly also lithologic, positions. Study of such distributions could help determine primary controls over soil development.

D. Climatic Diversity

Accurate information concerning the local climate is readily available from the Chapel Hill weather station, located less than one mile from the natural area. The area receives an average of 46.66 inches of precipitation each year, with July having the maximum average precipitation (5.20 inches) and November the minimum (2.82 inches). The area has 77 days a year with precipitation greater than 0.10 inch and 30 days a year with more than 0.50 inch. These days are spaced fairly evenly throughout the year, with five to eight days a month receiving more than 0.10 inch and two to four days a month receiving more than 0.50 inch.

The average mean annual temperature is 60.2°F, with July having the highest mean monthly temperature (78.7°F) and January the lowest (41.4°F). July also has the highest mean daily maximum temperature (90.2°F), whereas January has the lowest mean daily minimum temperature (31.4°F). The highest temperature recorded in 75 years is 107°F, during July; the lowest temperature recorded in 75 years is -6°F, during February. The hottest months of the year are June, July, and August, with averages of days with the temperature over 90°F at 92°, 98°, and 94°, respectively. The coldest months of the year are November, December, January, February, and March, with averages of days with the temperature below 32°F at 12°, 21°, 20°, 16°, and 12°, respectively. The area is essentially freeze-free from May through October.

The rate of precipitation in the Iron Mine Hill Natural Area does not differ from the local rate. Moisture availability varies with changes in the topographic position of a particular Community Type, the drainage pattern, and the degree of protection from water loss by evaporation. The overall trend is for lower slopes and valley floors to be moister than upper slopes and ridge and hill crests. Similarly, in temperature patterns, the upper slopes and hill and ridge crests receive more daily sun and are probably several degrees warmer on the average than the lower slopes and valley floors, which receive less overall direct sunlight. Drying winds, high winds, ice storms, and other damaging climatic phenomena also take their greatest toll on the higher, more exposed sites and do less damage to the more protected lower slopes and valley floors.

E. Hydrologic Diversity

No large bodies of standing water occur within the natural area. A few small farm ponds of insignificant size are present; however, no distinct aquatic plant communities have become established because of their use as sources of water for livestock. Nothing is known about the animal inhabitants of these ponds. No large streams flow through the natural area. Bolin Creek, the largest perennial stream in the area (Fig. A5, p. 166), flows not more than a foot or two deep and never wider than 20 feet. It can rise several feet during flash floods but not for extended periods of time because of the relatively steep stream gradient. No distinct aquatic plant communities have been located in the stream, and, as of now, no work has been done on the aquatic animals. The smaller streams range from spring-fed and perennial to ephemeral runoffs active during periods of high rainfall. All of the above hydrologic environments control plant distribution in the sense that the lands directly adjacent to them contain higher soil moisture values than the surrounding, drier, better drained lands.

The areas of greatest hydrologic interest in some way directly relate to flooding or a high or perched water table. In this natural area, soils of the Chewacla series (Fluvaquentic Dystrochrept), the Helena series (Aquic Hapludult), the Lignum series (Aquic Hapludult), and the Orange series (Albaquic Hapludalf) cover such locations. The Chewacla series occurs on long, flat areas parallel to the larger streams and is somewhat poorly drained, moderately permeable soil that originates in recent alluvium and commonly undergoes flooding for brief periods of time (Dunn, 1977). The Helena series occupies broad ridges of the uplands and is moderately well-drained, slowly permeable soil commonly underlain by a perched water table at 12 to 30 inches below the surface during wet seasons (Dunn, 1977). The Lignum series consists of moderately well-drained, slowly permeable soil found on interstream divides and around the head of drainageways. The slowly

Figure A5. *Drainage patterns in the Iron Mine Hill Natural Area.* Compiled from aerial photographs, Dunn (1977), and the Chapel Hill, N.C., 7.5 minute topographic quadrangle (1946).

permeable subsoil of this series allows a perched water table to develop at about 12 to 30 inches below the surface during wet seasons (Dunn, 1977). The *Acer rubrum*-mixed mesic hardwoods-oaks Community Type (No. 1, p. 168) exists over this soil. The Orange series combines features of moderately well-drained, slowly permeable soil on broad ridges and a slowly permeable subsoil that allows a perched water table to develop at 12 to 36 inches below the surface during wet seasons (Dunn, 1977).

Within the natural area, the Congaree series (Typic Udifluvent) and the Wilkes series (Typic Hapludalf) represent other soils in which water plays an important role. The Congaree series consists of well-drained, moderately permeable soil that forms from loamy alluvium on floodplains or at the base of slopes. The seasonal high water table is at a depth of about 30 to 48 inches late in winter and early in spring. The soil is flooded frequently but for brief periods (Dunn, 1977). The Wilkes series is characterized by well-drained, moderately slowly permeable soil and is found on narrow side slopes. The seasonal high water table is at a depth of more than 72 inches, but a water table perches above the subsoil in places during prolonged wet periods (Dunn, 1977). The *Fagus grandifolia*-mixed mesic hardwoods-oaks (No. 2, p. 168) and the *Liriodendron tulipifera*-mixed mesic hardwoods-oaks-hickories (No. 3, p. 168) communities, as well as the *Fagus grandifolia*-dominated communities (Nos. 4 and 5, p. 171), occupy this soil.

Not all the locations covered by the above-mentioned six hydrologically important soil series in the Iron Mine Hill Natural Area were visited for this investigation. The sites visited did support Community Types distinctly different from the surrounding typical oak-and-hickory-dominated forests of the Piedmont Uplands, and they clearly were controlled by the hydrologic characteristics of the soil. This indicates a definite relationship between these soils and the vegetation inhabiting them.

II. BIOTIC DIVERSITY OF THE IRON MINE HILL NATURAL AREA

A. Botanical Diversity

This section contains discussions concerning the 11 hardwood Community Types analyzed for this report (Fig. A1, p. 152). Each community summary deals with unique features of a particular Community Type and observations on controlling factors that govern the distribution of that Community Type. The summaries are arranged in what appears to be a natural progression from the more mesic to the more xeric Community Types (Table A2, p. 169). Distribution and concentration of canopy trees in each Community Type form the basis for this progression. Detailed vegetational analyses for each Community Type appear in the Community Diversity Summaries at the end of this report.

Community Type Summaries

1. Acer rubrum-mixed mesic hardwoods-oaks/? Community Type. This community distinctly differs from any others analyzed. Topographically it inhabits a small flat at the headwaters of a spring-fed creek. The area appears relatively wet, yet also moderately well drained. The soil, a Lignum silt loam, derives from the underlying weathered slatelike rocks and, according to Dunn (1977), contains a locally perched water table at least 12 to 30 inches below the ground surface during wet seasons.

This Community Type is vegetationally distinct because it represents the only climax community in which vines, prevailingly species of *Lonicera* and *Vitis*, cover the ground. The canopy contains the best development of *Acer rubrum*, *Liquidambar styraciflua*, and *Quercus phellos* in the entire Iron Mine Hill complex. Several species, including *Fagus grandifolia*, *Quercus rubra*, and the *Carya* spp. are noticeably absent, although they appear immediately adjacent to the flat on the gentle slopes surrounding the community. *Platanus occidentalis* also occurs and concentrates along the more open portions of the creek, where some cutting obviously has occurred in the past. This species occurs nowhere else in the complex as mature trees.

The spring-fed stream supplies most of the water to the community, with the remainder coming from drainage of the surrounding slopes, and keeps the soil of the small flat wet for a significant portion of the year. This moisture apparently accounts for the absence from the community of the upland oaks (except *Q. alba*) and all the *Carya* spp. Furthermore, the moisture allows more mesic, bottomland trees that prefer good drainage to become established.

2. Fagus grandifolia-mixed mesic hardwoods-oaks/? Community Type and *3. Liriodendron tulipifera-mixed mesic hardwoods, oaks, and hickories/? Community Type.* In the southeastern corner of the natural area, portions of Bolin Creek have developed a small, yet distinct, floodplain, which is only a few hundred feet wide and several feet deep at its greatest extent. The floodplain can be divided into two terracelike segments, both of which are well drained and show no evidence

Table A2. *Distribution and importance values of canopy species in the analyzed communities of the Iron Mine Hill hardwood complex*

	Community Types*										
Species	1	2	3	4	5	6	7	8	9	10	11
Platanus occidentalis	8.5										
Fraxinus pennsylvanica	22.2										
Quercus falcata var. *pagodaefolia*	7.3	7.4									
Quercus phellos	33.6		8.6								
Quercus phellos X *Q. falcata*		8.3									
Liquidambar styraciflua	67.6	45.4	43.5	8.1	8.7						
Acer saccharum ssp. *floridanum*		34.7					7.8				
Juniperus virginiana			7.7								
Nyssa sylvatica	16.3					14.8	7.5				
Pinus taeda	7.4		6.5			7.3	8.4	18.3	11.2		
Liriodendron tulipifera	23.5	37.4	78.0	16.5		28.2	10.6	38.8	29.9	8.3	
Acer rubrum	89.0	14.3	7.5	32.1	14.5	7.1	34.0			7.4	
Fagus grandifolia		77.6	42.0	159.3	138.8	6.3				9.4	
Quercus rubra		22.1	31.8	17.8	37.2	93.0	33.2	36.7	31.3	26.6	23.8
Quercus alba	24.3	34.5	35.8	56.4	61.6	82.5	78.3	114.1	67.2	35.9	8.6
Carya tomentosa		10.2	12.4		32.4	22.1	94.2	69.7	6.4	31.8	69.5
Carya glabra		7.1	22.0		7.7				6.4	80.5	28.7
Quercus velutina						14.8	7.8		43.9		
Quercus coccinea							9.1		22.1		
Fraxinus americana								6.5		23.1	
Quercus falcata var. *falcata*						6.7		7.6	58.1	26.5	
Quercus stellata						6.8	9.1	7.1		32.8	124.7
Carya ovalis									7.0		
Carya ovata									11.3	6.9	23.6
Pinus echinata									8.2		
Carya carolinae-septentrionalis										9.8	
Quercus marilandica											21.1

*Key to Community Types:
1. *Acer rubrum*-mixed mesic hardwoods and oaks/?
2. *Fagus grandifolia*-mixed mesic hardwoods and oaks/?
3. *Liriodendron tulipifera*-mixed mesic hardwoods, oaks, and hickories/?
4. *Fagus grandifolia*/?
5. *Fagus grandifolia*-mixed mesic oaks and hickories
6. *Quercus rubra-Quercus alba*/Mixed hardwoods/*Cornus florida*
7. *Carya tomentosa-Quercus alba*/Mixed hardwoods/Mixed hardwoods/*Viburnum rafinesquianum*
8. *Quercus alba*-mixed upland hardwoods/?
9. Mixed upland oaks/Mixed hardwoods/*Cornus florida*
10. *Carya glabra*-mixed upland oaks and hickories/?
11. *Quercus stellata*-mixed upland hickories and oaks/*Viburnum rafinesquianum*

of extended periods of standing water. The lower segment consists of a flat on the inside of a large curve in Bolin Creek. The ground surface of this flat does bear evidence of periodic but short-term flooding: low shrub and herb diversity and density and areas swept clean of leaf litter. This area is analyzed as a distinct Community Type. The upper segment occurs on slightly higher ground along a straighter stretch of Bolin Creek, just downstream from the flat. This segment shows little evidence of frequent flooding. A large part of this upper section has been cut over and now is dominated by a mixture of immature *Pinus taeda*, *Liquidambar styraciflua*, *Liriodendron tulipifera*, and minor numbers of other tree species. Portions of this forest close to the creek remain undisturbed and contain large, mature hardwood trees. This latter streamside community is analyzed as another distinct Community Type.

Both the lower flat and the upper terrace Community Types contain essentially the same canopy species (Table A2, p. 169), the major difference being variation in abundance of each species. Several species (*Liquidambar styraciflua*, *Acer rubrum*, *Quercus rubra*, *Q. phellos*, *Q. alba*, and *Carya tomentosa*) are present in roughly the same quantities in both Community Types. Two species (*Juniperus virginiana* and *Pinus taeda*) occupy the higher ground but not the lower flat. *Acer saccharum* ssp. *floridanum* (hereafter referred to as *A. saccharum*), *Nyssa sylvatica*, *Q. falcata* var. *pagodaefolia*, and a specimen that appears to be a hybrid between *Q. phellos* and *Q. falcata* var. *pagodaefolia* are found on the flat but not on the higher ground. *Fagus grandifolia* more commonly inhabits the flat than the higher terrace, whereas *L. tulipifera* and *C. glabra* more commonly inhabit the latter. These distribution changes appear to result from changes in moisture availability, the lower flat being wetter even though it is as well drained as the upper terrace. The upper terrace contains numerous small exposures of bedrock, suggesting that the ground around these patches of rock is covered with a thinner, rockier soil than the lower flat. This could account for the slightly drier aspect of the upper terrace canopy.

Overall, three species (*L. tulipifera*, *L. styraciflua*, and *A. saccharum*) are more abundant in these two Community Types than anywhere else in the Iron Mine Hill complex. Other species that are markedly lower in abundance on the valley floor include *Q. alba* and *C. tomentosa*. *Quercus rubra* occurs about as frequently on the valley floor as it does in most of the other Community Types in the complex.

The density and distribution of the species in these two Community Types result from the mesic environment provided by the sites' protection from adverse climate. This protection stems from this portion of the Bolin Creek valley being surrounded by 90- to 140-foot hills. The rich alluvial soils and the abundant moisture produced by the creek and from drainage off the surrounding hills aid in the development of this environment.

4. Fagus grandifolia/? Community Type and *5. Fagus grandifolia-mixed oaks-hickories Community Type.* These two communities, probably continuous at one time, presently are separated by a strip of forest of young pines, oaks, and hardwoods. *Fagus grandifolia* plays a dominant role in these two Community Types and in the adjacent valley floor Community Type (see Mixed mesic hardwood discussion, p. 168).

The *Fagus grandifolia* Community Type, found along a steep south-facing slope of Bolin Creek (Fig. A3, p. 158), grades abruptly upward into a mixed *Carya-Quercus* forest (these two genera dominate 87% of the canopy). The dominant species is *C. glabra* (see *Carya glabra*-mixed upland hickories and oaks Community Type discussion, No. 10, p. 177). *Fagus grandifolia* forms 53% of the canopy downslope and only 3% of the canopy upslope. This decrease occurs where slope degree changes abruptly and obviously marks a difference in erosive conditions and soils and moisture availability, possibly delimiting the upslope limit of the erosive control of Bolin Creek.

Two species, *F. grandifolia* and *Q. alba*, dominate the canopies of these two *F. grandifolia* Community Types. The *F. grandifolia* Community Type has importance values of 159 and 56 for the two species, respectively, and the *F. grandifolia*-mixed oaks-hickories Community Type has values of 139 and 62 for these two species, respectively. The only difference in the canopies of the two Community Types lies in the variation in minor canopy elements. In the small valley where the *F. grandifolia*-mixed oaks-hickories Community Type is found, *Carya* spp. become more abundant, whereas on the large south-facing slope of Bolin Creek and in the *F. grandifolia* Community Type, *L. tulipifera* and *A. rubrum* increase in importance.

The major differences in these two Community Types occur in the lower vegetational layers. The *F. grandifolia* Community Type, for which only the canopy was analyzed, has low-diversity, open, tall and short subcanopies, a sparse shrub layer of transgressive trees, and a sparse herb layer. On the other hand, the *F. grandifolia*-mixed oaks-hickories Community Type contains high-diversity, open, tall and short subcanopies, as well as high-diversity shrub and herb layers. All of these layers contain a mixture of upland, slope, and moist bottomland species (see Community Diversity Summary, No. 5, p. 202).

The *F. grandifolia*-mixed oaks-hickories Community Type occupies the narrow floor and lower, steep, southwest-facing slope of a small, south-flowing portion of a perennial tributary of Bolin Creek (see Fig. A3, p. 158). Hills ranging up to 100 feet above and surrounding the valley floor protect this narrow, relatively deep valley from adverse climatic conditions (winds, ice, and the like). The east-facing slope within this portion of the valley contains a few canopy-sized *F. grandifolia* on the lowermost 20 feet of the slope. This grades rapidly upward into a mixture dominated by *Q. alba*, *Q. rubra*, and *Carya* species. The canopy of the west-facing slope grades more gradually upward into a mixture containing a greater number of *Q. alba* and *Acer rubrum* and a lesser number of *F. grandifolia*. Upstream the valley widens, the slopes become gentler, the *F. grandifolia* decrease to only widely scattered individuals in the subcanopy, and the oaks and hickories become dominant (see No. 6, *Q. rubra-Q. alba*; No. 7, *C. tomentosa-Q. alba*; and No. 9, Mixed upland oaks Community Type discussions, pp. 172, 172, and 175). Downstream, after the creek changes direction and flows westward into the more open, finer-textured, flat floodplain of Bolin Creek, the forest changes into the cutover mixture of hardwoods already described in the Mixed mesic hardwoods discussion (Nos. 2 and 3, p. 168).

In the Iron Mine Hill Natural Area mature, canopy-sized *F. grandifolia* are restricted predominantly to the well-drained, mesic bottomland and lower slopes along Bolin Creek and along the more protected portions of its larger tributaries. Communities characterized by *F. grandifolia* abruptly change into the more typical oak-hickory communities of the Piedmont Uplands. Scattered canopy-sized individuals inhabit the more open, lower slopes and bottoms farther upstream along Bolin Creek and its tributaries. Isolated subcanopy-sized specimens inhabit the oak-hickory forests of the middle to upper slopes, and a few short subcanopy- and shrub-sized individuals inhabit the drier ridge crests dominated by *Pinus virginiana* and *P. echinata*. These specimens of *F. grandifolia* probably never will reach canopy height.

6. *Quercus rubra-Quercus alba/Mixed hardwoods/Cornus florida Community Type* and 7. *Carya tomentosa-Quercus alba/Mixed hardwoods/Mixed hardwoods/Viburnum rafinesquianum Community Type*. These two Community Types are associated closely and possibly could be considered segregates of a single, larger community, but because they are differentiated readily in the field and are controlled by different factors, they are treated separately. The two Community Types occur as zones alternately dominated by *Q. rubra* and *C. tomentosa* on an east-southeast-

facing slope along the same tributary of Bolin Creek as, but farther upstream from, the *F. grandifolia*-mixed oaks-hickories Community Type just discussed (Fig. A3, p. 158).

Along this slope the most obvious evidence that two Community Types exist is the difference in density in the shrub layers, both dominated by *Viburnum rafinesquianum* and *V. acerifolium*. Portions of the slope have small drainages running perpendicular to the overall slope direction and support a high-density shrub layer. In contrast, the flatter, interdrainage areas support a low-density shrub layer. By using this density difference as a means of recognizing different environmental conditions, each shrub area was analyzed separately. The resulting data (see Community Diversity Summaries, pp. 207 and 211) show corresponding changes in canopy and subcanopy composition between the two shrub zones.

Quercus rubra dominates the low-density shrub layers and *C. tomentosa* the high-density layers. *Quercus alba* is distributed evenly over the entire slope. A study of importance values (Table A2, p. 169) for *Q. alba*, *Q. rubra*, and *C. tomentosa* in these two zones shows that the values for *Q. alba* remain essentially the same, 78 versus 82. The values for *Q. rubra* and *C. tomentosa* become almost exactly reversed in the two Community Types, *Q. rubra* having values of 22 and 94 and *C. tomentosa* having values of 93 and 33, respectively.

Both of these Community Types, as stated above, occur on a gentle, east-southeast slope along a small tributary of Bolin Creek. Small, wet-weather drainages that feed perpendicularly into a stream at the base of the slope dissect this slope. The relief in these drainages ranges from only a few feet to over 15 feet. This dissection and relief give the slope a gentle undulatory appearance when viewed upslope. These small drainages control the distribution of *Q. rubra* and *C. tomentosa* and the density of the shrubs.

By comparing these two Community Types with the more xeric *Q. stellata*-dominated Community Type (No. 11, p. 178) farther upslope, one can resolve their environmental relationships. The *C. tomentosa*-*Q. alba* Community Type contains intermediate numbers of *C. tomentosa* and *Q. rubra* when compared to the *Q. rubra*-*Q. alba* and the *Q. stellata*-dominated Community Types. It contains fewer *Q. rubra* and more *C. tomentosa* than the *Q. rubra*-*Q. alba* Community Type and more *Q. rubra* and roughly an equal number of *C. tomentosa* as compared with the *Q. stellata*-dominated Community Type. The density of the dominant shrub, *Viburnum rafinesquianum*, in the *C. tomentosa*-*Q. alba* Community Type also lies intermediate between the other two Community Types, being more abundant in the *Q. stellata*-dominated

Community Type and less abundant in the *Q. rubra-Q. alba* Community Type. One other species that indicates changing environmental controls in this situation is *Cornus florida*. This species, which forms a closed, short subcanopy layer in the *Q. rubra-Q. alba* Community Type, has a scattered distribution in the *C. tomentosa-Q. alba* Community Type and is almost totally absent from the *Q. stellata*-dominated Community Type. When comparing the Community Diversity Summaries for these three Community Types (pp. 207, 211, and 223), similar trends in other species distributions can be detected.

These distributional trends demonstrate that the *Q. rubra-Q. alba* Community Type is more mesic than the *C. tomentosa-Q. alba* Community Type and that both are more mesic than the *Q. stellata*-dominated Community Type. The concentration of *C. tomentosa* and its associated tree and shrub species along the slopes of the small, wet-weather drainage systems suggests that these areas are better drained and less mesic than the flatter interdrainage divides where the *Q. rubra* concentrates.

The *Q. stellata*, *Q. coccinea*, and *Q. falcata* var. *falcata* (hereafter referred to as *Q. falcata*) located in these two Community Types mostly occupy the upper half of the slope, indicating a general decrease in moisture availability upslope. Because of the abrupt change from protected, mesic slope to unprotected, xeric hill top, this environment does not provide an opportunity for a Community Type composed predominantly of a mixture of these less mesic (but not xeric) oaks to develop. Instead, such a mixed upland oak Community Type is found upstream in an environmental situation intermediate between the two Community Types just discussed and the more xeric *Q. stellata* Community Type that occupies the exposed hill top above the *Q. rubra-Q. alba* and *C. tomentosa-Q. alba* Community Types.

One major concern in the maintenance of the Community Types on this slope is that, in the process of construction of a railroad line and a two-lane road along the base of the slope, the stream responsible for the development of the slope has been rerouted into a man-made channel. This channel, with steep banks up to 12 feet high, cuts into the base of the forested slope. This artificial movement of the stream into the slope has lowered the local base level of the small drainages feeding off the slope. This has caused, and will continue to cause, more rapid drainage in the small slope drainages and can hasten moisture loss and nutrient removal from the communities. It definitely will bring about more rapid erosion of the slope. The effects of these changes and of the increased openness of the Community Type at the base of the slope may not be apparent now, but over a long period of time they surely will alter severely the composition of the Community Types.

8. *Quercus alba-mixed upland hardwoods/? Community Type.* This Community Type occurs over weathered, fine-grained, acidic volcanic rocks rather than the coarse-grained, intermediate, plutonic diorite on which most of the other analyzed oak-and-hickory-dominated Community Types are found. The immediately noticeable difference in this Community Type, as compared with the others, is the lack of all *Carya* species except *C. tomentosa* and the large increase in the number of *Q. alba*. The lower base content of the underlying rocks and the resultant soils probably explain the absence of *Carya* species. The site is topographically higher on the valley slope than most of the other Community Types analyzed, thus making the site drier than the ones containing abundant *Q. rubra* and more mesic trees. Because of a series of shallow but very wide, wet-weather drainages in the site, enough moisture seems to be available to allow the *Q. alba* to outcompete the drier oaks. The overall even distribution of *Q. alba* over the community site, with a concentration of *Q. rubra* and *Liriodendron tulipifera* specimens in the center of the drainages and *Q. stellata* and *Q. falcata* on the interdrainage divides, supports this observation.

Variation in underlying lithologies supplies the most important difference in environmental factors. The weathered volcanics produce a better-drained, loamier, more acidic soil than that developed over diorite. These soil characteristics probably aid in the exclusion of the *Carya* species and allow better development of *Q. alba*.

9. *Mixed upland oaks/Mixed hardwoods/Cornus florida Community Type.* The Mixed upland oaks Community Type differs from the previous oak-and-hickory-dominated Community Types in the sense that the hickories have a much lower importance value in the canopy, whereas the more xeric oaks increase in value. Two major factors possibly explain this change: topographic position and slope profile. Topographic position represents the most obvious difference between the present Community Type and the others thus far discussed. The former lies close to the headwaters of a Bolin Creek tributary and occurs nearer the typical flat uplands than the previously described Community Types, which are found closer to larger streams and in more dissected terrain that offers increased protection from adverse climatic factors (Fig. A3, p. 158). This, plus the general concavity of the slope, tends to make the area drier and less favorable for the more mesic oaks, hickories, and other hardwood species.

Fagus grandifolia and *Acer saccharum* are both absent from this Community Type, and *A. rubrum* is present only in the lower vegetation layers. Seventy-five percent of the canopy consists of a mixture of *Quercus alba*, *Q. falcata*, *Q. velutina*, *Q. rubra*, and *Q. coccinea*. The remaining 25% consists of scattered individual *Pinus* and *Carya* specimens and of *L. tulipifera*. Several interesting segregations exist within this Community Type. *Quercus falcata* is more abundant on the drier, concave, upper slope than on the lower slope. Most of the *Carya* listed for the site grows on the lower portion of the slope, which opens onto a small, relatively flat, mesic area through which the stream flows. *Carya*, which develops best over intermediate to basic rocks (Radford, 1977, personal communication), possibly reduces in quantity at this site because of excessive leaching of nutrients from the upper, concave portion of the slope on which the Community Type is found. The accumulation of these nutrients on the lower slope through downslope drainage may allow scattered *Carya* trees to become established.

A variety of shrubs compose this Community Type, but they do not form a distinct layer as in the *Q. rubra*-*Q. alba*-, *C. tomentosa*-*Q. alba*-, or *Q. stellata*-dominated Community Types. *Viburnum* species occur as scattered individuals and as small clumps but do not develop nearly as well as in the aforementioned Community Types. The ericaceous shrubs, here represented by *Vaccinium vacillans*, *V. tenellum*, *V. stamineum*, and *Rhododendron nudiflorum*, are better developed at this site than at the others. The ericads are present as isolated populations up to 12 or more feet in diameter. Soil pits in the *Vaccinium* clumps reveal nothing unusual or different about the soil as compared with the areas lacking *Vaccinium*. The amount of light coming through the canopy and reaching the forest floor might control this patchy distribution because shrubs seem to inhabit the better-lighted portions of the Community Type.

Investigation of a *R. nudiflorum* clump, however, reveals a definite change in soil texture. The much lighter red soil under the *R. nudiflorum* has a much siltier, but less clayey, texture than the Ultic Hapludalf that dominates the site. Fragments of vein quartz in the soil under the *Rhododendron* suggest that this particular soil is localized and derived from a highly weathered quartz vein, whereas the Ultic Hapludalf is derived from diorite.

The larger oaks in the community are about 120 to 130 years old--approximately the same age as the specimens in the oak-hickory communities downstream. These Community Types therefore can be considered part of the same forest system. Nevertheless, it can be said safely that they do not represent seral stages of a single climax forest type but are, in themselves, climax communities in a complex forest system.

The pines that are found in the canopy of this community and the others thus far discussed are as large as the hardwoods but much younger--from 80 to 100 years old. It is interesting to note that the growth rate of some of the pines, determined from thickness of the tree rings in a core, has decreased by approximately half in the past 30 year. These pines probably established within the hardwood forest when the canopy opened, perhaps when a large hardwood tree fell. Pines from the surrounding cutover fields quickly seeded these open areas. The pine seedlings established themselves, grew rapidly, and quickly filled the canopy opening. As they approached canopy height, however, the surrounding mature hardwoods began to compete more directly with the pines. This increased competition, which started about 30 years ago in the Mixed upland oak Community Type, reduced the growth rate of the pines.

10. Carya glabra-mixed upland hickories and oaks/? Community Type. As with most Community Types in the complex, this one exhibits several features that distinguish it from the others. The major difference is the great increase in numbers of *Carya glabra* specimens, which constitute 25% of the canopy. These are three times more prevalent here than in any other Community Type of the complex and also represent the most abundant species in the canopy of this Community Type. *Carya tomentosa*, the dominant hickory in the oak-hickory forests found over diorite in this complex, is much less abundant, composing from one-half to one-third the importance value as it did in the other upland Community Types. The reason for this change is not known, although some basic factor controlling *Carya tomentosa* must differ at this site, either not allowing *C. tomentosa* to become established and thus allowing *C. glabra* to come in or allowing *C. glabra* directly to outcompete *C. tomentosa*.

The only readily observable major change in environmental factors is the topographic position of the Community Type. The Community Types in which *C. tomentosa* plays an important role occupy the smaller tributaries of Bolin Creek, mostly on more open, upper slopes. This Community Type, in which *C. glabra* occurs as an important canopy species, is found directly above the *F. grandifolia* Community Type discussed on page 171. These two Community Types stop and start at a definite break in a slope on the outside edge of a large curve in Bolin Creek, with the lower slope being steeper than the upper slope. This slope change together with the presence of a larger stream downslope possibly can create a set of conditions more favorable for *C. glabra* than for *C. tomentosa*, but these critical conditions are not readily discernible.

11. Quercus stellata-mixed upland hickories and oaks/Viburnum rafinesquianum Community Type. The *Quercus stellata*-mixed upland hickories and oaks Community Type occupies an undisturbed site on the upper slopes and hill crest above the previously discussed Community Types of *C. tomentosa-Q. alba* and *Q. rubra-Q. alba* (Fig. A3, p. 158). It grades rapidly into these other two Community Types on the east slope of the hill and is bordered on its south and north sides by a *Pinus virginiana-P. echinata* forest, which is only 30 feet wide on the north side. Beyond this is an old field now densely overgrown with young pines. These pine stands are much younger than the hardwoods in this Community Type and have grown up in old cutover areas. The pine forest contains abundant young hardwoods that eventually will replace the pines. The *Q. stellata*-mixed upland hickories and oaks Community Type continues westward along the hill top and upper slopes as a strip a hundred or more feet wide running through the surrounding pines.

This Community Type is notably older than the others in the complex, yet the trees are markedly smaller. The oldest trees, which are about 250 years old, are only half the diameter of the larger 125-150-year-old trees downslope in the *Q. rubra*, *Q. alba*, and *C. tomentosa* Community Types. Apparently the *Q. stellata*-dominated Community Type was left undisturbed when at one time the entire area, including the land downslope, was cleared for farming or timber. The large amount of rock exposed on the surface would make farming impossible. The shortness of the trees and the presence of branches all the way up the trunks possibly rendered the wood unsuitable for lumber. Overall poorer soil, lesser amounts of soil moisture, increased leaching of soil nutrients, and increased rockiness produce less favorable growing conditions.

Within the *Q. stellata*-dominated Community Type, large, semiburied boulders of amphibole-rich diorite cover most of the ground. Where these boulders are densest (associated with the shallowest soil), very little shrub growth appears. Where the boulders are fewest (associated with deeper soils) the shrubs, dominated by *Viburnum rafinesquianum*, are dense enough to make walking through the thicket difficult.

The trees in the Community Type have the general appearance of being in an open, exposed area because the canopy is open. Many branches of the canopy trees are within arm's reach, whereas in the other climax Community Types pronounced branching does not occur much below 30 feet. The trees in the Community Type are stunted, being about 40% shorter than the canopy trees downslope. Tree ring spacing, an average of 22 per inch as compared to 10-12 per inch for the trees downslope, shows that growing conditions on top of the hill are much less favorable than downslope. A subcanopy is virtually absent, with typical subcanopy species lacking in all of the vegetational layers. The overall aspect is one of openness, exposure, and poor growing conditions (poor in the sense of the poorest of the Community Types analyzed for this report).

Quercus stellata and *C. tomentosa*, the two dominant trees in this Community Type, most often are associated with dry growing conditions (Ashe, 1897; Radford et al., 1968). Other trees (such as *Q. marilandica*) which develop best in harsh, xerophytic situations, appear in the canopy of this Community Type. Some plants (such as *Viburnum rafinesquianum* and *Juniperus virginiana*), which develop best over intermediate to basic rocks (Radford et al., 1968), are common to abundant in this Community Type. Many trees typically found in moister situations, such as *Q. rubra*, *L. tulipifera*, *A. rubrum*, *A. saccharum*, and *F. grandifolia*, are either present in much smaller numbers or are totally absent. The shrubs respond likewise. *Viburnum acerifolium*, present in significant numbers downslope, is missing totally from this Community Type.

Carya ovata and *Viburnum prunifolium*, however, illustrate two discrepancies in this overall distributional trend. These two species most often are associated with low, wet to mesic environments (Radford et al., 1968). In this complex, *Carya ovata* occurs in two different situations: this species is well developed in the wet area at the base of the Mixed upland oaks Community Type (No. 9, p. 175); it also grows in the driest Community Type in the complex. *Viburnum prunifolium* is abundantly present along the bottomlands of Bolin Creek yet also occurs in this Community Type. In fact, several individual *V. prunifolium* specimens look more like small trees than shrubs. Why these two species exist in these two very different environments, yet are essentially absent in the intermediate Community Types, remains unanswered.

B. Zoological Diversity

No work has been done on the zoological diversity of the Iron Mine Hill Natural Area. Casual observations suggest that with the wide diversity of habitats, ranging from open fields and young pine lots, to mature pine forests and hardwood forests, to aquatic habitats, the area should provide habitats for a wide diversity of animal forms. Abundant bird and small mammal life was observed, although no rare species were encountered. Many small streams and wet areas, along with the abundant rock outcrops, supply numerous suitable habitats for reptiles and amphibians. The water in Bolin Creek is pure enough to support fish life and numerous invertebrate forms. Because the numerous fields and forests in different successional stages and the varied types of forest litter furnish a wide variety of habitats, insect and other invertebrate life should be highly diverse.

III. PIEDMONT HARDWOOD DIVERSITY

No comprehensive vegetational analysis taking into consideration all possible abiotic factors controlling vegetation of the hardwood forests in the Piedmont Province of the Appalachian Highlands has been undertaken yet. Numerous publications deal with Piedmont vegetation, and some of these have been regarded as comprehensive studies. Oosting (1942), for example, usually is quoted as the definitive work on the Piedmont, yet this work treats only 5,000 acres of forest land in the Duke Forest (Durham and Orange counties, North Carolina). Oosting homogenizes many of the local Community Types into generalized oak-hickory associations, justifying this homogenization as necessary when analyzing a forest from a forester's viewpoint, for lumbering purposes. Braun (1950) characterizes the Piedmont Province as the Atlantic Slope Section of her oak-pine forest region. She essentially summarizes the literature to the date of her publication and adds her personal observations. She, too, however, tends to homogenize the forests into upland oak-hickory forests, slope forests characterized by mixed hydrophytic hardwoods, and bottomland forests containing mixed hydrophytic and mesophytic hardwoods. Ashe (1897) probably made the most comprehensive survey of forest canopy vegetation in the Piedmont of North Carolina. He includes a tremendous amount of information on lithologic, pedologic, topographic, and hydrologic control of most of the common Piedmont canopy species, and he is able to recognize the association between changes in species associations and changes in certain abiotic factors. Radford et al. (1968) include ecological data on individual species, but, because of the nature of their publication, community analyses are not provided. Only within the last 13 years have articles concerned specifically with analytical studies of the Piedmont hardwood communities and the multitude of ecological factors that control the communities appeared (for example, Dayton, 1966; Nemeth, 1968; Sechrest and Cooper, 1970).

This literature reveals, and the analyses for this report further emphasize, that the Piedmont has complex geological, pedological, topographical, and hydrological variations. With these complexities come intricately mixed factors that control moisture and nutrient availability, provide protection from adverse climatic conditions, and ultimately develop an even more complex vegetational assemblage. This study and previous studies have just begun to touch upon the unraveling and understanding of these complexities, such as associations between various species and numerous abiotic environmental controls. Much work still needs to be done to determine habitat requirements for most species. An effort is being made in the ecosystematics program at the University of North Carolina at Chapel Hill to discover the relationships between community vegetational composition and distribution and abiotic factors. To facilitate this study the literature regarding the major species associations and what factors influence the distribution of these associations must be reviewed and summarized.

Ashe (1897) divides the original forest lands of the North Carolina Piedmont into three parallel belts. The Eastern Pine Belt is covered with soils derived from slates, sandstones, and gneisses, and it originally supported a forest containing a large proportion of pine. The middle belt, the Broadleaf Forest Belt, in which the Iron Mine Hill Natural Area is located, is covered with deep, loamy soils mainly derived from granitic rocks; it originally supported hardwood forests of first quality, with only a small percentage of pine present. The Western Pine Belt contains extensive areas of gneissic soils underlying a forest of smaller-sized hardwoods and more pines than the middle belt.

Ashe (1897) further subdivides these three belts into smaller areas, each characterized by specific soils derived from specific underlying lithologies. The soils and lithologies that Ahse (1897) found to dominate one particular region occur in smaller quantities in the other regions. Therefore, the major plant associations he found over the different soils and lithologies can be considered by examining only one of his three forest belts. The Iron Mine Hill Natural Area is located within the Broadleaf Forest Belt, and this belt contains the best-developed hardwood forests; therefore, this discussion concentrates on it.

Within the Broadleaf Forest Belt, Ashe (1897) distinguishes between upland forests growing in stiff red soils originating from hornblende-bearing rocks (Alfisols) and upland forests growing in loose gray loams originating from rocks containing large amounts of quartz (Ultisols). The following chart, derived from Ashe, lists the distribution of the dominant hardwood canopy species in relation to changes in topographic position and in soils.

	Stiff Red Clays	Loose Gray Loams
Ridge crests	*Quercus stellata*	*Quercus prinus*
Upper slopes (species listed in decreasing importance)	*Quercus velutina* *Quercus alba* *Carya tomentosa* *Carya ovalis* *Quercus stellata*	*Quercus alba* *Quercus falcata* *Quercus stellata* *Quercus velutina* *Carya tomentosa* *Quercus coccinea*
Lower slopes and steep north-facing slopes	species listed above in addition to: *Liriodendron tulipifera* *Quercus rubra* *Fraxinus americana* *Carya ovata*	species listed above in addition to: *Liriodendron tulipifera* *Fraxinus* spp. *Acer rubrum* *Quercus rubra*

Occasionally, abrupt changes in community composition may occur with abrupt changes in topography, soils, lithology, moisture availability, and/or exposure. In most cases, however, a zone of mixing (ecotone) joins adjacent communities because many environmental factors exhibit gradational change. In the above list, the canopy species that are most abundant on the upper slopes extend onto the lower slopes but generally decrease in importance, while the species most abundant on the lower slopes and steep north-facing slopes typically characterize these topographic situations and usually dominate the community. These lower-slope species also occur further upslope where localized microhabitats allow establishment, but they compose a lesser percentage of the canopy.

Ashe (1897) also describes what he calls the forests of the Piedmont Lowlands, these confined mostly to narrow strips of sediment deposited along streams. He found very few broad floodplain swamps in the Piedmont such as those typical of the adjacent Coastal Plain. Hardwoods dominate the climax lowland forests except along the eastern border of the Piedmont, where *Pinus taeda* occasionally forms a dominant portion of the canopy. Ashe differentiates six major lowland environments, with canopy species characteristic of each:

1. Open stream banks in full sunlight, usually lined with *Betula nigra* and *Salix nigra*.

2. Flats subject to frequent and periodic overflow, generally covered with compact growths of *B. nigra* and *S. nigra*, or flats in areas of prolonged inundations, with *Fraxinus* and *Ulmus* species.

3. Occasional broad flats in the eastern Piedmont, with soils that remain

moist or even wet but are inundated rarely, usually dominated by mixtures of *Quercus nigra*, *Q. lyrata*, *Q. phellos*, and fringed by *Q. alba* and *Q. velutina*.

4. Flats adjacent to larger and more slowly flowing streams usually containing silty and mud alluvium, with little organic matter (indicating good drainage), supporting stands of *Liquidambar styraciflua*, *Nyssa sylvatica*, *Carya cordiformis*, *Q. lyrata*, *Q. michauxii*, *Platanus occidentalis*, *Celtis occidentalis*, *Q. phellos*, *Q. nigra*, *Q. shumardii*, and *C. ovata*.

5. Hollows and borders of small streams, with sandy loams containing a large proportion of organic constituents, supporting a distinctive canopy association of *Fagus grandifolia*, *Q. rubra*, *Q. alba*, *Acer rubrum*, *A. saccharum*, and *Liriodendron tulipifera*. As these smaller streams open out into larger stream valleys or the soil becomes increasingly siltier, the trees common to the larger flats (No. 4, above) become more conspicuous. The trees normally on the sandy loam slowly drop out, *F. grandifolia* being the first to disappear, followed by *Q. rubra*, then *Q. alba*, and finally *L. tulipifera*.

6. Ashe describes an additional "lowland" type, consisting of mud or clay depressions on the crests of ridges and usually occurring in areas of shallow soils, often where slates constitute the country rock. These areas remain wet for some time after rainy weather (suggesting a perched water table) because little subsurface drainage takes place. During summer and autumn these areas become exceedingly dry. The depressions contain a curious mixture of *Q. phellos*, *Q. marilandica*, and *Q. stellata*.

With additional work, the species characteristic of each of Ashe's environments could be differentiated further by acquiring more detailed habitat requirements. Most forest communities contain a mixture of species. One species often dominates the canopy of a particular community, but rarely does a climax community consist of only one species. Early successional communities frequently are composed of dense stands of a single species, but by the time a community has matured, both biotically and abiotically, many microhabitats usually will have developed, enabling different species to establish themselves in their own particular niches. For example, *Quercus alba* plays a very conspicuous role in most of the Piedmont upland hardwood communities. In some sites, were it not for the presence of occasional microhabitats suitable for other species, it very possibly could form solid, mature stands. Small drainages within the community support *L. tulipifera* and *N. sylvatica*, whereas deeper, moister soils can support specimens of *Q. rubra*. Shallow, rocky soil permits a few representatives of *Q. coccinea* and

Q. falcata to grow. Pockets of more basic soil developed from soil richer in hornblende and/or plagioclase, and other related factors, allow *Carya* spp. to become established. If a large area exhibits all the environmental characters capable of controlling species diversity uniformly throughout the area, one would expect to find extensive stands of the species best suited to that particular set of characters. The great diversity within the abiotic portion of nature, and the capacity of this diversity to appear within very short distances, however, allow and almost force an equally diverse biotic situation in nature.

In a study of forest communities of Duke Forest, a part of Ashe's Broadleaf Forest Belt, Oosting (1942) finds several major climax community associations. In some aspects these consolidate Ashe's associations, whereas in other aspects they differentiate Ashe's communities into finer divisions. Oosting finds one association commonly dominated by *Q. stellata* on sites with inferior exposures, on drier ridges and knolls, and over poorer soils. *Quercus marilandica*, *C. glabra*, and *C. tomentosa* also are associated with these environments. *Quercus alba* is of minor importance, and *Q. velutina* is usually absent. Small numbers of *Q. rubra*, *Q. coccinea*, and *Q. falcata* are sometimes present.

The more favorable upland sites are dominated by mixtures of *Q. alba*, *Q. velutina*, *Q. rubra*, and various *Carya* species (not differentiated by Oosting). These sites also could contain the other oaks and hickories in lesser numbers and also occasional specimens of more mesic species, such as *Nyssa sylvatica*, *A. rubrum*, and *L. tulipifera*. Oosting (1942) finds south-facing bluffs very dry and populated by open stands of *Q. stellata*, *Q. marilandica*, *Juniperus virginiana*, and *Pinus virginiana*. North-facing slopes, however, differ substantially. The bluff tops support vegetation related to the surrounding upland forest, typically a *Quercus-Carya* mixture. The bluff's talus slopes and lower slopes, frequently overlooking a stream, support a more mesic mixture, containing *Q. phellos*, *Q. prinus*, *Q. alba*, *Q. rubra*, upland *Carya* spp., *C. ovata*, *C. cordiformis*, *L. tulipifera*, *P. occidentalis*, *F. grandifolia*, *A. saccharum*, *A. rubrum*, *Juglans nigra*, and *Prunus serotina*.

The lowland communities analyzed by Oosting (1942) contain mixtures of mesic trees, usually with specimens of *Q. phellos*, *L. styraciflua*, and *Q. falcata* more numerous than those of the other species present, including *Q. alba*, *P. taeda*, *C. ovata*, *Q. lyrata*, *N. sylvatica*, *C. carolinae-septentrionalis*, *A. rubrum*, *A. saccharum*, *Celtis occidentalis*, *Q. velutina*, *Q. stellata*, and *Ulmus americana*.

As does Ashe, Oosting lists certain associations that, with more detailed analyses and differentiation into smaller, more distinct associations, could have resulted in more specific Community Types. If specific controlling factors for individual species had been observed in greater detail, a much better understanding of these distributions would have resulted.

Braun (1950) generally agrees with Oosting's divisions of the upland forests. She further divides Oosting's bottomland forests into (1) typical streamside environments with *B. nigra*, *S. nigra*, *Populus deltoides*, *Platanus occidentalis*, and *L. styraciflua* being most abundant; and (2) old, wide, bottomland flats containing *L. styraciflua*, *Q. phellos*, *U. alata*, *U. americana*, *A. rubrum*, *L. tulipifera*, *Fraxinus* spp., *Q. nigra*, and *Celtis laevigata*.

The hardwood canopy species of the Piedmont can be arranged along numerous gradients, including moisture (wet to dry), soil orders (Entisols to Ultisols), rock chemistries (acidic to basic), soil textures (clayey to sandy), topographic positions (lowlands to uplands), and many others. Detailed quantitative work is required to place different species into their proper position along most of these gradients. This work has not been done as yet. Generalizations, but not exact positionings, can be made for many of these gradients. The topographic gradient, however, has been studied, and species distributions along this gradient are readily observable in the field. Most major works on the Piedmont, including Ashe (1897), Oosting (1942), Braun (1950), Dayton (1966), Nemeth (1968), and Sechrest and Cooper (1970), concentrate on this topographic gradient. Thus arranging the more common Piedmont canopy trees along a topoenvironmental gradient with some degree of confidence is possible.

Nine distinct yet intergrading topoenvironments are recognized here and with additional detailed work these nine easily could be subdivided. As an example, see the section on Swift Creek in this publication (p. 229) for further subdivision of floodplain swamp communities along Swift Creek. The hardwoods listed in Table A3 (p. 186) include the most common trees in these nine environments. The environments are:

1. Bottomland along streams in which the soil usually is saturated with standing or slowly flowing water for much of the year. This could be regarded as a swamp environment.

2. Bottomland along streams with soil that is saturated infrequently but is generally wet most of the year and poorly drained.

Table A3. *Topoenvironmental distribution of hardwood trees in the Piedmont of North Carolina*

	Topoenvironment*								
Species	1	2	3	4	5	6	7	8	9
Populus deltoides	X								
Ulmus alata	X	X							
Quercus michauxii	X	X	X						
Quercus lyrata	X	X	X						
Liquidambar styraciflua	X	X		X					
Acer rubrum	X	X	X	X					
Quercus falcata var. *pagodaefolia*	X	X	X	X					
Nyssa sylvatica	X	X		X	X				
Quercus shumardii	X	X	X	X	X				
Ulmus americana	X	X	X	X	X				
Quercus nigra		X	X						
Quercus laurifolia		X	X						
Betula nigra		X	X						
Salix nigra		X	X						
Acer saccharinum		X	X						
Quercus phellos		X	X	X					
Platanus occidentalis		X	X	X					
Fraxinus americana		X	X	X	X	X			
Acer negundo			X						
Oxydendrum arboreum			X	X	X	X			
Ulmus rubra				X	X				
Magnolia acuminata				X	X				
Magnolia tripetala				X	X				
Fraxinus pennsylvanica				X	X				
Morus rubra				X	X				
Celtis laevigata				X	X				
Juglans nigra				X	X				
Carya cordiformis				X	X				
Fagus grandifolia				X	X	X			
Liriodendron tulipifera				X	X	X			
Acer saccharum				X	X	X			
Quercus alba				X	X	X	X	X	
Carya tomentosa				X	X	X	X	X	X
Tilia heterophylla					X	X			
Quercus rubra					X	X	X		
Carya carolinae-septentrionalis					X	X	X		
Carya ovata					X	X	X	X	X
Carya glabra						X	X	X	
Quercus velutina							X	X	
Quercus coccinea							X	X	

Table A3. *(continued)*

Species	Topoenvironment* 1	2	3	4	5	6	7	8	9
Quercus falcata var. *falcata*							X	X	
Carya pallida							X	X	X
Carya ovalis							X	X	X
Celtis occidentalis								X	X
Quercus marilandica								X	X
Quercus stellata								X	X
Quercus prinus								X	X

*Key to types of topoenvironment:

1. Swamps--saturated soils.
2. Wet, poorly drained bottomland.
3. Stream margins, swamp margins.
4. Low, rich, well-balanced bottomland.
5. Lower slopes along large streams, hollows, and borders of small streams.
6. Midslopes.
7. Well-drained upper slopes of large streams, upper reaches of small streams.
8. Interstream divides, flat uplands.
9. Ridge crests and rocky hill tops.

3. Stream and swamp margins in more open sunlight than the rest of the surrounding bottomlands. Because these areas usually are associated with a break or change in slope, they are often better drained than the surrounding forest. A slightly coarser texture than the surrounding soils frequently characterizes the soil of these areas, especially if a levee borders the stream margins.

4. Low, rich bottomlands, sometimes seasonally flooded but generally well drained, and underlain by a moist but not saturated soil. Generally these occur on upper terraces of larger streams and sometimes include the entire floodplain of small, high-gradient streams. On these floodplains local depressions (such as oxbows) reach or approach the water table and support communities similar to those found in (1) or (2).

5. Lower slopes, especially north-facing slopes along larger streams, plus the borders of small streams and sometimes the entire slope of the surrounding small valley. These valleys, although usually well drained, remain moist and

nutrient-rich because they accumulate water and nutrients moving downslope from adjacent uplands. Situated in the valleys of the Piedmont, these areas are protected from the full brunt of many adverse climatic factors.

6. The midportion of most Piedmont valley slopes, often acting as a transition zone between lower and upper slope communities. This slope position frequently supports its own distinctive community, composed of a mixture of elements from two different communities usually easily distinguished. The abundance of particular microhabitats usually determines the locally dominant species.

7. Open, somewhat protected, well-drained upper slopes along larger streams and open, gentle slopes along the upper reaches of small upland streams. These environments often support similar communities because of the overall corresponding controlling factors.

8. Very broad, relatively flat interstream divides between large streams and uppermost, very gently sloping to nearly level upper reaches of smaller streams. Such areas are drier and more exposed than those discussed in (7). In the eastern Piedmont, fine-grained metavolcanic and metasedimentary rocks often underlie these extensive upland flats. Weathering of these rocks produces soils that are often very iron-rich and relatively dry clay loams. Perched water tables frequently develop in the more clay-rich soils wherever shallow depressions occur; in these areas a distinctive community can develop (see discussion concerning Ashe, 1897, on p. 181).

9. Rocky ridges, south-facing bluffs, and hill tops. These represent the most xeric topoenvironments in the eastern Piedmont. They often are exposed, open to adverse climatic conditions, well to excessively drained, and very rocky. They frequently are dominated by species that are not usually a major component in any of the other environments.

As can be seen in Table A3 (p. 186), a large number of species occur in each of these topoenvironments, and each species is usually present in several of the environments. Every species listed for a given environment usually does not grow at any given site. The species composition found in a particular topoenvironment changes as other controlling factors change, for example, nutrient availability, soil texture, depth of soil, amount of exposed rock, and drainage patterns. Ideally, each species adapts best to a particular range for each of its controlling ecologic factors. Each species listed in Table A3 becomes a probable canopy dominant in only one or two of its listed topoenvironments--the ones that have controlling factors most favorable to it. The other environments in which the species occur contain small, localized microenvironments in which the set of controlling factors is favorable for the maturity of the species.

Clearly, position along a topographic gradient is not an entity in itself but rather is one of many factors that aid in controlling species distributions. Topographic positioning constitutes a readily apparent, external consequence of a large number of additional, intricately interrelated vegetational controls encompassing many facets of climate, geology, hydrology, and pedology. Much additional work is necessary before a complete understanding of the many environmental factors that control plant and animal distributions and densities can be attained. The investigation undertaken for this report hopefully will provide one additional step toward that understanding.

IV. CONCLUSIONS

The Iron Mine Hill Natural Area shows definite significance on a regional level and perhaps on a national level. The area is in some jeopardy and could be destroyed with expansion of the Horace Williams Airport and/or expansion of the nearby cities of Chapel Hill and Carrboro. Although no threatened or endangered plant species inhabit the complex, the entire complex might be considered an endangered entity. Much of the Piedmont Province has been or is being cut over for a variety of reasons, and many areas have been cut several times. A mature, relatively undisturbed complex of climax hardwood Community Types is becoming increasingly rare in the Piedmont. Such a Community Type complex, when discovered, should be set aside and preserved so that the natural diversity of this region can be recognized, studied, and more fully understood.

This area displays exceptional botanical diversity. Thus far, the climax hardwood forest Community Types contain 34 species of angiosperms and gymnosperms in the canopy and subcanopy. This total includes almost 70% of the canopy species commonly found in the Piedmont of North Carolina (Table A3, p. 186). These species segregate into numerous Community Types composed of different combinations and concentrations of trees, shrubs, and herbs. These segregations result from variations in moisture, nutrients, and local climate, which are in turn controlled by variations in topography, soils, and underlying lithologies.

The natural area is relatively small (3.0 square miles) and contains land ranging from recently cleared fields to mature, climax hardwood forest. Natural pine lots range from recently invaded fields to pine stands containing trees well over 100 years old. This range of successional stages in such a small area allows for intense, detailed successional studies over a diversity of soils, topography, and lithologies.

The proximity of the natural area to the Triassic Durham Basin, now a topographic lowland, resulted in the development of a terrain that ranges from highly dissected to relatively flat upland over a very short distance. The area is situated geologically in the Slate Belt of the North Carolina Piedmont. Within the area rock types differ tremendously: (1) acidic to ultramafic plutonic igneous complexes; (2) fine-grained, water-deposited tuffs to volcanic breccias containing clasts up to 0.5 foot in diameter; (3) metamorphics ranging from slates to chlorite schists; and (4) a small iron deposit. This topographic and lithologic diversity causes the development of a wide variety of soils. The area contains four soil orders and 16 soil series, representing 77% of all the soil series presently recognized in Orange County.

The biotic and abiotic diversity present in this area makes the land worthy of preservation. The numerous uses to which the land now is being subjected require that a variety of management recommendations be proposed. First, and most important, the mature hardwood complex should be left undisturbed to preserve a part of the hardwood diversity of the Piedmont Province. With its complex changes in species distributions and concentrations, this area offers a great opportunity to investigate the microhabitat requirements of many of the Piedmont trees and shrubs. Second, the area is excellent for conducting successional studies over a variety of soils, lithologies, and landforms. The boundaries of the natural area are drawn purposefully to include roads and actively farmed land (Map A1, p. 150). To understand all stages of succession, one must see how the land was disturbed originally and how this disturbance injured or aided the original recolonization by natural components of the biosphere.

Controlled investigations are very important for a thorough understanding of succession on disturbed land. This natural area contains a wide variety of successional stages. Perhaps the entire area should be left undisturbed and allowed to reach a climax stage, thus allowing close monitoring of the successional stages. The final result would yield a much clearer picture of succession in the eastern Piedmont of North Carolina.

To develop this area into a truly scientific and educational facility, additional recommendations need to be made. Plant and animal colonization of disturbed land depends not only on how the land was disturbed and for how long but also on what lived on the land before that time and what lived adjacent to it at the time of colonization. Much of this natural area still is owned privately; some of it still is farmed, and portions of it still are cut. A fantastic opportunity exists here to establish a research natural area where disturbances still are permitted but closely monitored. This natural area offers an opportunity to study factors that control succession.

It is recommended that the remaining portions of the natural area not yet biotically and abiotically analyzed be studied, with the landowners' permission, so that a comprehensive data base for the area can be established. Future disturbances could be monitored and successional studies carried out in a more meaningful manner than ever before. This type of study would lead to a much better understanding of succession and what controls it over a long period of time.

Bibliography

Ashe, W. W. 1897. Forests of North Carolina. Pages 139-224 *in* G. Pinchot and W. W. Ashe. Timber trees and forests of North Carolina. North Carolina Geological Survey Bulletin 6. Raleigh.

Braun, E. L. 1950. Deciduous forests of eastern North America. Hafner Press, New York.

Dayton, B. R. 1966. The relationship of vegetation to Iredell and other Piedmont soils in Granville County, North Carolina. J. Elisha Mitchell Sci. Soc. 82: 108-18.

Dunn, J. 1977. Soil survey of Orange County, North Carolina. Soil Conservation Service. U.S. Department of Agriculture.

Mann, V. I., T. G. Clarke, L. D. Hayes, and D. S. Kirstein. 1965. Geology of the Chapel Hill Quadrangle, North Carolina. North Carolina Geological Survey Special Publication 1. Raleigh.

Nemeth, J. C. 1968. The hardwood vegetation and soils of Hill Demonstration Forest, Durham County, North Carolina. J. Elisha Mitchell Sci. Soc. 84: 482-91.

Oosting, H. J. 1942. An ecological analysis of the plant communities of the Piedmont, North Carolina. Amer. Midl. Naturalist 28:1-126.

Radford, A. E., H. E. Ahles, and C. R. Bell. 1968. Manual of the vascular flora of the Carolinas. The University of North Carolina Press, Chapel Hill.

Sechrest, C. G., and A. W. Cooper. 1970. An analysis of the vegetation and soils of upland hardwood stands in the Piedmont and Coastal Plain of Moore County, North Carolina. Castanea 35:26-57.

Stuckey, J. L. 1965. North Carolina: Its geology and mineral resources. North Carolina State University Print Shop, Raleigh.

Natural Area Diversity Summary

CLIMATE: A. Mesothermal; AA. Warm temperate. B. Cool and moderately moist yearly, moderately hot and moderately wet summers, moderately warm and moderately dry winters, with an average freeze-free period; BB. Cool and moist to moderately moist yearly, very hot and moderately wet summers, moderately cold and moderately wet winters, with an average to long freeze-free period. C. Similar to local climate; CC. Similar to natural area climate.

SOILS: A. (1, 8) Ultisol, A. (2-7, 9-11) Alfisol; AA. (1, 8) Udult, AA. (2-7, 9-11) Udalf. B. (1, 8) Hapludult, B. (2-7, 9-11) Hapludalf; BB. (1) Aquic Hapludult, BB. (8) Typic Hapludult, BB. (2-5) Typic Hapludalf, BB. (6, 7, 9-11) Ultic Hapludalf. C. (1) Clayey, mixed, thermic Aquic Hapludult, C. (8) Clayey, kaolinitic, thermic Typic Hapludult, C. (2-5) Mixed, thermic, shallow Typic Hapludalf, C. (6, 7, 9-11) Fine, mixed, thermic Ultic Hapludalf; CC. (1) Lignum silt loam, CC. (8) Herndon silt loam, CC. (2-5) Wilkes gravelly loam, CC. (6, 7, 9-11) Enon loam.

GEOLOGY: A. (1-3, 8) Metamorphic, A. (4-7, 9-11) Igneous; AA. (1, 8) Mixed metasedimentary and metavolcanic, AA. (2, 3) Metavolcanic, AA. (4-7, 9-11) Plutonic. B. (1-3, 8) Acidic, B. (4-7, 9-11) Intermediate; BB. Water-deposited sediments and fine volcanics, BB. (2, 3) Undetermined origin, BB. (4-7, 9-11) Pluton. C. (1-11) Mixed plutonic, metasedimentary, and metavolcanic rocks of varying composition and degree of exposure; CC. (1) Folded, fine-grained, tuffaceous slates and volcanic tuffs with no surface exposure, CC. (2) Folded, rhyolitic, welded, tuffaceous breccia buried under recent fluvial sediments, CC. (3) Folded, welded, tuffaceous rhyolitic breccia exposed as scattered, fresh boulders but mostly buried under recent fluvial sediments, CC. (4) Diorite exposed as fresh, widely scattered, large boulders, CC. (5) Diorite exposed as large, scattered, fresh, isolated boulders and piles of boulders, CC. (6, 7) Plagioclase porphyritic diorite occurring as very scattered, isolated, highly weathered, small boulders, CC. (8) Folded, tuffaceous slates and rhyolitic tuffs not exposed on the surface, CC. (9) Undeformed, plagioclase porphyritic diorite exposed as weathered, small- to medium-sized boulders concentrated in large patches at the top and bottom of the slope, CC. (10) Diorite exposed as isolated, fresh, large boulders, CC. (11) Undeformed, amphibole porphyritic diorite exposed as small to large, weathered, isolated, and piled boulders.

HYDROLOGY: A. Terrestrial; AA. (1-3) Wet, AA. (4, 6-11) Dry-mesic, AA. (5) Dry-mesic and wet. B. (1-11) Fresh; BB. (1-3) Temporarily saturated, BB. (4, 6-11) Permanently exposed, BB. (5) Permanently exposed and temporarily saturated. C. All the hydrologic characteristics listed in CC.; CC. (1) Moderately well drained, wetted by a seasonally perched water table, occasional flooding, and rains, CC. (2, 3) Well drained, wetted by rains, a seasonally perched water table, and occasional flooding by Bolin Creek, CC. (4, 6-8, 10) Well drained, wetted by rains and downslope drainage, CC. (5) Well drained, some portions wetted by rains and downslope drainage and other portions additionally wetted by a seasonally perched water table and occasional flooding by a small stream, CC. (9) Well drained, wetted by rains and slight downslope drainage, CC. (11) Well drained, wetted by rains.

TOPOGRAPHY: A. Plains; AA. Irregular plain to irregular plain with slight relief. B. Low hills and shallow to relatively deep valleys; BB. (1-5, 9) Water-eroded stream valley, BB. (6-8, 10, 11) Water-eroded hill. C. All the topographic characteristics listed in CC.; CC. (1) Water-deposited, slightly sheltered valley flat containing abandoned channels along a small, perennial, southwest-flowing stream, CC. (2) Sheltered, smooth, level flat on the inside of a curve of a northwest-southeast-flowing stream, CC. (3) Sheltered, irregular to rocky, level valley floor along a northwest-southeast-flowing stream, CC. (4) Sheltered, south-facing, irregular to rocky, constant, strongly sloping, lower slope along a northwest-southeast-flowing perennial stream, CC. (5) Sheltered, small, flat valley floor containing exposed rock, small pools, and abandoned channels and gentle to very steep, convex, west-facing, rocky lower- to midslope along a small, perennial south-flowing stream, CC. (6) Open, east-southeast-facing, gently sloping, irregular to undulatory lower- and midportion of a constant hill slope, CC. (7) Open, east-southeast-facing, undulatory, gentle, constant midslope of a north-south-oriented hill facing a north-south-flowing perennial stream, CC. (8) Open, undulatory, southeast-facing, constant upper slope of a low, north- to south-aligned hill, CC. (9) Open, west-southwest-facing, smooth to bouldery, gentle, concave, lower slope along the upper reaches of a small, perennial stream, CC. (10) Open, irregular to slightly rocky, south-southwest-facing, constant, gentle midslope of an overall convex hill slope overlooking a relatively deep, protected, northwest-southeast-oriented small stream valley, CC. (11) Open, variously exposed, bouldery, flat but gently sloping hill crest and gentle, convex, extreme upper slopes.

PHYSIOGRAPHY: A. Appalachian Highlands; AA. Piedmont Province. B. Piedmont Uplands; BB. Iron Mine Hill complex. C. (1, 4, 5, 9) Stream valley, C. (2, 3) Bolin Creek Valley, C. (6-8, 10, 11) Hill; CC. (1) Valley flat, underlain by Paleozoic tuffs and slates, CC. (2, 3) Valley floor, underlain by Paleozoic volcanic rhyolites, CC. (4) Lower valley slope, underlain by Paleozoic igneous diorite, CC. (5) Valley floor and lower- to midslopes, underlain by Paleozoic igneous diorite, CC. (6, 7, 10) Midslope, underlain by Paleozoic igneous diorite, CC. (8) Upper slope, underlain by early Paleozoic tuffs and slates, CC. (9) Lower slope along upper reaches of stream, underlain by Paleozoic igneous diorite, CC. (11) Hill crest and extreme upper slopes, underlain by Paleozoic igneous diorite.

BIOLOGY:

1. A. TERRESTRIAL BROADLEAF FOREST SYSTEM;
   AA. Large, deliquescent, deciduous trees/?. B. Maple-hardwood-oak flat; BB. Sapindales-mixed hardwoods-Fagales. C. *Acer rubrum*-mixed mesic hardwoods and oaks; CC. *Acer rubrum*-mixed mesic hardwoods and oaks/?.
2. A. TERRESTRIAL BROADLEAF FOREST SYSTEM;
   AA. Large, deliquescent, deciduous trees/?. B. Beech-mixed hardwood flat; BB. Fagales-mixed hardwoods. C. *Fagus grandifolia*-mixed mesic hardwoods and oaks; CC. *Fagus grandifolia*-mixed mesic hardwoods and oaks/?.
3. A. TERRESTRIAL BROADLEAF FOREST SYSTEM;
   AA. Large, deliquescent, deciduous trees/?. B. Tulip tree-mixed hardwood flat; BB. Magnoliales-mixed hardwoods-Fagales-Juglandales/?. C. *Liriodendron tulipifera*-mixed mesic hardwoods, oaks, and hickories; CC. *Liriodendron tulipifera*-mixed mesic hardwoods, oaks, and hickories/?.
4. A. TERRESTRIAL BROADLEAF FOREST SYSTEM;
   AA. Large, deliquescent, deciduous trees/?. B. Beech slope; BB. Fagales/?. C. *Fagus grandifolia*; CC. *Fagus grandifolia*/?.
5. A. TERRESTRIAL BROADLEAF FOREST SYSTEM;
   AA. Large, deliquescent, deciduous trees. B. Beech-oak-hickory cove;

BB. Fagales-Juglandales. C. *Fagus grandifolia*-mixed mesic oaks and hickories; CC. *Fagus grandifolia*-mixed mesic oaks and hickories.

6. A. TERRESTRIAL BROADLEAF FOREST SYSTEM;
AA. Large, deliquescent, deciduous trees/Large, deliquescent, deciduous trees/Medium, deliquescent, deciduous trees. B. Oak slope; BB. Fagales/Mixed hardwoods/Cornales. C. *Quercus rubra-Quercus alba*; CC. *Quercus rubra-Quercus alba*/Mixed hardwoods/*Cornus florida*.
7. A. TERRESTRIAL BROADLEAF FOREST SYSTEM;
AA. Large, deliquescent, deciduous trees/Transgressive, deliquescent, deciduous trees/Medium, deliquescent, deciduous trees/Tall, rhizomatous, deciduous shrubs. B. Oak-hickory slope; BB. Juglandales-Fagales/Mixed hardwoods/Mixed hardwoods/Dipsacales. C. *Carya tomentosa-Quercus alba*; CC. *Carya tomentosa-Quercus alba*/Mixed hardwoods/Mixed hardwoods/*Viburnum rafinesquianum*.
8. A. TERRESTRIAL BROADLEAF FOREST SYSTEM;
AA. Large, deliquescent, deciduous trees/?. B. Oak-hardwood slope; BB. Fagales-mixed hardwoods/?. C. *Quercus alba*-mixed upland hardwoods; CC. *Quercus alba*-mixed upland hardwoods/?.
9. A. TERRESTRIAL BROADLEAF FOREST SYSTEM;
AA. Large, deliquescent, deciduous trees/Large, deliquescent, deciduous trees/Medium, deliquescent, deciduous trees. B. Oak slope; BB. Fagales/Mixed hardwoods/Cornales. C. Mixed upland oaks; CC. Mixed upland oaks/Mixed hardwoods/*Cornus florida*.
10. A. TERRESTRIAL BROADLEAF FOREST SYSTEM;
AA. Large, deliquescent, deciduous trees/Tall, rhizomatous, deciduous shrubs. B. Oak-hickory hill crest; BB. Fagales-Juglandales/Dipsacales. C. *Quercus stellata*-mixed upland hickories and oaks; CC. *Quercus stellata*-mixed upland hickories and oaks/*Viburnum rafinesquianum*.
11. A. TERRESTRIAL BROADLEAF FOREST SYSTEM;
AA. Large, deliquescent, deciduous trees/?. B. Oak-hickory slope; BB. Juglandales-Fagales/?. C. *Carya glabra*-mixed upland oaks and hickories; CC. *Carya glabra*-mixed upland oaks and hickories/?.

Community Diversity Summary 1

Terrestrial broadleaf forest system	Maple-hardwood-oak flat
Large, deliquescent, deciduous trees	Sapindales-mixed hardwoods-Fagales

ACER RUBRUM-MIXED MESIC HARDWOODS AND OAKS/?
Acer rubrum-mixed mesic hardwoods and oaks

CLIMATE: A. Mesothermal; AA. Warm temperate. B. Cool and moderately moist yearly, moderately hot and moderately wet summers, moderately warm and moderately dry winters, with an average freeze-free period; BB. Cool and moist to moderately moist yearly, very hot and moderately wet summers, moderately cold and moderately wet winters, with an average to long freeze-free period. C. Similar to local climate; CC. Similar to natural area climate.

SOILS: A. Ultisol; AA. Udult. B. Hapludult; BB. Aquic Hapludult. C. Clayey, mixed, thermic Aquic Hapludult; CC. Lignum silt loam. No data for soil horizons.

GEOLOGY: A. Metamorphic; AA. Mixed metavolcanic and metasedimentary. B. Acidic; BB. Water-deposited sediments and volcanics. C. Mixed plutonic, metasedimentary, and metavolcanic rocks of varying composition and degree of exposure; CC. Folded, fine-grained, tuffaceous slates and volcanic tuffs with no surface exposure.
HYDROLOGY: A. Terrestrial; AA. Wet. B. Fresh; BB. Temporarily saturated. C. Refer to Natural Area Diversity Summary; CC. Moderately well drained, wetted by a seasonally perched water table, occasional flooding, and fresh rains.
TOPOGRAPHY: A. Plains; AA. Irregular plain to irregular plain with slight relief. B. Low hills and shallow to relatively deep valleys; BB. Water-eroded stream valley. C. Refer to Natural Area Diversity Summary; CC. Water-deposited, slightly sheltered valley flat containing abandoned channels along a small, perennial, southwest-flowing stream.
PHYSIOGRAPHY: A. Appalachian Highlands; AA. Piedmont Province. B. Piedmont Uplands; BB. Iron Mine Hill complex. C. Stream valley; CC. Valley flat, underlain by Paleozoic tuffs and slates.

CANOPY: Height, dbh, and age not determined.
CANOPY DOMINANTS: *Acer rubrum*, mixed mesic hardwoods, oaks.
CANOPY PHYSIOGNOMY: Large, deliquescent, deciduous trees.

Canopy Analysis

Species	I.V.	Rel. Den.	Rel. Dom.	Rel. Freq.
Acer rubrum	89.0	30.0	30.9	28.1
Liquidambar styraciflua	67.6	25.0	20.7	21.9
Quercus phellos	33.6	10.0	14.3	9.4
Quercus alba	24.3	7.5	10.6	6.2
Liriodendron tulipifera	23.5	7.5	5.6	9.4
Fraxinus pennsylvanica	22.2	7.5	5.3	9.4
Nyssa sylvatica	16.3	5.0	5.0	6.3
Platanus occidentalis	8.5	2.5	2.9	3.1
Pinus taeda	7.4	2.5	1.8	3.1
Quercus falcata var. *pagodaefolia*	7.3	2.5	1.7	3.1

Number of points: 10 d: Not determined Number of individuals/acre: Not determined

CANOPY SPECIES PRESENT, BUT NOT IN ANALYSIS: None.

SUBCANOPY: Not analyzed.

SHRUB LAYER: Not analyzed.

HERB LAYER: Not analyzed.

COMMUNITY REFERENCES: None.

COMMUNITY DOCUMENTATION: Area analyzed by Lee J. Otte for A. E. Radford's Botany 235 class, UNC, Chapel Hill. Specimens deposited in the NCU Herbarium by Lee J. Otte and Deborah K. S. Otte in 1978.

COMMUNITY ECOLOGICAL CHARACTERIZATION: Vegetationally: Sapindalean-mixed hardwood-Fagalean/? terrestrial broadleaf forest system with a closed canopy of large, deliquescent, deciduous trees. Climatically: Warm temperate, mesothermal climate, similar to the local yearly cool and moist to moderately moist climate, which has very hot and moderately wet summers, moderately cold and moderately wet winters, and an average to long freeze-free period. Pedologically: Clayey, mixed, thermic Aquic Hapludult, Lignum silt loam soil. Geologically: Folded, acidic, water-deposited metavolcanic tuffs and metasedimentary tuffaceous slates with no surface exposure. Hydrologically: Seasonally wet, moderately well drained, wet, terrestrial system wetted by fresh rains, occasional flooding, and a seasonally perched water table. Topographically: Water-deposited, slightly sheltered valley flat, containing small, abandoned channels, in a stream valley of a small, perennial, southwest-flowing stream in a landscape of low hills and shallow to relatively deep valleys on an irregular plain with slight relief. Temporally and Spatially: Climax stage of a lithosere on a flat in a stream valley, underlain by Paleozoic tuffs and slates, in the Iron Mine Hill complex in the Uplands of the Piedmont Province of the Appalachian Highlands.

Community Diversity Summary 2

Terrestrial broadleaf forest system
Large, deliquescent, deciduous trees

Beech-mixed mesic hardwood flat
Fagales-mixed hardwoods

FAGUS GRANDIFOLIA-MIXED MESIC HARDWOODS AND OAKS/?
Fagus grandifolia-mixed mesic hardwoods and oaks

CLIMATE: A. Mesothermal; AA. Warm temperate. B. Cool and moderately moist yearly, moderately hot and moderately wet summers, moderately warm and moderately dry winters, with an average freeze-free period; BB. Cool and moist to moderately moist yearly, very hot and moderately wet summers, moderately cold and moderately wet winters, with an average to long freeze-free period. C. Similar to local climate; CC. Similar to natural area climate.

SOILS: A. Alfisol; AA. Udalf. B. Hapludalf; BB. Typic Hapludalf. C. Mixed, thermic, shallow Typic Hapludalf; CC. Wilkes gravelly loam. No data for soil horizons.

GEOLOGY: A. Metamorphic; AA. Metavolcanic. B. Acidic; BB. Undetermined origin. C. Mixed plutonic, metasedimentary, and metavolcanic rocks of varying compositions and degree of exposure; CC. Folded, rhyolitic, welded, tuffaceous breccia buried under recent fluvial sediments.

HYDROLOGY: A. Terrestrial; AA. Wet. B. Fresh; BB. Temporarily saturated. C. Refer to Natural Area Diversity Summary; CC. Well drained, wetted by rains, a seasonally perched water table, and occasional flooding by Bolin Creek.

TOPOGRAPHY: A. Plains; AA. Irregular plain to irregular plain with slight relief. B. Low hills and shallow to relatively deep valleys; BB. Water-eroded stream valley. C. Refer to Natural Area Diversity Summary; CC. Sheltered, smooth, level flat on the inside of a curve on a northwest-southeast-flowing stream.
PHYSIOGRAPHY: A. Appalachian Highlands; AA. Piedmont Province. B. Piedmont Uplands; BB. Iron Mine Hill complex. C. Bolin Creek valley; CC. Valley floor, underlain by Paleozoic volcanic rhyolites.

CANOPY: Height, dbh, and age not determined.
CANOPY DOMINANTS: *Fagus grandifolia*, mixed mesic hardwoods, and *Quercus* spp.
CANOPY PHYSIOGNOMY: Large, deliquescent, deciduous trees.

Canopy Analysis

Species	I.V.	Rel. Den.	Rel. Dom.	Rel. Freq.
Fagus grandifolia	77.6	25.0	31.6	21.0
Liquidambar styraciflua	45.4	15.0	12.4	18.0
Liriodendron tulipifera	37.4	12.5	9.9	15.0
Acer saccharum var. *floridanum*	34.7	15.0	13.7	6.0
Quercus alba	34.5	10.0	12.5	12.0
Quercus rubra	22.1	7.5	5.6	9.0
Acer rubrum	14.3	5.0	3.3	6.0
Carya tomentosa	10.2	2.5	4.7	3.0
Quercus falcata X *Quercus phellos*	8.3	2.5	2.8	3.0
Quercus falcata var. *pagodaefolia*	7.4	2.5	1.9	3.0
Carya glabra	7.1	2.5	1.6	3.0

Number of points: 10 d: Not determined Number of individuals/acre: Not determined

CANOPY SPECIES PRESENT, BUT NOT IN ANALYSIS: *Nyssa sylvatica, Pinus taeda.*

SUBCANOPY: Not analyzed.

SHRUB LAYER: Not analyzed.

HERB LAYER: Not analyzed.

COMMUNITY REFERENCES: None.

COMMUNITY DOCUMENTATION: Area analyzed by Lee J. Otte for A. E. Radford's Botany 235 class, UNC, Chapel Hill. Specimens deposited in the NCU Herbarium by Lee J. Otte and Deborah K. S. Otte in 1978.

COMMUNITY ECOLOGICAL CHARACTERIZATION: Vegetationally: Fagalean-mixed hardwood/? terrestrial broadleaf forest system with a closed canopy of large, deliquescent, deciduous trees. Climatically: Warm temperate, mesothermal climate, similar to the local cool and moist to moderately moist yearly climate, which has very hot and moderately wet summers, moderately cold and moderately wet winters, and an average to long freeze-free period. Pedologically: Mixed, thermic, shallow Typic Hapludalf, Wilkes gravelly loam soil. Geologically: Folded, acidic, welded, tuffaceous, rhyolitic breccia buried under recent fluvial sediments. Hydrologically: Temporarily saturated, well-drained, wet terrestrial system wetted by fresh rains, occasional flooding by Bolin Creek, and a seasonally perched water table. Topographically: Sheltered, smooth, level, water-deposited flat on the inside of a curve on a generally northwest-southeast-flowing stream in a water-eroded stream valley in a landscape of low hills and shallow to relatively deep valleys on an irregular plain with slight relief. Temporally and Spatially: Climax stage of a lithosere on a flat along Bolin Creek, underlain by Paleozoic volcanic rhyolites, in the Iron Mine Hill complex in the Uplands of the Piedmont Province of the Appalachian Highlands.

Community Diversity Summary 3

Terrestrial broadleaf forest system
Large, deliquescent, deciduous trees/?

Tulip tree-mixed hardwood flat
Magnoliales-mixed hardwoods-Fagales-Juglandales/?

LIRIODENDRON TULIPIFERA-MIXED MESIC HARDWOODS, OAKS, AND HICKORIES/?
Liriodendron tulipifera-mixed mesic hardwoods, oaks, and hickories

CLIMATE: A. Mesothermal; AA. Warm temperate. B. Cool and moderately moist yearly, moderately hot and moderately wet summers, moderately warm and moderately dry winters, with an average freeze-free period; BB. Cool and moist to moderately moist yearly, very hot and moderately wet summers, moderately cold and moderately wet winters, with an average to long freeze-free period. C. Similar to local climate; CC. Similar to natural area climate.

SOILS: A. Alfisol; AA. Udalf. B. Hapludalf; BB. Typic Hapludalf. C. Mixed, thermic, shallow Typic Hapludalf; CC. Wilkes gravelly loam. No data for soil horizons.

GEOLOGY: A. Metamorphic; AA. Metavolcanic. B. Acidic; BB. Undetermined origin. C. Mixed plutonic, metasedimentary, and metavolcanic rocks of varying compositions and degree of exposure; CC. Folded, welded, tuffaceous, rhyolitic breccia exposed as scattered, fresh boulders, but mostly buried under recent fluvial sediments.

HYDROLOGY: A. Terrestrial; AA. Wet. B. Fresh; BB. Temporarily saturated. C. Refer to Natural Area Diversity Summary; CC. Well drained, wetted by rains, a seasonally perched water table, and occasional flooding by Bolin Creek.

TOPOGRAPHY: A. Plains; AA. Irregular plain to irregular plain with slight relief. B. Low hills and shallow to relatively deep valleys; BB. Water-eroded stream valley. C. Refer to Natural Area Diversity Summary; CC. Sheltered, irregular to rocky, level valley floor along a generally northwest-southeast-flowing stream.

PHYSIOGRAPHY: A. Appalachian Highlands; AA. Piedmont Province. B. Piedmont Uplands; BB. Iron Mine Hill complex. C. Bolin Creek valley; CC. Valley floor, underlain by Paleozoic volcanic rhyolites.

CANOPY: Height, dbh, and age not determined.

CANOPY DOMINANTS: *Liriodendron tulipifera*, mixed mesic hardwoods, oaks, and hickories.

CANOPY PHYSIOGNOMY: Large, deliquescent, deciduous trees.

Canopy Analysis

Species	I.V.	Rel. Den.	Rel. Dom.	Rel. Freq.
Liriodendron tulipifera	77.95	25.0	34.2	18.75
Liquidambar styraciflua	43.50	17.5	10.4	15.60
Fagus grandifolia	42.00	15.0	11.4	15.60
Quercus alba	35.80	10.0	13.3	12.50
Quercus rubra	31.80	10.0	12.4	9.38
Carya glabra	21.96	7.5	5.1	9.36
Carya tomentosa	12.42	5.0	4.3	3.12
Quercus phellos	8.62	2.5	3.0	3.12
Juniperus virginiana	7.72	2.5	2.1	3.12
Acer rubrum	7.53	2.5	1.9	3.12
Pinus taeda	6.52	2.5	1.9	3.12

Number of points: 10 d: Not determined Number of individuals/acre: Not determined

CANOPY SPECIES PRESENT, BUT NOT IN ANALYSIS: None.

SUBCANOPY: Not analyzed.

SHRUB LAYER: Not analyzed.

HERB LAYER: Not analyzed.

COMMUNITY REFERENCES: None.

COMMUNITY DOCUMENTATION: Area analyzed by Lee J. Otte for A. E. Radford's Botany 235 class, UNC, Chapel Hill. Specimens deposited in the NCU Herbarium by Lee J. Otte and Deborah K. S. Otte in 1978.

COMMUNITY ECOLOGICAL CHARACTERIZATION: Vegetationally: Magnolialean-mixed hardwood-Fagalean-Juglandalean/? terrestrial broadleaf forest system with a closed canopy of large, deliquescent, deciduous trees. Climatically: Warm temperate, mesothermal climate, similar to the local yearly cool and moist to moderately moist climate, which has very hot and moderately wet summers, moderately cold and moderately wet winters, and an average to long freeze-free period. Pedologically: Mixed, thermic, shallow Typic Hapludalf, Wilkes gravelly loam soil. Geologically: Folded, acidic, welded, tuffaceous, rhyolitic breccia exposed as small, in place outcrops, but mostly buried by recent fluvial sediments. Hydrologically: Temporarily saturated, well-drained, wet terrestrial system wetted by rains, occasional flooding by Bolin Creek, and a seasonally perched water table. Topographically: Sheltered, irregular to rocky, level valley floor along a generally northwest-southeast-flowing stream in a water-eroded valley in a landscape of low hills and shallow to relatively deep valleys on an irregular plain with slight relief. Temporally and Spatially: Climax stage of a lithosere on the valley floor along Bolin Creek, underlain by Paleozoic volcanic rhyolites, in the Iron Mine Hill complex in the Uplands of the Piedmont Province of the Appalachian Highlands.

Community Diversity Summary 4

Terrestrial broadleaf forest system	Beech slope
Large, deliquescent, deciduous trees/?	Fagales/?

FAGUS GRANDIFOLIA/?
Fagus grandifolia

CLIMATE: A. Mesothermal; AA. Warm temperate. B. Cool and moderately moist yearly, moderately hot and moderately wet summers, moderately warm and moderately dry winters, with an average freeze-free period; BB. Cool and moist to moderately moist yearly, very hot and moderately wet summers, moderately cold and moderately wet winters, with an average to long freeze-free period. C. Similar to local climate; CC. Similar to natural area climate.

SOILS: A. Alfisol; AA. Udalf. B. Hapludalf; BB. Typic Hapludalf. C. Loamy, mixed, thermic, shallow Typic Hapludalf; CC. Wilkes gravelly loam. No data for soil horizons.

GEOLOGY: A. Igneous; AA. Plutonic. B. Intermediate; BB. Pluton. C. Mixed plutonic, metasedimentary, and metavolcanic rocks of varying composition and degree of exposure; CC. Diorite exposed as fresh, widely scattered, large boulders.

HYDROLOGY: A. Terrestrial; AA. Dry-mesic. B. Fresh; BB. Permanently exposed. C. Refer to Natural Area Diversity Summary; CC. Well drained, wetted by rains and downslope drainage.

TOPOGRAPHY: A. Plains; AA. Irregular plain to irregular plain with slight relief. B. Low hills and shallow to relatively deep valleys; BB. Water-eroded stream valley. C. Refer to Natural Area Diversity Summary; CC. Sheltered, south-facing, irregular to rocky, constant, strongly sloping, lower slope along a northwest-southeast-flowing perennial stream.

PHYSIOGRAPHY: A. Appalachian Highlands; AA. Piedmont Province. B. Piedmont Uplands; BB. Iron Mine Hill complex. C. Stream valley; CC. Lower valley slope, underlain by Paleozoic igneous diorite.

CANOPY: Height, dbh, and age not determined.
CANOPY DOMINANT: *Fagus grandifolia*.
CANOPY PHYSIOGNOMY: Large, deliquescent, deciduous trees.

Canopy Analysis

Species	I.V.	Rel. Den.	Rel. Dom.	Rel. Freq.
Fagus grandifolia	159.3	57.5	61.8	40.0
Quercus alba	56.4	17.5	18.9	20.0
Acer rubrum	32.1	10.0	6.1	16.0
Quercus rubra	17.8	5.0	4.8	8.0
Liriodendron tulipifera	16.5	5.0	3.5	8.0
Quercus stellata	9.7	2.5	3.2	4.0
Liquidambar styraciflua	8.1	2.5	1.6	4.0

Number of points: 10   d: Not determined   Number of individuals/acre: Not determined

CANOPY SPECIES PRESENT, BUT NOT IN ANALYSIS: None.

SUBCANOPY: Not analyzed.

SHRUB LAYER: Not analyzed.

HERB LAYER: Not analyzed.

COMMUNITY REFERENCES: None

COMMUNITY DOCUMENTATION: Area analyzed by Lee J. Otte for A. E. Radford's Botany 235 class, UNC, Chapel Hill. Specimens deposited in the NCU Herbarium by Lee J. Otte and Deborah K. S. Otte in 1978.

COMMUNITY ECOLOGICAL CHARACTERIZATION: Vegetationally: Fagalean/? terrestrial broadleaf forest system with a closed canopy of large, deliquescent, deciduous trees. Climatically: Warm temperate, mesothermal climate, similar to the local yearly cool and moist to moderately moist climate, which has very hot and moderately wet summers, moderately cold and moderately wet winters, and an average to long freeze-free period. Pedologically: Mixed, thermic, shallow Typic Hapludalf, Wilkes gravelly loam soil. Geologically: Undeformed, intermediate, igneous pluton of diorite, exposed as fresh, widely scattered, large boulders. Hydrologically: Permanently exposed, well-drained, dry-mesic, terrestrial system wetted by fresh rains and downslope drainage. Topographically: Sheltered, south-facing, irregular to rocky, constant, strongly sloping, lower slope along a northwest-southeast-flowing perennial stream in a water-eroded valley in a landscape of low hills and shallow to relatively deep valleys on an irregular plain with slight relief. Temporally and Spatially: Climax stage of a lithosere on a lower stream valley slope, underlain by Paleozoic igneous diorite, in the Iron Mine Hill complex in the Uplands of the Piedmont Province of the Appalachian Highlands.

Community Diversity Summary 5

Terrestrial broadleaf forest system
Large, deliquescent, deciduous trees

Beech-oak-hickory cove
Fagales-Juglandales

FAGUS GRANDIFOLIA-MIXED MESIC OAKS AND HICKORIES
Fagus grandifolia-mixed mesic oaks and hickories

CLIMATE: A. Mesothermal; AA. Warm temperate. B. Cool and moderately moist yearly, moderately hot and moderately wet summers, moderately warm and moderately dry winters, with an average freeze-free period; BB. Cool and moist to moderately moist yearly, very hot and moderately wet summers, moderately cold and moderately wet winters, with an average to long freeze-free period. C. Similar to local climate; CC. Similar to natural area climate.

SOILS: A. Alfisol; AA. Udalf. B. Hapludalf; BB. Typic Hapludalf. C. Mixed, thermic, shallow Typic Hapludalf; CC. Wilkes gravelly loam.

GEOLOGY: A. Igneous; AA. Plutonic. B. Intermediate; BB. Pluton. C. Mixed plutonic, metasedimentary, and metavolcanic rocks of varying compositions and degree of exposure; CC. Diorite exposed as large, scattered, fresh, isolated boulders and piles of boulders.

HYDROLOGY: A. Terrestrial; AA. Dry-mesic and wet. B. Fresh; BB. Permanently exposed and temporarily saturated. C. Refer to Natural Area Diversity Summary; CC. Well drained, some portions wetted by rains and downslope drainage and other portions wetted by a seasonally perched water table and occasional flooding by a small stream.

TOPOGRAPHY: A. Plains; AA. Irregular plain to irregular plain with slight relief. B. Low hills and shallow to relatively deep valleys; BB. Water-eroded stream valley. C. Refer to Natural Area Diversity Summary; CC. Sheltered, small channels and gentle to very steep, convex, west-facing, rocky lower to midvalley slope along a small, perennial, south-flowing stream.

PHYSIOGRAPHY: A. Appalachian Highlands; AA. Piedmont Province. B. Piedmont Uplands; BB. Iron Mine Hill complex. C. Stream valley; CC. Valley floor and lower to midslopes, underlain by Paleozoic igneous diorite.

CANOPY: Height, dbh, and age not determined.
CANOPY DOMINANTS: *Fagus grandifolia*, mixed *Quercus* and *Carya* spp.
CANOPY PHYSIOGNOMY: Large, deliquescent, deciduous trees.

Canopy Analysis

Species	I.V.	Rel. Den.	Rel. Dom.	Rel. Freq.
Fagus grandifolia	138.8	52.5	50.3	36.0
Quercus alba	61.6	17.5	22.5	21.6
Quercus rubra	37.2	10.0	12.8	14.4
Carya tomentosa	32.4	10.0	8.0	14.4
Acer rubrum	14.5	5.0	2.1	7.4
Liquidambar styraciflua	8.7	2.5	2.6	3.6
Carya glabra	7.7	2.5	1.6	3.6

Number of points: 10 d: Not determined Number of individuals/acre: Not determined

CANOPY SPECIES PRESENT, BUT NOT IN ANALYSIS: *Liriodendron tulipifera.*

TALL SUBCANOPY: Height, dbh, and age not determined.
SUBCANOPY DOMINANTS: None.
SUBCANOPY PHYSIOGNOMY: Not applicable.

Tall Subcanopy Analysis

Species	1 C.S	2 C.S	3 C.S	4 C.S	5 C.S	6 C.S	7 C.S	8 C.S	9 C.S	10 C.S
Transgressive large trees										
Acer rubrum	-	-	-	-	4.1	-	-	-	-	-
Carya tomentosa	2.1	-	3.1	-	-	-	5.1	-	-	-
Fagus grandifolia	-	-	1.1	4.1	-	-	-	4.1	-	-
Quercus alba	-	-	5.1	-	-	-	-	-	-	-

Number of relevés: 10 Relevé size: 5 m X 5 m

TALL SUBCANOPY SPECIES PRESENT, BUT NOT IN ANALYSIS: *Oxydendrum arboreum.*

SHORT SUBCANOPY: Height, dbh, and age not determined.
SUBCANOPY DOMINANTS: None.
SUBCANOPY PHYSIOGNOMY: Not applicable.

Short Subcanopy Analysis

Species	1 C.S	2 C.S	3 C.S	4 C.S	5 C.S	6 C.S	7 C.S	8 C.S	9 C.S	10 C.S
Transgressive trees										
Acer rubrum	-	-	-	-	-	-	3.1	-	2.1	1.1
Carya glabra	-	-	-	-	1.1	-	-	-	-	-
Carya tomentosa	-	-	-	-	1.1	-	-	-	-	-
Fagus grandifolia	-	2.1	-	-	-	-	-	-	2.1	-
Fraxinus americana	-	1.1	-	-	-	-	-	-	-	-

Short Subcanopy Analysis (continued)

Species	1 C.S	2 C.S	3 C.S	4 C.S	5 C.S	6 C.S	7 C.S	8 C.S	9 C.S	10 C.S
Juniperus virginiana	-	-	1.1	-	-	-	-	-	-	-
Nyssa sylvatica	-	-	3.1	4.1	4.1	1.1	3.1	-	-	-
Oxydendrum arboreum	1.1	-	4.1	4.1	1.1	+	-	2.1	3.1	3.1
Quercus alba	2.1	-	-	-	-	-	-	-	-	-
Medium trees										
Carpinus caroliniana	2.1	-	-	-	3.1	-	-	-	-	-
Cercis canadensis	-	-	-	-	1.1	-	-	-	-	-
Cornus florida	2.1	2.1	-	3.1	-	5.1	2.1	2.1	-	2.1

Number of relevés: 10 Relevé size: 5 m X 5 m

SHORT SUBCANOPY SPECIES PRESENT, BUT NOT IN ANALYSIS: TRANSGRESSIVE TREE: *Ulmus americana*.

SHRUB LAYER DOMINANTS: None.
SHRUB LAYER PHYSIOGNOMY: Not applicable.

Shrub Analysis

Species	GF	1 C.S	2 C.S	3 C.S	4 C.S	5 C.S	6 C.S	7 C.S	8 C.S	9 C.S	10 C.S
Transgressive trees											
Acer rubrum	H2	+	+	1.1	-	-	1.1	2.1	+	+	+
Carpinus caroliniana	H2	1.1	1.1	1.1	-	-	-	-	-	-	-
Carya glabra	H2	-	-	-	-	-	-	-	+	-	-
Carya tomentosa	H2	1.1	1.1	1.1	-	-	-	+	-	-	+
Cercis canadensis	H2	1.1	-	-	1.1	+	+	-	+	-	-
Cornus florida	H2	1.1	1.1	-	-	1.1	2.1	-	-	+	-
Fagus grandifolia	H2	1.1	-	+	1.1	1.1	-	-	-	+	+
Fraxinus americana	H2	1.1	1.1	+	-	-	-	-	-	-	-
Liriodendron tulipifera	H2	+	+	-	-	-	-	1.1	-	-	-
Nyssa sylvatica	H2	-	-	1.1	-	-	-	-	-	-	-
Oxydendrum arboreum	H2	1.1	-	-	1.1	-	-	-	-	1.1	1.1
Prunus serotina	H2	+	+	+	+	-	+	+	+	+	-
Quercus alba	H2	1.1	1.1	1.1	+	-	+	-	-	-	+
Quercus rubra	H2	1.1	1.1	+	+	-	+	+	-	-	-
Sassafras albidum	H2	-	-	-	+	-	-	+	+	+	-
Normal shrubs											
Aesculus sylvatica	H7	1.1	-	-	-	-	-	-	-	-	-
Chionanthus virginicus	H11	1.1	-	+	1.1	+	1.1	1.1	2.1	-	-
Ilex ambigua var. *montana*	H7	-	-	-	-	-	-	-	-	-	1.1
Rhododendron nudiflorum	H11	+	-	3.1	-	+	-	-	-	-	-
Rubus argutus	H11	-	-	-	-	+	-	-	-	-	-
Viburnum acerifolium	H14	-	+	-	+	-	1.1	1.1	1.1	1.1	-
Viburnum rafinesquianum	H11	2.1	2.1	-	1.1	2.1	2.1	2.1	2.1	+	+

Shrub Analysis (continued)

Species	GF	1 C.S	2 C.S	3 C.S	4 C.S	5 C.S	6 C.S	7 C.S	8 C.S	9 C.S	10 C.S
Tall dwarf shrubs											
Euonymus americanus	H11	+	+	1.1	-	+	+	+	+	-	-
Viburnum prunifolium	H11	-	-	-	-	1.1	-	+	+	-	-
Normal dwarf shrubs											
Vaccinium tenellum	H11	1.1	-	-	-	-	-	-	-	-	-
Vaccinium vacillans	H11	-	-	-	-	-	+	-	-	-	+
Vines											
Campsis radicans	H16	-	-	2.1	-	-	-	-	-	-	-
Lonicera japonica	H19	-	-	-	-	-	+	-	-	-	-
Lonicera sempervirens	H19	-	-	-	1.1	-	-	+	+	-	-
Parthenocissus quinquefolia	H18	-	-	+	-	+	-	-	-	+	-
Rhus radicans	H16	-	-	-	+	-	-	-	-	-	-
Vitis aestivalis	H18	-	-	1.1	-	-	+	-	+	1.1	-
Vitis rotundifolia	H18	1.1	+	-	-	1.1	-	+	+	1.1	-

Number of relevés: 10 Relevé size: 5 m X 5 m

SHRUB SPECIES PRESENT, BUT NOT IN ANALYSIS: TRANSGRESSIVE TREES: *Diospyros virginiana, Juniperus virginiana, Morus rubra*; NORMAL SHRUBS: *Cornus amomum, Sambucus canadensis*; TALL DWARF SHRUBS: *Hypericum* sp., *Viburnum rufidulum*; VINES: *Smilax rotundifolia*.

HERB LAYER DOMINANTS: None.
HERB LAYER PHYSIOGNOMY: Not applicable.

Herb Analysis

Species	GF	1 C.S	2 C.S	3 C.S	4 C.S	5 C.S	6 C.S	7 C.S	8 C.S	9 C.S	10 C.S
Tall forbs											
Desmodium nudiflorum	H7	+	-	-	1.1	-	1.1	1.1	-	1.1	-
Desmodium sp.	H7	-	1.1	-	+	-	-	-	-	-	-
Elephantopus tomentosus	H13	-	1.1	-	-	-	-	-	-	-	-
Polygonatum biflorum	H11	-	-	-	1.1	-	-	-	+	-	-
Prenanthes altissima	H7	-	-	-	-	1.1	-	-	-	-	-
Smilacina racemosa	H11	-	-	-	+	-	-	+	1.1	-	-
Solidago caesia	H11	-	1.1	-	+	1.1	-	-	-	-	-
Zizia aurea	H7	-	1.1	-	-	1.1	-	-	-	-	-
Medium forbs											
Chimaphila maculata	H11	-	-	-	+	-	-	-	-	+	-
Goodyera pubescens	H13	-	-	-	-	-	+	-	-	-	-
Hexastylis arifolia	H11	+	+	-	-	+	+	-	+	-	-
Tipularia discolor	H6	-	-	-	-	-	-	-	-	+	-
Trillium sp.	H11	-	-	-	-	-	-	-	-	-	-

Herb Analysis (continued)

Species	GF	1 C.S	2 C.S	3 C.S	4 C.S	5 C.S	6 C.S	7 C.S	8 C.S	9 C.S	10 C.S
Small forbs											
Hepatica americana	H11	-	1.1	-	1.1	+	-	-	-	-	-
Houstonia caerulea	H11	-	-	-	-	+	-	-	-	-	-
Iris cristata	H11	-	-	-	+	-	-	-	-	-	-
Viola papilionacea	H11	-	1.1	-	-	-	-	-	-	-	-
Viola sp.	H11	-	-	-	+	-	-	-	-	-	-
Medium graminoid											
Panicum sp.	H11	-	1.1	-	-	+	-	-	+	-	-
Tall fern											
Polystichum acrostichoides	H11	-	1.1	-	-	1.1	-	+	-	-	-

Number of relevés: 10 Relevé size: 1 m X 1 m

HERB SPECIES PRESENT, BUT NOT IN ANALYSIS: TALL FORBS: *Aster infirmus, Aureolaria virginica, Coreopsis major, Desmodium rotundifolium, Elephantopus carolinianus, Heuchera americana, Hieracium venosum, Prunella vulgaris, Scutellaria elliptica, Scutellaria integrifolia, Thaspium barbinode*; MEDIUM FORBS: *Antennaria plantaginifolia, Chrysogonum virginianum, Cunila origanoides*; SMALL FORBS: *Houstonia purpurea, Oxalis* sp., *Potentialla canadensis*; TALL GRAMINOIDS: *Leersia virginica*; TALL FERN: *Osmunda regalis* var. *spectabilis*; MEDIUM FERNS: *Asplenium platyneuron, Polypodium polypodioides*; VINE: *Dioscorea villosa*.

COMMUNITY REFERENCES: None.

COMMUNITY DOCUMENTATION: Not applicable.

COMMUNITY ECOLOGICAL CHARACTERIZATION: Vegetationally: Fagalean-Juglandalean terrestrial broadleaf forest system with a closed canopy of large, deliquescent, deciduous trees, an open tall subcanopy of large, deliquescent, deciduous trees, an open, short subcanopy of medium, deliquescent, deciduous trees and transgressive, deliquescent, deciduous trees, an open shrub layer of mixed, deciduous shrubs and transgressive, deliquescent, deciduous trees, and an open herb layer of abundant, mixed, mesophytic ferns, graminoids, and forbs. Climatically: Warm temperate, mesothermal climate, similar to the local yearly cool and moist to moderately moist climate, which has very hot and moderately wet summers, moderately cold and moderately wet winters, and an average to long freeze-free period. Pedologically: Mixed, thermic, shallow Typic Hapludalf, Wilkes gravelly loam soil. Geologically: Undeformed, intermediate, igneous pluton of diorite exposed as large, scattered, fresh, isolated boulders and piles of boulders. Hydrologically: Permanently exposed and temporarily saturated, well-drained, wet to dry-mesic terrestrial system, with some portions wetted by fresh rains and downslope drainage and other portions wetted by a seasonally perched water table and occasional flooding by a small stream. Topographically: Sheltered, small, flat valley floor containing scattered exposed rock, small pools, and abandoned channels, and gentle to very steep, convex, west-facing, rocky lower- to midvalley slope along a small, perennial, south-flowing stream. Temporally and Spatially: Climax stage of a lithosere on the floor and lower to midslopes of a stream valley, underlain by Paleozoic igneous diorite, in the Iron Mine Hill complex in the Uplands of the Piedmont Province of the Appalachian Highlands.

Community Diversity Summary 6

Terrestrial broadleaf forest system	Oak slope
Large, deliquescent, deciduous trees/ Large, deliquescent, deciduous trees/ Medium, deliquescent, deciduous trees	Fagales/Mixed hardwoods/Cornales

QUERCUS RUBRA-QUERCUS ALBA/MIXED HARDWOODS/*CORNUS FLORIDA*
Quercus rubra-Quercus alba

CLIMATE: A. Mesothermal; AA. Warm temperate. B. Cool and moderately moist yearly, moderately hot and moderately wet summers, moderately warm and moderately dry winters, with an average freeze-free period; BB. Cool and moist to moderately moist yearly, very hot and moderately wet summers, moderately cold and moderately wet winters, with an average to long freeze-free period. C. Similar to local climate; CC. Similar to natural area climate.

SOILS: A. Alfisol; AA. Udalf. B. Hapludalf; BB. Ultic Hapludalf. C. Fine, mixed, thermic Ultic Hapludalf; CC. Enon loam.

A_0: 0-6 cm, decayed leaf litter.
A_1: 6-30 cm, rocky, clayey sand, reddish yellow, pH 6.0.
B_1: 30-41 cm, clayey sand, light red, pH 6.0.
B_2: 41+ cm, sandy clay, red, pH 6.0.

GEOLOGY: A. Igneous; AA. Plutonic. B. Intermediate; BB. Pluton. C. Mixed plutonic, metasedimentary, and metavolcanic rocks of varying composition and degree of exposure; CC. Plagioclase porphyritic diorite, occurring as very scattered, isolated, highly weathered, small boulders.

HYDROLOGY: A. Terrestrial; AA. Dry-mesic. B. Fresh; BB. Permanently exposed. C. Refer to Natural Area Diversity Summary; CC. Well drained, wetted by fresh rains and downslope drainage.

TOPOGRAPHY: A. Plains; AA. Irregular plain to irregular plain with slight relief. B. Low hills and shallow to relatively deep valleys; BB. Water-eroded hill. C. Refer to Natural Area Diversity Summary; CC. Open, east-southeast-facing, gently sloping, irregular to undulatory lower- and midportion of a constant hill slope.

PHYSIOGRAPHY: A. Appalachian Highlands; AA. Piedmont Province. B. Piedmont Uplands; BB. Iron Mine Hill complex. C. Hill; CC. Midslope, underlain by Paleozoic igneous diorite.

CANOPY: *Quercus alba*--Height 30 m, dbh 61 cm, age 120 years.
Quercus rubra--Height 24 m, dbh 48 cm, age 95 years.

CANOPY DOMINANTS: *Quercus rubra, Quercus alba.*
CANOPY PHYSIOGNOMY: Large, deliquescent, deciduous trees.

Canopy Analysis

Species	I.V.	Rel. Den.	Rel. Dom.	Rel. Freq.
Quercus rubra	93.0	32.5	36.0	24.5
Quercus alba	82.5	25.0	30.0	27.5
Liriodendron tulipifera	38.2	12.5	13.7	12.0
Carya tomentosa	22.1	7.5	5.6	9.0
Nyssa sylvatica	15.2	5.0	4.2	6.0
Quercus velutina	14.8	5.0	3.8	6.0
Pinus taeda	7.3	2.5	1.8	3.0
Acer rubrum	7.1	2.5	1.6	3.0
Quercus stellata	6.8	2.5	1.3	3.0
Quercus falcata	6.7	2.5	1.2	3.0
Fagus grandifolia	6.3	2.5	0.8	3.0

Number of Points: 10 d: 18.5 ft Number of individuals/acre: 129

CANOPY SPECIES PRESENT, BUT NOT IN ANALYSIS: LARGE TREE: *Liquidambar styraciflua.*

TALL SUBCANOPY: Height, dbh, and age not determined.
SUBCANOPY DOMINANTS: Mixed hardwoods.
SUBCANOPY PHYSIOGNOMY: Large, deliquescent, deciduous trees.

Tall Subcanopy Analysis

Species	GF	1 C.S	2 C.S	3 C.S	4 C.S	5 C.S
Large trees						
Acer rubrum	H2	-	-	5.1	5.1	-
Carya tomentosa	H2	2.1	-	-	-	-
Liriodendron tulipifera	H2	-	-	3.1	-	1.1
Oxydendrum arboreum	H2	-	-	-	-	5.1
Quercus alba	H2	5.1	-	-	-	-

Number of relevés: 5 Relevé size: 5 m X 5 m

TALL SUBCANOPY SPECIES PRESENT, BUT NOT IN ANALYSIS: None.

SHORT SUBCANOPY: Height, dbh, and age not determined.
SUBCANOPY DOMINANT: *Cornus florida.*
SUBCANOPY PHYSIOGNOMY: Medium, deliquescent, deciduous trees.

Short Subcanopy Analysis

Species	GF	1 C.S	2 C.S	3 C.S	4 C.S	5 C.S
Large trees						
Acer rubrum	H2	-	4.1	-	2.1	3.1
Nyssa sylvatica	H2	-	1.1	-	-	-
Oxydendrum arboreum	H2	-	2.1	-	2.1	3.1
Medium trees						
Cornus florida	H2	5.1	5.1	5.1	5.1	5.1

Number of relevés: 5 Relevé size: 5 m X 5 m

SHORT SUBCANOPY SPECIES PRESENT, BUT NOT IN ANALYSIS: None.

SHRUB LAYER DOMINANTS: None.
SHRUB LAYER PHYSIOGNOMY: Not applicable.

Shrub Analysis

Species	GF	1 C.S	2 C.S	3 C.S	4 C.S	5 C.S
Transgressive trees						
Acer rubrum	H2	2.1	1.1	1.1	+.1	1.1
Carya glabra	H2	+.1	-	-	-	1.1
Carya tomentosa	H2	+.1	1.1	+.1	1.1	1.1
Cornus florida	H2	1.1	-	-	2.1	-
Diospyros virginiana	H2	-	1.1	-	-	1.1
Fagus grandifolia	H2	-	-	1.1	-	-
Fraxinus americana	H2	-	-	1.1	-	1.1
Juniperus virginiana	H2	-	-	-	-	+.1
Liriodendron tulipifera	H2	-	1.1	-	+.1	-
Nyssa sylvatica	H2	2.1	1.1	-	+.1	-
Oxydendrum arboreum	H2	-	1.1	2.1	-	1.1
Prunus serotina	H2	1.1	+.1	+.1	1.1	-
Quercus alba	H2	-	-	2.1	+.1	+.1
Quercus rubra	H2	+.1	+.1	-	1.1	1.1
Quercus velutina	H2	-	-	-	1.1	1.1
Sassafras albidum	H2	1.1	-	-	-	-
Tall shrubs						
Viburnum prunifolium	H11	-	+.1	-	-	-
Viburnum rafinesquianum	H11	4.3	3.3	2.3	2.3	3.3
Normal shrubs						
Euonymus americanus	H11	+.1	-	+.1	+.1	+.1
Viburnum acerifolium	H11	2.3	-	3.3	4.3	-
Vines						
Lonicera sempervirens	H19	1.1	1.1	-	-	-
Parthenocissus quinquefolia	H18	-	-	-	-	+.1
Smilax bona-nox	H18	-	-	-	-	+.1
Vitis aestivalis	H18	1.1	1.1	1.1	+.1	-

Number of relevés: 5 Relevé size: 5 m X 5 m

SHRUB SPECIES PRESENT, BUT NOT IN ANALYSIS: TALL DWARF SHRUB: *Hypericum* sp.; VINES: *Rhus radicans*, *Smilax glauca*.

HERB LAYER DOMINANTS: None.
HERB LAYER PHYSIOGNOMY: Not applicable.

Herb Analysis

Species	GF	1 C.S	2 C.S	3 C.S	4 C.S	5 C.S
Tall forbs						
Desmodium nudiflorum	H11	-	1.1	-	-	-
Tipularia discolor	H5	-	-	-	-	1.1
Medium forbs						
Chimaphila maculata	H11	+.1	1.1	1.1	-	-
Hexastylis arifolia	H11	1.1	-	1.1	1.1	1.1

Number of relevés: 5 Relevé size: 1 m X 1 m

HERB SPECIES PRESENT, BUT NOT IN ANALYSIS: TALL FORBS: *Desmodium rotundifolium*, *Gentiana villosa*; MEDIUM FORB: *Monotropa hypopithys*; TALL GRAMINOIDS: *Carex rosea*, *Carex* sp., *Panicum* sp.; VINE: *Dioscorea villosa*.

COMMUNITY REFERENCES: None.

COMMUNITY DOCUMENTATION: Area analyzed by Lee J. Otte for A. E. Radford's Botany 235 class, UNC, Chapel Hill. Specimens deposited in the NCU Herbarium by Lee J. Otte and Deborah K. S. Otte in 1978.

COMMUNITY ECOLOGICAL CHARACTERIZATION: Vegetationally: Fagalean/Mixed hardwood/ Cornalean terrestrial broadleaf forest system with a closed canopy of large, deliquescent, deciduous trees, a closed tall subcanopy of large, deliquescent, deciduous trees, a closed short subcanopy of medium, deliquescent, deciduous trees, an open shrub layer containing transgressive trees and mixed shrubs, and a sparse herb layer containing widely scattered forbs and graminoids. Climatically: Warm temperate, mesothermal climate, similar to the local yearly cool and moist to moderately moist climate, which has very hot and moderately wet summers and moderately cold and moderately wet winters, and an average to long freeze-free period. Pedologically: Fine, mixed, thermic Ultic Hapludalf, Enon loam soil. Geologically: Undeformed, intermediate, igneous pluton of plagioclase porphyritic diorite, exposed as very scattered, isolated, highly weathered, small boulders. Hydrologically: Permanently exposed, well-drained, dry-mesic, terrestrial system, wetted by fresh rains and downslope drainage. Topographically: Open, east-southeast-facing, irregular to undulatory, gentle, constant midslope of a water-eroded hill facing a north-south-flowing, small, perennial stream in a landscape of low hills and shallow to relatively deep valleys on an irregular plain with slight relief. Temporally and Spatially: A climax community of a lithosere on the midslope of a hill, underlain by Paleozoic igneous diorite, in the Iron Mine Hill complex in the Uplands of the Piedmont Province of the Appalachian Highlands.

Community Diversity Summary 7

Terrestrial broadleaf forest system
Large, deliquescent, deciduous trees/
Large, deliquescent, deciduous trees/
Transgressive, deliquescent, deciduous trees-medium, deliquescent, deciduous trees/Tall, rhizomatous, deciduous shrubs

Oak-hickory slope
Juglandales-Fagales/Mixed hardwoods/Mixed hardwoods/
Dipsacales

CARYA TOMENTOSA-QUERCUS ALBA/MIXED HARDWOODS/
MIXED HARDWOODS/*VIBURNUM RAFINESQUIANUM*
Carya tomentosa-Quercus alba

CLIMATE: A. Mesothermal; AA. Warm temperate. B. Cool and moderately moist yearly, moderately hot and moderately wet summers, moderately warm and moderately dry winters, with an average freeze-free period; BB. Cool and moist to moderately moist yearly, very hot and moderately wet summers, moderately cold and moderately wet winters, with an average to long freeze-free period. C. Similar to local climate; CC. Similar to natural area climate.
SOILS: A. Alfisol; AA. Udalf. B. Hapludalf; BB. Ultic Hapludalf. C. Fine, mixed, thermic Ultic Hapludalf; CC. Enon loam. No data for soil horizons.
GEOLOGY: A. Igneous; AA. Plutonic. B. Intermediate; BB. Pluton. C. Mixed plutonic, metavolcanic, and metasedimentary rocks of varying composition and degree of exposure; CC. Plagioclase porphyritic diorite exposed only as very scattered, isolated, highly weathered small boulders.
HYDROLOGY: A. Terrestrial; AA. Dry-mesic. B. Fresh; BB. Permanently exposed. C. Refer to Natural Area Diversity Summary; CC. Well drained, wetted by rains and downslope drainage.
TOPOGRAPHY: A. Plains; AA. Irregular plain to irregular plain with slight relief. B. Low hills and shallow to relatively deep valleys; BB. Water-eroded hill. C. Refer to Natural Area Diversity Summary; CC. Open, east-southeast-facing, undulatory, gentle, constant midslope of a north-south-aligned hill facing a north-south-flowing, small, perennial stream.
PHYSIOGRAPHY: A. Appalachian Highlands; AA. Piedmont Province. B. Piedmont Uplands; BB. Iron Mine Hill complex. C. Hill; CC. Midslope, underlain by Paleozoic igneous diorite.

CANOPY: *Quercus alba*--Height 29 m, dbh 61 cm, age 120 years.
Carya tomentosa--Height 24 m, dbh 42 cm, age 100 years.
CANOPY DOMINANTS: *Carya tomentosa, Quercus alba.*
CANOPY PHYSIOGNOMY: Large, deliquescent, deciduous trees.

Canopy Analysis

Species	I.V.	Rel. Den.	Rel. Dom.	Rel. Freq.
Carya tomentosa	94.2	35.0	30.3	28.9
Quercus alba	78.3	25.0	21.2	32.1
Acer rubrum	34.0	12.5	15.3	6.2
Quercus rubra	33.2	10.0	12.2	11.0
Liriodendron tulipifera	10.6	2.5	3.0	5.1
Quercus stellata	9.1	2.5	3.0	3.6
Quercus coccinea	9.1	2.5	3.0	3.6
Pinus taeda	8.4	2.5	3.0	2.9
Quercus velutina	7.8	2.5	3.0	2.3
Acer saccharum var. *floridanum*	7.8	2.5	3.0	2.3
Nyssa sylvatica	7.5	2.5	3.0	2.0

Number of points: 10 d: 16.0 ft Number of individuals/acre: 172

CANOPY SPECIES PRESENT, BUT NOT IN ANALYSIS: None.

TALL SUBCANOPY: Height, dbh, and age not determined.
SUBCANOPY DOMINANTS: Mixed hardwoods.
SUBCANOPY PHYSIOGNOMY: Large, deliquescent, deciduous trees.

Tall Subcanopy Analysis

Species	GF	1 C.S	2 C.S	3 C.S	4 C.S	5 C.S
Transgressive large trees						
Acer rubrum	H2	-	-	5.1	-	2.1
Carya tomentosa	H2	-	-	-	-	5.1
Oxydendrum arboreum	H2	2.1	1.1	-	-	-
Quercus alba	H2	-	-	2.1	5.1	-

Number of relevés: 5 Relevé size: 5 m X 5 m

TALL SUBCANOPY SPECIES PRESENT, BUT NOT IN ANALYSIS: LARGE TREE: *Acer saccharum* var. *floridanum*.

SHORT SUBCANOPY: Height, dbh, and age not determined.
SUBCANOPY DOMINANTS: Mixed hardwoods.
SUBCANOPY PHYSIOGNOMY: Transgressive, deliquescent, deciduous trees and medium, deliquescent, deciduous trees.

Short Subcanopy Analysis

Species	GF	1 C.S	2 C.S	3 C.S	4 C.S	5 C.S
Transgressive trees						
Acer rubrum	H2	-	5.1	-	3.1	5.1
Nyssa sylvatica	H2	-	-	1.1	2.1	-
Oxydendrum arboreum	H2	5.1	-	-	3.1	3.1
Medium trees						
Cornus florida	H2	4.1	4.1	3.1	-	-

Number of relevés: 5 Relevé size: 5 m X 5 m

SHORT SUBCANOPY SPECIES PRESENT, BUT NOT IN ANALYSIS: TRANSGRESSIVE TALL TREE: *Fraxinus pennsylvanica*.

SHRUB LAYER DOMINANT: *Viburnum rafinesquianum*.
SHRUB LAYER PHYSIOGNOMY: Tall, rhizomatous, deciduous shrubs.

Shrub Analysis

Species	GF	1 C.S	2 C.S	3 C.S	4 C.S	5 C.S
Transgressive trees						
Acer rubrum	H2	2.1	1.1	1.1	2.1	-
Acer saccharum var. *floridanum*	H2	+.1	-	-	-	-
Carya glabra	H2	-	-	-	+.1	1.1
Carya tomentosa	H2	+.1	1.1	+.1	2.1	1.1
Cornus florida	H2	2.1	3.1	2.1	1.1	-
Diospyros virginiana	H2	-	-	-	-	+.1
Fraxinus americana	H2	1.1	-	1.1	1.1	-
Juniperus virginiana	H2	-	2.1	1.1	-	-
Liriodendron tulipifera	H2	-	-	2.1	2.1	-
Nyssa sylvatica	H2	-	-	2.1	3.1	-
Oxydendrum arboreum	H2	2.1	-	-	-	3.1
Prunus serotina	H2	1.1	-	-	+.1	+.1
Quercus alba	H2	+.1	1.1	+.1	-	-
Quercus rubra	H2	1.1	1.1	+.1	1.1	1.1
Quercus velutina	H2	-	-	1.1	-	-
Transgressive small tree						
Cercis canadensis	H2	-	-	-	-	2.1
Transgressive giant shrub						
Chionanthus virginicus	H11	-	2.1	-	-	2.1
Tall shrubs						
Viburnum prunifolium	H11	1.1	-	-	-	-
Viburnum rafinesquianum	H11	4.3	5.4	5.4	5.4	5.4
Viburnum rufidulum	H11	-	-	3.1	-	-
Normal shrubs						
Euonymus americanus	H11	+.1	2.1	1.3	1.1	+.1
Viburnum acerifolium	H11	4.3	5.4	5.4	1.1	-

Shrub Analysis (continued)

Species	GF	1 C.S	2 C.S	3 C.S	4 C.S	5 C.S
Tall dwarf shrubs						
Vaccinium stamineum	H11	-	1.1	1.1	1.1	-
Vaccinium tenellum	H11	1.1	-	-	-	-
Vines						
Lonicera sempervirens	H19	+.1	-	-	+.1	-
Vitis aestivalis	H18	-	+.1	-	-	-
Vitis rotundifolia	H18	-	2.1	1.1	-	-

Number of relevés: 5 Relevé size: 5 m X 5 m

SHRUB SPECIES PRESENT, BUT NOT IN ANALYSIS: TRANSGRESSIVE LARGE TREE: *Morus rubra*; TALL DWARF SHRUB: *Hypericum* sp.; VINES: *Rhus radicans, Smilax glauca*.

HERB LAYER DOMINANTS: None.
HERB LAYER PHYSIOGNOMY: Not applicable.

Herb Analysis

Species	GF	1 C.S	2 C.S	3 C.S	4 C.S	5 C.S
Tall forbs						
Desmodium nudiflorum	H11	-	-	2.1	-	-
Galium circaezans	H11	-	-	+.1	-	-
Polygonatum biflorum	H11	1.1	-	1.1	1.1	-
Medium forbs						
Chimaphila maculata	H6	1.1	-	-	-	+.1
Goodyera pubescens	H6	1.1	-	-	-	-
Hexastylis arifolia	H11	-	1.1	-	2.1	-

Number of relevés: 5 Relevé size: 1 m X 1 m

HERB SPECIES PRESENT, BUT NOT IN ANALYSIS: TALL FORBS: *Desmodium rotundifolium, Heuchera americana, Uvularia perfoliata*; VINE: *Dioscorea villosa*.

COMMUNITY REFERENCES: None.

COMMUNITY DOCUMENTATION: Area analyzed by Lee J. Otte for A. E. Radford's Botany 235 class, UNC, Chapel Hill. Specimens deposited in the NCU Herbarium by Lee J. Otte and Deborah K. S. Otte in 1978.

COMMUNITY ECOLOGICAL CHARACTERIZATION: Vegetationally: Juglandalean-Fagalean/ Mixed hardwood/Mixed hardwood/Dipsacalean terrestrial broadleaf forest system with a closed canopy of large, deliquescent, deciduous trees, a closed tall subcanopy of large, deliquescent, deciduous trees, a closed short subcanopy of transgressive, deliquescent, deciduous trees and medium, deliquescent, deciduous trees, a closed shrub layer of tall, rhizomatous, deciduous shrubs, and a sparse herb layer containing widely scattered forbs. Climatically: Warm temperate, mesothermal climate, similar to the local yearly cool and moist to moderately moist climate, which has very hot and moderately wet summers, moderately cold and moderately wet winters, and an average to long freeze-free period. Pedologically: Fine, mixed, thermic Ultic Hapludalf, Enon loam soil. Geologically: Undeformed, intermediate, igneous pluton of plagioclase porphyritic diorite, exposed as very scattered, isolated, highly weathered, small boulders. Hydrologically: Permanently exposed, well-drained, dry-mesic, terrestrial system, wetted by fresh rains and downslope drainage. Topographically: Open, east-southeast-facing, undulatory, gentle, constant midslope of a north-south-aligned hill, facing a north-south-flowing, small, perennial stream, in a landscape of low hills and shallow to relatively deep valleys on an irregular plain with slight relief. Temporally and Spatially: A climax community of a lithosere on the midslope of a hill, underlain by Paleozoic igneous diorite, in the Iron Mine Hill complex in the Uplands of the Piedmont Province of the Appalachian Highlands.

Community Diversity Summary 8

Terrestrial broadleaf forest system	Oak-hardwood slope
Large, deliquescent, deciduous trees/?	Fagales-mixed hardwoods/?

QUERCUS ALBA-MIXED UPLAND HARDWOODS/?
Quercus alba-Mixed upland hardwoods

CLIMATE: A. Mesothermal; AA. Warm temperate. B. Cool and moderately moist yearly, moderately hot and moderately wet summers, moderately warm and moderately dry winters, with an average freeze-free season; BB. Cool and moist to moderately moist yearly, very hot and moderately wet summers, moderately cold and moderately wet winters, with an average to long freeze-free season. C. Similar to local climate; CC. Similar to natural area climate.

SOILS: A. Ultisol; AA. Udult. B. Hapludult; BB. Typic Hapludult. C. Clayey, kaolinitic, thermic Typic Hapludult; CC. Herndon silt loam.

GEOLOGY: A. Metamorphic; AA. Mixed metasedimentary and metavolcanic. B. Acidic; BB. Waterlain, fine-grain sediments and tuffs. C. Mixed plutonic, metasedimentary, and metavolcanic rocks of varying composition and degree of exposure; CC. Folded, tuffaceous slates and rhyolitic tuffs not exposed on the surface.

HYDROLOGY: A. Terrestrial; AA. Dry-mesic. B. Fresh; BB. Permanently exposed. C. Refer to Natural Area Diversity Summary; CC. Well drained, wetted by rains and downslope drainage.

TOPOGRAPHY: A. Plains; AA. Irregular plain to irregular plain with slight relief. B. Low hills and shallow to relatively deep valleys; BB. Water-eroded hill. C. Refer to Natural Area Diversity Summary; CC. Open, undulatory, southeast-facing, constant upper slope of a low north- to south-aligned hill.

PHYSIOGRAPHY: A. Appalachian Highlands; AA. Piedmont Province. B. Piedmont Uplands; BB. Iron Mine Hill complex. C. Hill; CC. Upper slope, underlain by early Paleozoic tuffs and slates.

CANOPY: Height, dbh, and age not determined.
CANOPY DOMINANTS: *Quercus alba*, mixed upland hardwoods.
CANOPY PHYSIOGNOMY: Large, deliquescent, deciduous trees.

Canopy Analysis

Species	I.V.	Rel. Den.	Rel. Dom.	Rel. Freq.
Quercus alba	114.1	37.5	43.6	33.0
Carya tomentosa	69.7	27.5	19.1	23.1
Liriodendron tulipifera	38.8	10.0	15.6	13.2
Quercus rubra	36.7	12.5	11.0	13.2
Pinus taeda	18.3	5.0	6.7	6.6
Quercus falcata var. *falcata*	7.6	2.5	1.8	3.3
Quercus stellata	7.1	2.5	1.3	3.3
Fraxinus americana	6.5	2.5	0.8	3.3

Number of points: 10 d: Not determined Number of individuals/acre: Not determined

CANOPY SPECIES PRESENT, BUT NOT IN ANALYSIS: *Acer rubrum, Liquidambar styraciflua, Quercus velutina.*

SUBCANOPY: Not analyzed.

SHRUB LAYER: Not analyzed.

HERB LAYER: Not analyzed.

COMMUNITY REFERENCES: None.

COMMUNITY DOCUMENTATION: Area analyzed by Lee J. Otte for A. E. Radford's Botany 235 class, UNC, Chapel Hill. Specimens deposited in the NCU Herbarium by Lee J. Otte and Deborah K. S. Otte in 1978.

COMMUNITY ECOLOGICAL CHARACTERIZATION: Vegetationally: Fagalean-mixed upland hardwood/? terrestrial broadleaf forest system with a closed canopy dominated by large, deliquescent, deciduous trees. Climatically: Warm temperate, mesothermal climate, similar to the local yearly cool and moist to moderately moist climate, which has very hot and moderately wet summers, moderately cold and moderately wet winters, and an average to long freeze-free period. Pedologically: Clayey, kaolinitic, thermic Typic Hapludult, Herndon silt loam soil. Geologically: Folded, acidic, mixed metasedimentary and meta-volcanic, waterlain tuffaceous slates and rhyolitic tuffs not exposed on the surface. Hydrologically: Permanently exposed, well-drained, dry-mesic terrestrial system, wetted by fresh rains and downslope drainage. Topographically: Open, undulatory, southeast-facing constant upper slope of a low north- to south-aligned hill in a water-eroded landscape of low hills and shallow to relatively deep valleys on an irregular plain with slight relief. Temporally and Spatially: A climax stage of a lithosere on the upper slope of a hill, underlain by early Paleozoic tuffs and slates, in the Iron Mine Hill complex in the Uplands of the Piedmont Province of the Appalachian Highlands.

Community Diversity Summary 9

Terrestrial broadleaf forest system
Large, deliquescent, deciduous trees/
Large, deliquescent, deciduous trees/
Medium, deliquescent, deciduous trees

Oak slope
Fagales/Mixed hardwoods/Cornales

MIXED UPLAND OAKS/MIXED HARDWOODS/*CORNUS FLORIDA*
Mixed upland oaks

CLIMATE: A. Mesothermal; AA. Warm temperate. B. Cool and moderately moist yearly, moderately hot and moderately wet summers, moderately warm and moderately dry winters, with an average freeze-free period; BB. Cool and moist to moderately moist yearly, very hot and moderately wet summers, moderately cold and moderately wet winters, with an average to long freeze-free period. C. Similar to local climate; CC. Similar to natural area climate.

SOILS: A. Alfisol; AA. Udalf. B. Hapludalf; BB. Ultic Hapludalf. C. Fine, mixed, thermic Ultic Hapludalf; CC. Enon loam.
A_0: 0-8 cm, recent leaf litter.
A_1: 8-13 cm, sandy, decayed leaf litter, gray to black.
A_2: 13-51 cm, rocky, fine, sandy clay, yellowish brown, pH 6.0.
B_1: 51+ cm, packed sandy clay, more clay-rich with depth, red, pH 6.5.

GEOLOGY: A. Igneous; AA. Plutonic. B. Intermediate; BB. Pluton. C. Mixed plutonic, metasedimentary, and metavolcanic rocks of varying composition and degree of exposure; CC. Undeformed, plagioclase porphyritic diorite exposed as weathered, small- to medium-sized boulders concentrated in large patches at the top and bottom of the slope.

HYDROLOGY: A. Terrestrial; AA. Dry-mesic. B. Fresh; BB. Permanently exposed. C. Refer to Natural Area Diversity Summary; CC. Well drained, wetted by rains and slight downslope drainage.

TOPOGRAPHY: A. Plains; AA. Irregular plain to irregular plain with slight relief. B. Low hills and shallow to relatively deep valleys; BB. Water-eroded stream valley. C. Refer to Natural Area Diversity Summary; CC. Open, west-southwest-facing, smooth to bouldery, gentle, concave lower slope along the upper reaches of a small, perennial stream.

PHYSIOGRAPHY: A. Appalachian Highlands; AA. Piedmont Province. B. Piedmont Uplands; BB. Iron Mine Hill complex. C. Stream valley; CC. Lower slope along upper reaches of stream, underlain by Paleozoic igneous diorite.

CANOPY: *Quercus alba*--Height 26 m, dbh 63 cm, age 125 years.
Quercus coccinea--Height 24 m, dbh 46 cm, age 90 years.
CANOPY DOMINANTS: Mixed *Quercus* spp.
CANOPY PHYSIOGNOMY: Large, deliquescent, deciduous trees.

Canopy Analysis

Species	I.V.	Rel. Den.	Rel. Dom.	Rel. Freq.
Quercus alba	67.2	22.5	18.7	26.0
Quercus falcata	58.1	20.0	15.7	22.4
Quercus velutina	43.9	15.0	15.7	13.2
Quercus rubra	31.3	10.0	12.5	8.8
Liriodendron tulipifera	26.9	7.5	9.4	10.0
Quercus coccinea	22.1	7.5	9.4	5.2
Carya ovata	11.3	5.0	3.1	3.2
Pinus taeda	11.2	2.5	3.1	5.6
Pinus echinata	8.2	2.5	3.1	2.6
Carya ovalis	7.0	2.5	3.1	1.4
Carya tomentosa	6.4	2.5	3.1	0.8
Carya glabra	6.4	2.5	3.1	0.8

Number of points: 10 d: 14.0 ft Number of individuals/acre: 225

CANOPY SPECIES PRESENT, BUT NOT IN ANALYSIS: None.

TALL SUBCANOPY: Height, dbh, and age not determined.
SUBCANOPY DOMINANTS: Mixed hardwoods.
SUBCANOPY PHYSIOGNOMY: Large, deliquescent, deciduous trees.

Tall Subcanopy Analysis

Species	GF	1 C.S	2 C.S	3 C.S	4 C.S	5 C.S	6 C.S
Large trees							
Acer rubrum	H2	-	-	2.1	-	-	4.1
Carya tomentosa	H2	-	-	4.1	-	-	-
Liquidambar styraciflua	H2	5.1	-	-	1.1	-	-
Liriodendron tulipifera	H2	-	-	-	-	2.1	-
Oxydendrum arboreum	H2	-	3.1	3.1	-	5.1	3.1
Quercus alba	H2	5.1	3.1	5.1	-	3.1	5.1
Quercus rubra	H2	5.1	-	-	-	-	-

Number of relevés: 6 Relevé size: 5 m X 5 m

TALL SUBCANOPY SPECIES PRESENT, BUT NOT IN ANALYSIS: LARGE TREE: *Juniperus virginiana*.

SHORT SUBCANOPY: Height, dbh, and age not determined.
SUBCANOPY DOMINANT: *Cornus florida*.

SUBCANOPY PHYSIOGNOMY: Medium, deliquescent, deciduous trees.

Short Subcanopy Analysis

Species	GF	1 C.S	2 C.S	3 C.S	4 C.S	5 C.S	6 C.S
Medium tree							
Cornus florida	H2	4.1	5.1	5.1	5.1	5.1	5.1

Number of relevés: 6 Relevé size: 5 m X 5 m

SHORT SUBCANOPY SPECIES PRESENT, BUT NOT IN ANALYSIS: TRANSGRESSIVES: *Acer rubrum, Oxydendrum arboreum, Quercus alba.*

SHRUB LAYER DOMINANTS: None.
SHRUB LAYER PHYSIOGNOMY: Not applicable.

Shrub Analysis

Species	GF	1 C.S	2 C.S	3 C.S	4 C.S	5 C.S	6 C.S
Transgressive trees							
Acer rubrum	H2	5.1	5.1	-	3.1	4.1	3.1
Carya tomentosa	H2	+.1	-	+.1	2.1	+.1	1.1
Fraxinus americana	H2	2.1	-	1.1	-	-	1.1
Juniperus virginiana	H1	-	-	-	-	1.1	-
Nyssa sylvatica	H2	-	-	-	4.1	-	-
Oxydendrum arboreum	H2	-	3.1	-	-	3.1	2.1
Prunus serotina	H2	+.1	-	-	-	+.1	1.1
Quercus alba	H2	-	-	+.1	1.1	-	-
Quercus rubra	H2	+.1	-	-	1.1	-	-
Transgressive shrub							
Chionanthus virginicus	H11	1.1	-	1.1	-	-	-
Tall shrubs							
Aesculus sylvatica	H6	-	-	1.1	-	-	-
Rhododendron nudiflorum	H11	-	2.3	3.3	-	-	-
Viburnum prunifolium	H11	+.1	-	1.1	+.1	+.1	-
Viburnum rafinesquianum	H11	4.3	+.1	1.1	1.1	1.1	2.1
Normal shrub							
Euonymus americanus	H11	2.1	+.1	2.1	-	1.1	+.1
Tall dwarf shrubs							
Vaccinium stamineum	H11	2.3	-	1.1	-	+.1	1.1
Vaccinium tenellum	H11	-	3.3	-	-	3.3	-
Vaccinium vacillans	H11	-	3.3	-	-	-	-
Vines							
Lonicera sempervirens	H19	-	-	-	-	-	+.1
Vitis rotundifolia	H18	-	-	-	2.1	-	-

Number of relevés: 6 Relevé size: 5 m X 5 m

SHRUB SPECIES PRESENT, BUT NOT IN ANALYSIS: TRANSGRESSIVE TREES: *Cornus florida, Ilex opaca*; TALL SHRUB: *Viburnum rufidulum*.

HERB LAYER DOMINANTS: None.
HERB LAYER PHYSIOGNOMY: Not applicable.

Herb Analysis

Species	GF	1 C.S	2 C.S	3 C.S	4 C.S	5 C.S	6 C.S
Medium forbs							
Chimaphila maculata	H11	-	-	-	-	1.1	2.1
Hexastylis arifolia	H11	-	-	-	-	1.1	1.1

Number of relevés: 6 Relevé size: 1 m X 1 m

HERB SPECIES PRESENT, BUT NOT IN ANALYSIS: TALL FORBS: *Desmodium nudiflorum, Tipularia discolor*; TALL GRAMINOIDS: *Carex* sp., *Panicum* sp.

COMMUNITY REFERENCES: None.

COMMUNITY DOCUMENTATION: Area analyzed by Lee J. Otte for A. E. Radford's Botany 235 class, UNC, Chapel Hill. Specimens deposited in the NCU Herbarium by Lee J. Otte and Deborah K. S. Otte in 1978.

COMMUNITY ECOLOGICAL CHARACTERIZATION: Vegetationally: Fagalean/Mixed hardwood/ Cornalean terrestrial broadleaf forest system with a closed canopy of large, deliquescent, deciduous trees, a closed tall subcanopy of large, deliquescent, deciduous trees, a closed short subcanopy of medium, deliquescent, deciduous trees, a sparse shrub layer of scattered, transgressive, deliquescent, deciduous trees and mixed, tall dwarf to tall, deciduous shrubs, and a sparse herb layer of scattered forbs and graminoids. Climatically: Warm temperate, mesothermal climate, similar to the local yearly cool and moist to moderately moist climate, which has very hot and moderately wet summers, moderately cold and moderately wet winters, and an average to long freeze-free period. Pedologically: Fine, mixed, thermic Ultic Hapludalf, Enon loam soil. Geologically: Undeformed, intermediate, igneous pluton of plagioclase porphyritic diorite exposed as weathered, small- to medium-sized boulders concentrated in large patches at the top and bottom of the slope. Hydrologically: Permanently exposed, well-drained, dry-mesic, terrestrial system, wetted by fresh rains and some downslope drainage. Topographically: Open, west-southwest-facing, smooth to bouldery, gentle, concave lower slope along the upper reaches of a small, perennial stream in a water-eroded stream valley in a landscape of very low hills and shallow to relatively deep valleys on an irregular plain with slight relief. Temporally and Spatially: A climax community of a lithosere on the lower slopes of a stream valley, underlain by Paleozoic igneous diorite, in the Iron Mine Hill complex in the Uplands of the Piedmont Province of the Appalachian Highlands.

Community Diversity Summary 10

Terrestrial broadleaf forest system — Oak-hickory slope
Large, deliquescent, deciduous trees/? — Juglandales-Fagales/?

CARYA GLABRA-MIXED UPLAND OAKS AND HICKORIES/?
Carya glabra-Mixed upland oaks and hickories

CLIMATE: A. Mesothermal; AA. Warm temperate. B. Cool and moderately moist yearly, moderately hot and moderately wet summers, moderately warm and moderately dry winters, with an average freeze-free period; BB. Cool and moist to moderately moist yearly, very hot and moderately wet summers, moderately cold and moderately wet winters, with an average to long freeze-free period. C. Similar to local climate; CC. Similar to natural area climate.
SOILS: A. Alfisol; AA. Udalf. B. Hapludalf; BB. Ultic Hapludalf. C. Fine, mixed, thermic Ultic Hapludalf; CC. Enon loam. No data for soil horizons.
GEOLOGY: A. Igneous; AA. Plutonic. B. Intermediate; BB. Pluton. C. Mixed plutonic, metasedimentary, and metavolcanic rocks of varying composition and degree of exposure; CC. Diorite exposed as isolated, fresh, large boulders.
HYDROLOGY: A. Terrestrial; AA. Dry-mesic. B. Fresh; BB. Permanently exposed. C. Refer to Natural Area Diversity Summary; CC. Well drained, wetted by rains and downslope drainage.
TOPOGRAPHY: A. Plains; AA. Irregular plain to irregular plain with slight relief. B. Low hills and shallow to relatively deep valleys; BB. Water-eroded hill. C. Refer to Natural Area Diversity Summary; CC. Open, irregular to slightly rocky, south-southwest-facing, constant, gentle midslope of an overall convex hill slope overlooking a relatively deep, protected, northwest-southeast-oriented small stream valley.
PHYSIOGRAPHY: A. Appalachian Highlands; AA. Piedmont Province. B. Piedmont Uplands; BB. Iron Mine Hill complex. C. Hill; CC. Midslope, underlain by Paleozoic igneous diorite.

CANOPY: Height, dbh, and age not determined.
CANOPY DOMINANTS: *Carya glabra*, mixed *Quercus* and *Carya* spp.
CANOPY PHYSIOGNOMY: Large, deliquescent, deciduous trees.

Canopy Analysis

Species	I.V.	Rel. Den.	Rel. Dom.	Rel. Freq.
Carya glabra	80.5	30.0	31.3	19.2
Quercus alba	35.9	12.5	10.6	12.8
Quercus stellata	32.8	10.0	10.0	12.8
Carya tomentosa	31.8	12.5	9.7	9.6
Quercus rubra	26.6	7.5	9.5	9.6
Quercus falcata	26.5	7.5	9.4	9.6
Fraxinus americana	23.1	7.5	6.0	9.6
Carya carolinae-septentrionalis	9.8	2.5	4.1	3.2

Canopy Analysis (continued)

Species	I.V.	Rel. Den.	Rel. Dom.	Rel. Freq.
Fagus grandifolia	9.4	2.5	3.7	3.2
Liriodendron tulipifera	8.3	2.5	2.6	3.2
Acer rubrum	7.4	2.5	1.7	3.2
Carya ovata	6.9	2.5	1.2	3.2

Number of points: 10 d: Not determined Number of individuals/acre: Not determined

CANOPY SPECIES PRESENT, BUT NOT IN ANALYSIS: None.

SUBCANOPY: Not analyzed.

SHRUB LAYER: Not analyzed.

HERB LAYER: Not analyzed.

COMMUNITY REFERENCES: None.

COMMUNITY DOCUMENTATION: Area analyzed by Lee J. Otte for A. E. Radford's Botany 235 class, UNC, Chapel Hill. Specimens deposited in the NCU Herbarium by Lee J. Otte and Deborah K. S. Otte in 1978.

COMMUNITY ECOLOGICAL CHARACTERIZATION: Vegetationally: Juglandalean-Fagalean/? terrestrial broadleaf forest system with a closed canopy of large, deliquescent, deciduous trees. Climatically: Warm temperate, mesothermal climate, similar to the local yearly cool and moist to moderately moist climate, which has very hot and moderately wet summers, moderately cold and moderately wet winters, and an average to long freeze-free period. Pedologically: Fine, mixed, thermic Ultic Hapludalf, Enon loam soil. Geologically: Undeformed, intermediate, igneous pluton of diorite, exposed as isolated, fresh, large boulders. Hydrologically: Permanently exposed, well-drained, dry-mesic, terrestrial system, wetted by rains and downslope drainage. Topographically: Open, irregular to slightly rocky, south-southwest-facing, constant, gentle midslope of an overall convex hill slope, overlooking a relatively deep, protected, northwest-southeast-oriented small stream valley in a landscape of low hills and shallow to relatively deep valleys on an irregular plain with slight relief. Temporally and Spatially: A climax community of a lithosere on the midslope of a hill, underlain by Paleozoic igneous diorite, in the Iron Mine Hill complex in the Uplands of the Piedmont Province of the Appalachian Highlands.

Community Diversity Summary 11

Terrestrial broadleaf forest system
Large, deliquescent, deciduous trees/
Tall, rhizomatous, deciduous shrubs

Oak-hickory hill crest
Fagales-Juglandales/Dipsacales

QUERCUS STELLATA-MIXED UPLAND HICKORIES AND OAKS/*VIBURNUM RAFINESQUIANUM*
Quercus stellata-mixed upland hickories and oaks

CLIMATE: A. Mesothermal; AA. Warm temperate. B. Cool and moderately moist yearly, moderately hot and moderately wet summers, moderately warm and moderately dry winters, with an average freeze-free period; BB. Cool and moist to moderately moist yearly, very hot and moderately wet summers, moderately cold and moderately wet winters, with an average to long freeze-free period. C. Similar to local climate; CC. Similar to natural area climate.

SOILS: A. Alfisol; AA. Udalf. B. Hapludalf; BB. Ultic Hapludalf. C. Fine, mixed, thermic Ultic Hapludalf; CC. Enon loam.
A_0: 0-10 cm, leaf litter and decayed organics, gray.
A_1: 10-43 cm, leached, loosely packed, rocky, clayey silt, yellow, pH 6.0.
B_1: 43+ cm, packed, silty clay, yellowish brown, increasingly red with depth, pH 6.0.

GEOLOGY: A. Igneous; AA. Plutonic. B. Intermediate; BB. Pluton. C. Mixed plutonic, metasedimentary, and metavolcanic rocks of varying composition and degree of exposure; CC. Undeformed, amphibole porphyritic diorite exposed as small to large, weathered, isolated, and piled boulders.

HYDROLOGY: A. Terrestrial; AA. Dry-mesic. B. Fresh; BB. Permanently exposed. C. Refer to Natural Area Diversity Summary; CC. Well drained, wetted by rains.

TOPOGRAPHY: A. Plains; AA. Irregular plain to irregular plain with slight relief. B. Low hills and shallow to relatively deep valleys; BB. Water-eroded hill. C. Refer to Natural Area Diversity Summary; CC. Open, variously exposed, bouldery, flat but gently sloping hill crest and gentle, convex, extreme upper slopes.

PHYSIOGRAPHY: A. Appalachian Highlands; AA. Piedmont Province. B. Piedmont Uplands; BB. Iron Mine Hill complex. C. Hill; CC. Hill crest and extreme upper slopes, underlain by Paleozoic igneous diorite.

CANOPY: *Quercus stellata*--Height 13 m, dbh 34 cm, age 150 years.
Carya ovata--Height 18 m, dbh 38 cm, age 187 years.
Carya tomentosa--Height 15 m, dbh 50 cm, age 220 years.

CANOPY DOMINANTS: *Quercus stellata*, mixed upland *Carya* and *Quercus* spp.

CANOPY PHYSIOGNOMY: Large, deliquescent, deciduous trees.

Canopy Analysis

Species	I.V.	Rel. Den.	Rel. Dom.	Rel. Freq.
Quercus stellata	124.7	47.5	46.5	30.7
Carya tomentosa	69.5	22.5	20.1	26.9
Carya glabra	28.7	7.5	9.6	11.7

Canopy Analysis (continued)

Species	I.V.	Rel. Den.	Rel. Dom.	Rel. Freq.
Quercus rubra	23.8	7.5	4.8	11.4
Carya ovata	23.6	5.0	10.9	7.7
Quercus marilandica	21.1	7.5	5.9	7.7
Quercus alba	8.6	2.5	2.2	3.9

Number of points: 10 d: 12.5 ft Number of individuals/acre: 282

CANOPY SPECIES PRESENT, BUT NOT IN ANALYSIS: None.

SUBCANOPY: No well-defined subcanopy present.

SHRUB LAYER DOMINANT: *Viburnum rafinesquianum*.
SHRUB LAYER PHYSIOGNOMY: Tall, rhizomatous, deciduous shrubs.

Shrub Analysis

Species	GF	1 C.S	2 C.S	3 C.S	4 C.S	5 C.S	6 C.S	7 C.S
Transgressives								
Acer rubrum	H2	-	-	-	-	4.1	1.1	-
Carya glabra	H2	-	4.1	5.1	3.1	-	-	3.1
Carya tomentosa	H2	-	-	-	3.1	3.1	2.1	2.1
Cornus florida	H2	-	-	1.1	5.1	-	2.1	-
Fraxinus americana	H2	+.1	-	-	-	-	5.1	+.1
Juniperus virginiana	H1	2.1	5.1	-	5.1	2.1	-	1.1
Prunus serotina	H2	-	-	-	-	1.1	-	-
Quercus alba	H2	4.1	-	3.1	-	-	-	2.1
Quercus marilandica	H2	-	1.1	-	-	-	-	-
Quercus phellos	H2	-	-	-	1.1	-	-	-
Quercus rubra	H2	4.1	1.1	-	-	-	1.1	5.1
Quercus velutina	H2	4.1	1.1	-	1.1	1.1	-	-
Tall shrubs								
Viburnum prunifolium	H11	5.3	-	-	2.1	2.1	2.1	-
Viburnum rafinesquianum	H11	5.4	5.4	5.4	5.4	4.4	4.4	3.3
Normal shrub								
Euonymus americanus	H11	-	+.1	-	-	-	-	-
Tall dwarf shrubs								
Vaccinium stamineum	H11	-	1.1	-	-	3.3	1.1	5.4
Vaccinium tenellum	H11	-	-	-	-	5.4	-	-
Vines								
Lonicera japonica	H19	-	-	1.1	-	-	-	-
Vitis rotundifolia	H18	-	-	-	-	2.1	-	-

Number of relevés: 7 Relevé size: 5 m X 5 m

SHRUB SPECIES PRESENT, BUT NOT IN ANALYSIS: TRANSGRESSIVE: *Ulmus alata*; PARASITIC SHRUB: *Phoradendron serotinum*; VINES: *Lonicera sempervirens*, *Smilax bona-nox*, *Smilax glauca*.

HERB LAYER DOMINANTS: None.
HERB LAYER PHYSIOGNOMY: Not applicable.
HERB ANALYSIS: Only a species presence list completed. Species list--TALL FORBS: *Chimaphila maculata*, *Desmodium paniculatum*, *Tephrosia* sp.; TALL GRAMINOIDS: *Carex* sp., *Panicum* spp.

COMMUNITY REFERENCES: None.

COMMUNITY DOCUMENTATION: Area analyzed by Lee J. Otte for A. E. Radford's Botany 235 class, UNC, Chapel Hill. Specimens deposited in the NCU Herbarium by Lee J. Otte and Deborah K. S. Otte in 1978.

COMMUNITY ECOLOGICAL CHARACTERIZATION: Vegetationally: Fagalean-Juglandalean/ Dipsacalean terrestrial broadleaf forest system with a closed canopy of large, deliquescent, deciduous trees, a closed shrub layer of tall, rhizomatous, deciduous shrubs, and a sparse herb layer containing widely scattered forbs and graminoids. Climatically: Warm temperate, mesothermal climate, similar to the local yearly cool and moist to moderately moist climate, which has very hot and moderately wet summers, moderately cold and moderately wet winters, and an average to long freeze-free period. Pedologically: Fine, mixed, thermic Ultic Hapludalf, Enon loam soil. Geologically: Undeformed, intermediate, igneous pluton of amphibole porphyritic diorite exposed as frequent isolated and piled, small to large, weathered boulders. Hydrologically: Permanently exposed, well-drained, dry-mesic, terrestrial system wetted by fresh rains. Topographically: Open, variously exposed, bouldery, flat but gently sloping hill crest and gentle, convex, extreme upper slopes on a water-eroded hill in a landscape of very low hills and shallow to relatively deep valleys on an irregular plain with slight relief. Temporally and Spatially: A climax community of a lithosere on the crest and upper slopes of a hill, underlain by Paleozoic igneous diorite, in the Iron Mine Hill complex in the Uplands of the Piedmont Province of the Appalachian Highlands.

Master Species Presence List

ANGIOSPERMS

Araliales
  Apiaceae
    Thaspium barbinode (Michaux) Nuttall (5*)
    Zizia aurea (L.) W. D. J. Koch (5)

Aristolochiales
  Aristolochiaceae
    Hexastylis arifolia (Michaux) Small (5-7, 9)

*Refer to Community Types. See the end of the Master Species Presence List.

Asterales
Asteraceae
Antennaria plantaginifolia (L.) Richardson (5)
Aster infirmus Michaux (5)
Chrysogonum virginianum L. (5)
Coreopsis major Walter (5)
Elephantopus carolinianus Willd. (5)
E. tomentosus L. (5)
Hieracium venosum L. (5)
Prenanthes altissima L. (5)
Solidago caesia L. (5)

Celastrales
Aquifoliaceae
Ilex ambigua var. *montana* (T. & G.) Ahles (5)
I. opaca Aiton (9)
Celastraceae
Euonymus americanus L. (5-7, 9, 11)

Cornales
Cornaceae
Cornus amomum Miller (5)
C. florida L. (5-7, 9, 11)
Nyssaceae
Nyssa sylvatica Marshall (1, 2, 5-7, 9)

Cyperales
Cyperaceae
Carex rosea Schkuhr (6)
Carex spp. (6, 9, 11)
Poaceae
Leersia virginica Willd. (5)
Panicum spp. (5, 6, 9, 11)

Dipsacales
Caprifoliaceae
Lonicera japonica Thunberg (5, 11)
L. sempervirens L. (5-7, 9, 11)
Sambucus canadensis L. (5)
Viburnum acerifolium L. (5-7)
V. prunifolium L. (5-7, 9, 11)
V. rafinesquianum Schultes
V. rufidulum Raf. (5, 7, 9)

Ebenales
Ebenaceae
Diospyros virginiana L. (5-7)

Ericales
Ericaceae
Chimaphila maculata (L.) Pursh (5-7, 9, 11)
Monotropa hypopithys L. (6)
Oxydendrum arboreum (L.) DC. (5-7, 9)
Rhododendron nudiflorum (L.) Torrey (5, 9)
Vaccinium stamineum L. (7, 9, 11)
V. tenellum Aiton (5, 7, 9, 11)
V. vacillans Torrey (5, 9)

Fagales
Betulaceae
Carpinus caroliniana Walter (5)
Fagaceae
Fagus grandifolia Ehrhart (2-6, 10)
Quercus alba L. (1-11)
Q. coccinea Muenchh. (7, 9)
Q. falcata var. *falcata* Michaux (6, 8-10)
Q. falcata var. *pagodaefolia* Ell. (1, 2)
Q. falcata X *Q. phellos* (2)
Q. marilandica Muenchh. (11)
Q. phellos L. (1-3, 11)
Q. rubra L. (2-11)
Q. stellata Wang. (4, 6-11)
Q. velutina Lam. (6-9, 11)

Gentianales
Gentianaceae
Gentiana villosa L. (6)

Geraniales
Oxalidaceae
Oxalis sp. (5)

Hamamelidales
Hamamelidaceae
Liquidambar styraciflua L. (1-6, 8, 9)
Platanaceae
Platanus occidentalis L. (1)

Juglandales
Juglandaceae
Carya carolinae-septentrionalis (Ashe) Engler & Graebner (10)
Carya glabra (Miller) Sweet (2, 3, 5-7, 9-11)
C. ovalis (Wang.) Sargeant (9, 10)

C. ovata (Miller) K. Koch (10, 11)
C. tomentosa (Poiret) Nuttall (2, 3, 5, 6-11)

Lamiales
Lamiaceae
Cunila origanoides (L.) Britton (5)
Prunella vulgaris L. (5)
Scutellaria elliptica Muhl. (5)
S. integrifolia L. (5)

Liliales
Dioscoreaceae
Dioscorea villosa L. (5-7)
Iridaceae
Iris cristata Aiton (5)
Liliaceae
Polygonatum biflorum (Walter) Ell. (5, 7)
Smilacina racemosa (L.) Desf. (5)
Trillium sp. L. (5)
Uvularia perfoliata L. (7)
Smilacaceae
Smilax bona-nox L.
S. glauca Walter (6, 7, 11)
S. rotundifolia L. (5)

Magnoliales
Lauraceae
Sassafras albidum (Nuttall) Nees (5, 6)
Magnoliaceae
Liriodendron tulipifera L. (1-10)

Orchidales
Orchidaceae
Goodyera pubescens (Willd.) R. Brown (5, 7)
Tipularia discolor (Pursh) Nuttall (5, 6, 9)

Ranunculales
Ranunculaceae
Hepatica americana (DC.) Ker.

Rhamnales
Vitaceae
Parthenocissus quinquefolia (L.) Planchon (5, 6)
Vitis aestivalis Michaux (5-7)
V. rotundifolia Michaux (5, 7, 9, 11)

Rosales
Fabaceae
Cercis canadensis L. (5, 7)
Desmodium nudiflorum (L.) DC. (5-7, 9)
D. paniculatum (L.) DC. (11)
D. rotundifolium DC. (5-7)
D. sp. (5)
Tephrosia sp. (11)
Rosaceae
Potentilla canadensis L. (5)
Prunus serotina Ehrhart (5-7, 9, 11)
Rubus argutus Link (5)
Saxifragaceae
Heuchera americana L. (5, 7)

Rubiales
Rubiaceae
Galium circaezans Michaux (7)
Houstonia caerulea L. (5)
H. purpurea L. (5)

Santalales
Loranthaceae
Phoradendron serotinum (Raf.) M. C. Johnson (11)

Sapindales
Aceraceae
Acer rubrum L. (1-11)
A. saccharum ssp. *floridanum* (Chapman) Desmarais (2, 7)
Anacardiaceae
Rhus radicans L. (5-7)
Hippocastanaceae
Aesculus sylvatica Bartram (5, 9)

Scrophulariales
Bignoniaceae
Campsis radicans (L.) Seemann (5)
Oleaceae
Chionanthus virginicus L. (5, 7, 9)
Fraxinus americana L. (5-11)
F. pennsylvanica Marshall (1, 11)
Scrophulariaceae
Aureolaria virginica (L.) Pennell (5)

Theales
Hypericaceae
Hypericum sp. (5-7)

Urticales
Moraceae
Morus rubra L. (5, 7)
Ulmaceae
Ulmus alata Michaux (11)
U. americana L. (5)

Violales
Violaceae
Viola papilionacea Pursh (5)
V. sp. (5)

GYMNOSPERMS

Coniferales
Cupressaceae
Juniperus virginiana L. (3, 5-7, 9, 11)
Pinaceae
Pinus echinata Miller (9)
P. taeda L. (1-3, 6-9)

FERNS

Filicales
Aspidiaceae
Polystichum acrostichoides (Michaux) Schott (5)
Aspleniaceae
Asplenium platyneuron (L.) Oakes (5)
Osmundaceae
Osmunda regalis var. *spectabilis* (Willd.) Gray (5)
Polypodiaceae
Polypodium polypodioides (L.) Watt (5)

Iron Mine Hill Communities

1. *Acer rubrum*-mixed mesic hardwoods and oaks/?
2. *Fagus grandifolia*-mixed mesic hardwoods and oaks/?
3. *Liriodendron tulipifera*-mixed mesic hardwoods, oaks, and hickories/?
4. *Fagus grandifolia*/?
5. *Fagus grandifolia*-mixed mesic oaks and hickories
6. *Quercus rubra*-*Quercus alba*/Mixed hardwoods/*Cornus florida*
7. *Carya tomentosa*-*Quercus alba*/Mixed hardwoods/Mixed hardwoods/*Viburnum rafinesquianum*
8. *Quercus alba*-mixed upland hardwoods/?
9. Mixed upland oaks/Mixed hardwoods/*Cornus florida*
10. *Carya glabra*-mixed upland oaks and hickories/?
11. *Quercus stellata*-mixed upland hickories and oaks/*Viburnum rafinesquianum*

B

Swift Creek Swamp Forest Natural Area*

Edgecombe County, North Carolina; ca. 3 miles (4.8 km) east-northeast of Battleboro, N.C., along SSR 1404 0.5 mile (0.8 km) east of its junction with SSR 1411, and extending to the Atlantic Coast Line Railroad tracks.
36°04'N, 77°41'W; Whitakers, N.C., quad. 7 1/2', 1961
Wicomico Terrace; Atlantic Coastal Plain; Coastal Plain
75'-85'; 23 m-26 m
ca. 3,200 acres (1,300 hectares)

Ownership: Local agricultural companies and private. The largest portion, consisting of most of the area south of the creek, is owned by the M. C. Braswell Corporation, operating out of Battleboro. Other landowners include Robert Merritt, Jr., and J. N. "Sandy" Taylor.

Administration: Same as ownership.

Land use: Hunting by the Pope Gun Club and others, fishing in Swift Creek, lumbering at times over most of the area, some farming in small areas within the swamp that have been cleared and drained.

Land classification: Unknown.

Danger to integrity: Further lumbering in at least parts of the forest seems likely although not imminent. Robert Merritt, Jr., on whose land the least disturbed portions of the swamp occur, says that he "does not plan to cut." Further clearing and draining for farmland also seem likely. In addition, road-building and small fires have come dangerously close to destroying the only known location in the Carolinas for *Ranunculus flabellaris* (Water Crowfoot).

Publicity sensitivity: Little or none.

Significance and protection priority: Appears to be regionally significant; site is in some jeopardy.

Reasons for priority rating:

Endangered or threatened species and communities: Presence of endangered disjunct *Ranunculus flabellaris*--only known population in the Carolinas.

General diversity: An excellent collection of characteristic swamp forest species; the natural area contains 45 canopy species, including 13 oak species. Excellent example of climax zonation, the canopy species tending to occupy characteristic niches within the swamp. Unique community composition; the swamp oak communities contain much greater than average diversity of oak species. No mention is found in the literature of such diversity.

---

*Data compiled and report written by Alan S. Weakley in 1977 and 1978. Revised and edited by Alan S. Weakley, Lee J. Otte, and Deborah K. Strady Otte in 1979.

Natural features condition: Excellent stand of swamp forest trees, much of it undisturbed for many years. The following large trees have been noted:

Taxodium distichum (Bald Cypress)	52" (13.2 dm) dbh
Quercus palustris (Pin Oak)	47" (11.9 dm)
Nyssa aquatica (Water Gum)	45" (11.4 dm)
Ulmus americana (American Elm)	42" (10.7 dm)
Quercus falcata var. *pagodaefolia* (Cherrybark Oak)	40" (10.2 dm)
Quercus hemisphaerica (Laurel Oak)	39" (9.9 dm)
Quercus lyrata (Overcup Oak)	39" (9.9 dm)
Liquidambar styraciflua (Sweet-gum)	37" (9.4 dm)
Acer rubrum (Red Maple)	36" (9.1 dm)
Quercus bicolor (Swamp White Oak)	27" (6.9 dm)
Populus heterophylla (Swamp Cottonwood)	25" (6.4 dm)
Pinus taeda (Loblolly Pine)	24" (6.1 dm)
Carya aquatica (Water Hickory)	21" (5.3 dm)
Diospyros virginiana (Persimmon)	20" (5.1 dm)
Carya ovata (Shagbark Hickory)	19" (4.8 dm)

Distribution features: Presence of transition communities and zones; the swamp proper contains at least five significantly different communities. Presence of excellent continua; all the swamp species occupy relatively specific niches and are present or abundant in only one or two communities but phase in and out across the spectrum from hydric to mesic.

Humanistic features: The Swift Creek Swamp has high aesthetic value. The very large trees, high canopy, many lianas, and sparse lower layers of vegetation make it a striking and unusual area to walk through in the dry season. The scientific research value is also high; the very high diversity of canopy species and the opportunity to study their associations and the environmental parameters governing their distribution within the swamp are outstanding.

Productivity: The biomass and productivity of the swamp communities are very high; the following animals have breeding grounds or important territories within the natural area: River Otter, mink, muskrat, raccoon, deer, Pileated Woodpecker, Swainson's Warbler, Prothonotary Warbler, Wood Duck, Water Moccasin, freshwater clams, and mussels.

Management recommendations: The natural area should be preserved partly as wilderness and partly managed, goals compatible with its extensive size. Purposes for management include endangered species preservation (*Ranunculus flabellaris*), scientific research, and educational resource (for further study of the numerous and distinctive swamp forest communities, particularly those communities associated with the bottomland oaks, to determine the environmental parameters involved in the "sorting out" of species in the swamp), possible unique communities preservation, wildlife resource, and recreation.

Data sources:
Dan Harrell, employee, M. C. Braswell Corporation, Battleboro, N.C.
Robert Merritt, Jr., owner, Merritt Farms, Battleboro, N.C.
Albert E. Radford, Department of Botany, UNC-Chapel Hill.

General scientific references: None.

General documentation and authentication: Primary data were collected in June 1977 by A. E. Radford's Ecosystematics class (Botany 235) at UNC-Chapel Hill. Subsequent visits were made in July, August, September, October, and November of 1977 and in February, April, May, and July of 1978. Specimens deposited in the NCU Herbarium by Alan S. Weakley in 1977 and 1978.

Discussion

The Swift Creek Swamp Forest is located about three miles (4.8 km) east-northeast of Battleboro, North Carolina, and eleven miles (17.7 km) northeast of the small city of Rocky Mount, North Carolina. It lies in the extreme upper Coastal Plain--the Piedmont bluffs begin less than ten miles (16 km) to the west. The part of the forest examined occupies a strip 2,500 yards (2,285 m) wide and stretches five miles (8 km) along Swift Creek in a broad, flat, alluvial valley. Surrounding land is largely agricultural--corn, tobacco, cotton, soybeans, and peanuts are the principal crops.

The climate is warm temperate, the annual mean precipitation moderately moist, the freeze-free period long (U.S. Department of the Interior, 1970). The rainfall is greatest in the summer, but increased evaporation in this season makes it effectively the driest season in the swamp. In the summer and fall one can walk about in the swamp in areas that are covered by a foot to two feet (3-6 dm) of standing water in winter and early spring.

The soils of Swift Creek Natural Area are diverse, and this diversity is related closely to that of hydroperiod and vegetation. The pools are underlain by a Cumulic Humaquept of the Johnston series, a soil with a layer of muck overlying a relatively impervious loamy fine sand. The flat is a Typic Albaqualf of the Meggett series, a very wet, basic soil with an impervious clay layer. A slightly higher flat is characterized by an Aquic Hapludult of the Altavista series, a saturated soil, but somewhat coarser textured and more permeable. The remaining higher areas are underlain by Typic Hapludults of the Wickham and Gritney series; these soils are more mesic, in general better drained because of permeability and runoff, and more acidic than the swamp soils. The presence of a clay layer in most of the swamp accounts for the standing water even in the driest summers. (See Steila, 1976; Soil Survey Staff, 1975; and Community Diversity Summaries.) A high correlation is found to exist between soil type and plant community. Variations in texture and minor topography within a soil type delimit microhabitats and consequently microcommunities.

The entire swamp is underlain by Recent, unconsolidated, alluvial deposits, overlying the Yorktown formation (North Carolina Department of Conservation and Development, 1958), which contains marl fragments. The resultant high pH's (to 8.0) and nutrient availability, conbined with adequate moisture, provide excellent conditions for plant growth. This may explain in part the high diversity of canopy species and the presence of certain basophiles, such as *Ostrya virginiana* (Hop-hornbeam), *Quercus bicolor* (Swamp White Oak), and *Carya ovata* (Shagbark Hickory).

Despite its name, Swift Creek is a low-gradient, slow, meandering, brown-water stream, rarely more than thirty feet (9 m) wide and three feet (9 dm) deep. It drops less than a foot per mile (3 dm per km) as it winds through the swamp. In fact, the variation in elevation throughout the six-square-mile (15.5 km^2) area is only about ten feet (3 m), but these minute irregularities of topography produce profound changes in vegetation. A transect across the swamp might produce elevation differences of only a foot (3 dm), but the lowest areas would be flooded permanently and the highest rarely or never flooded. Because climate and geology are constants for the entire swamp and variations in soil and hydrology are for the most part regulated by changes in topography, the elevation in any given area largely controls the vegetation. Texture, and therefore drainage, is also of primary importance but is a less critical factor than frequency of inundation. Only fairly minor variations in species distribution along an isohydric line* occur, and these are almost always occasioned by changes in soil texture. Of any two locations on an isohydric line, the coarser textured will be relatively more mesic, the finer textured more hydric, with corresponding changes in species present. This brings about a shift in a species' water tolerance--a hydrophilic species may be able to live in a clayey substrate with a hydroperiod of 220 days, but under sandier conditions may require a 250-day hydroperiod. Other factors are operating also, however. Some species have more absolute preferences for soil texture and have a preference for a fairly specific hydroperiod and soil texture; these species obviously will not respond to the "shift effect." These various factors working together through time produce the characteristic distribution of species through the swamp and are therefore directly responsible for the composition of

---

*Isohydric lines are the edges of flooding at successive times during the season. They are thus measures of equal periods of inundation and of elevation relative to the creek.

the various communities. Figure B1 is a qualitative representation of the hydroperiod (moisture) requirements of canopy species found at Swift Creek. Figure B2 shows texture preferences for selected canopy species, and Figure B3 combines Figures B1 and B2. The various shapes in Figure B3 reveal interesting factors in the environmental parameters of a species: a horizontal oval indicates a species for which texture is more critical than hydroperiod, a vertical oval indicates a species for which hydroperiod is the more important factor, and a diagonal oval (upper left-lower right) shows approximately equal importance, with the "shift effect" working. Quantification of these factors would be an excellent topic of future study.

The swamp forest also bears a distinctive and interesting fauna. Abundant mammal species include deer, muskrats, and raccoons. Raccoons occur particularly along the banks of Swift Creek, where they feed on freshwater mussels and clams, leaving behind mounds of empty shells. Area trappers also report good populations of mink, River Otter, and weasel. Birds commonly nesting within the swamp include Pileated Woodpecker, Wood Duck, Green Heron, Red-eyed Vireo, and Prothonotary and Swainson's Warblers; the two warblers are of particular interest because their primary nesting sites are limited to swamp forests and they are rather uncommon. The abundance of Pileated Woodpeckers is related to their need for extensive forested land, a condition met at Swift Creek but increasingly rarely in the Southeast. The only common reptile encountered is the Cottonmouth or Water Moccasin, whose abundance around the deep pools demands precaution. An unusual amphibian, the Neuse River Waterdog, a giant aquatic salamander averaging 6-9 inches (15-23 cm) and endemic to the Neuse and Tar River systems, probably occurs in Swift Creek (Conant, 1975). Freshwater mussels and clams abound in Swift Creek and, because these tend to require unpolluted water with a good nutrient balance, are probably indicative of a rich aquatic fauna.

The history of disturbance within the Swift Creek Natural Area is varied. Much of the land owned by the Braswell Corporation regularly has been cut selectively, but retains a relatively undisturbed aspect. Many less accessible areas deep in the swamp along the creek have not been cut even selectively in at least sixty years and probably longer. For the purposes of community analysis, tracts of land were selected that showed the least signs of disturbance. All communities discussed, with the possible exceptions of Community Types 4 and 6, are considered to be in a climax condition. Even where disturbance has occurred, replacement is rapid because of the very favorable conditions for timber growth (Dean, 1969).

A. DISCUSSION OF COMMUNITIES

Community Type 1, the deep pool community, occurs in permanently flooded depressions in the bottomland flat. The soil is a fine clay Humaquept, with considerable organic material toward the top. These impermeable depressions retain whatever water they receive, either from the winter inundation, periodic flooding across the levee in other seasons, or rainfall, and thus dry out only in the driest summers. This very rare loss of water is necessary for the continuation of a forested climax because *Nyssa aquatica* (Water Gum) and *Taxodium distichum* (Bald Cypress) seeds must germinate while not submerged and grow to a sufficient height to avoid total inundation in subsequent seasons (Demaree, 1932; Shunk, 1939).

The water is stained a dark brown color because of leaching from the abundant

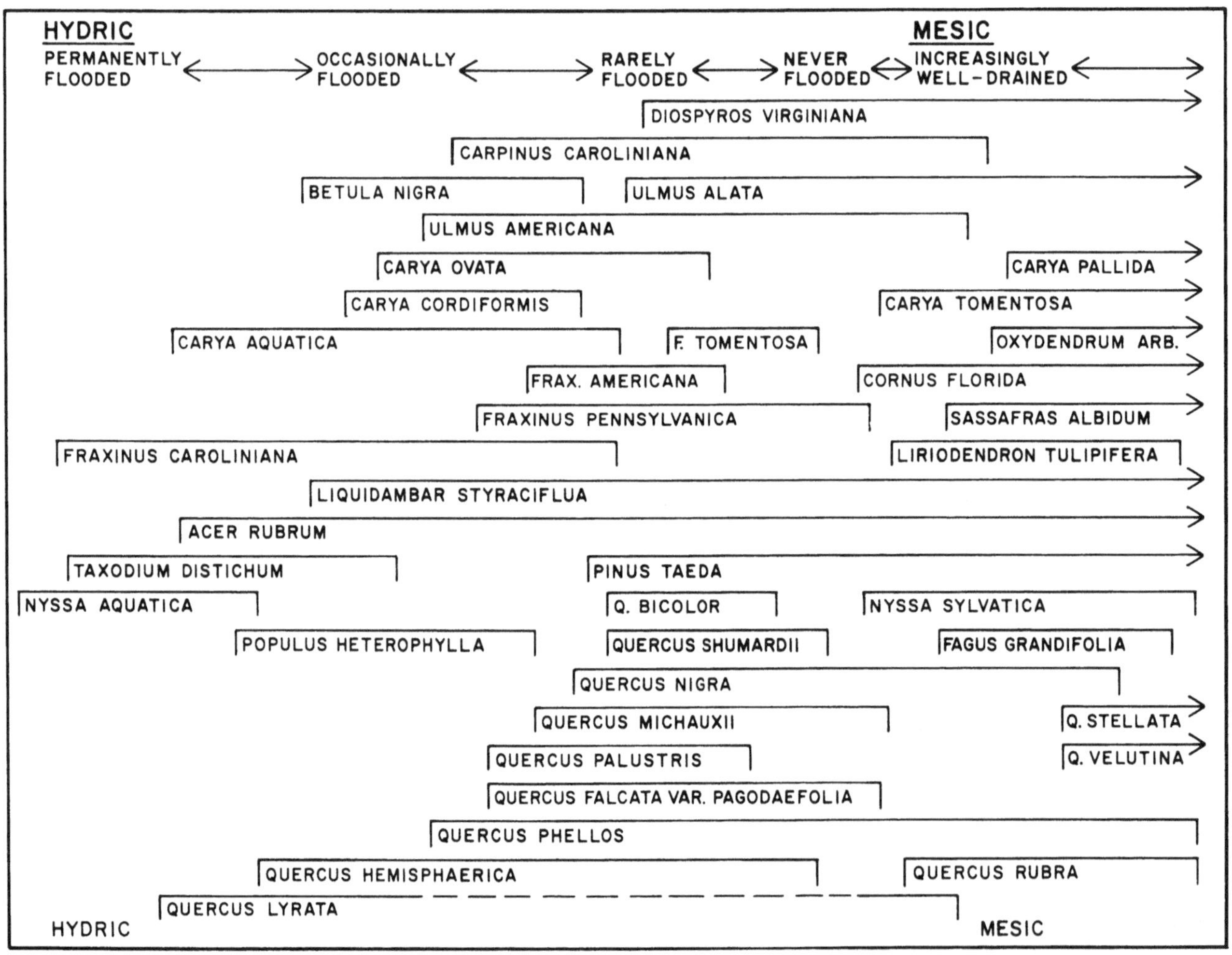

Figure B1. *Moisture tolerances of canopy species.*

organic material that collects in the pools. Typically the pools are interconnected in long chains and the water flows slowly from one pool to another and eventually into Swift Creek. During the winter floods, the pools are connected more strongly to the creek, and fish spread out into the pools, where they frequently are stranded during summer droughts. The presence of fish accounts for the abundance of Water Moccasins and raccoons around the pools in late spring and summer.

The differences between the interlinked pools and the creek should be emphasized. The flow of water is much slower in the pools, and it carries almost no alluvial material--deposition is almost entirely of organic material. For this reason, the well-developed levee of the creek is totally absent. The substrate texture of the pools is mucky to clayey, as opposed to silty or even sandy in the

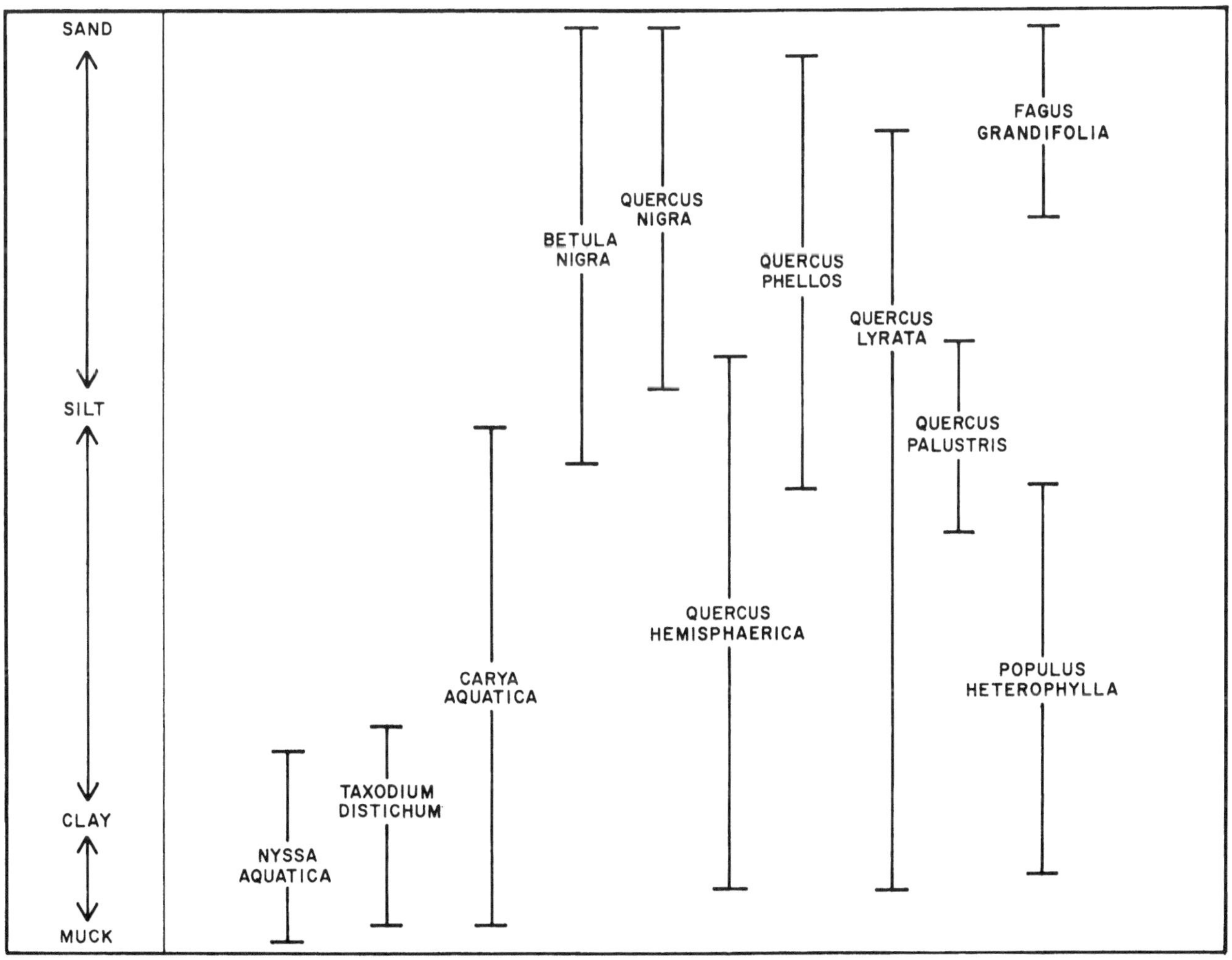

Figure B2. *Soil texture tolerances of selected canopy species.*

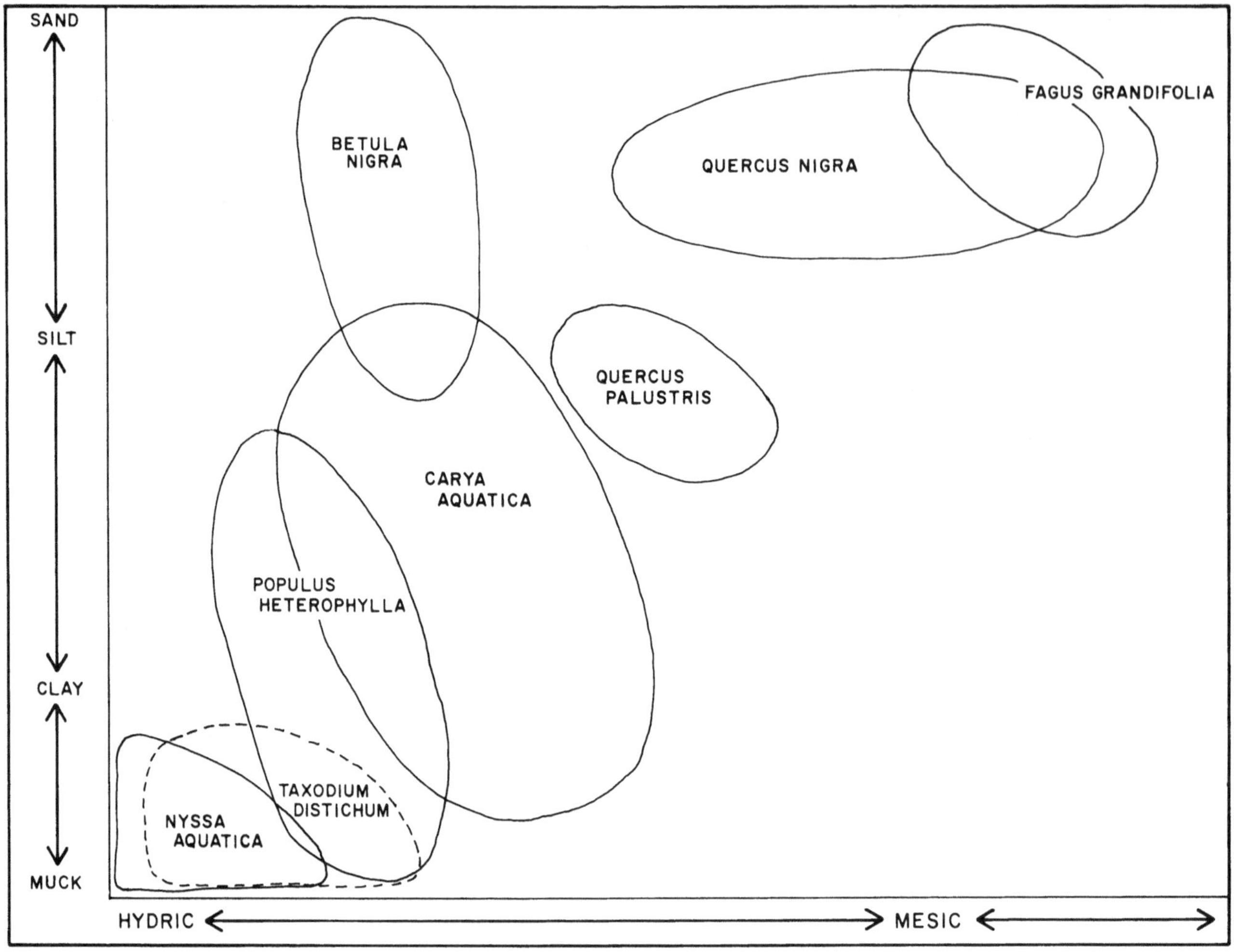

Figure B3. *Moisture and soil texture tolerances of selected canopy species.*

creek. Most important, the rare drying out of the pools allows the presence of a forest community, whereas Swift Creek is totally without trees. The greater oxygen supply in the creek caused by the more rapid water flow and the lesser degree of leaching, however, makes for a richer aquatic fauna.

The deepest pools are dominated by *Nyssa aquatica* with large basal swellings; dbh's range up to 46 inches (114 cm), although diameters above the swelling are smaller--to 26 inches (65 cm). In shallower waters other canopy species appear: *Taxodium distichum* and an occasional *Acer rubrum* (Red Maple). The *Taxodium* have typical cone buttresses and knees (Kurz and Demaree, 1934).

Fraxinus caroliniana (Water Ash) makes up the subcanopy, along with small specimens of the canopy species. Reproduction of the canopy trees is effective; it is therefore predicted that the community will maintain itself, barring any permanent changes in water level. The *Fraxinus* occur in slightly raised areas and on submerged tree trunks and occasionally reach canopy size.

Only two species of aquatics are found in the pools: *Proserpinaca palustris* (Mermaid-weed) and *Ranunculus flabellaris* (Water Crowfoot). The latter is an endangered peripheral and occurs in only one pool on the edge of State Secondary Road 1404. The sunlight allowed in by the roadcut is probably crucial in maintenance of the population; it is thus not a true species of the deep swamp. Further research should be done on its environmental requirements. Several other herb species occur in this community in very specialized niches--*Pilea pumila* (Clearweed) grows on rotting, fallen trees above the water level, and *Polypodium polypodioides* (Resurrection Fern) grows on the slanting, swollen bases of the *Nyssa aquatica*, also above high-water mark, along with numerous mosses and liverworts.

Community Type 2, the pool bank community, actually represents only slight differences in conditions from Community Type 1, but even this very minor variation produces significant changes in vegetation. All conditions are as described above except that the slightly higher areas are not covered with standing water for several months in the summer.

The canopy has a much greater diversity: *Taxodium distichum, Populus heterophylla* (Swamp Cottonwood), and *Acer rubrum* are almost as important and prevalent as *Nyssa aquatica*. In addition, the most hydric oaks are sparsely present: *Q. lyrata* (Overcup Oak), *Q. hemisphaerica* (Laurel Oak), and *Q. falcata* var. *pagodaefolia* (Cherrybark Oak). *Liquidambar styraciflua* (Sweet-gum), *Fraxinus pennsylvanica* (Red Ash), and *Carya aquatica* (Water Hickory), all of which are more typical of more mesic communities within the swamp, are scattered sporadically where slightly better drainage occurs.

The subcanopy once again consists primarily of *Fraxinus caroliniana*, with transgressives of the canopy species. Canopy reproduction here is excellent--the greater canopy diversity than in Community Type 1 probably results from the greater period of soil exposure allowing slightly less water-tolerant species to establish themselves. The lower vegetation layers, almost totally absent in Community Type 1, are sparsely present here, also resulting from the periods of exposure. A few lianas, shrubs, vines, and herbs (see Community Diversity Summary 2) have been able to establish themselves, often on fallen tree trunks or on the higher areas of the ground. The most prevalent herb species are tall, forb perennials: *Pilea pumila, Boehmeria cylindrica* (False Nettle), *Sium suave*

(Water Parsnip), and *Saururus cernuus* (Lizard's Tail). The first three of these species flower in this area in July and August and fruit in September and October (Radford et al., 1968). This period of flowering and fruiting appears to be adaptive to the seasonally flooded habitat--the plants do not begin vegetative growth until exposed, flower and fruit in late summer and early fall, the driest seasons in the swamp, and complete their fertile cycle by the time of the late fall-winter inundation. *Saururus cernuus*, listed as blooming in May through July (Radford et al., 1968), in this habitat seems typically to delay its flowering to July or even August, also a result of the late start in vegetative growth caused by flooding. The presence of these particular species in the pool banks community is not by chance.

Where a tree has fallen a persistent mound of earth with little organic matter remains. This microhabitat hosts an interesting collection of typically mesic species such as *Asplenium platyneuron* (Ebony Spleenwort), *Mitchella repens* (Partridgeberry), *Viola papilionacea* (Wood Violet), and *Rhus radicans* (Poison Ivy).

Community Type 3, the low flat community, typically grades into the pool bank community (Type 2) at its lower edges. It is slightly higher and is flooded only occasionally for most of the year, usually when Swift Creek rises over its low levee, depositing water in the flat beyond. The water in the flat slowly subsides because of slow flow into the pools and thence to the creek, slow flow directly into the creek at erosional low places in the levee, gradual seepage, and evaporation. The water table is probably never more than six inches (15 cm) or so below ground level. Slight variations in surface topography and in soil texture occasion variations in species present.

The canopy in the low flat community is very diverse, and, in undisturbed areas, of spectacular size. *Fraxinus pennsylvanica* and *Acer rubrum* are the two most abundant trees, although they rarely reach the large sizes of some of the other species. *Liquidambar styraciflua* is common and frequently of very large size, in some very localized areas forming nearly pure stands of specimens two to three feet (6-9 dm) in diameter. No real Sweet-gum community has been found yet within Swift Creek Natural Area, although one may exist; such a community does occur in the Congaree Swamp, South Carolina, and at Great Lake, North Carolina. Most characteristic of the low flat community is the great diversity of swamp oak species: in order of importance, *Quercus michauxii* (Swamp Chestnut Oak), *Q. hemisphaerica*, *Q. lyrata*, *Q. falcata* var. *pagodaefolia*, *Q. bicolor*, *Q. palustris* (Pin Oak), *Q. shumardii* (Swamp Red Oak), and *Q. nigra* (Water Oak).

The individual species of oak do not occur uniformly throughout the community. *Quercus lyrata* tends to occupy low depressions, *Q. hemisphaerica* and *Q. falcata* var. *pagodaefolia* the low to medium areas, *Q. palustris* and *Q. shumardii* intermediate areas of a siltier texture, and *Q. michauxii*, *Q. bicolor*, and *Q. nigra* intermediate to high areas of siltier textures. It should be emphasized that "low" and "high" are relative terms and that the change in elevation between the lowest and highest areas in this community is no more than three or four inches (7 or 10 cm). *Carya aquatica* is another relatively important species of this community, with *C. ovata* and *C. cordiformis* (Bitternut Hickory) occurring rarely on the best drained sites. *Ulmus americana* (American Elm) occupies intermediate, silty sites. *Taxodium distichum*, *Nyssa aquatica*, and *Populus heterophylla*, indicator species of the previous two communities, are sporadic in the lowest, most hydric sites.

A striking feature of the meso-hydric swamp forest communities (the present community and the following two) is the many, very large vines climbing into the canopy. The term "liana" has been loosely used in this report to refer to any of these high-climbing vines, even those like *Rhus radicans* and *Decumaria barbara* (Climbing Hydrangea) that are not free-hanging. Many lianas, of eight species, are present in this community, although they do not form as dense a layer as in the following two communities.

An open subcanopy layer is present, consisting for the most part of transgressives of the canopy species. *Fraxinus caroliniana* is present in the wettest depressions, and a few small specimens of *Fraxinus americana* and *Ulmus alata* (Winged Elm) are found, although none of canopy height is seen. *Crataegus viridis* (Hawthorn) occurs in medium to high areas; one specimen is thirteen inches (31 cm) in diameter and thirty feet (9 m) tall.

Scattered shrubs of the following species are present, mostly in the drier areas: *Ilex decidua* (Possum Haw), *Euonymus americanus* (Burning Bush), *Itea virginica* (Virginia Willow), and *Arundinaria gigantea* (Cane). The cane is very locally abundant on the highest spots, but nowhere approaches the vast stands found in some swamp forests (Braun, 1950). The herb species are mostly the same as in the pool bank community (Type 2), but are more abundant in the low flat community because inundation is less frequent. On some higher areas *Saururus cernuus* is a local dominant, but overall the herb layer is very sparse.

Community Type 4, the low-medium flat community, is in many respects very similar to the low flat community (Type 3) in topographic, hydrologic, and biotic features. It is slightly higher on the average than the low flat, but is more irregular and therefore contains greater diversity of microtopographical situations. The soil texture also is more varied, from clay to coarse silt. The area seems to have been more recently cut than the low flat areas, perhaps because of its greater accessibility. The *Pinus taeda* (Loblolly Pine) probably got its start when the area was disturbed, but very well may persist as a component of the climax canopy, especially given occasional openings in the canopy caused by windthrow that allows further germination of pine seedlings. *Pinus taeda* does seem able to maintain itself as a climax species in bottomland situations such as the Congaree Swamp.

The canopy in the low-medium flat community is very similar to that of the low flat community, with a few notable exceptions. *Pinus taeda*, absent on the low flat, is the third most important species here. *Fraxinus pennsylvanica*, the most important species on the low flat, is very rare in this area. Among the oaks, *Quercus palustris* and *Q. nigra* are more common, and *Q. phellos* (Willow Oak) is present, because of their preference for siltier, better-drained soil.

The canopy supports a layer of lianas. The subcanopy also is well developed, consisting of younger individuals of the canopy species as well as *Carpinus caroliniana* (Ironwood) and *Crataegus viridis* on better-drained sites, and *Fraxinus caroliniana* in the least well-drained sites. The shrub layer is very similar to that in the low flat: a few scattered specimens of *Ilex decidua, Itea virginica, Viburnum dentatum* (Viburnum), *Euonymus americanus,* and an occasional patch of *Arundinaria gigantea.*

Saururus cernuus dominates the herb layer, with scattered individuals of other species. An interesting commentary on the importance of minute changes in elevation is the distribution of *Saururus*. In the previous, slightly more hydric community, *Saururus* occurred on the very slightly higher ridges. In this community, just an inch or two (3-5 cm) greater in elevation, it occurs in the low depressions and finds the "ridges" too mesic.

Community Type 5, the levee community, occurs in the Swift Creek Natural Area in a narrow band along the creek. It is characterized topographically by a low, silty levee about ten feet (3 m) broad, rising two feet (6 dm) above the normal low-water level of Swift Creek. It varies somewhat in width and height, at times entirely filling a tight loop in the creek, at other times being replaced by a sloping, muddy seepage area from the flat. The combination of proximity to the creek yet highness compared to surrounding areas gives the levee a very different pattern of flooding; it is probably never flooded for an extended period of time. When flooding does occur, the water is sloughed off fairly quickly into the flat or back into the creek, and the somewhat coarser texture also allows some drainage. The levee contains many significant microhabitats--the actual creek bank, the top of the levee, the slope to the flat, and the low, mucky seepages. Species distinctive of each of these situations occur.

For this reason, the canopy is very diverse, with twenty species present. Typical species of the actual creek bank are *Betula nigra* (River Birch), *Carya aquatica, Carpinus caroliniana,* and *Quercus lyrata*. Low seepages into the creek, closely resembling the pool banks, are characterized by common pool bank community trees: *Taxodium distichum, Populus heterophylla, Nyssa aquatica,* and *Fraxinus caroliniana*. More common in the community are the true levee species: *Quercus hemisphaerica, Acer rubrum, Ulmus americana, Carya aquatica, Liquidambar styraciflua, Quercus michauxii, Q. palustris, Carya ovata,* and *Quercus phellos*. These species prefer rare flooding, coarser-textured soil, and relatively good drainage.

The liana layer finds one of its greatest developments in this community, with very large specimens of eight species present. Four-inch (10 cm) diameters with heights into the canopy are common for *Anisostichus capreolata* (Cross-vine), *Campsis radicans* (Trumpet Creeper), *Parthenocissus quinquefolia* (Virginia Creeper), *Rhus radicans*, and *Vitis* spp. (Grape Vine).

The subcanopy contains young specimens of most canopy species as well as several smaller species. *Carpinus caroliniana* and *Ilex opaca* (American Holly) are found on the levee in fairly well-drained situations. *Fraxinus caroliniana* occupies sinks and pans in and around the levee proper. Shrubs are scattered infrequently, only *Ilex decidua* and *Arundinaria gigantea* being commonly encountered. A shrubby, stoloniferous form of *Rhus radicans* is very common and locally dominant; it has been included in the herb layer in the community description because its distribution is linked with that of the herbs--where one is present the other is absent, so that the two taken together form a layer. The herbs are mixed, with local patches of *Commelina virginica, Saururus cernuus,* other tall herbs, *Viola papilionacea* and *V. palmata* var. *triloba* (Violet) being present.

Community Type 6, the high flat community, is not really a swamp forest community because it never is flooded. It is included, however, as a part of the Swift Creek Natural Area and an interesting example of the vegetation of a more mesic area only very slightly higher than the five communities already discussed. This community is probably a late successional stage resulting from past clearing (other clearing has been maintained nearby). The community occupies old sandy terraces approximately a foot (3 dm) above the level of nearby swamp forest communities.

The largest canopy trees are *Pinus taeda*, with specimens ranging up to two feet (6 dm) in diameter. *Liquidambar styraciflua, Nyssa sylvatica* (Black Gum), and *Acer rubrum* are other important canopy trees, but their smaller size indicates that they likely germinated in the pine forest and only now are reaching canopy size. This idea is supported by the subcanopy layer of transgressives that eventually will reach canopy size: *Quercus michauxii, Q. rubra* (Red Oak), *Q. phellos, Fraxinus pennsylvanica,* and others. Ultimately, it seems likely that a mixed oak-mixed hardwood canopy will develop, although *Pinus taeda* may well remain a part of the climax community; reproduction is ongoing at present, despite a dense canopy.

An open shrub layer is present, particularly in areas of greater disturbance. Important species are *Myrica cerifera* (Wax Myrtle) and *Viburnum dentatum*. Also indicative of the disturbed condition of this community is the heavy layer of *Lonicera japonica* (Japanese Honeysuckle) that effectively prevents the presence of herb species except in occasional spots.

The more mesic nature of the community is accentuated by the presence of several mesic species totally absent from the swamp forest communities, including *Quercus rubra, Nyssa sylvatica, Cornus florida* (Flowering Dogwood), *Fagus grandifolia* (Beech), *Liriodendron tulipifera* (Tulip Tree), *Sassafras albidum* (Sassafras), *Vaccinium corymbosum* (Highbush Blueberry), *Vitis rotundifolia* (Muscadine), and *Lonicera japonica*.

The last Community Type (7) surveyed is found on the gentle, sandy slopes surrounding the Swift Creek Swamp Forest basin and can be called the mesic slope community. It forms the boundary line between the natural area and the mostly cultivated upland.

Fagus grandifolia is the most important tree, with many large specimens found. The remainder of the canopy consists primarily of mesophytic trees: *Quercus alba* (White Oak), *Pinus taeda, Quercus phellos, Nyssa sylvatica, Acer rubrum,* and *Carpinus caroliniana*. Some more hydrophytic trees are present, mostly in disturbed areas, such as *Quercus nigra* and *Q. lyrata*, as well as a few xerophytic species, such as *Carya tomentosa* (Mockernut Hickory), *Carya pallida* (Pale Hickory), *Quercus velutina* (Black Oak), and *Q. stellata* (Post Oak). This mixture of species, apparently more or less randomly distributed within the community, probably is related to the proximity of this well-drained habitat to a large wetland; hydrophytic species are present as waifs or carryovers in a habitat much drier than typical for the species.

The subcanopy is a mixture of canopy species transgressives and common mesophytic subcanopy species: *Cornus florida, Oxydendrum arboreum* (Sourwood), *Carpinus caroliniana,* and *Ilex opaca*. Unimportant scattered layers of shrubs, vines, and herbs are present. On the whole, Community Type 7 is a fairly typical mesic community of the Upper Coastal Plain or Piedmont, but with the influence of the nearby swamp apparent in the species present.

B. OVERVIEW OF THE SWIFT CREEK SWAMP COMMUNITIES

It was deemed impractical to attempt to map the communities in the swamp because the boundaries between them are nebulous and even if defined sharply are so convoluted within the natural area that a map would be merely a jumble of lines. Instead, a representative profile across the swamp is provided (Figure B4), from which the relative areas covered by each community can be judged roughly.

In the true swamp forest or bottomland communities (Types 1 to 5), replacement of canopy trees is largely dependent on gaps in the canopy allowing germination and rapid growth of transgressive trees. The general absence in the swamp of lower layers of vegetation is occasioned by the dense shade and, in the more hydric communities, by the surface water. The levee community (Type 5) has the most herbaceous and other low-layer vegetation owing to light from the break in the canopy above the creek and infrequent inundation. Where a tree falls, dense competition and growth of lower layers occur until the dense shade provided by the canopy is restored. The prevalence of lianas in the swamp forest also is related to this problem; their ability to grow with the tree species and therefore maintain their foliage in the upper layer gives them the "best of both worlds"--abundant sunlight as well as high-nutrient, high-moisture soil. The lianas probably also get their start in openings and then establish the same growth rate as the trees that support them. This hypothesis is backed up by the frequent absence of any leaves to a height of 50 feet (15 m) or more. Also, the lianas frequently fall free from the canopy to their root systems, with no support.

In general, the greatest diversity within the swamp forest communities is in the woody species that can take advantage of both the sunlight and the high-nutrient, high-moisture soil--the canopy trees and lianas. Over half of the species present in the first five communities belong to one of these groups--30 canopy species and nine liana species out of a total of seventy-four. This high diversity of woody vegetation is one of the factors that make the swamp forest communities so intriguing.

C. NOTES ON SELECTED SPECIES

This section is meant as a further explanation of the data from a species biology rather than a community viewpoint. Obviously the two are inseparable because the individual species biologies of the forty-odd tree species determine the canopy composition of any community within the swamp forest. Most tree species present

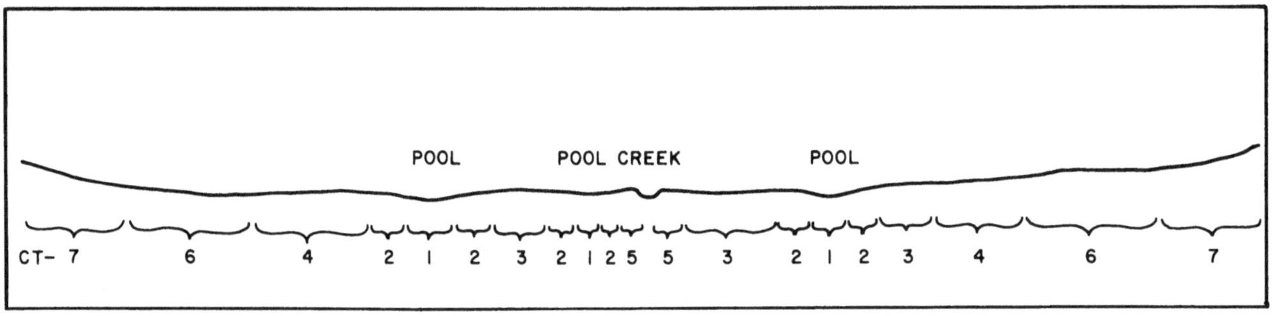

Figure B4. *Transect across Swift Creek communities.*

	PENFOUND, 1952	BRAUN, 1950	HOTCHKISS & STEWART, 1947	WINTERS et al., 1938
CT-1 POOL CT-2 POOL BANKS	DEEP FRESHWATER	SWAMP FOREST	——— ———	CYPRESS-TUPELO
CT-3 LOW FLAT CT-4 LOW-MEDIUM FLAT CT-5 LEVEE	(CELTIS-ULMUS-FRAXINUS) ACER-LIQUIDAMBAR-QUERCUS	HARDWOOD BOTTOMS (CANE RIDGE)	RIVER SWAMP	RED GUM-WATER OAKS MIXED OAK-MIXED HARDWOOD
CT-6 HIGH SANDY FLAT	———	LOBLOLLY PINE-HARDWOODS	———	PINE-HARDWOOD
CT-7 MESIC SLOPE	———	MESOPHYTIC MIXED HARDWOODS	———	———

Figure B5. *A comparison chart of some swamp forest classifications.*

at Swift Creek seem to have rather narrow environmental tolerances; thus, a very good separation usually occurs down to a community and even microcommunity level (as should be evident from the community discussions). No community is homogeneous in composition. The following qualitative assessments of the environmental parameters controlling a given taxon's distribution within the swamp are given only as a suggestion of the possibilities of this sort of study.

Quercus spp. (Oaks). The Swift Creek natural area is an excellent area for the study of oaks, with thirteen present (see the species presence list). The most important factor in the environmental separation of the oaks is moisture. This variable by itself only roughly separates one species from another in ecological preference (Figure B1). The addition of another variable, soil texture, gives much more information. For swamp species, these two factors alone give almost the entire picture. *Quercus lyrata* is the most hydric of the oaks present, found most typically on pool banks, in low depressions in the flat, and on the creek bank--all locations in which surface water is present almost year-round. These are also for the most part fine-textured soils, but hydroperiod is the more critical parameter. In disturbed areas, *Q. lyrata* can occur in mesic areas, but this seems related to their proximity to the swamp. *Quercus hemisphaerica* is the second most hydric oak at Swift Creek and the most abundant. Its wettest stations occur on usually flooded pool banks, and it is most common on the flat and on the levee. It tends to prefer silty rather than mucky or clayey soils. *Quercus palustris* also in general follows this pattern, but is even more particular to the levee. *Quercus phellos* is less hydric, being limited to silty to sandy soil and at most rare flooding. *Quercus nigra* can occur in areas with fairly long hydroperiods but seems to require the relatively good drainage of the siltiest and sandiest spots within each community. Its greatest abundance within the natural area is in the mesic slope community, perhaps as a result of disturbance. *Quercus michauxii* and *Q. falcata* var. *pagodaefolia* both have a wide amplitude of tolerance for hydroperiod and substrate texture, ranging from low flat to high sandy flat and from silty clay to silty sand. The two least frequently encountered oaks are *Q. shumardii* and *Q. bicolor*; these both seem to fall in the intermediate mesohydric range, although not enough specimens were seen to be able to characterize them ecologically with confidence. Although each of the wetland oak species has its individual preferences as to substrate and hydroperiod, it will be seen (Figure B1, p. 234) that in the mesohydric (infrequently flooded) part of the swamp the ranges of almost all of the oaks overlap. The low flat, the low-medium flat, and the levee each display eight *Quercus* species, each tending to occupy a fairly distinct microhabitat.

In the most xeric part of the Swift Creek area, the mesic slope, four more species of *Quercus* appear: *Q. alba, Q. rubra, Q. stellata,* and *Q. velutina.* These will not be discussed here because they are not swamp species and are present here only at one extreme of their ecological amplitude.

Carya spp. (Hickories). Five species of hickory are present in the natural area, three in the swamp itself. By far the most abundant and the most hydric is *C. aquatica.* It always is found near standing water, liking best the relatively well-drained levee but also present occasionally on pool banks and near intermittent pools in the flat. *Carya ovata* seems similar in general pattern of distribution but is apparently more particular, being almost restricted to the levee community (Type 5). It also is found occasionally in silty locations near intermittent pools in the flat. *Carya ovata* and *C. aquatica* seem to have similar hydroperiod tolerances, but *C. ovata* requires the better drainage of siltier textured soils. *Carya cordiformis* is present but not common in the low and low-medium flat communities, where it tolerates poorer drainage than *C. ovata*, but with a typically shorter hydroperiod. The remaining two hickories, *C. pallida* and *C. tomentosa*, are found only in the mesic slope community, from which they range to more xeric conditions.

Fraxinus spp. (Ashes). All four species of ash known to occur in the Carolinas are present at Swift Creek. The commonest is *F. caroliniana*, present as a subcanopy dominant in the two most hydric communities and extending occasionally into the other three swamp communities in the wettest spots. Typically it reaches only the subcanopy, but occasional specimens are found eight inches (20 cm) in diameter and of canopy height. *Fraxinus pennsylvanica* is also common, but restricted in habitat. It is abundant and large in the low flat, but only occasional in less hydric communities. *Fraxinus americana* (American Ash) is rare in the swamp and seems to have approximately the same ecological parameters as *F. pennsylvanica*, although it is too infrequent to draw definite conclusions. *Fraxinus tomentosa* (Pumpkin Ash) is rare and occasional on the levee only and may be limited to this well-drained habitat.

Populus heterophylla. Swamp Cottonwood has hydroperiod requirements very similar to those of *Carya aquatica* and is thus found in the same communities (pool banks, low flat, low-medium flat, and levee). Its soil texture preference is very different, however; it prefers mucky and clayey soils with poor drainage. It is abundant in the pool bank community and present in the other communities in stations with similar hydroperiod but decreasingly common as the texture coarsens.

Rhus radicans. This species seems to be present in two very distinct forms in the swamp--the first a large, high-climbing vine, the second a low, stoloniferous shrub. The first form would seem to be the typical variety of *Rhus radicans* and is encountered frequently in the mesohydric communities in the swamp. The second form, for which no satisfactory nomenclature was found, occurs only on the levee and grows in large stoloniferous patches. Usually the plants are about two feet (6 dm) tall with a simple, woody stem reminiscent of *Rhus toxicodendron* (Poison Oak).

D. A SHORT SUMMARY OF SWAMP FOREST LITERATURE APPLICABLE TO SWIFT CREEK

The literature on swamp forest communities for the most part falls into two categories--generalized classifications and detailed studies of specific locations. The literature was found to be very sparse and frequently of little applicability to small Atlantic Coastal Plain swamps such as Swift Creek; more work has been done in the Mississippi River swamps and in the extensive swamps in the lower Atlantic Coastal Plain, such as the Dismal and Okefinokee. These large swamps represent a special case, more significant for their differences than their similarities to the more frequent river or stream swamps.

Penfound (1952) defines a swamp as "a woody community occurring in an area where the soil is usually saturated or covered with surface water for one or more months of the growing season." This seems a reasonable definition and fairly simple to apply, but most writers on the subject do not define "swamp" at all, apparently reserving the term "swamp" for nearly permanently flooded areas.

Braun (1950, p. 295) states: "Most data on the composition of bottomland forests do not separate the communities of the several bottomland habitats. All hardwoods are considered together, thus developmental sequences and correlations with site factors are obscured." This statement can be taken as a summary of the current literature on wooded wetlands. Very little critical observation of the distinctive communities present under very slightly varying conditions in the swamp has been made. Figure B5 shows how the seven communities described in this report fit into several classification schemes. It can be seen immediately that problems are present--some investigators use cover type, others topography or hydrography, and so forth. One of the most detailed classifications, that of Winters et al. (1938), fails to correlate the cover types in any direct way to hydroperiod, texture, or any other characteristic, thereby limiting the usefulness and pertinence of the study. Other classifications are limited by their broad generalizations (Hotchkiss and Stewart, 1947; Braun, 1950; Penfound, 1952). Other more specific studies are worthwhile as far as they go--some pertain to factors affecting swamp plants (Harper, 1905; Demaree, 1932; Shunk, 1939; Henry, 1970), some to a particular swamp forest (Beaven and Oosting, 1939; Brown, 1943; Thieret, 1971). Almost no work has addressed itself to the environmental parameters of the distribution of the various communities. This study has attempted to take the environmental approach toward one specific swamp forest. What is now needed is to analyze the correlations between plant communities and environmental factors such as hydroperiod, soil, elevation, and geographical locations in other swamp forests. This holistic view ultimately should provide a catalog of swamp forest communities and the environmental parameters determining their distributions.

Bibliography

Barker, E. D. 1922. A note from Okefinokee. Torreya 22(6):104-6.

Beaven, G. F., and H. J. Oosting. 1939. Pocomoke Swamp: A study of a cypress swamp on the eastern shore of Maryland. Bull. Torrey Bot. Club 66:367-89.

Bergman, H. F. 1920. The relation of aeration to the growth and activity of roots and its influence on the ecesis of plants in swamps. Ann. Bot. (London) 34:13-33.

Braun, E. L. 1950. Deciduous forests of eastern North America. Hafner Press, New York.

Brown, C. A. 1943. Vegetation and lake level correlations at Catahoula Lake, Louisiana. Geogr. Rev. (New York) 33:435-45.

Buell, M., and W. A. Wistendahl. 1955. Flood plain forests of the Raritan River. Bull. Torrey Bot. Club 82:463-72.

Conant, R. 1975. A field guide to reptiles and amphibians of eastern and central North America. Houghton Mifflin, Boston.

Cypert, E. 1961. The effects of fires in the Okefinokee Swamp in 1954 and 1955. Amer. Midl. Naturalist 66:485-503.

Dean, G. W. 1969. Forests and forestry in the Dismal Swamp. Virginia J. Sci. 20:166-73.

Demaree, D. 1932. Submersing experiments with *Taxodium*. Ecology 13:258-62.

Fernald, M. L. 1950. Gray's manual of botany. 8th ed. D. Van Nostrand, New York.

Gemborys, S. R., and E. J. Hodgkins. 1971. Forests of small stream bottoms in the Coastal Plain of southwestern Alabama. Ecology 52:70-84.

Hall, T. F., and W. T. Penfound. 1939. A phytosociological analysis of a cypress-gum swamp in southeastern Louisiana. Amer. Midl. Naturalist 21:378-95.

________. 1943. Cypress-gum communities in the Blue Girth Swamp near Selma, Alabama. Ecology 24:208-17.

Harper, R. M. 1905. Further observations on *Taxodium*. Bull. Torrey Bot. Club 32:105-15.

Henry, E. F. 1970. Soils of the Dismal Swamp of Virginia. Virginia J. Sci. 21:41.

Hotchkiss, N., and R. E. Stewart. 1947. Vegetation of Patuxent Refuge, Maryland. Amer. Midl. Naturalist 38:1-75.

Kurz, H., and D. Demaree. 1934. Cypress buttresses and knees in relation to water and air. Ecology 15:36-41.

Lee, W. D. 1955. The soils of North Carolina: Their formation, identification, and use. Tech. Bull. No. 115. North Carolina Agricultural Experiment Station, Raleigh.

Lowe, E. N. 1921. Plants of Mississippi: A list of flowering plants and ferns. Mississippi State Geological Survey. Bull. No. 17. Hederman Bros., Jackson, Mississippi.

Mueller-Dombois, D., and H. Ellenberg. 1974. Aims and methods of vegetation ecology. John Wiley & Sons, New York.

North Carolina Department of Conservation and Development. 1958. Geologic map of North Carolina. Raleigh.

Penfound, W. T. 1952. Southern swamps and marshes. Bot. Rev. (Lancaster) 18:413-46.

________, and T. F. Hall. 1939. A phytosociological analysis of a tupelo gum forest near Huntsville, Alabama. Ecology 20:358-64.

________, and E. S. Hathaway. 1938. Plant communities of the marshlands of southeastern Louisiana. Ecol. Monogr. 8:1-56.

Pessin, L. S. 1933. Forest associations in the uplands of the lower Gulf Coastal Plain. Ecology 14:1-14.

Putnam, J. A., G. M. Furnival, and J. S. McKnight. 1960. Management and inventory of southern hardwoods. Handbook 181. U.S. Department of Agriculture, Washington, D.C.

Radford, A. E. 1976. Vegetation--habitats--floras--natural areas in the southeastern United States: Field data and information. University of North Carolina Student Stores, Chapel Hill.

________. 1977. A natural area and diversity classification system. A standardized scheme for basic inventory of species, community, and habitat diversity. University of North Carolina Student Stores, Chapel Hill.

________, H. E. Ahles, and C. R. Bell. 1968. Manual of the vascular flora of the Carolinas. University of North Carolina Press, Chapel Hill.

Radford, A. E., L. J. Otte, and D. K. S. Otte. 1978. Natural heritage classification and information systems: Ecological diversity classification and inventory. University of North Carolina Student Stores, Chapel Hill.

Radford, A. E., W. C. Dickison, J. R. Massey, and C. R. Bell. 1974. Vascular plant systematics. Harper and Row, New York.

Shunk, I. V. 1939. Oxygen requirements for germination of seeds of *Nyssa aquatica*, tupelo gum. Science 90:565-66.

Small, J. K. 1933. An Everglade cypress swamp. J. New York Bot. Gard. 34:261-67.

Soil Conservation Service. In press. Soil survey of Edgecombe County, North Carolina. U.S. Department of Agriculture, Washington, D.C.

Soil Survey Staff. 1972. Soil series of the United States, Puerto Rico, and the Virgin Islands: Their taxonomic classification. U.S. Department of Agriculture, Washington, D.C.

________. 1975. Soil taxonomy: A basic system of soil classification for making and interpreting soil surveys. Agriculture Handbook No. 436. Soil Conservation Service, U.S. Department of Agriculture, Washington, D.C.

Steila, D. 1976. The geography of soils: Formation, distribution, and management. Prentice-Hall, Englewood Cliffs, New Jersey.

Thieret, J. W. 1971. Quadrat study of a bottomland forest in St. Martin Parish, Louisiana. Castanea 36:174-81.

U.S. Department of the Interior. 1970. The national atlas of the United States of America. U.S. Government Printing Office, Washington, D.C.

Viosca, P. 1928. Louisiana wetlands and the value of their wildlife and fishery resources. Ecology 9:216-30.

Wells, B. W. 1928. Plant communities of the Coastal Plain of North Carolina and their successional relations. Ecology 9:230-42.

________. 1942. Ecological problems of the southeastern United States Coastal Plain. Bot. Rev. (Lancaster) 8:533-61.

Winters, R. K., J. A. Putnam, and I. F. Eldredge. 1938. Forest resources of the North Louisiana Delta. Misc. Publ. 308. U.S. Department of Agriculture, Washington, D.C.

Wright, A. H., and A. A. Wright. 1932. The habitats and composition of the vegetation of Okefinokee Swamp, Georgia. Ecol. Monogr. 2:109-232.

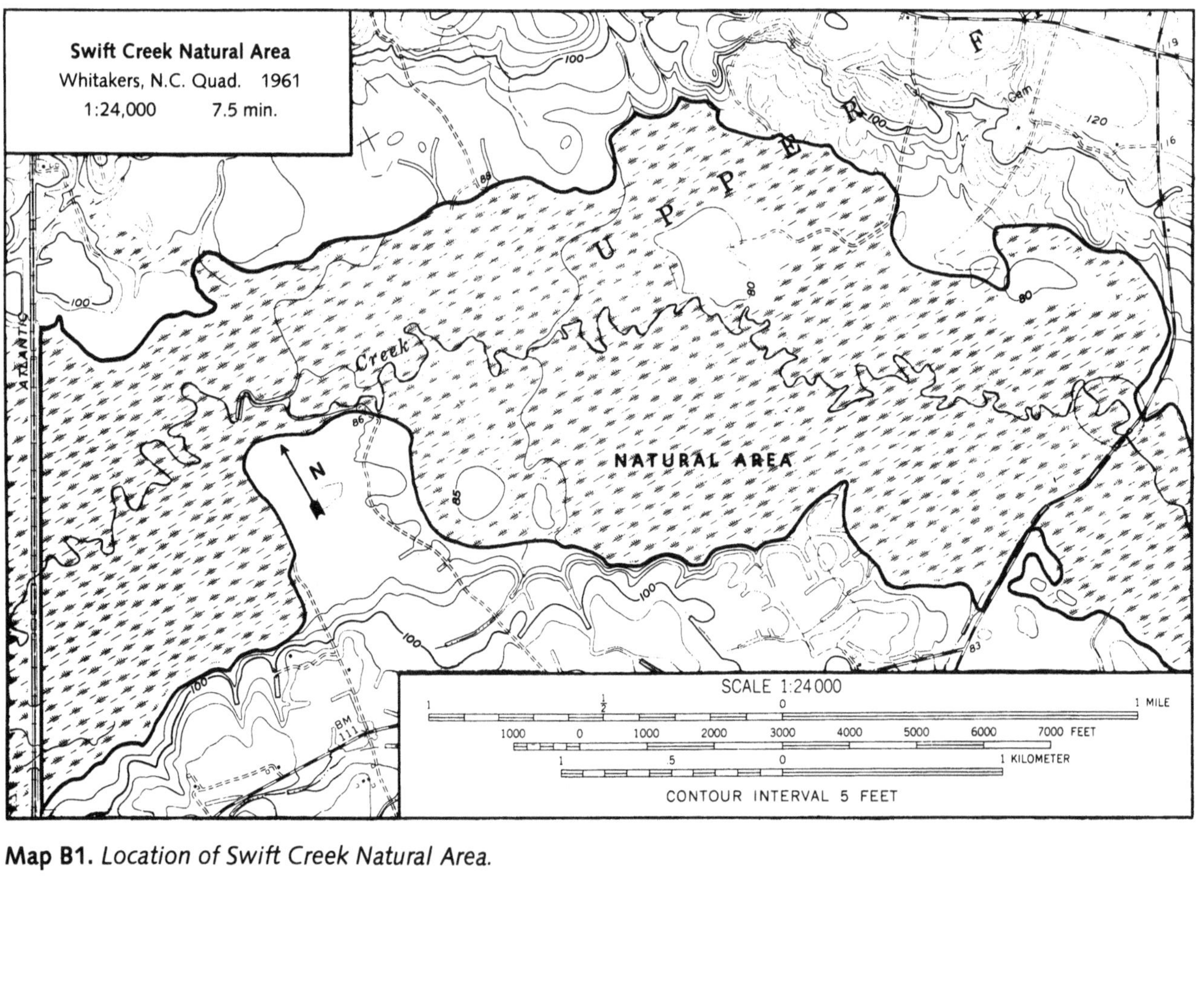

Map B1. *Location of Swift Creek Natural Area.*

Natural Area Diversity Summary

CLIMATE: A. Mesothermal; AA. Warm temperate. B. Cool and moderately moist yearly, with moderately hot and moderately wet summers, moderately warm and moderately dry winters, and a long freeze-free period; BB. Similar to sectional climate. C. Cooler and wetter than local climate; CC. Similar to natural area climate.

SOILS: A. (1-2) Inceptisol, (3, 5) Alfisol, (4, 6-7) Ultisol; AA. (1-2) Aquept, (3, 5) Aqualf, (4, 6-7) Udult. B. (1-2) Humaquept, (3, 5) Albaqualf, (4, 6-7) Hapludult; BB. (1-2) Cumulic Humaquept, (3, 5) Typic Albaqualf, (4) Aquic Hapludult, (6-7) Typic Hapludult. C. (1-2) Coarse-loamy, siliceous, acid, thermic, (3, 5) Fine, mixed, thermic, (4, 6) Fine-loamy, mixed, thermic, (7) Clayey, mixed, thermic; CC. (1-2) Johnston, (3, 5) Meggett, (4) Altavista, (6) Wickham, (7) Gritney.

GEOLOGY: A. Sedimentary; AA. (1-2) Mixed clastic and organic, (3-7) Clastic. B. (1-2) Siliceous and organic, (3-7) Siliceous; BB. (1-4) Floodplain swamp deposit, (5) Levee deposit, (6) Stream terrace deposit, (7) Undetermined origin. C. (1-2) Mixed fine clastics and organics, (3-7) Mixed fine clastics; CC. (1-2) Unconsolidated, unaligned, water-saturated muck, underlain by a low-porosity, low-permeability clay, (3) Unbedded, unconsolidated, low-porosity, low-permeability clay, (4) Unbedded, unaligned, unconsolidated, low-permeability silt, (5) Unbedded, unconsolidated, highly porous and permeable mixtures of silt and sand, (6) Unbedded, unconsolidated, highly porous and permeable mixture of silt and sand, (7) Unbedded, unconsolidated, highly porous and permeable mixture of silt and clay.

HYDROLOGY: A. (1-5) Palustrine, (6-7) Terrestrial; AA. (1) Aqueous, (2-5) Interaqueous, (6-7) Dry. B. Fresh; BB. (1) Permanently flooded, (2) Seasonally to intermittently flooded, (3-5) Intermittently flooded, (6-7) Permanently exposed. C. Variously drained and wetted; CC. (1-3) Very poorly drained, wetted by fresh rains, flooding, and a high water table, (4) Poorly drained, wetted by fresh rains, flooding, and a high water table, (5) Moderately poorly drained, wetted by fresh rains and occasional flooding, (6) Moderately well drained, wetted by fresh rains, (7) Well drained, wetted by fresh rains and downslope drainage.

TOPOGRAPHY: A. Plain; AA. Flat plain. B. Broad, east-west-running, shallow river valley; BB. (1-6) Water-deposited and organically accumulated floodplain, (7) Valley slopes. C. Entire floodplain and adjoining valley slopes; CC. (1) Open, nearly level bottom of shallow, concave depressions on an alluvial flat, (2) Open, nearly level borders of shallow, concave depressions on an alluvial flat, (3-4) Open, nearly level, slightly irregular, alluvial flat, (5) Open, nearly level, irregularly mounded, streamside slope and crest of a levee, (6) Open, nearly level, slightly irregular flat on an upper stream terrace, (7) Open, north-, northeast-, south-, and southwest-facing, gently sloping to sloping, convex-concave, entire valley slope.

PHYSIOGRAPHY: A. Atlantic Plain; AA. Coastal Plain. B. Embayed Section; BB. Tar River drainage system. C. (1-6) Swift Creek floodplain, (7) Swift Creek valley slope; CC. (1-2) Shallow depressions on an alluvial flat, underlain by Recent alluvium, (3-4) Alluvial flat, underlain by Recent alluvium, (5) Streamside levee, underlain by Recent alluvium, (6) Upper stream terrace, underlain by Recent alluvium, (7) Valley slope, underlain by Miocene (?) sediments.

BIOLOGY:

1. A. WETLAND BROADLEAF FOREST SYSTEM;
AA. Large, deliquescent, deciduous trees/Medium, deliquescent, deciduous trees. B. Tupelo Gum swamp forest; BB. Canopy: *Taxodium distichum*/ Subcanopy: *Nyssa aquatica*--Cornales/Scrophulariales. C. *Nyssa aquatica*; CC. *Nyssa aquatica/Fraxinus caroliniana*.

2. A. WETLAND BROADLEAF-NEEDLELEAF FOREST SYSTEM;
AA. Large, deliquescent and excurrent, deciduous trees/Medium, deliquescent, deciduous trees. B. Mixed swamp forest; BB. Canopy: *Nyssa aquatica, Taxodium distichum, Populus heterophylla, Acer rubrum, Quercus hemisphaerica*/Subcanopy: *Nyssa aquatica*--Hardwood-cypress/Scrophulariales. C. Mixed bottomland trees; CC. Mixed bottomland trees/*Fraxinus caroliniana*.

3. A. WETLAND BROADLEAF FOREST SYSTEM;
AA. Large, deliquescent, deciduous trees. B. Mixed oak-mixed hardwood swamp forest; BB. Canopy: *Fraxinus pennsylvanica, Acer rubrum, Liquidambar styraciflua, Quercus michauxii, Quercus hemisphaerica*/Subcanopy: *Acer rubrum, Quercus hemisphaerica*/Herb: *Saururus cernuus*--Mixed hardwoods. C. Mixed bottomland oaks-mixed bottomland hardwoods; CC. Mixed bottomland oaks-mixed bottomland hardwoods.

4. A. WETLAND BROADLEAF FOREST SYSTEM;
AA. Large, deliquescent, deciduous trees/Medium, deliquescent, deciduous trees/Tall, rhizomatous, rhizocarpic forbs//Large, tendril- and root-climbing, deciduous lianas. B. Red Maple-mixed hardwood swamp forest; BB. Canopy: *Quercus hemisphaerica, Pinus taeda, Liquidambar styraciflua, Ulmus americana, Quercus palustris*/Subcanopy: *Acer rubrum, Quercus hemisphaerica, Carpinus caroliniana*//Liana: *Anisostichus capreolata, Berchemia scandens, Campsis radicans, Decumaria barbara, Rhus radicans*--Sapindales-mixed hardwoods/Mixed hardwoods/Piperales//Mixed lianas. C. *Acer rubrum*-mixed bottomland hardwoods; CC. *Acer rubrum*-mixed bottomland hardwoods/Mixed bottomland hardwoods/*Saururus cernuus*//Mixed lianas.

5. A. WETLAND BROADLEAF FOREST SYSTEM;
AA. Large, deliquescent, deciduous trees/Medium, deliquescent, deciduous trees/Tall dwarf, stoloniferous, deciduous shrubs-small to tall, rhizomatous, rhizocarpic herbs//Large, tendril- and root-climbing, deciduous lianas. B. Mixed hardwood-mixed oak swamp forest; BB. Canopy: *Quercus hemisphaerica, Acer rubrum, Ulmus americana, Carya aquatica, Liquidambar styraciflua, Quercus michauxii, Betula nigra*/Subcanopy: *Fraxinus caroliniana, Carpinus caroliniana, Acer rubrum*/Herb: *Commelina virginica, Viola papilionacea, Saururus cernuus, Panicum* sp.//Liana: *Anisostichus capreolata, Campsis radicans, Smilax rotundifolia*--Mixed hardwoods/Mixed hardwoods/Sapindales-mixed forb perennials//Mixed lianas. C. Mixed bottomland hardwoods-mixed bottomland oaks; CC. Mixed bottomland hardwoods-mixed bottomland oaks/Mixed bottomland hardwoods/*Rhus radicans*-mixed forb perennials//Mixed lianas.

6. A. TERRESTRIAL NEEDLELEAF-BROADLEAF FOREST SYSTEM;
AA. Large, excurrent, evergreen trees-large, deliquescent, deciduous trees/Medium, deliquescent, deciduous trees//Low, spread-climbing, evergreen vines. B. Loblolly Pine-mixed bottomland hardwood forest; BB. Canopy: *Liquidambar styraciflua, Nyssa sylvatica, Acer rubrum*/ Subcanopy: *Quercus michauxii, Quercus phellos, Quercus rubra, Liquidambar styraciflua*/Shrub: *Myrica cerifera, Viburnum dentatum*--Coniferales-mixed hardwoods/Mixed hardwoods/Dipsacales. C. *Pinus taeda*-

mixed bottomland hardwoods; CC. *Pinus taeda*-mixed bottomland hardwoods/ Mixed bottomland oaks//*Lonicera japonica*.

7. A. TERRESTRIAL BROADLEAF FOREST SYSTEM;
AA. Large, deliquescent, deciduous trees/Medium, deliquescent, deciduous trees. B. Beech-mixed hardwood forest; BB. Canopy: *Quercus alba, Liquidambar styraciflua*/Subcanopy: *Ilex opaca, Quercus alba, Fagus grandifolia, Oxydendrum arboreum, Sassafras albidum*/Vine: *Vitis rotundifolia*/Herb: *Panicum* sp.--Fagales-mixed hardwoods/Mixed hardwoods. C. *Fagus grandifolia* mixed hardwoods; CC. *Fagus grandifolia*-mixed hardwoods/Mixed hardwoods.

Community Diversity Summary 1

Wetland broadleaf forest system
Large, deliquescent, deciduous trees/
Medium, deliquescent, deciduous trees

Tupelo gum swamp forest
Canopy: *Taxodium distichum*/
Subcanopy: *Nyssa aquatica*--
Cornales/Scrophulariales

NYSSA AQUATICA/FRAXINUS CAROLINIANA
Nyssa aquatica

CLIMATE: A. Mesothermal; AA. Warm temperate. B. Cool and moderately moist yearly, with moderately hot and moderately wet summers, moderately warm and moderately dry winters, and a long freeze-free period; BB. Similar to sectional climate. C. Cooler and wetter than local climate; CC. Similar to natural area climate.

SOILS: A. Inceptisol; AA. Aquept. B. Humaquept; BB. Cumulic Humaquept. C. Coarse-loamy, siliceous, acid, thermic; CC. Johnston.
A horizon: 0 to 56 cm, black mucky loam, pH 4.6.
C horizon: 56 to 140+ cm, light gray loamy fine sand, with some shells toward bottom, pH 5.4 to 7.0.

GEOLOGY: A. Sedimentary; AA. Mixed clastic and organic. B. Siliceous and organic; BB. Floodplain swamp deposit. C. Mixed fine clastics and organics; CC. Unconsolidated, unaligned, water-saturated muck, underlain by a low porosity, low permeability clay.

HYDROLOGY: A. Palustrine; AA. Aqueous. B. Fresh; BB. Permanently flooded. C. Variously drained and wetted; CC. Very poorly drained, wetted by fresh rains, flooding, and a high water table.

TOPOGRAPHY: A. Plain; AA. Flat plain. B. Broad, east-west-running, shallow river valley; BB. Water-deposited and organically accumulated floodplain. C. Entire floodplain and adjoining valley slopes; CC. Open, nearly level bottom of shallow, concave depressions on an alluvial flat.

PHYSIOGRAPHY: A. Atlantic Plain; AA. Coastal Plain. B. Embayed Section; BB. Tar River drainage system. C. Swift Creek floodplain; CC. Shallow depressions on an alluvial flat, underlain by Recent alluvium.

CANOPY: *Nyssa aquatica*--Height 26 m, dbh 114 cm (65 cm above the bottleneck), age not determined.

Taxodium distichum--Height 36 m, dbh 79 cm, age not determined.
CANOPY DOMINANT: *Nyssa aquatica*.
CANOPY PHYSIOGNOMY: Large, deliquescent, deciduous trees.

Canopy Analysis

Species	I.V.	Rel. Den.	Rel. Dom.	Rel. Freq.
Nyssa aquatica	225.4	82.7	78.4	64.3
Taxodium distichum	46.3	7.7	17.2	21.4
Acer rubrum	15.9	7.7	1.1	7.1
Fraxinus caroliniana	12.3	1.9	3.3	7.1

Number of relevés: 9 Relevé size: 10 m X 10 m Number of individuals/ acre: 256

CANOPY SPECIES PRESENT, BUT NOT IN ANALYSIS: None.

SUBCANOPY: *Fraxinus caroliniana*--Height 5 m, dbh 12 cm, age not determined.
SUBCANOPY DOMINANT: *Fraxinus caroliniana*.
SUBCANOPY PHYSIOGNOMY: Medium, deliquescent, deciduous trees.

Subcanopy Analysis

Species	GF	1 C.S	2 C.S	3 C.S	4 C.S	5 C.S	6 C.S	7 C.S	8 C.S	9 C.S
Large trees										
Nyssa aquatica	H2	-	-	2.1	-	2.1	-	-	-	4.1
Acer rubrum	H2	-	-	-	-	4.1	-	-	-	-
Medium trees										
Fraxinus caroliniana	H2	5.1	5.1	-	4.1	-	5.1	5.1	-	5.1
Ilex opaca	H2	-	1.1	-	-	-	-	-	-	-

Number of relevés: 9 Relevé size: 5 m X 5 m

SUBCANOPY SPECIES PRESENT, BUT NOT IN ANALYSIS: None.

SHRUB LAYER DOMINANTS: None.
SHRUB LAYER PHYSIOGNOMY: Not applicable.

Shrub Analysis

Species	GF	1 C.S	2 C.S	3 C.S	4 C.S	5 C.S	6 C.S	7 C.S	8 C.S	9 C.S
Vine										
Smilax rotundifolia	H18	-	2.1	-	-	1.1	-	-	-	-

Number of relevés: 9 Relevé size: 5 m X 5 m

SHRUB SPECIES PRESENT, BUT NOT IN ANALYSIS: None.

HERB LAYER DOMINANTS: None.
HERB LAYER PHYSIOGNOMY: Not applicable.

Herb Analysis

Species	GF	1 C.S	2 C.S	3 C.S	4 C.S	5 C.S	6 C.S	7 C.S	8 C.S	9 C.S
Medium forb										
Proserpinaca palustris	H31	-	2.1	-	1.1	-	-	-	-	-
Small fern										
Polypodium polypodioides	H11	-	1.1	1.1	-	-	2.1	-	-	-

Number of relevés: 9 Relevé size: 1 m X 1 m

HERB SPECIES PRESENT, BUT NOT IN ANALYSIS: TALL FORB: *Pilea pumila*; MEDIUM FORB: *Ranunculus flabellaris*.

COMMUNITY REFERENCES: None.

COMMUNITY DOCUMENTATION: Area analyzed in 1977 and 1978 by Alan S. Weakley for A. E. Radford's Botany 235 class, UNC, Chapel Hill. Specimens deposited in the NCU Herbarium by Alan S. Weakley in 1978.

COMMUNITY ECOLOGICAL CHARACTERIZATION: Vegetationally: Cornalean/Scrophularialean wetland broadleaf forest system with a closed canopy dominated by large, deliquescent, deciduous trees, a closed subcanopy dominated by medium, deliquescent, deciduous trees, a sparse vine layer of normal, tendril-climbing, deciduous vines, and a sparse herb layer containing scattered aquatic forbs and epiphytic ferns. Climatically: Warm temperate, mesothermal climate, cooler and wetter than the local and sectional cool and moderately moist yearly climate, which has moderately hot and moderately wet summers, moderately warm and moderately dry winters, and a long freeze-free period. Pedologically: Coarse-loamy, siliceous, acid, thermic Cumulic Humaquept, Johnston soil. Geologically: Mixed clastic and organic, siliceous and organic, sedimentary floodplain swamp deposit, consisting of unconsolidated, unaligned, water-saturated muck, underlain by a low-porosity, low-permeability clay. Hydrologically: Very poorly drained, permanently flooded, aqueous, palustrine system, wetted by fresh rains, flooding, and a high water table. Topographically: Open, nearly level bottom of shallow, concave depressions on an alluvial flat on a water-deposited and organically accumulated floodplain in a broad, east-west-running portion of a shallow river valley situated on a flat plain. Temporally and Spatially: A climax community of a hydropelosere in shallow depressions on an alluvial flat on the Swift Creek floodplain, underlain by Recent alluvium, in the Tar River drainage system in the Embayed Section of the Coastal Plain Province of the Atlantic Plain.

Community Diversity Summary 2

Wetland broadleaf-needleleaf forest system
Large, deliquescent and excurrent, deciduous trees/Medium, deliquescent, deciduous trees

Mixed swamp forest
Canopy: *Nyssa aquatica, Taxodium distichum, Populus heterophylla, Acer rubrum, Quercus hemisphaerica*/Subcanopy: *Nyssa aquatica*--Hardwood-cypress/ Scrophulariales

MIXED BOTTOMLAND TREES/*FRAXINUS CAROLINIANA*
Mixed bottomland trees

CLIMATE: A. Mesothermal; AA. Warm temperate. B. Cool and moderately moist yearly, with moderately hot and moderately wet summers, moderately warm and moderately dry winters, and a long freeze-free period; BB. Similar to sectional climate. C. Cooler and wetter than local climate; CC. Similar to natural area climate.

SOILS: A. Inceptisol; AA. Aquept. B. Humaquept; BB. Cumulic Humaquept. C. Coarse-loamy, siliceous, acid, thermic; CC. Johnston.
A horizon: 0 to 56 cm, black mucky loam, pH 4.6.
C horizon: 56 to 140+ cm, light gray loamy fine sand, with some shells toward bottom, pH 5.4 to 7.0.

GEOLOGY: A. Sedimentary; AA. Mixed clastic and organic. B. Siliceous and organic; BB. Floodplain swamp deposit. C. Mixed fine clastics and organics; CC. Unconsolidated, unaligned, water-saturated muck, underlain by a low porosity, low-permeability clay.

HYDROLOGY: A. Palustrine; AA. Interaqueous. B. Fresh; BB. Seasonally to intermittently flooded. C. Variously drained and wetted; CC. Very poorly drained, wetted by fresh rains, flooding, and a high water table.

TOPOGRAPHY: A. Plain; AA. Flat plain. B. Broad, east-west-running, shallow river valley; BB. Water-deposited and organically accumulated floodplain. C. Entire floodplain and adjoining valley slopes; CC. Open, nearly level borders of shallow, concave depressions on an alluvial flat.

PHYSIOGRAPHY: A. Atlantic Plain; AA. Coastal Plain. B. Embayed Section; BB. Tar River drainage system. C. Swift Creek floodplain; CC. Shallow depressions on an alluvial flat, underlain by Recent alluvium.

CANOPY: *Taxodium distichum*--Height 32 m, dbh 80 cm, age not determined.
Populus heterophylla--Height 27 m, dbh 39 cm, age not determined.
CANOPY DOMINANTS: Mixed bottomland trees.
CANOPY PHYSIOGNOMY: Large, deliquescent and excurrent, deciduous trees.

Canopy Analysis

Species	I.V.	Rel. Den.	Rel. Dom.	Rel. Freq.
Nyssa aquatica	69.8	29.4	21.4	19.0
Taxodium distichum	55.2	8.8	32.1	14.3
Populus heterophylla	45.9	14.7	21.7	9.5

Canopy Analysis (continued)

Species	I.V.	Rel. Den.	Rel. Dom.	Rel. Freq.
Acer rubrum	40.3	17.6	8.4	14.3
Quercus hemisphaerica	29.5	11.8	3.4	14.3
Fraxinus caroliniana	16.3	5.9	0.9	9.5
Quercus lyrata	11.7	2.9	4.0	4.8
Liquidambar styraciflua	11.7	2.9	4.0	4.8
Quercus falcata var. *pagodaefolia*	10.0	2.9	2.3	4.8
Fraxinus pennsylvanica	9.5	2.9	1.8	4.8

Number of relevés: 4 Relevé size: 10 m X 10 m Number of individuals/ acre: 320

CANOPY SPECIES PRESENT, BUT NOT IN ANALYSIS: LARGE TREE: *Carya aquatica*.

SUBCANOPY: *Fraxinus caroliniana*--Height 4.5 m, dbh 11 cm, age not determined.
SUBCANOPY DOMINANT: *Fraxinus caroliniana*.
SUBCANOPY PHYSIOGNOMY: Medium, deliquescent, deciduous trees.

Subcanopy Analysis

Species	GF	1 C.S	2 C.S	3 C.S	4 C.S
Large trees					
Acer rubrum	H2	-	-	-	1.1
Morus rubra	H2	-	+.1	-	-
Nyssa aquatica	H2	1.1	-	2.1	1.1
Medium tree					
Fraxinus caroliniana	H2	5.1	2.1	5.1	4.1
Small tree					
Crataegus viridis	H2	-	1.1	-	-

Number of relevés: 4 Relevé size: 5 m X 5 m

SUBCANOPY SPECIES PRESENT, BUT NOT IN ANALYSIS: SMALL TREE: *Ilex opaca*.

SHRUB LAYER DOMINANTS: None.
SHRUB LAYER PHYSIOGNOMY: Not applicable.

Shrub Analysis

Species	GF	1 C.S	2 C.S	3 C.S	4 C.S
Normal shrub					
Crataegus marshallii	H7	1.1	-	-	-

Shrub Analysis (continued)

Species	GF	1 C.S	2 C.S	3 C.S	4 C.S
Vines					
Rhus radicans	H16	-	1.1	-	-
Smilax rotundifolia	H18	-	1.1	-	-

Number of relevés: 4 Relevé size: 5 m X 5 m

SHRUB AND VINE SPECIES PRESENT, BUT NOT IN ANALYSIS: None.

HERB LAYER DOMINANTS: None.
HERB LAYER PHYSIOGNOMY: Not applicable.

Herb Analysis

Species	GF	1 C.S	2 C.S	3 C.S	4 C.S
Tall forb					
Pilea pumila	H7	1.1	-	-	-
Small forb					
Viola papilionacea	H11	-	1.1	-	-
Medium fern					
Asplenium platyneuron	H11	-	1.1	-	-

Number of relevés: 4 Relevé size: 1 m X 1 m

HERB SPECIES PRESENT, BUT NOT IN ANALYSIS: TALL FORBS: *Boehmeria cylindrica, Orontium aquaticum, Saururus cernuus, Sium suave*; MEDIUM FORB: *Proserpinaca palustris*; SMALL FORBS: *Hydrocotyle umbellata, Mitchella repens*.

LIANA DOMINANTS: None.
LIANA PHYSIOGNOMY: Not applicable.

Liana Analysis

Species	GF	1 C.S	2 C.S	3 C.S	4 C.S
Rhus radicans	H16	-	-	-	1.1
Vitis aestivalis	H18	-	1.1	-	-

Number of relevés: 4 Relevé size: 5 m X 5 m

LIANA SPECIES PRESENT, BUT NOT IN ANALYSIS: None.

COMMUNITY REFERENCES: None.

COMMUNITY DOCUMENTATION: Area analyzed in 1977 and 1978 by Alan S. Weakley for A. E. Radford's Botany 235 class, UNC, Chapel Hill. Specimens deposited in the NCU Herbarium by Alan S. Weakley in 1978.

COMMUNITY ECOLOGICAL CHARACTERIZATION: Vegetationally: Hardwood-Cypress/ Scrophularialean wetland broadleaf-needleleaf forest system with a closed canopy of mixed large, deliquescent and excurrent, deciduous trees, a closed subcanopy of medium, deliquescent, deciduous trees, a sparse shrub layer of normal, rhizomatous, deciduous shrubs, a sparse vine layer of normal, tendril- and root-climbing, deciduous vines, a sparse herb layer of scattered forbs and ferns, and a sparse liana layer of large, root- and tendril-climbing, deciduous lianas. Climatically: Warm temperate, mesothermal climate, cooler and wetter than the local and sectional cool and moderately moist yearly climate, which has moderately hot and moderately wet summers, moderately warm and moderately dry winters, and a long freeze-free period. Pedologically: Coarse-loamy, siliceous, acid, thermic Cumulic Humaquept, Johnston soil. Geologically: Mixed clastic and organic, siliceous and organic, sedimentary floodplain swamp deposit, consisting of unconsolidated, unaligned, water-saturated muck, underlain by a low-porosity, low-permeability clay. Hydrologically: Very poorly drained, seasonally to intermittently flooded, interaqueous palustrine system, wetted by fresh rains, flooding, and a high water table. Topographically: Open, nearly level borders of shallow, concave depressions on an alluvial flat on a water-deposited and organically accumulated floodplain in a broad, east-west-running portion of a shallow river valley situated on a flat plain. Temporally and Spatially: A climax community of a hydropelosere on the borders of shallow depressions on an alluvial flat on the Swift Creek floodplain, underlain by Recent alluvium, in the Tar River drainage system in the Embayed Section of the Coastal Plain Province of the Atlantic Plain.

Community Diversity Summary 3

Wetland broadleaf forest system
Large, deliquescent, deciduous trees

Mixed oak-mixed hardwood swamp forest
Canopy: *Fraxinus pennsylvanica, Acer rubrum, Liquidambar styraciflua, Quercus michauxii, Quercus hemisphaerica*/Subcanopy: *Acer rubrum, Quercus hemisphaerica*/Herb: *Saururus cernuus*--Mixed oaks-mixed hardwoods

MIXED BOTTOMLAND OAKS-MIXED BOTTOMLAND HARDWOODS
Mixed bottomland oaks-mixed bottomland hardwoods

CLIMATE: A. Mesothermal; AA. Warm temperate. B. Cool and moderately moist yearly, with moderately hot and moderately wet summers, moderately warm and moderately dry winters, and a long freeze-free period; BB. Similar to sectional climate. C. Cooler and wetter than local climate; CC. Similar to natural area climate.

SOILS: A. Alfisol; AA. Aqualf. B. Albaqualf; BB. Typic Albaqualf. C. Fine, mixed, thermic; CC. Meggett.

A horizon: 0 to 16 cm, dark gray fine sandy loam, pH 5.7.

B horizon: 16 to 110 cm, gray sandy clay with mottles, pH 7.1.

C horizon: 110 to 150+ cm, light gray loamy fine sand with marl and shell fragments, pH 8.0.

GEOLOGY: A. Sedimentary; AA. Clastic. B. Siliceous; BB. Floodplain swamp deposit. C. Mixed fine clastics; CC. Unbedded, unconsolidated, low-porosity, low-permeability clay.

HYDROLOGY: A. Palustrine; AA. Interaqueous. B. Fresh; BB. Intermittently flooded. C. Variously drained and wetted; CC. Very poorly drained, wetted by fresh rains, flooding, and a high water table.

TOPOGRAPHY: A. Plain; AA. Flat plain. B. Broad, east-west-running, shallow river valley; BB. Water-deposited and organically accumulated floodplain. C. Entire floodplain and adjoining valley slopes; CC. Open, nearly level, slightly irregular, alluvial flat.

PHYSIOGRAPHY: A. Atlantic Plain; AA. Coastal Plain. B. Embayed Section; BB. Tar River drainage system. C. Swift Creek floodplain; CC. Alluvial flat, underlain by Recent alluvium.

CANOPY: *Taxodium distichum*--Height 36 m, dbh 160 cm, age not determined.
Quercus michauxii--Height 29 m, dbh 81 cm, age not determined.
Liquidambar styraciflua--Height 33 m, dbh 79 cm, age not determined.
Fraxinus pennsylvanica--Height 27 m, dbh 46 cm, age not determined.

CANOPY DOMINANTS: Mixed bottomland oaks-mixed bottomland hardwoods.

CANOPY PHYSIOGNOMY: Large, deliquescent, deciduous trees.

Canopy Analysis

Species	I.V.	Rel. Den.	Rel. Dom.	Rel. Freq.
Fraxinus pennsylvanica	74.3	31.9	19.1	23.3
Acer rubrum	62.3	28.6	12.0	21.7
Liquidambar styraciflua	29.3	6.6	12.7	10.0
Quercus michauxii	22.7	4.4	13.3	5.0
Quercus hemisphaerica	21.1	7.7	3.4	10.0
Carya aquatica	20.1	5.5	6.3	8.3
Quercus lyrata	19.6	2.2	14.1	3.3
Quercus falcata var. *pagodaefolia*	12.8	4.4	3.4	5.0
Taxodium distichum	10.0	1.1	7.4	1.7
Ulmus americana	8.2	2.2	2.7	3.3
Quercus bicolor	7.4	1.1	4.6	1.7
Populus heterophylla	3.1	1.1	0.3	1.7
Quercus palustris	3.1	1.1	0.3	1.7
Quercus shumardii	3.0	1.1	0.2	1.7
Quercus nigra	2.9	1.1	0.1	1.7

Number of relevés: 16   Relevé size: 10 m X 10 m   Number of individuals/acre: 270

CANOPY SPECIES PRESENT, BUT NOT IN ANALYSIS: LARGE TREES: *Carya cordiformis*, *Carya ovata*.

SUBCANOPY: *Crataegus viridis*--Height 11 m, dbh 31 cm, age not determined.
SUBCANOPY DOMINANTS: None.
SUBCANOPY PHYSIOGNOMY: Not applicable.

Subcanopy Analysis

Species	GF	1 C.S	2 C.S	3 C.S	4 C.S	5 C.S	6 C.S	7 C.S	8 C.S	9 C.S	10 C.S	11 C.S	12 C.S	13 C.S	14 C.S	15 C.S	16 C.S
Large trees																	
Acer rubrum	H2	-	5.1	3.1	-	2.1	-	2.1	1.1	3.1	-	4.1	-	-	-	3.1	1.1
Fraxinus pennsylvanica	H2	-	2.1	-	-	3.1	-	-	-	-	-	-	-	-	-	-	1.1
Quercus falcata var. *pagodaefolia*	H2	-	-	-	-	-	-	-	-	-	-	-	1.1	-	-	-	-
Quercus hemisphaerica	H2	3.1	-	-	-	-	1.1	-	-	-	4.1	-	-	-	-	3.1	-
Quercus michauxii	H2	-	-	2.1	-	-	-	-	-	-	-	-	-	-	-	-	-
Quercus shumardii	H2	-	-	-	-	-	-	-	1.1	-	-	-	-	-	-	-	-
Medium tree																	
Fraxinus caroliniana	H2	-	-	-	-	-	-	-	-	-	-	-	1.1	-	-	-	-
Small tree																	
Crataegus viridis	H2	-	-	-	5.1	-	-	-	-	-	-	-	-	-	-	-	-

Number of relevés: 16   Relevé size: 5 m X 5 m

SUBCANOPY SPECIES PRESENT, BUT NOT IN ANALYSIS: LARGE TREES: *Fraxinus americana*, *Quercus palustris*, *Ulmus alata*, *Ulmus americana*.

SHRUB LAYER DOMINANTS: None.
SHRUB LAYER PHYSIOGNOMY: Not applicable.

Shrub Analysis

Species	GF	1 C.S	2 C.S	3 C.S	4 C.S	5 C.S	6 C.S	7 C.S	8 C.S	9 C.S	10 C.S	11 C.S	12 C.S	13 C.S	14 C.S	15 C.S	16 C.S
Tall shrub																	
Ilex decidua	H7	-	-	-	3.1	-	-	-	-	2.1	-	-	-	-	-	-	-
Normal shrubs																	
Arundinaria gigantea	H11	2.1	-	-	1.1	-	-	-	-	-	-	-	-	-	-	-	-
Itea virginica	H11	-	-	-	-	-	+.1	-	-	-	-	-	-	-	-	-	-

Shrub Analysis (continued)

Species	GF	1 C.S	2 C.S	3 C.S	4 C.S	5 C.S	6 C.S	7 C.S	8 C.S	9 C.S	10 C.S	11 C.S	12 C.S	13 C.S	14 C.S	15 C.S	16 C.S
Vine																	
Smilax rotundifolia	H18	-	1.1	-	-	-	-	-	-	-	2.1	-	-	-	-	-	-

Number of relevés: 16 Relevé size: 5 m X 5 m

SHRUB SPECIES PRESENT, BUT NOT IN ANALYSIS: NORMAL SHRUB: *Euonymus americanus.*

HERB LAYER DOMINANTS: None.
HERB LAYER PHYSIOGNOMY: Not applicable.

Herb Analysis

Species	GF	1 C.S	2 C.S	3 C.S	4 C.S	5 C.S	6 C.S	7 C.S	8 C.S	9 C.S	10 C.S	11 C.S	12 C.S	13 C.S	14 C.S	15 C.S	16 C.S
Tall forbs																	
Pilea pumila	H7	-	-	1.1	-	-	-	-	-	-	-	-	-	-	-	-	-
Saururus cernuus	H11	-	-	-	-	-	-	2.1	-	3.1	-	-	5.1	-	-	-	-
Sium suave	H11	-	-	-	1.1	-	-	-	1.1	-	-	-	-	-	-	-	-
Small forbs																	
Hydrocotyle umbellata	H11	-	-	-	-	-	-	-	1.1	-	-	-	-	-	-	-	-
Viola papilionacea	H11	-	-	-	-	-	-	-	-	1.1	-	-	-	1.1	-	-	-

Number of relevés: 16 Relevé size: 1 m X 1 m

HERB SPECIES PRESENT, BUT NOT IN ANALYSIS: TALL FORBS: *Lycopus virginicus, Boehmeria cylindrica*; MEDIUM FORB: *Proserpinaca palustris*; MEDIUM GRAMINOID: *Carex louisianica*; TALL FERN: *Onoclea sensibilis*; MEDIUM FERN: *Asplenium platyneuron.*

LIANA DOMINANTS: None.
LIANA PHYSIOGNOMY: Not applicable.

Liana Analysis

Species	GF	1 C.S	2 C.S	3 C.S	4 C.S	5 C.S	6 C.S	7 C.S	8 C.S	9 C.S	10 C.S	11 C.S	12 C.S	13 C.S	14 C.S	15 C.S	16 C.S
Anisostichus capreolata	H18	-	-	-	-	-	-	-	-	-	-	-	5.1	-	4.1	-	-
Berchemia scandens	H19	-	-	-	-	-	-	-	2.1	-	-	-	-	-	-	-	-

Liana Analysis (continued)

Species	GF	1 C.S	2 C.S	3 C.S	4 C.S	5 C.S	6 C.S	7 C.S	8 C.S	9 C.S	10 C.S	11 C.S	12 C.S	13 C.S	14 C.S	15 C.S	16 C.S
Campsis radicans	H16	-	-	-	-	-	-	-	3.1	-	-	-	-	-	-	-	-
Decumaria barbara	H16	-	-	3.1	-	-	-	-	-	-	-	-	-	-	-	-	-
Parthenocissus quinquefolia	H18	-	-	-	-	-	-	-	-	-	-	-	-	-	2.1	-	-
Rhus radicans	H16	-	-	-	-	-	-	1.1	-	-	-	-	-	-	-	-	-
Smilax rotundifolia	H18	-	1.1	-	-	-	2.1	-	-	1.1	-	-	-	-	-	-	-
Vitis aestivalis	H18	-	-	-	-	-	-	-	-	-	2.1	-	-	-	-	-	-

Number of relevés: 16   Relevé size: 5 m X 5 m

LIANA SPECIES PRESENT, BUT NOT IN ANALYSIS: None.

COMMUNITY REFERENCES: None.

COMMUNITY DOCUMENTATION: Area analyzed in 1977 and 1978 by Alan S. Weakley for A. E. Radford's Botany 235 class, UNC, Chapel Hill. Specimens deposited in the NCU Herbarium by Alan S. Weakley in 1978.

COMMUNITY ECOLOGICAL CHARACTERIZATION: Vegetationally: Mixed oak-mixed hardwood broadleaf forest system, with a closed canopy of large, deliquescent, deciduous trees, an open subcanopy of medium, deliquescent, deciduous trees, a sparse shrub layer of mixed shrubs, a sparse vine layer of normal, tendril-climbing, deciduous vines, a sparse herb layer of mixed, scattered forbs, and an open liana layer of large, root- and tendril-climbing, deciduous lianas. Climatically: Warm temperate, mesothermal climate, cooler and wetter than the local and sectional cool and moderately moist yearly climate, which has moderately hot and moderately wet summers, moderately warm and moderately dry winters, and a long freeze-free period. Pedologically: Fine, mixed, thermic Typic Albaqualf, Meggett soil. Geologically: Clastic, siliceous, sedimentary, floodplain swamp deposit, consisting of unbedded, unconsolidated, low-porosity, low-permeability clay. Hydrologically: Very poorly drained, intermittently flooded, interaqueous palustrine system, wetted by fresh rains, flooding, and a high water table. Topographically: Open, nearly level, slightly irregular, alluvial flat on a water-deposited and organically accumulated floodplain in a broad, east-west-running portion of a shallow river valley situated on a flat plain. Temporally and Spatially: A climax community of a pelopsammosere on an alluvial flat on the Swift Creek floodplain, underlain by Recent alluvium, in the Tar River drainage system in the Embayed Section of the Coastal Plain Province of the Atlantic Plain.

Community Diversity Summary 4

Wetland broadleaf forest system
Large, deliquescent, deciduous trees/
Medium, deliquescent, deciduous trees/
Tall, rhizomatous, rhizocarpic forbs//
Large, tendril- and root-climbing, deciduous lianas

Red Maple-mixed hardwood swamp forest
Canopy: *Quercus hemisphaerica, Pinus taeda, Liquidambar styraciflua, Ulmus americana, Quercus palustris*/Subcanopy: *Acer rubrum, Quercus hemisphaerica, Carpinus caroliniana*// Liana: *Anisostichus capreolata, Berchemia scandens, Campsis radicans, Decumaria barbara, Rhus radicans*--Sapindales-mixed hardwoods/ Mixed hardwoods/Piperales// Mixed lianas

ACER RUBRUM-MIXED BOTTOMLAND HARDWOODS/MIXED BOTTOMLAND HARDWOODS/ *SAURURUS CERNUUS*//MIXED LIANAS

Acer rubrum-mixed bottomland hardwoods

CLIMATE: A. Mesothermal; AA. Warm temperate. B. Cool and moderately moist yearly, with moderately hot and moderately wet summers, moderately warm and moderately dry winters, and a long freeze-free period; BB. Similar to sectional climate. C. Cooler and wetter than local climate; CC. Similar to natural area climate.

SOILS: A. Ultisol; AA. Udult. B. Hapludult; BB. Aquic Hapludult. C. Fine-loamy, mixed, thermic; CC. Altavista.
A horizon: 0 to 22 cm, brown fine sandy loam, pH 5.1.
B horizon: 22 to 110 cm, brownish yellow sandy clay loam, pH 5.3.
C horizon: 110 to 160+ cm, light gray to brown sandy loam, pH 5.7.

GEOLOGY: A. Sedimentary; AA. Clastic. B. Siliceous; BB. Floodplain stream deposit. C. Mixed fine clastics; CC. Unbedded, unaligned, unconsolidated, low-permeability silt.

HYDROLOGY: A. Palustrine; AA. Interaqueous. B. Fresh; BB. Intermittently flooded. C. Variously drained and wetted; CC. Poorly drained, wetted by fresh rains, occasional flooding, and a high water table.

TOPOGRAPHY: A. Plain; AA. Flat plain. B. Broad, east-west-running, shallow river valley; BB. Water-deposited and organically accumulated floodplain. C. Entire floodplain and adjoining valley slopes; CC. Open, nearly level, slightly irregular, alluvial flat.

PHYSIOGRAPHY: A. Atlantic Plain; AA. Coastal Plain. B. Embayed Section; BB. Tar River drainage system. C. Swift Creek floodplain; CC. Alluvial flat, underlain by Recent alluvium.

CANOPY: *Acer rubrum*--Height 29 m, dbh 90 cm, age not determined.
Pinus taeda--Height 30 m, dbh 62 cm, age not determined.
Quercus hemisphaerica--Height 21 m, dbh 35 cm, age not determined.
CANOPY DOMINANTS: *Acer rubrum*-mixed bottomland hardwoods.
CANOPY PHYSIOGNOMY: Large, deliquescent, deciduous trees.

Canopy Analysis

Species	I.V.	Rel. Den.	Rel. Dom.	Rel. Freq.
Acer rubrum	76.7	24.3	33.3	19.1
Quercus hemisphaerica	32.9	12.9	9.4	10.6
Pinus taeda	28.9	7.1	15.4	6.4
Liquidambar styraciflua	28.4	11.4	6.4	10.6
Ulmus americana	22.6	7.1	4.9	10.6
Quercus palustris	20.4	7.1	4.8	8.5
Quercus falcata var. *pagodaefolia*	17.2	7.1	3.7	6.4
Quercus lyrata	15.5	4.3	9.1	2.1
Quercus michauxii	12.3	4.3	1.6	6.4
Quercus phellos	10.8	2.9	3.6	4.3
Quercus nigra	10.4	2.9	3.2	4.3
Carya aquatica	9.0	1.4	2.0	2.1
Fraxinus caroliniana	8.0	2.9	0.8	4.3
Quercus shumardii	6.0	2.9	1.0	2.1
Populus heterophylla	4.5	1.4	1.0	2.1

Number of relevés: 10 Relevé size: 10 m X 10 m Number of individuals/acre: 305

CANOPY SPECIES PRESENT, BUT NOT IN ANALYSIS: LARGE TREES: *Carya cordiformis*, *Fraxinus pennsylvanica*, *Nyssa aquatica*, *Taxodium distichum*.

SUBCANOPY: *Quercus hemisphaerica*--Height 13 m, dbh 16 cm, age not determined.
SUBCANOPY DOMINANTS: Mixed bottomland hardwoods.
SUBCANOPY PHYSIOGNOMY: Medium, deliquescent, deciduous trees.

Subcanopy Analysis

Species	GF	1 C.S	2 C.S	3 C.S	4 C.S	5 C.S	6 C.S	7 C.S	8 C.S	9 C.S	10 C.S
Large trees											
Acer rubrum	H2	2.1	-	-	4.1	4.1	2.1	1.1	4.1	-	2.1
Carya aquatica	H2	-	-	1.1	-	-	-	-	-	-	-
Fraxinus pennsylvanica	H2	-	-	-	-	-	2.1	-	-	-	-
Liquidambar styraciflua	H2	-	-	-	2.1	-	-	-	-	-	-
Nyssa aquatica	H2	2.1	-	-	-	-	-	-	-	-	-
Quercus falcata var. *pagodaefolia*	H2	3.1	-	2.1	-	-	-	-	-	-	-
Quercus hemisphaerica	H2	2.1	5.1	2.1	-	-	-	2.1	2.1	-	-
Quercus lyrata	H2	3.1	-	-	-	-	2.1	-	-	-	-
Quercus phellos	H2	-	-	-	-	-	-	1.1	-	-	-
Populus heterophylla	H2	-	-	-	-	-	1.1	-	-	-	-
Ulmus americana	H2	-	-	-	-	-	-	2.1	-	-	-

Subcanopy Analysis (continued)

Species	GF	1 C.S	2 C.S	3 C.S	4 C.S	5 C.S	6 C.S	7 C.S	8 C.S	9 C.S	10 C.S
Medium trees											
Carpinus caroliniana	H2	-	4.1	5.1	-	-	-	-	2.1	-	-
Crataegus viridis	H2	1.1	-	-	-	-	-	-	-	-	-
Fraxinus caroliniana	H2	4.1	-	-	-	1.1	-	-	-	-	-

Number of relevés: 10 Relevé size: 5 m X 5 m

SUBCANOPY SPECIES PRESENT, BUT NOT IN ANALYSIS: LARGE TREES: *Quercus michauxii*, *Quercus shumardii*.

SHRUB LAYER DOMINANTS: None.
SHRUB LAYER PHYSIOGNOMY: Not applicable.

Shrub Analysis

Species	GF	1 C.S	2 C.S	3 C.S	4 C.S	5 C.S	6 C.S	7 C.S	8 C.S	9 C.S	10 C.S
Tall shrub											
Ilex decidua	H7	-	-	-	-	-	-	-	-	-	5.2
Normal shrub											
Itea virginica	H11	-	-	3.2	-	-	2.1	-	-	-	-
Vine											
Smilax rotundifolia	H18	3.1	-	-	-	-	-	-	-	-	-

Number of relevés: 10 Relevé size: 5 m X 5 m

SHRUB SPECIES PRESENT, BUT NOT IN ANALYSIS: NORMAL SHRUBS: *Arundinaria gigantea*, *Euonymus americanus*, *Viburnum dentatum*.

HERB LAYER DOMINANT: *Saururus cernuus*.
HERB LAYER PHYSIOGNOMY: Tall, rhizomatous, rhizocarpic forbs.

Herb Analysis

Species	GF	1 C.S	2 C.S	3 C.S	4 C.S	5 C.S	6 C.S	7 C.S	8 C.S	9 C.S	10 C.S
Tall forbs											
Lycopus virginicus	H7	-	-	1.1	-	-	-	-	-	-	-
Saururus cernuus	H11	1.1	3.1	-	3.1	2.1	5.1	2.1	4.1	-	3.1
Small forbs											
Hydrocotyle umbellata	H11	-	-	-	-	1.1	-	-	-	-	-

Herb Analysis (continued)

Species	GF	1 C.S	2 C.S	3 C.S	4 C.S	5 C.S	6 C.S	7 C.S	8 C.S	9 C.S	10 C.S
Viola palmata var. *triloba*	H11	-	-	-	-	-	-	-	-	2.1	-
Viola papilionacea	H11	-	-	-	-	-	-	-	-	3.1	-
Medium graminoids											
Panicum sp.	H7	-	-	-	-	-	-	-	1.1	-	-

Number of relevés: 10 Relevé size: 1 m X 1 m

HERB SPECIES PRESENT, BUT NOT IN ANALYSIS: TALL FORBS: *Arisaema triphyllum, Boehmeria cylindrica, Pilea pumila, Sium suave*; TALL FERN: *Onoclea sensibilis.*

LIANA DOMINANTS: Mixed lianas.
LIANA PHYSIOGNOMY: Large, tendril- and root-climbing, deciduous lianas.

Liana Analysis

Species	GF	1 C.S	2 C.S	3 C.S	4 C.S	5 C.S	6 C.S	7 C.S	8 C.S	9 C.S	10 C.S
Anisostichus capreolata	H18	-	-	-	-	-	-	4.1	-	-	-
Berchemia scandens	H19	-	4.1	2.1	-	-	-	-	-	-	-
Campsis radicans	H16	-	-	-	4.1	-	-	-	-	-	-
Decumaria barbara	H16	-	-	-	-	-	-	-	-	-	4.1
Parthenocissus quinquefolia	H18	1.1	1.1	-	-	-	-	-	-	-	-
Rhus radicans	H16	-	2.1	-	-	2.1	-	4.1	-	-	1.1
Smilax rotundifolia	H18	-	1.1	3.1	-	-	-	-	1.1	-	-
Vitis aestivalis	H18	-	-	-	2.1	-	4.1	-	-	-	-
Vitis labrusca	H18	-	2.1	-	-	-	-	-	-	-	-

Number of relevés: 10 Relevé size: 5 m X 5 m

LIANA SPECIES PRESENT, BUT NOT IN ANALYSIS: None.

COMMUNITY REFERENCES: None.

COMMUNITY DOCUMENTATION: Area analyzed in 1977 and 1978 by Alan S. Weakley for A. E. Radford's Botany 235 class, UNC, Chapel Hill. Specimens deposited in the NCU Herbarium by Alan S. Weakley in 1978.

COMMUNITY ECOLOGICAL CHARACTERIZATION: Vegetationally: Sapindalean-mixed hardwood/Mixed hardwood/Piperalean//Mixed liana wetland broadleaf forest system, with a closed canopy dominated by large, deliquescent, deciduous trees, a closed subcanopy of medium, deliquescent, deciduous trees, a sparse shrub layer of mixed, normal to tall, rhizomatous, deciduous shrubs, a sparse vine layer of normal, tendril-climbing, deciduous vines, a closed herb layer dominated by a tall, rhizomatous, rhizocarpic forb, and a closed liana layer of large, root- and tendril-climbing, deciduous lianas. Climatically: Warm temperate, mesothermal climate, cooler and wetter than the local and sectional cool and moderately moist yearly climate, which has moderately hot and moderately wet summers, moderately warm and moderately dry winters, and a long freeze-free period. Pedologically: Fine-loamy, mixed, thermic Aquic Hapludult, Altavista soil. Geologically: Clastic, siliceous, sedimentary, floodplain stream deposit, consisting of unbedded, unaligned, low-permeability, unconsolidated silt. Hydrologically: Poorly drained, intermittently flooded, interaqueous palustrine system, wetted by fresh rains, occasional flooding, and a high water table. Topographically: Open, nearly level, slightly irregular, alluvial flat on a water-deposited and organically accumulated floodplain in a broad, east-west-running portion of a shallow river valley situated on a flat plain. Temporally and Spatially: A climax community of a pelopsammosere on an alluvial flat on the Swift Creek floodplain, underlain by Recent alluvium, in the Tar River drainage system in the Embayed Section of the Coastal Plain Province of the Atlantic Plain.

Community Diversity Summary 5

Wetland broadleaf forest system
Large, deliquescent, deciduous trees/ Medium, deliquescent, deciduous trees/ Tall dwarf, stoloniferous, deciduous shrubs-small to tall, rhizomatous, rhizocarpic herbs//Large, tendril- and root-climbing, deciduous lianas

Mixed hardwood-mixed oak swamp forest
Canopy: *Quercus hemisphaerica, Acer rubrum, Ulmus americana, Carya aquatica, Liquidambar styraciflua, Quercus michauxii, Betula nigra*/Subcanopy: *Fraxinus caroliniana, Carpinus caroliniana, Acer rubrum*/Herb: *Commelina virginica, Viola papilionacea, Saururus cernuus, Panicum* sp.//Liana: *Anisostichus capreolata, Campsis radicans, Smilax rotundifolia*--Mixed hardwoods-mixed oaks/ mixed hardwoods/Sapindales-mixed forb perennials//Mixed lianas

MIXED BOTTOMLAND HARDWOODS-MIXED BOTTOMLAND OAKS/
MIXED BOTTOMLAND HARDWOODS/*RHUS RADICANS*-MIXED FORB PERENNIALS//MIXED LIANAS
Mixed bottomland hardwoods-mixed bottomland oaks

CLIMATE: A. Mesothermal; AA. Warm temperate. B. Cool and moderately moist yearly, with moderately hot and moderately wet summers, moderately warm and moderately dry winters, and a long freeze-free period; BB. Similar to sectional climate. C. Cooler and wetter than local climate; CC. Similar to natural area climate.

SOILS: A. Alfisol; AA. Aqualf. B. Albaqualf; BB. Typic Albaqualf. C. Fine, mixed, thermic; CC. Meggett.

A horizon: 0 to 16 cm, dark gray fine sandy loam, pH 5.7.

B horizon: 16 to 110 cm, gray sandy clay with mottles, pH 7.1.

C horizon: 110 to 150+ cm, light gray loamy fine sand with marl and shell fragments, pH 8.0.

GEOLOGY: A. Sedimentary; AA. Clastic. B. Siliceous; BB. Levee deposit. C. Mixed fine clastics; CC. Unbedded, unconsolidated mixtures of silt and clay.

HYDROLOGY: A. Palustrine; AA. Interaqueous. B. Fresh; BB. Intermittently flooded. C. Variously drained and wetted; CC. Moderately poorly drained, wetted by fresh rains and occasional flooding.

TOPOGRAPHY: A. Plain; AA. Flat plain. B. Broad, east-west-running, shallow river valley; BB. Water-deposited and organically accumulated floodplain. C. Entire floodplain and adjoining valley slopes; CC. Open, nearly level, irregularly mounded, streamside slope and crest of a levee.

PHYSIOGRAPHY: A. Atlantic Plain; AA. Coastal Plain. B. Embayed Section; BB. Tar River drainage system. C. Swift Creek floodplain; CC. Streamside levee, underlain by Recent alluvium.

CANOPY: *Quercus hemisphaerica*--Height 33 m, dbh 97 cm, age not determined.

Acer rubrum--Height 24 m, dbh 47 cm, age not determined.

Ulmus americana--Height 29 m, dbh 90 cm, age not determined.

Quercus falcata var. *pagodaefolia*--Height 33 m, dbh 102 cm, age not determined.

CANOPY DOMINANTS: Mixed bottomland hardwoods-mixed bottomland oaks.

CANOPY PHYSIOGNOMY: Large, deliquescent, deciduous trees.

Canopy Analysis

Species	I.V.	Rel. Den.	Rel. Dom.	Rel. Freq.
Quercus hemisphaerica	35.9	9.1	17.9	8.9
Acer rubrum	33.3	13.0	7.8	12.5
Ulmus americana	31.6	7.8	13.1	10.7
Carya aquatica	27.6	11.7	7.0	8.9
Liquidambar styraciflua	24.5	7.8	7.8	8.9
Quercus michauxii	18.8	7.8	5.6	5.4
Betula nigra	18.8	9.1	4.3	5.4
Quercus lyrata	17.1	6.5	1.7	8.9
Quercus palustris	15.3	3.9	6.0	5.4
Quercus falcata var. *pagodaefolia*	15.2	1.3	12.1	1.8
Taxodium distichum	13.4	5.2	4.6	3.6
Carya ovata	9.3	2.6	3.1	3.6
Carpinus caroliniana	9.0	3.9	1.5	3.6
Quercus phellos	7.1	2.6	0.9	3.6
Diospyros virginiana	6.1	1.3	3.0	1.8
Fraxinus pennsylvanica	5.6	2.6	1.2	1.8
Nyssa aquatica	5.3	1.3	2.2	1.8
Populus heterophylla	3.2	1.3	0.1	1.8
Fraxinus tomentosa	3.2	1.3	0.1	1.8

Number of relevés: 10 Relevé size: 10 m X 10 m Number of individuals/ acre: 445

CANOPY SPECIES PRESENT, BUT NOT IN ANALYSIS: LARGE TREE: *Quercus nigra*.

SUBCANOPY: *Carpinus caroliniana*--Height 16 m, dbh 17 cm, age not determined. *Fraxinus caroliniana*--Height 14 m, dbh 8 cm, age not determined.
SUBCANOPY DOMINANTS: Mixed bottomland hardwoods.
SUBCANOPY PHYSIOGNOMY: Medium, deliquescent, deciduous trees.

Subcanopy Analysis

Species	GF	1 C.S	2 C.S	3 C.S	4 C.S	5 C.S	6 C.S	7 C.S	8 C.S	9 C.S
Large trees										
Acer rubrum	H2	1.1	2.1	2.1	1.1	1.1	-	-	-	-
Carya ovata	H2	-	-	-	2.1	-	-	-	-	-
Fraxinus americana	H2	2.1	-	-	-	-	-	-	-	-
Fraxinus tomentosa	H2	-	-	-	-	-	-	-	2.1	-
Quercus hemisphaerica	H2	1.1	-	-	-	4.1	1.1	-	-	-
Quercus lyrata	H2	-	-	-	-	-	-	-	-	2.1
Quercus michauxii	H2	-	1.1	-	1.1	-	2.1	-	2.1	-
Taxodium distichum	H1	-	3.1	-	-	-	-	-	-	-
Medium trees										
Carpinus caroliniana	H2	3.1	-	-	1.1	5.1	4.1	-	-	-
Fraxinus caroliniana	H2	3.1	5.1	1.1	-	2.1	2.1	3.1	1.1	2.1
Ilex opaca	H2	-	-	-	2.1	-	-	-	-	-
Ulmus alata	H2	-	-	2.1	2.1	-	-	-	3.1	-
Small tree										
Crataegus viridis	H2	-	-	-	-	-	-	3.1	-	4.1

Number of relevés: 9 Relevé size: 5 m X 5 m

SUBCANOPY SPECIES PRESENT, BUT NOT IN ANALYSIS: MEDIUM TREE: *Ostrya virginiana*.

SHRUB LAYER DOMINANTS: *Rhus radicans* (co-dominant with herb layer).
SHRUB LAYER PHYSIOGNOMY: Tall dwarf, stoloniferous, deciduous shrub.

Shrub Analysis

Species	GF	1 C.S	2 C.S	3 C.S	4 C.S	5 C.S	6 C.S	7 C.S	8 C.S	9 C.S
Tall shrubs										
Ilex decidua	H6	-	2.1	4.1	-	-	-	-	2.1	-
Normal shrubs										
Arundinaria gigantea	H10	-	-	-	-	-	-	1.1	1.1	2.1
Cornus stricta	H6	-	-	-	-	-	-	-	-	1.1

Shrub Analysis (continued)

Species	GF	1 C.S	2 C.S	3 C.S	4 C.S	5 C.S	6 C.S	7 C.S	8 C.S	9 C.S
Itea virginica	H10	1.1	-	-	-	-	-	-	-	-
Rhus radicans	H13	1.1	4.1	-	5.1	2.1	-	-	4.1	-
Vines										
Smilax rotundifolia	H10	1.1	-	-	-	1.1	-	-	-	-

Number of relevés: 9   Relevé size: 5 m X 5 m

SHRUB SPECIES PRESENT, BUT NOT IN ANALYSIS: NORMAL SHRUBS: *Crataegus marshallii, Ilex verticillata, Viburnum dentatum, Viburnum prunifolium*; TALL DWARF SHRUBS: *Hypericum stans.*

HERB LAYER DOMINANTS: Mixed perennial herbs (co-dominant with shrub layer).
HERB LAYER PHYSIOGNOMY: Small to tall, rhizomatous, rhizocarpic herbs.

Herb Analysis

Species	GF	1 C.S	2 C.S	3 C.S	4 C.S	5 C.S	6 C.S	7 C.S	8 C.S	9 C.S
Tall forbs										
Arisaema triphyllum	H5	+.1	-	-	-	-	-	-	-	-
Boehmeria cylindrica	H10	-	-	-	-	-	-	-	-	1.1
Commelina virginica	H10	-	-	-	-	1.1	4.1	-	-	-
Impatiens capensis	H10	-	-	-	-	-	1.1	-	-	-
Pilea pumila	H10	-	-	-	-	-	-	1.1	-	-
Saururus cernuus	H10	1.1	-	-	-	3.1	-	-	-	-
Small forbs										
Viola palmata var. *triloba*	H10	-	-	1.1	-	-	1.1	-	-	-
Viola papilionacea	H10	-	1.1	2.1	-	1.1	1.1	2.1	-	-
Very small forbs										
Mitchella repens	H9	-	-	-	1.1	-	-	-	-	-
Tall graminoids										
Carex lurida	H7	-	-	-	1.1	-	-	-	-	-
Medium graminoids										
Panicum sp.	H7	-	-	1.1	1.1	+.1	2.1	1.1	1.1	-
Tall ferns										
Onoclea sensibilis	H10	-	-	-	-	-	-	+.1	-	-

Number of relevés: 9   Relevé size: 1 m X 1 m

HERB SPECIES PRESENT, BUT NOT IN ANALYSIS: TALL FORBS: *Justicia ovata, Orontium aquaticum, Sabatia angularis*; SMALL FORBS: *Hydrocotyle umbellata*; MEDIUM FERNS: *Asplenium platyneuron.*

LIANA DOMINANTS: Mixed lianas.
LIANA PHYSIOGNOMY: Large, tendril- and root-climbing, deciduous lianas.

Liana Analysis

Species	GF	1 C.S	2 C.S	3 C.S	4 C.S	5 C.S	6 C.S	7 C.S	8 C.S	9 C.S
Anisostichus capreolata	H17	-	4.1	-	-	-	3.1	-	-	-
Campsis radicans	H15	1.1	3.1	-	3.1	2.1	-	-	-	2.1
Parthenocissus quinquefolia	H17	-	-	1.1	-	-	-	4.1	-	-
Rhus radicans	H15	-	-	2.1	-	-	2.1	-	4.1	-
Smilax rotundifolia	H17	2.1	-	-	2.1	1.1	3.1	-	-	2.1
Vitis aestivalis	H17	-	2.1	-	-	-	-	-	-	-

Number of relevés: 9 Relevé size: 5 m X 5 m

LIANA SPECIES PRESENT, BUT NOT IN ANALYSIS: LIANAS: *Berchemia scandens, Vitis labrusca*.

COMMUNITY REFERENCES: None.

COMMUNITY DOCUMENTATION: Area analyzed in 1977 and 1978 by Alan S. Weakley for A. E. Radford's Botany 235 class, UNC, Chapel Hill. Specimens deposited in the NCU Herbarium by Alan S. Weakley in 1978.

COMMUNITY ECOLOGICAL CHARACTERIZATION: Vegetationally: Mixed hardwood-Fagalean/Mixed hardwood/Sapindalean-mixed forb perennial//Mixed liana wetland broadleaf forest system, with a closed canopy of large, deliquescent, deciduous trees, a closed subcanopy of medium, deliquescent, deciduous trees, an open shrub layer of tall dwarf, stoloniferous, deciduous shrubs, an open herb layer of small to tall, rhizomatous, rhizocarpic herbs, and a closed liana layer of large, tendril- and root-climbing, deciduous lianas. Climatically: Warm temperate, mesothermal climate, cooler and wetter than the local and sectional cool and moderately moist yearly climate, which has moderately hot and moderately wet summers, moderately warm and moderately dry winters, and a long freeze-free period. Pedologically: Fine, mixed, thermic Typic Albaqualf, Meggett soil. Geologically: Siliceous, clastic, sedimentary levee deposit, consisting of unbedded, unconsolidated mixtures of silt and clay. Hydrologically: Moderately poorly drained, intermittently flooded, interaqueous palustrine system, wetted by fresh rains and occasional flooding. Topographically: Open, nearly level, irregularly mounded, streamside slope and crest of a levee, on a water-deposited and organically accumulated floodplain in a broad, east-west-running portion of a shallow river valley situated on a flat plain. Temporally and Spatially: A climax community of a pelopsammosere on a levee on the Swift Creek floodplain, underlain by Recent alluvium, in the Tar River drainage system in the Embayed Section of the Coastal Plain Province of the Atlantic Plain.

Community Diversity Summary 6

Terrestrial needleleaf-broadleaf forest system
Large, excurrent, evergreen trees-large, deliquescent, deciduous trees/Medium, deliquescent, deciduous trees//Low, spread-climbing, evergreen vines

Loblolly Pine-mixed bottomland hardwood forest
Canopy: *Liquidambar styraciflua, Nyssa sylvatica, Acer rubrum*/ Subcanopy: *Quercus michauxii, Quercus phellos, Quercus rubra, Liquidambar styraciflua*/Shrub: *Myrica cerifera, Viburnum dentatum*--Coniferales-mixed hardwoods/Mixed hardwoods// Dipsacales

PINUS TAEDA-MIXED BOTTOMLAND HARDWOODS/MIXED BOTTOMLAND OAKS//*LONICERA JAPONICA*
Pinus taeda-mixed bottomland hardwoods

CLIMATE: A. Mesothermal; AA. Warm temperate. B. Cool and moderately moist yearly, with moderately hot and moderately wet summers, moderately warm and moderately dry winters, and a long freeze-free period; BB. Similar to sectional climate. C. Cooler and wetter than local climate; CC. Similar to natural area climate.

SOILS: A. Ultisol; AA. Udult. B. Hapludult; BB. Typic Hapludult. C. Fine-loamy, mixed, thermic; CC. Wickham.
A horizon: 0 to 19 cm, brown sandy loam, pH 5.2.
B1 horizon: 19 to 28 cm, reddish yellow sandy loam, pH 5.4.
B2 horizon: 28 to 83 cm, yellowish red sandy clay loam, pH 6.0.
C horizon: 83 to 140+ cm, brown loamy sand, pH 6.0.

GEOLOGY: A. Sedimentary; AA. Clastic. B. Siliceous; BB. Stream terrace deposit. C. Mixed fine clastics; CC. Unbedded, unconsolidated, highly porous and permeable mixtures of silt and sand.

HYDROLOGY: A. Terrestrial; AA. Dry. B. Fresh; BB. Permanently exposed. C. Variously drained and wetted; CC. Moderately well drained, wetted by fresh rains.

TOPOGRAPHY: A. Plain; AA. Flat plain. B. Broad, east-west-running, shallow river valley; BB. Water-deposited and organically accumulated floodplain. C. Entire floodplain and adjoining valley slopes; CC. Open, nearly level, slightly irregular flat on an upper stream terrace.

PHYSIOGRAPHY: A. Atlantic Plain; AA. Coastal Plain. B. Embayed Section; BB. Tar River drainage system. C. Swift Creek floodplain; CC. Upper stream terrace, underlain by Recent alluvium.

CANOPY: *Pinus taeda*--Height 30 m, dbh 50 cm, age not determined.
Ulmus americana--Height 31 m, dbh 53 cm, age not determined.

CANOPY DOMINANTS: *Pinus taeda*-mixed bottomland hardwoods.

CANOPY PHYSIOGNOMY: Large, excurrent, evergreen trees-large, deliquescent, deciduous trees.

Canopy Analysis

Species	I.V.	Rel. Den.	Rel. Dom.	Rel. Freq.
Pinus taeda	122.0	35.3	60.6	26.1
Liquidambar styraciflua	79.3	33.3	15.6	30.4
Nyssa sylvatica	29.5	11.8	4.7	13.0
Acer rubrum	20.9	5.9	6.3	8.7
Quercus phellos	15.8	5.9	1.2	8.7
Ulmus americana	15.6	1.9	9.4	4.3
Ulmus alata	10.1	3.9	1.9	4.3
Quercus rubra	6.5	1.9	0.3	4.3

Number of relevés: 7 Relevé size: 10 m X 10 m Number of individuals/ acre: 294

CANOPY SPECIES PRESENT, BUT NOT IN ANALYSIS: LARGE TREES: *Fagus grandifolia, Liriodendron tulipifera.*

SUBCANOPY: *Quercus phellos*--Height 12 m, dbh 12 cm, age not determined. *Quercus rubra*--Height 13 m, dbh 11 cm, age not determined. *Quercus michauxii*--Height 10 m, dbh 8 cm, age not determined.
SUBCANOPY DOMINANTS: Mixed bottomland oaks.
SUBCANOPY PHYSIOGNOMY: Medium, deliquescent, deciduous trees.

Subcanopy Analysis

Species	GF	1 C.S	2 C.S	3 C.S	4 C.S	5 C.S	6 C.S	7 C.S
Large trees								
Acer rubrum	H2	-	2.1	2.1	-	3.1	-	-
Fraxinus pennsylvanica	H2	-	2.1	-	-	-	-	-
Liquidambar styraciflua	H2	-	-	-	-	3.1	-	5.1
Nyssa sylvatica	H2	-	3.1	-	-	-	-	-
Pinus taeda	H1	-	-	-	2.1	-	5.1	-
Quercus falcata var. *pagodaefolia*	H2	1.1	-	-	-	-	-	-
Quercus hemisphaerica	H2	1.1	-	-	-	-	-	-
Quercus michauxii	H2	-	4.1	4.1	4.1	-	-	-
Quercus phellos	H2	3.1	1.1	-	-	3.1	2.1	-
Quercus rubra	H2	-	-	-	-	-	2.1	3.1
Ulmus alata	H2	-	-	3.1	-	-	-	-
Medium trees								
Cornus florida	H2	-	-	-	3.1	-	-	-
Diospyros virginiana	H2	2.1	-	-	-	-	-	-
Ilex opaca	H2	-	-	-	2.1	-	-	-

Number of relevés: 7 Relevé size: 5 m X 5 m

SUBCANOPY SPECIES PRESENT, BUT NOT IN ANALYSIS: None.

SHRUB LAYER DOMINANT: *Lonicera japonica*.
SHRUB LAYER PHYSIOGNOMY: Low, spread-climbing, evergreen vines.

Shrub Analysis

Species	GF	1 C.S	2 C.S	3 C.S	4 C.S	5 C.S	6 C.S	7 C.S
Transgressive trees								
Fagus grandifolia	H2	-	-	-	-	-	-	1.1
Liriodendron tulipifera	H2	-	-	-	2.1	-	-	-
Sassafras albidum	H2	-	-	-	-	-	1.1	-
Large shrubs								
Crataegus viridis	H7	3.1	-	-	-	-	-	-
Myrica cerifera	H11	-	-	-	-	-	5.1	5.1
Normal shrubs								
Vaccinium corymbosum	H11	-	1.1	-	-	-	-	-
Viburnum dentatum	H11	-	-	-	2.1	2.1	-	1.1
Vines								
Lonicera japonica	H17	5.1	1.1	-	5.1	4.1	-	2.1
Parthenocissus quinquefolia	H18	+.1	+.1	-	-	-	-	-
Vitis rotundifolia	H18	-	+.1	-	-	-	-	-

Number of relevés: 7 Relevé size: 5 m X 5 m

SHRUB SPECIES PRESENT, BUT NOT IN ANALYSIS: TRANSGRESSIVE TREES: *Carya tomentosa*, *Morus rubra*, *Prunus serotina*; NORMAL SHRUB: *Arundinaria gigantea*; VINES: *Smilax rotundifolia*, *Vitis aestivalis*.

HERB LAYER DOMINANTS: None.
HERB LAYER PHYSIOGNOMY: Not applicable.

Herb Analysis

Species	GF	1 C.S	2 C.S	3 C.S	4 C.S	5 C.S	6 C.S	7 C.S
Medium ferns								
Asplenium platyneuron	H11	1.1	-	-	-	-	-	-
Onoclea sensibilis	H11	-	-	5.1	-	-	-	-

Number of relevés: 7 Relevé size: 1 m X 1 m

HERB SPECIES PRESENT, BUT NOT IN ANALYSIS: VERY SMALL FORB: *Mitchella repens*; SMALL GRAMINOID: *Panicum* sp.; TALL FERNS: *Athyrium asplenioides*, *Osmunda cinnamomea*.

COMMUNITY REFERENCES: None.

COMMUNITY DOCUMENTATION: Area analyzed in 1977 and 1978 by Alan S. Weakley for A. E. Radford's Botany 235 class, UNC, Chapel Hill. Specimens deposited in the NCU Herbarium by Alan S. Weakley in 1978.

COMMUNITY ECOLOGICAL CHARACTERIZATION: Vegetationally: Coniferalean-mixed hardwood/Mixed hardwood/Dipsacalean terrestrial needleleaf-broadleaf forest system, with a closed canopy of large, excurrent, evergreen trees and large, deliquescent, deciduous trees, a closed subcanopy of medium, deliquescent, deciduous trees, an open shrub layer of normal to large, rhizomatous, deciduous and evergreen shrubs, a closed vine layer of low, spread-climbing, evergreen vines, and a sparse herb layer of scattered, medium to tall, rhizomatous ferns. Climatically: Warm temperate, mesothermal climate, cooler and wetter than the local and sectional cool and moderately moist yearly climate, which has moderately hot and moderately wet summers, moderately warm and moderately dry winters, and a long freeze-free period. Pedologically: Fine-loamy, mixed, thermic Typic Hapludult, Wickham soil. Geologically: Siliceous, clastic, sedimentary, stream terrace deposit, consisting of unbedded, unconsolidated, highly porous and permeable mixtures of silt and sand. Hydrologically: Moderately well-drained, permanently exposed, dry, terrestrial system wetted by fresh rains. Topographically: Open, nearly level, slightly irregular flat on an upper stream terrace of a water-deposited and organically accumulated floodplain in a broad, east-west-running portion of a shallow stream valley located on a flat plain. Temporally and Spatially: A late successional stage community of a psammosere on an upper stream terrace on the Swift Creek floodplain, underlain by Recent alluvium, in the Tar River drainage system in the Embayed Section of the Coastal Plain Province of the Atlantic Plain.

Community Diversity Summary 7

Terrestrial broadleaf forest system
Large, deliquescent, deciduous trees/ Medium, deliquescent, deciduous trees

Beech-mixed hardwood forest
Canopy: *Quercus alba, Liquidambar styraciflua*/ Subcanopy: *Ilex opaca, Quercus alba, Fagus grandifolia, Oxydendrum arboreum, Sassafras albidum*/Vine: *Vitis rotundifolia*/Herb: *Panicum* sp.--Fagales-mixed hardwoods/Mixed hardwoods

FAGUS GRANDIFOLIA-MIXED HARDWOODS/MIXED HARDWOODS
Fagus grandifolia-mixed hardwoods

CLIMATE: A. Mesothermal; AA. Warm temperate. B. Cool and moderately moist yearly, with moderately hot and moderately wet summers, moderately warm and moderately dry winters, and a long freeze-free period; BB. Similar to sectional climate. C. Cooler and wetter than local climate; CC. Similar to natural area climate.

SOILS: A. Ultisol; AA. Udult. B. Hapludult; BB. Typic Hapludult. C. Clayey, mixed, thermic; CC. Gritney.

A horizon: 0 to 13 cm, brown fine sandy loam, pH 5.0.

B1 horizon: 13 to 56 cm, reddish yellow clay with mottles, pH 5.0.

B2 horizon: 56 to 120 cm, yellowish red to gray clay with mottles, pH 5.4.

C horizon: 120 to 150+ cm, yellow, red, or gray sandy clay loam, pH 5.4.

GEOLOGY: A. Sedimentary; AA. Clastic. B. Siliceous; BB. Undetermined origin. C. Mixed fine clastics; CC. Unbedded, unconsolidated, highly porous and permeable mixture of silt and clay.

HYDROLOGY: A. Terrestrial; AA. Dry. B. Fresh; BB. Permanently exposed. C. Variously drained and wetted; CC. Well drained, wetted by fresh rains and downslope drainage.

TOPOGRAPHY: A. Plain; AA. Flat plain. B. Broad, east-west-running, shallow river valley; BB. Valley slopes. C. Entire floodplain and adjoining valley slopes; CC. Open, north-, northeast-, south-, and southwest-facing, gently sloping to sloping, convexo-concave, entire valley slope.

PHYSIOGRAPHY: A. Atlantic Plain; AA. Coastal Plain. B. Embayed Section; BB. Tar River drainage system. C. Swift Creek valley slope; CC. Valley slope, underlain by Miocene (?) sediments.

CANOPY: *Fagus grandifolia*--Height 19 m, dbh 64 cm, age not determined.

Quercus alba--Height 17 m, dbh 30 cm, age not determined.

CANOPY DOMINANTS: *Fagus grandifolia*-mixed hardwoods.

CANOPY PHYSIOGNOMY: Large, deliquescent, deciduous trees.

Canopy Analysis

Species	I.V.	Rel. Den.	Rel. Dom.	Rel. Freq.
Fagus grandifolia	125.1	42.4	54.1	28.6
Quercus alba	49.9	21.2	14.4	14.3
Liquidambar styraciflua	24.6	6.1	9.0	9.5
Pinus taeda	19.6	6.1	4.0	9.5
Quercus lyrata	11.4	3.0	3.6	4.8
Carya tomentosa	10.3	3.0	2.5	4.8
Quercus nigra	10.3	3.0	2.5	4.8
Quercus stellata	10.3	3.0	2.5	4.8
Quercus phellos	10.3	3.0	2.5	4.8
Nyssa sylvatica	10.3	3.0	2.5	4.8
Acer rubrum	9.0	3.0	1.2	4.8
Carpinus caroliniana	9.0	3.0	1.2	4.8

Number of relevés: 7 Relevé size: 10 m X 10 m Number of individuals/acre: 192

CANOPY SPECIES PRESENT, BUT NOT IN ANALYSIS: None.

SUBCANOPY: *Ilex opaca*--Height 11 m, dbh 13 cm, age not determined.

Oxydendrum arboreum--Height 13 m, dbh 13 cm, age not determined.

SUBCANOPY DOMINANTS: Mixed hardwoods.

SUBCANOPY PHYSIOGNOMY: Medium, deliquescent, deciduous trees.

Subcanopy Analysis

Species	GF	1 C.S	2 C.S	3 C.S	4 C.S	5 C.S	6 C.S	7 C.S
Large trees								
Acer rubrum	H2	-	-	-	3.1	-	-	-
Fagus grandifolia	H2	-	-	-	4.1	3.1	-	-
Liquidambar styraciflua	H2	-	-	-	-	-	2.1	-
Liriodendron tulipifera	H2	-	-	-	-	-	3.1	-
Quercus alba	H2	-	4.1	-	-	3.1	-	-
Quercus rubra	H2	-	-	1.1	2.1	-	-	-
Medium trees								
Carpinus caroliniana	H2	5.1	-	-	-	-	-	-
Cornus florida	H2	-	-	2.1	-	2.1	-	-
Ilex opaca	H2	-	-	-	-	3.1	3.1	5.1
Oxydendrum arboreum	H2	-	3.1	-	-	2.1	-	-
Sassafras albidum	H2	-	3.1	2.1	-	-	-	-

Number of relevés: 7 Relevé size: 5 m X 5 m

SUBCANOPY SPECIES PRESENT, BUT NOT IN ANALYSIS: LARGE TREES: *Nyssa sylvatica, Quercus velutina.*

SHRUB LAYER DOMINANTS: None.
SHRUB LAYER PHYSIOGNOMY: Not applicable.

Shrub Analysis

Species	GF	1 C.S	2 C.S	3 C.S	4 C.S	5 C.S	6 C.S	7 C.S
Transgressive trees								
Carya pallida	H2	-	-	-	-	2.1	-	-
Carya tomentosa	H2	-	-	-	-	-	-	2.1
Diospyros virginiana	H2	+.1	-	-	-	-	-	-
Liriodendron tulipifera	H2	-	-	-	-	-	1.1	-
Oxydendrum arboreum	H2	1.1	-	-	-	-	-	-
Prunus serotina	H2	-	-	-	+.1	-	-	-
Quercus alba	H2	-	-	-	-	-	-	+.1
Quercus michauxii	H2	-	-	-	-	-	1.1	-
Sassafras albidum	H2	-	-	-	1.1	-	-	-
Normal shrubs								
Clethra alnifolia	H11	-	-	-	-	-	2.1	3.1
Vaccinium corymbosum	H11	-	-	-	-	1.1	-	-
Vines								
Campsis radicans	H16	-	1.1	-	-	-	-	-
Vitis rotundifolia	H18	-	-	3.1	2.1	-	3.1	1.1

Number of relevés: 7 Relevé size: 5 m X 5 m

SHRUB SPECIES PRESENT, BUT NOT IN ANALYSIS: SMALL TREE: *Aralia spinosa*; TALL SHRUB: *Crataegus viridis*; NORMAL SHRUB: *Euonymus americanus*; VINES: *Gelsemium sempervirens, Lonicera japonica, Parthenocissus quinquefolia, Smilax rotundifolia, Vitis aestivalis.*

HERB LAYER DOMINANTS: None.
HERB LAYER PHYSIOGNOMY: Not applicable.

Herb Analysis

Species	GF	1 C.S	2 C.S	3 C.S	4 C.S	5 C.S	6 C.S	7 C.S
Tall forbs								
Pilea pumila	H7	3.1	-	-	-	-	-	-
Solidago sp.	H7	2.1	-	-	-	-	-	-
Medium forb								
Cypripedium acaule	H6	-	-	-	-	-	1.1	-
Small forb								
Chimaphila maculata	H7	-	-	-	-	-	1.1	-
Very small forb								
Mitchella repens	H10	-	-	-	-	-	1.1	-
Small graminoid								
Panicum sp.	H7	3.1	-	2.1	2.1	-	-	-
Tall ferns								
Athyrium asplenioides	H11	-	-	1.1	-	-	-	2.1
Osmunda cinnamomea	H11	-	-	-	-	-	-	2.1

Number of relevés: 7 Relevé size: 1 m X 1 m

HERB SPECIES PRESENT, BUT NOT IN ANALYSIS: None.

COMMUNITY REFERENCES: None.

COMMUNITY DOCUMENTATION: Area analyzed in 1977 and 1978 by Alan S. Weakley for A. E. Radford's Botany 235 class, UNC, Chapel Hill. Specimens deposited in the NCU Herbarium by Alan S. Weakley in 1978.

COMMUNITY ECOLOGICAL CHARACTERIZATION: Vegetationally: Fagalean-mixed hardwood/ Mixed hardwood terrestrial broadleaf forest system, with a closed canopy of large, deliquescent, deciduous trees, a closed subcanopy of medium, deliquescent, deciduous trees, a sparse shrub layer of mixed shrubs and transgressives, a sparse vine layer of tendril-climbing, deciduous vines, and a sparse herb layer of mixed herbs. Climatically: Warm temperate, mesothermal climate, cooler and wetter than the local and sectional cool and moderately moist yearly climate, which has moderately hot and moderately wet summers, moderately warm and moderately dry winters, and a long freeze-free period. Pedologically: Clayey, mixed, thermic Typic Hapludult, Gritney soil. Geologically: Siliceous, clastic, sedimentary deposit of undetermined origin, consisting of a highly porous and permeable mixture of silt and clay.

Hydrologically: Well-drained, permanently exposed, dry, terrestrial system, wetted by fresh rains and downslope drainage. Topographically: Open north-, northeast-, south-, southwest-facing, gently sloping to sloping, convexo-concave, entire valley slope along a broad, east-west-running portion of a shallow stream valley located on a flat plain. Temporally and Spatially: A climax community of a psammosere on a valley slope along Swift Creek, underlain by Miocene (?) sediments, in the Tar River drainage system in the Embayed Section of the Coastal Plain Province of the Atlantic Plain.

Master Species Presence List

ANGIOSPERMS

Arales
Araceae
Arisaema triphyllum (L.) Schott (4, 5†)
Orontium aquaticum L. (2, 5)

Alismatales
Alismataceae
Alisma subcordatum Raf. (8)

Asterales
Asteraceae
Solidago sp. (7)

Caryophyllales
Phytolaccaceae
Phytolacca americana L. (8)

Celastrales
Aquifoliaceae
Ilex decidua Walter 294‡ (3-5)
I. opaca Aiton 387 (2, 5-7)
I. verticillata (L.) Gray 361 (5)
Celastraceae
Euonymus americanus L. (3, 4, 7)

Commelinales
Commelinaceae
Commelina virginica L. 399 (5, 8)

Cornales
Cornaceae
Cornus florida L. 283, 391 (6, 7)
C. stricta Lam. (5, 8)
Nyssaceae
Nyssa aquatica L. 373 (1, 2, 4, 5)
N. sylvatica Marshall 385 (6, 7)

Cyperales
Cyperaceae
Carex louisianica Bailey (3, 8)
C. lurida Wahlenberg (5, 8)
Cyperus haspan L. (8)
Scirpus cyperinus (L.) Kunth (8)
S. divaricatus Ell. 402 (3, 4, 8)
Poaceae
Arundinaria gigantea (Walter) Muhl. 271 (3-6)
Panicum sp. (4-7)

Dipsacales
Caprifoliaceae
Lonicera japonica Thunberg (6-8)
Sambucus canadensis L. (8)
Viburnum dentatum L. 398 (4-6)
V. prunifolium L. (5)

Ebenales
Ebenaceae
Diospyros virginiana L. 390 (5-8)

---

*Nomenclature follows Radford, Ahles, and Bell (1968).

†Refers to Community Types. See end of Master Species Presence List.

‡Collection numbers of Alan S. Weakley, Department of Botany, University of North Carolina, Chapel Hill.

Ericales
  Clethraceae
    Clethra alnifolia L. 382 (7)
  Ericaceae
    Chimaphila maculata (L.) Pursh (7)
    Oxydendrum arboreum (L.) DC. 377 (7)
    Vaccinium corymbosum L. 386 (6, 7)

Fagales
  Betulaceae
    Alnus serrulata (Aiton) Willd. (8)
    Betula nigra L. 367 (5)
    Carpinus caroliniana Walter 296, 352 (4, 5, 7)
    Ostrya virginiana (Miller) K. Koch 295 (5)
  Fagaceae
    Fagus grandifolia Ehrhard 378 (6, 7)
    Quercus alba L. 401 (7)
    Q. bicolor Willd. 298 (3)
    Q. falcata var. *pagodaefolia* Ell. 376 (2-6)
    Q. hemisphaerica Bartram 304 (2-6)
    Q. lyrata Walter 299, 358 (2-5, 7)
    Q. michauxii Nuttall 297 (3-6)
    Q. nigra L. 380 (3-5, 7)
    Q. palustris Muenchh. 303, 368 (3-5)
    Q. phellos L. 273 (4-7)
    Q. rubra L. 300, 381 (6, 7)
    Q. shumardii Buckley 301, 359 (3-5)
    Q. stellata Wang. (7)
    Q. velutina Lam. 379 (7)

Gentianales
  Gentianaceae
    Sabatia angularis (L.) Pursh 392 (5, 8)
  Loganiaceae
    Gelsemium sempervirens (L.) Aiton f. 384 (7)

Geraniales
  Balsaminaceae
    Impatiens capensis Meerb. (5)

Haloragales
  Haloragaceae
    Proserpinaca palustris L. 357 (1-3)

Hamamelidales
  Hamamelidaceae
    Liquidambar styraciflua L. 289, 360 (2-7)

Juglandales
  Juglandaceae
    Carya aquatica (Michaux f.) Nuttall 274, 355 (2-5)
    C. cordiformis (Wang.) K. Koch 374 (3, 4)
    C. ovata (Miller) K. Koch 375 (3, 5)
    C. pallida (Ashe) Engler & Graebner 383 (7)
    C. tomentosa (Poiret) Nuttall (6, 7)

Lamiales
  Lamiaceae
    Lycopus virginicus L. (3)

Liliales
  Liliaceae
    Smilax rotundifolia L. 351 (1-7)

Magnoliales
  Lauraceae
    Sassafras albidum (Nuttall) Nees. 287, 389 (6, 7)
  Magnoliaceae
    Liriodendron tulipifera L. 395 (6, 7)

Myricales
  Myricaceae
    Myrica cerifera L. 394 (6)

Myrtales
  Melastomataceae
    Rhexia mariana L. (8)

Orchidales
  Orchidaceae
    Cypripedium acaule Aiton (7)

Piperales
  Saururaceae
    Saururus cernuus L. 364 (2-5)

Polygonales
  Polygonaceae
    Polygonum hydropiperoides var. *opelousanum* (Riddell *ex* Small) Stone 348 (8)

Ranunculales
  Ranunculaceae
    Ranunculus flabellaris Raf. (1)

Rhamnales
  Rhamnaceae
    Berchemia scandens K. Koch 277, 370 (3-5)
  Vitaceae
    Parthenocissus quinquefolia (L.) Planchon 285 (4-7)
    Vitis aestivalis Michaux 281 (2-7)
    V. labrusca L. (3-5)
    V. rotundifolia Michaux 280 (6, 7)

Rosales
  Rosaceae
    Crataegus marshallii Eggl. 369 (2, 5)
    C. viridis L. 298, 366 (2-5, 7)
    Prunus serotina Ehrhart (6, 7)
  Saxifragaceae
    Decumaria barbara L. (3, 4)
    Itea virginica L. 292 (3-5)

Rubiales
  Rubiaceae
    Cephalanthus occidentalis L. (8)
    Diodia virginiana L. 349 (8)
    Mitchella repens L. (2, 5-7)

Salicales
  Salicaceae
    Populus heterophylla L. 356 (2-5)
    Salix nigra Marshall (8)

Sapindales
  Aceraceae
    Acer rubrum L. 275, 279 (1-7)
  Anacardiaceae
    Rhus radicans L. (3-5)

Scrophulariales
  Acanthaceae
    Justicia ovata (Walter) Lindau 365 (5)
  Bignoniaceae
    Anisostichus capreolata (L.) Bureau 293 (3-5)
    Campsis radicans (L.) Seemarin 353 (3-5, 7)
  Oleaceae
    Fraxinus americana L. (3, 5)
    F. caroliniana Miller 352 (1-5)
    F. pennsylvanica Marshall 278 (3-6)
    F. tomentosa Michaux f. (5)
    Ligustrum sinense Lour. 282 (8)

Theales
  Hypericaceae
    Hypericum stans (Michaux) P. Adams & Robson 400 (5)

Typhales
  Sparganiaceae
    Sparganium americanum Nuttall 350 (8)
  Typhaceae
    Typha latifolia L. (8)

Umbellales
  Apiaceae
    Hydrocotyle umbellata L. (2-5)
    Sium suave Walter (2-4)
  Araliaceae
    Aralia spinosa L. 388 (7, 8)

Urticales
  Moraceae
    Morus rubra L. 286 (2, 6, 8)
  Ulmaceae
    Celtis laevigata Willd. 276 (8)
    Ulmus alata Michaux (3, 5, 6)
    U. americana L. 284, 396 (3-6)
  Urticaceae
    Boehmeria cylindrica (L.) Swartz 272, 362 (2-5)
    Pilea pumila (L.) Gray (2-5, 7)

Violales
  Violaceae
    Viola palmata var. *triloba* (Schweinitz) Ging. *ex* DC. (3-5)
    V. papilionacea Pursh (2-5)

GYMNOSPERMS

Coniferales
  Pinaceae
    Pinus taeda L. 393 (4, 6, 7)
  Taxodiaceae
    Taxodium distichum (L.) Richard 291 (1-5)

FERNS AND FERN ALLIES

Filicales
  Aspidiaceae
    Athyrium asplenioides (Michaux) A. A. Eaton 372 (6, 7)
    Onoclea sensibilis L. 363 (3-6)

Aspleniaceae
Asplenium platyneuron (L.) Oakes
371 (2, 3, 5, 6)
Polypodiaceae
Polypodium polypodioides (L.) Watt (1)

Osmundales
Osmundaceae
Osmunda cinnamomea L. (6, 7)

Swift Creek Communities

1. *Nyssa aquatica/Fraxinus caroliniana*
   deep swamp (pools), nearly permanently flooded
2. Mixed bottomland trees/*Fraxinus caroliniana*
   pool banks, seasonally to intermittently flooded
3. Mixed bottomland oaks-mixed bottomland hardwoods
   low flat, intermittently flooded
4. *Acer rubrum*-mixed bottomland hardwoods/Mixed bottomland hardwoods/*Saururus cernuus*//Mixed lianas
   low-medium flat, intermittently flooded
5. Mixed bottomland hardwoods-mixed bottomland oaks/Mixed bottomland hardwoods/*Rhus radicans*-mixed forb perennials//Mixed lianas
   stream bank (levee), intermittently flooded (rarely), well drained
6. *Pinus taeda*-mixed bottomland hardwoods/Mixed bottomland oaks//*Lonicera japonica*
   high flat, permanently exposed, silty to sandy
7. *Fagus grandifolia*-mixed hardwoods/Mixed hardwoods
   slope, permanently exposed, silty to sandy
8. Marsh, roadsides, disturbed areas, ditches

C

Carolina Beach State Park Natural Area*

New Hanover County, North Carolina; ca. 1.0 mile north of Carolina Beach on U.S. 421; ca. 0.2 mile southwest on Dow Road; ca. 0.6 mile west on state park road leading to parking lot at "Fly-trap Trail"; ca. 200 ft. south on the state park interpretive trail
34°02'45"N, 77°54'45"W; Carolina Beach, N.C., quad. 7.5 min., 1970
Atlantic Plain; Coastal Plain; Sea Island Section
10-15 ft; 2.9-4.8 m
ca. 5 acres

Ownership: State of North Carolina, Department of Natural Resources and Community Development, Division of Parks and Recreation.

Administration: Division of Parks and Recreation.

Land use: State park interpretive and hiking area.

Land classification: Unknown.

Danger to integrity: None presently known.

Publicity sensitivity: None presently known.

Significance and protection priority: Appears to be locally significant; site is protected.

Reasons for priority rating: An example of a savannah dominated by *Pinus palustris/Aristida stricta*. Presence of a threatened endemic insectivorous plant, *Dionaea muscipula*, as well as other insectivorous species occurring in local depressions and drainage ditches. Strong correlation between savannah and two Spodosol series sandwiched between a pocosin on an Aeric Paleaquult and a *Pinus palustris/Quercus laevis* community on a Spodic Quartzipsamment.

Management recommendations: Natural area should be managed as an educational and scientific resource with special attention paid to the populations of *Dionaea muscipula* and other insectivorous plants. This can be accomplished by periodic burning of the natural area and restricting visitors from harming the plant populations.

Data sources: None.

General references: None.

General documentation and authentication: Site analyzed by John B. Taggart on 16-17 October 1978. Specimens deposited in the NCU Herbarium by John B. Taggart in 1978.

---

*Report compiled by John B. Taggart for A. E. Radford's Botany 235 class. Report edited by Lee J. Otte and Deborah K. Strady Otte.

Discussion

The term "savannah" has long been commonly used to describe open grasslands with little or no canopy cover (Merriam, 1963). This concept has been applied primarily to certain tropical vegetation, particularly in Africa (Dyksterhuis, 1957). Beard (1953) defines savannahs in tropical North America as "communities comprising a virtually continuous, ecologically dominant stratum of more or less xeromorphic herbaceous plants of which grasses and sedges are the principal components and with scattered trees, shrubs, or palms sometimes present." Oosting (1956) generalizes the southeastern United States range in the following: "From the pitch pine barrens of New Jersey through loblolly pine and longleaf and slash pines in the more southern states, fire maintains pine dominance, usually in open stands, called savannahs, with the highly combustible wire grass (*Aristida stricta*) a common ground cover." Wells (1928) refers to southeastern savannahs as upland grass-sedge bogs with vegetation controlled by topography, soils, and fire. Numerous descriptions by early North Carolina botanists refer to the floral beauty and composition of the state's savannah areas. An excellent summary of these writings is given in Wilson (1977).

From available literature on savannahs, a generalized regime of habitat factors may be derived (Thomas, 1976): Climate--subtropical to tropical with a dry (winter) season alternating with a wet (summer) season; Topography--gently undulating to flat, sites are typically well drained at the surface with little relief; Soils--a sandy-sandy loam surface horizon overlying a layer preventing drainage, the impermeability of the subsurface horizon (possibly due to hardpan, bedrock, ironstone, clay, or a spodic horizon) will maintain a partly or wholly saturated soil solum during the rainy season, while during the dry season the perched water table may disappear completely; Fire--essential for community establishment and maintenance, particularly in regard to expansion into adjacent areas such as pocosin ecotones.

In considering the above environmental factors and the high diversity of species referred to in the literature that follows, one is left with the general conclusion that savannah species have wide amplitudes of habitat tolerance and can associate in a multitude of combinations. Kologiski (1977a) analyzed the plant communities of the Green Swamp, Brunswick County, North Carolina, and delineated three major and three minor savannah community types. Wells (1928) analyzed a savannah near Burgaw, North Carolina, and recognized various associes and consocies within the main cover class. Wilson (1977) performed a detailed study on savannahs of the pine plantations in the Croatan National Forest. The complexity of the vegetation was such that Community Cover Classes were the most functional descriptive units.

COMMUNITY DESCRIPTION

The Longleaf Pine (*Pinus palustris*)/Wire Grass (*Aristida stricta*) savannah is a cover type rapidly disappearing from the North Carolina Coastal Plain because of extensive land clearing, timbering, and general fire suppression (Carter, 1977). Because the savannah at Carolina Beach State Park is protected, it will be preserved as a remnant of these formerly extensive Coastal Plain grasslands.

The significance of the natural area lies in the community, species, and soil factors present. The *Pinus palustris*/*Aristida stricta* community type represents a specific unit within the savannah spectrum (see Figures C1 and C2). The community contains a variety of species adapted to the dry-wet moisture regimes, the most significant species being the threatened, exploited, endemic Venus' Fly Trap (*Dionaea muscipula*), as well as another insectivorous species (although not threatened or endangered), *Utricularia biflora*. A final important diversity element on a community basis is the correlation of the community type with two series of Spodosols (Leon and Rimini) situated between a pocosin on an Ultisol (Lynchburg) and a sand ridge community on an Entisol (Kureb). A transect of soils and possible community types through these areas is illustrated in Figure C3. Further discussion of these soils will be made later.

The savannah at Carolina Beach State Park falls into the drier portion of the matrix given in Figure C2. The frequency of fire (set intentionally) and the depth to the water table in the Rimini and Leon stands are the controlling factors. With increased moisture, the herb layer would be co-dominated by mixed herbs, but if one proceeds toward the Kureb sands, *Aristida stricta* dominance decreases and a xeric sand ridge community is encountered. Mixed herb occurrence is much greater over Leon soil than Rimini, but the overall sample did not reach 25% cover. Also of interest is the obvious difference in height of the graminoid dominant *Aristida stricta*: three to seven dm over Rimini, six to ten dm over Leon. Other species indicative of a microhabitat regime within the savannah are *Pterocaulon*

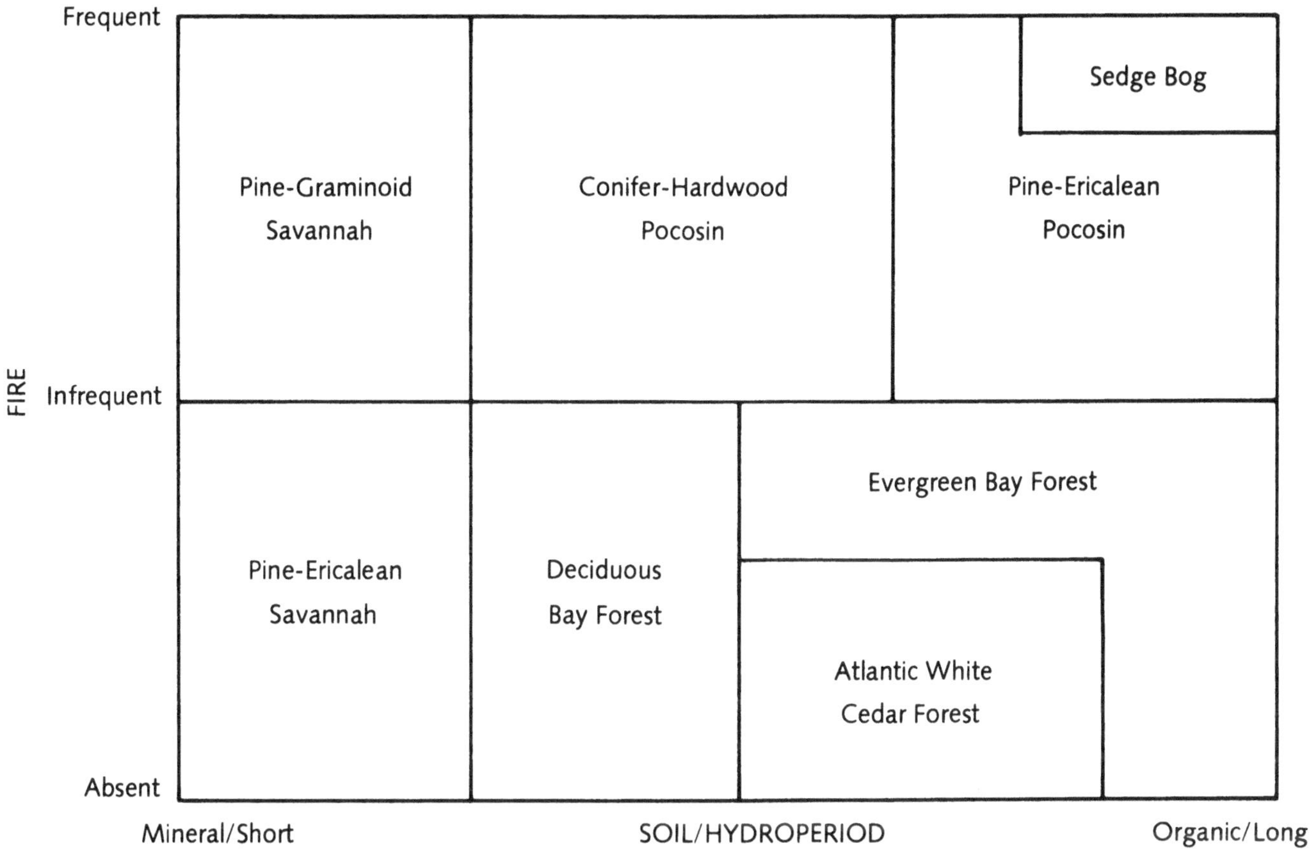

Figure C1. *Relationships between the dominant Green Swamp vegetation types (Community Classes) and soil/ hydroperiod and fire frequency.* From Kologiski (1977b).

pycnostachyum and *Polygonella polygama* in dry areas and *Lachnocaulon minus*, *Xyris torta*, and *Seymeria cassioides* in moister areas of local depressions.

Distribution of the open canopy dominant, Longleaf Pine, appears to be fairly uniform throughout the savannah, with intermixing of Pond Pine (*Pinus serotina*) near the pocosin edge. Toward the sand ridge community, there are scattered small trees of Turkey Oak (*Quercus laevis*) and Twin Oak (*Q. virginiana*). Establishment and maintenance of the *P. palustris* population was evidenced from random specimens of juvenile Longleafs ranging from grass stage (approximately one to five years) to transgressive trees two to six m in height.

The ecotones between the communities contain species whose habitat requirements are broad enough to fill these niches. To the pocosin side, a shrub layer of *Ilex glabra* and *Lyonia lucida* occurs, with larger shrubs and small trees

(*Cyrilla racemiflora*, *Magnolia virginiana*, and *Persea borbonia*) on the edge and in the pocosin proper. Local depressions and ditches at this ecotone are populated by insectivorous species as well as local occurrences of *Sphagnum* species. Progressing from the savannah to the sand ridge community, *Gaylussacia dumosa* increases while *Aristida stricta* is replaced gradually by *Cladonia* sp., *Selaginella arenicola*, and *Polygonella polygama*.

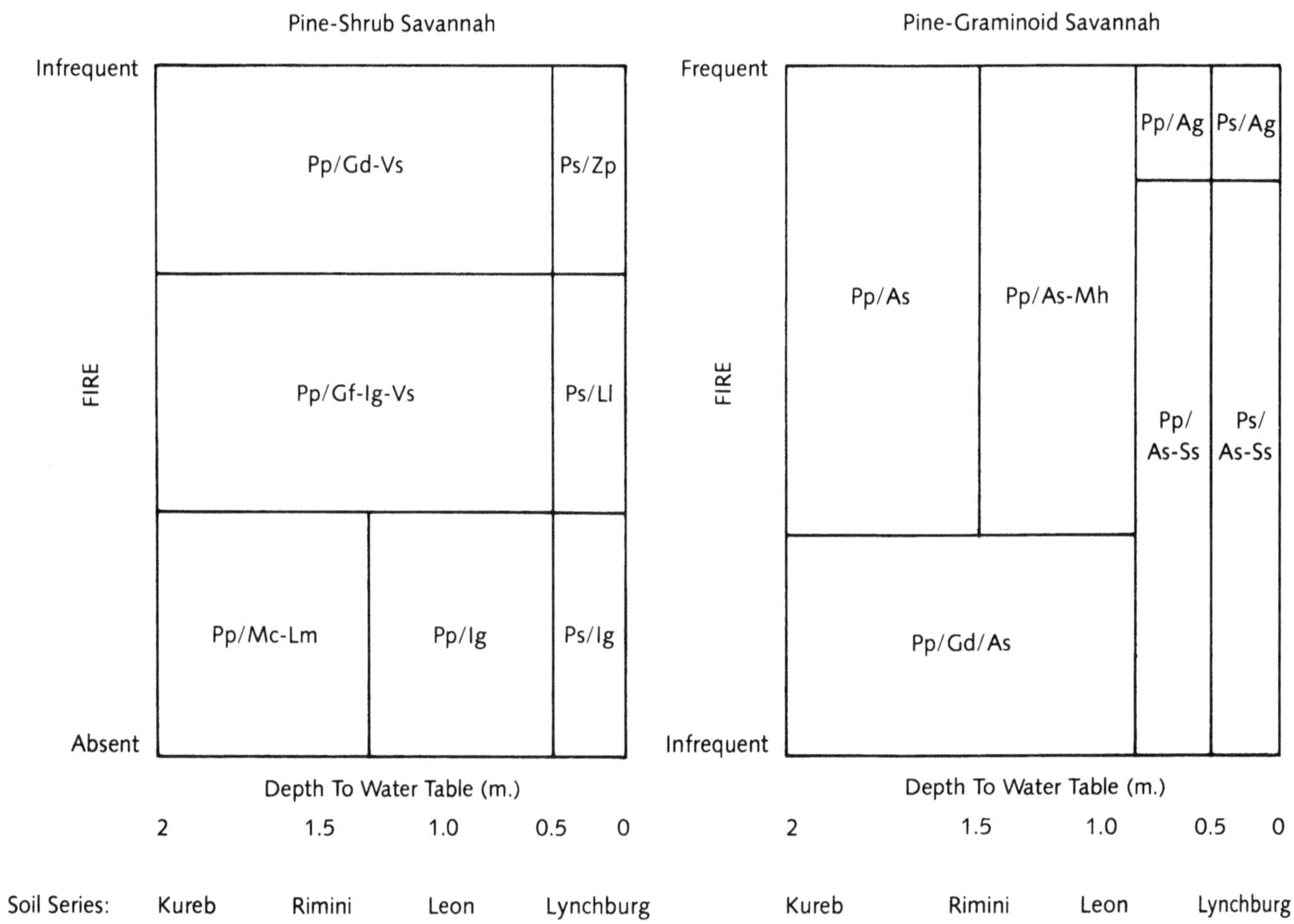

Ag = *Arundinaria gigantea*
As = *Aristida stricta*
Gd = *Gaylussacia dumosa*
Gf = *Gaylussacia frondosa*
Ig = *Ilex glabra*
Ll = *Lyonia lucida*
Lm = *Lyonia mariana*

Mc = *Myrica cerifera*
Mh = Mixed herbs
Pp = *Pinus palustris*
Ps = *Pinus serotina*
Ss = Sedge species
Vs = *Vaccinium* spp.
Zp = *Zenobia pulverulenta*

Figure C2. *Relationships between dominant species of the Green Swamp, fire frequency, depth to water table, and soil series.* Adapted from Kologiski (1977a).

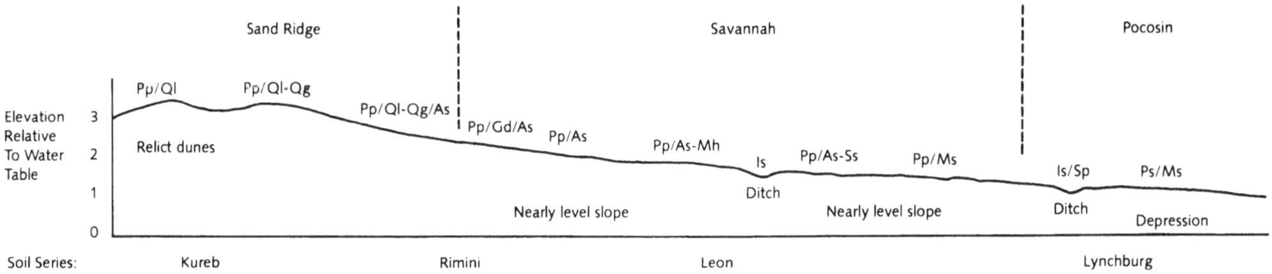

As = *Aristida stricta*
Gd = *Gaylussacia dumosa*
Mh = Mixed herbs
Pp = *Pinus palustris*
Qg = *Quercus geminata*
Ql = *Quercus laevis*

As = *Aristida stricta*
Is = Insectivorous species
Mh = Mixed herbs
Ms = Mixed evergreen shrubs
Pp = *Pinus palustris*
Ps = *Pinus serotina*
Ss = Sedge species
Sp = *Sphagnum* species

Figure C3. *Generalized sand ridge-savannah-pocosin transect.* Adapted from Wilson (1977).

Water table depth relative to the soil series is a very critical factor in this community. Because of the limitation of the soil auger length during sampling, only an estimation of water table depth could be made. Combining field observations with a recent soil survey of New Hanover County (Soil Conservation Service, 1977), the water table for the Rimini series (Grossarenic Entic Haplohumod) fluctuates between 1 and 2 m below the surface throughout the year. The Leon series (Aeric Haplaquod) is relatively wetter and fluctuates between 1.0 and less than 0.5 m. The Kureb (Spodic Quartzipsamment) series has the driest moisture regime with a water table depth greater than 2 m; the Lynchburg series (Aeric Paleaquult) has a moisture-retaining clay horizon that generally keeps the water table between 1.0 m and the surface. The values used in Figure C1 and C2 are therefore averages of these yearly fluctuations.

The relief of the natural area is a helpful guideline for estimating the extent of the savannah. Combining information from the topography map (Map C1) with field observations, the pocosin exists between 1.5 and 3 m above sea level, the savannah at 3 to 4.5 m and the sand ridge community above 4.5 m. Broadening or narrowing of the savannah "belt" generally can be accounted for by minor elevation differences (see Map C1), with best development of the community over the flat areas of Leon soil on the west side of the natural area.

The winter 1978 burn at Carolina Beach State Park has significantly increased savannah acreage and diversity (H. S. Jackson, personal communication). Such burning helps maintain the community by eliminating woody shrub and tree invaders, with the subsequent opening of the strata to herbaceous species (Vogl, 1972; Komarek, 1974). Soil modification by burning includes loss of ground litter and topsoil organic matter, resulting in a net loss of soil moisture and increased ground temperatures (Daubenmire, 1968). Christensen (1977) found that fires initially enriched savannah soil nutrient conditions, although certain compounds were lost to the atmosphere; however, the amount of the limiting element nitrogen was not altered significantly. The Longleaf Pine population generally is not affected by occasional fires if fuel levels are not high. *Pinus palustris* has evolved to tolerate most burns by its thick bark in mature adults and the "grass stage" growth habit of seedlings for five years prior to stem development (Chapman, 1932).

SUMMARY

The savannah at Carolina Beach State Park is an example of the *Pinus palustris/ Aristida stricta* community type. It is a pyric climax of a psammosere existing in a topographic and edaphic zone intermediate between hydric pocosin and xeric sand ridge habitats. The presence of the threatened, exploited, endemic *Dionaea muscipula* and the two Spodosol series are further elements of natural significance.

The natural area is locally significant and will be managed as an educational and scientific resource. Periodic controlled burnings will maintain the community and expand its area toward the pocosin; however, no more expansion toward the dry sand ridge will occur because of low soil moisture. Without fire, invasion of oaks and *Gaylussacia dumosa* from the sand ridge and evergreen shrubs (such as *Ilex glabra* and *Lyonia lucida*) from the pocosin gradually will overtop the savannah plants, with the eventual loss of the community type.

Because the property is owned and managed by the state of North Carolina, Division of Parks and Recreation, a permanent management plan will be instituted to maintain this savannah community. In addition, the endangered populations of *Dionaea muscipula* and other insectivorous species will be monitored carefully, and visitors will be informed of the park regulation pertaining to these species.

Further studies on savannahs must be carried out to better understand these diverse and beautiful areas. The complexity of environmental factors combined with the number of possible community types certainly has not been done justice by the paucity of work performed. Some of the most famous sites (such as Burgaw savannah) have been destroyed. The remaining areas should be inventoried and analyzed so that some may be preserved as living examples and not just memories.

Bibliography

Beard, J. S. 1933. The savannah vegetation of northern tropical America. Ecol. Monogr. 23:195-215.

Carter, J. 1977. Weymouth Woods management plan--Section I. Division of Parks and Recreation, Raleigh, North Carolina.

Chapman, H. S. 1932. Is the Longleaf type a climax? Ecology 13:328-34.

Christensen, N. L. 1977. Fire and soil--plant nutrient relations in a Pine-Wiregrass savannah of the Coastal Plain of North Carolina. Oecologia 31: 27-44.

Daubenmire, R. 1968. Ecology of fire in grasslands. Advances Ecol. Res. 5: 209-66.

Dyksterhuis, E. J. 1957. The savannah concept and its use. Ecology 38:435-42.

Jackson, H. S. 1978. Superintendent of Carolina Beach State Park. Personal communication of 17 October to J. Taggart.

Kologiski, R. L. 1977a. Phytosociology of the Green Swamp, North Carolina. Ph.D. dissertation. Department of Botany, North Carolina State University, Raleigh.

Kologiski, R. L. 1977b. The phytosociology of the Green Swamp, North Carolina. Agricultural Experiment Station Tech. Bull. No. 250. Raleigh, North Carolina.

Komarek, E. V. 1974. Effects of fire on temperate forests and related ecosystems. Pp. 251-77 *in* T. T. Kozlowski and C. E. Ahlgren, eds. Fire and ecosystems. Academic Press, New York.

Merriam, C. G., and Committee. 1963. Webster's seventh new collegiate dictionary. H. O. Houghton and Co., Cambridge, Massachusetts.

Oosting, H. J. 1956. The study of plant communities. W. H. Freeman & Co., San Francisco.

Soil Conservation Service. 1977. Soil Survey of New Hanover County, North Carolina. United States Department of Agriculture, Washington, D.C.

Thomas, W. 1976. Southeastern Coastal Plain savannah. Pp. 178-79 *in* A. E. Radford. Vegetation--habitats--floras--natural areas in the southeastern United States: Field data and information. University of North Carolina Student Stores, Chapel Hill.

Vogl, R. J. 1972. Fire in southeast grasslands. Proceedings, Tall Timbers Fire Ecology Conference 12:175-98.

Wells, B. W. 1928. A southern upland grass-sedge bog: An ecological study. Agricultural Experiment Station Tech. Bull. No. 32:1-75. Raleigh, North Carolina.

Wilson, E. J. 1977. A floristic study of the "savannas" on pine plantations in the Croatan National Forest. M.A. thesis. Botany Department, University of North Carolina, Chapel Hill.

Map C1. *Location of Carolina Beach State Park Natural Area.*

Natural Area Diversity Summary

CLIMATE: A. Mesothermal; AA. Warm temperate. B. Warm and moist yearly, moderately wet summers, moderately warm and moderately dry winters, with a very long freeze-free period; BB. Same as sectional climate. C. Similar to local climate; CC. Similar to natural area climate.
SOILS: A. Spodosol, Spodosol; AA. Aquod, Humod. B. Haplaquod, Haplohumod; BB. Aeric Haplaquod, Entic Haplohumod. C. Sandy, siliceous, thermic Aeric Haploquod; sandy, siliceous, thermic Entic Haplohumod; CC. Leon, Rimini.
GEOLOGY: A. Sedimentary; AA. Clastic. B. Siliceous; BB. Nearshore marine deposit. C. Mixture of course to fine sands; CC. Porous, unconsolidated, mixed quartz sands, exposed on the surface as soil constituents.
HYDROLOGY: A. Terrestrial; AA. Dry-xeric to wet. B. Fresh; BB. Permanently exposed to intermittently wet. C. Variously drained and wetted; CC. Well-drained to moderately poorly drained, wetted by fresh rains and by a seasonally high water table.
TOPOGRAPHY: A. Plains; AA. Flat plain. B. Marine-deposited coastal spit; BB. Beach ridges and swales on southwest portion of spit. C. Same as local landform; CC. Smooth, open, nearly level, constant middle slope between a swale and beach ridge.
PHYSIOGRAPHY: A. Atlantic Plain; AA. Coastal Plain. B. Sea Island Downwarp; BB. Lower Cape Fear Peninsula. C. Abandoned beach ridges and swales; CC. Slope between a ridge and swale, underlain by Recent marine quartz sands.
BIOLOGY: A. Terrestrial needleleaf forest system; AA. Large, excurrent, evergreen trees/Tall, cespitose, rhizocarpic graminoids. B. Pine-grass savannah; BB. Coniferales/Poaceae. C. *Pinus palustris*; CC. *Pinus palustris/Aristida stricta*.

Community Diversity Summary

Terrestrial needleleaf forest system
Large, excurrent, evergreen trees/Tall, cespitose, rhizocarpic graminoids

Pine-grass savannah
Coniferales/Poaceae

PINUS PALUSTRIS/ARISTIDA STRICTA
Pinus palustris

CLIMATE: A. Mesothermal; AA. Warm temperate. B. Warm and moist yearly, moderately hot and moderately wet summers, moderately warm and moderately dry winters, with a very long freeze-free period; BB. Same as sectional climate. C. Similar to local climate; CC. Similar to natural area climate.
SOILS: A. Spodosol; AA. Aquod. B. Haplaquod; BB. Aeric Haplaquod. C. Sandy, siliceous, thermic Aeric Haplaquod; CC. Leon.
A_0: 0-3 cm, sandy leaf litter, dark gray, pH 4.5.
A_1: 3-28 cm, sand, light gray, pH 4.5.
B_{1h}: 28-37 cm, sand, reddish brown, pH 4.5.
B_{2h}: 37-46 cm, sand, dark reddish brown, pH 4.5.
C: 46+ cm, sand, light gray, pH 4.5.

A. Spodosol; AA. Humod. B. Haplohumod; BB. Entic Haplohumod. C. Sandy, siliceous, thermic Entic Haplohumod; CC. Rimini.

A_0: 0-4 cm, sand with some leaf litter, gray, pH 4.5.

A_1: 4-42 cm, sand, dark gray, pH 4.5.

A_2: 42-50 cm, sand, white, pH 5.0.

B_{1h}: 50+ cm, sand, dark reddish brown with gray streaks, pH 4.5.

GEOLOGY: A. Sedimentary; AA. Clastic. B. Siliceous; BB. Nearshore marine deposit. C. Mixture of coarse to fine sands; CC. Porous, unconsolidated, mixed quartz sands, exposed on the surface as soil constituents.

HYDROLOGY: A. Terrestrial; AA. Dry-xeric to wet. B. Fresh; BB. Permanently exposed to intermittently wet. C. Variously drained and wetted; CC. Well-drained to moderately poorly drained, wetted by fresh rains and by a seasonally high water table.

TOPOGRAPHY: A. Plains; AA. Flat plain. B. Marine-deposited coastal spit; BB. Beach ridges and swales on southwest portion of spit. C. Same as local landform; CC. Smooth, open, nearly level, constant middle slope between a swale and beach ridge.

PHYSIOGRAPHY: A. Atlantic Plain; AA. Coastal Plain. B. Sea Island Downwarp; BB. Lower Cape Fear Peninsula. C. Abandoned beach ridges and swales; CC. Slope between a ridge and swale, underlain by Recent marine sands.

CANOPY: *Pinus palustris*--Height 11.2 m, dbh 9.29 m, age 35 years. *Pinus serotina*--Height 9.8 m, dbh 0.22 m, age 48 years.

CANOPY DOMINANT: *Pinus palustris*.

CANOPY PHYSIOGNOMY: Large, excurrent, evergreen trees.

Canopy Analysis

Species	I.V.	Rel. Den.	Rel. Dom.	Rel. Freq.
Pinus palustris	184.2	62.5	71.7	50.0
Pinus serotina	116.0	37.5	28.3	50.0

Number of points: 6 d: 8.3 m Number of individuals/acre: 64.9

CANOPY SPECIES PRESENT, BUT NOT IN ANALYSIS: None.

SUBCANOPY: Height, dbh, and age not determined.

SUBCANOPY DOMINANTS: None.

SUBCANOPY PHYSIOGNOMY: Not applicable.

SUBCANOPY ANALYSIS: Only species presence list completed. Species list--*Quercus virginiana, Quercus laevis*.

SHRUB LAYER DOMINANTS: None.

SHRUB LAYER PHYSIOGNOMY: Not applicable.

Shrub Analysis

Species	GF	1 C.S	2 C.S	3 C.S	4 C.S	5 C.S	6 C.S
Transgressive trees							
Pinus palustris	H1	1.1	-	-	1.1	-	2.1
Quercus virginiana	H2	1.1	-	-	-	1.1	-
Quercus laevis	H2	-	1.1	-	-	-	-
Transgressive tall shrub							
Myrica cerifera	H7	1.1	-	-	-	-	1.1
Normal shrubs							
Ilex glabra	H11	2.2	-	-	-	-	-
Lyonia lucida	H11	-	-	-	-	-	1.1
Tall dwarf shrubs							
Gaylussacia dumosa	H11	2.2	1.1	3.2	-	-	-
Gaylussacia frondosa	H11	-	-	2.1	1.1	-	2.1
Hypericum reductum	H11	2.2	2.1	-	-	2.1	2.2
Low dwarf shrub							
Vaccinium crassifolium	H11	3.1	2.2	2.1	-	-	3.2
Vine							
Smilax auriculata	H19	-	-	-	1.1	-	-

Number of relevés: 6 Relevé size: 5 m X 5 m

SHRUB SPECIES PRESENT, BUT NOT IN ANALYSIS: TRANSGRESSIVE TREES: *Cyrilla racemiflora, Magnolia virginiana, Nyssa sylvatica* var. *biflora, Persea borbonia, Pinus serotina, Sassafras albidum*; TALL SHRUB: *Arundinaria gigantea.*

HERB LAYER DOMINANT: *Aristida stricta.*
HERB LAYER PHYSIOGNOMY: Tall, cespitose, rhizocarpic graminoids.

Herb Analysis

Species	GF	1 C.S	2 C.S	3 C.S	4 C.S	5 C.S	6 C.S
Tall forbs							
Agalinis purpurea	H11	-	-	-	5.1	-	-
Pterocaulon pycnostachyum	H11	1.2	-	3.2	-	2.1	-
Rhexia mariana	H11	-	-	-	2.1	2.1	-
Seymeria cassioides	H7	-	-	-	4.1	-	-
Trilisa paniculata	H11	3.3	-	2.1	-	-	-
Xyris torta	H6	-	2.3	-	2.3	-	-
Medium forb							
Polygala lutea	H11	-	-	-	2.2	-	2.2
Small forb							
Lachnocaulon minus	H13	-	-	-	2.3	1.3	2.4

Herb Analysis (continued)

Species	GF	1 C.S	2 C.S	3 C.S	4 C.S	5 C.S	6 C.S
Tall graminoids							
Aristida stricta	H8	4.4	5.4	5.4	5.3	4.3	5.4
Juncus scirpoides	H11	-	-	2.3	-	-	-
Panicum hemitomon	H11	-	-	-	2.2	2.1	-
Scleria pauciflora	H11	-	2.2	-	-	-	-

Number of relevés: 6 Relevé size: 1 m X 1 m

HERB SPECIES PRESENT, BUT NOT IN ANALYSIS: TALL FORBS: *Aster paludosus, A. tortifolius, Carphephorus bellidifolius, Eupatorium pilosum, Heterotheca nervosa, Liatris graminifolia, Polygonella polygama*; SMALL FORBS: *Dionaea muscipula, Utricularia biflora*; TALL GRAMINOID: *Andropogon scoparius.*

COMMUNITY REFERENCES: None.

COMMUNITY DOCUMENTATION: Area analyzed by John B. Taggart for A. E. Radford's Botany 235 class, UNC, Chapel Hill. Specimens deposited in the NCU Herbarium by John B. Taggart in 1979.

COMMUNITY ECOLOGICAL CHARACTERIZATION: Vegetationally: Coniferalean/Poaceous terrestrial needleleaf forest system with an open canopy of large, excurrent, evergreen trees, a sparse subcanopy of small, deliquescent, deciduous trees, an open shrub layer of mixed shrubs and transgressive trees, and a closed herb layer of tall, cespitose, rhizocarpic graminoids. Climatically: Warm temperate, mesothermal climate, similar to the sectional and local warm and moist yearly climate, which has moderately hot and moderately wet summers, moderately warm and moderately dry winters, and a very long freeze-free period. Pedologically: Sandy, siliceous, thermic Aeric Haplaquod, Leon soil and sandy, siliceous, thermic Grossarenic Entic Haplohumod, Rimini soil. Geologically: Clastic, sedimentary, nearshore marine deposit, consisting of siliceous, porous, unconsolidated, mixed quartz sands exposed on the surface as soil constituents. Hydrologically: A well-drained to moderately poorly drained, permanently exposed to intermittently wet, dry-xeric to wet terrestrial system, wetted by fresh rains and by a seasonally high water table. Topographically: Smooth, open, nearly level, constant middle slope between a swale and beach ridge in an area of abandoned beach ridges and swales on the southwest portion of a large, marine-deposited coastal spit along the seaward edge of a flat plain. Temporally and Spatially: A pyroclimax of a psammosere on a slope between a ridge and swale in an area of abandoned beach ridges and swales, underlain by Recent marine sands, on the Lower Cape Fear Peninsula in the Sea Island Section of the Coastal Plain Province of the Atlantic Plain.

Master Species Presence List

ANGIOSPERMS

Asterales
   Asteraceae
      Aster paludosus Aiton 328*
      A. tortifolius Michaux 307
      Carphephorus bellidifolius (Michaux) T. & G. 311
      Eupatorium pilosum Walter 327
      Heterotheca nervosa (Willd.) Shinners 321
      Liatris graminifolia Willd. 329
      Pterocaulon pycnostachyum (Walter *ex* J. F. Gmelin) Cassini 310

Celastrales
   Aquifoliaceae
      Ilex glabra (L.) Gray 330

Commelinales
   Xyridaceae
      Xyris torta Smith 326

Cornales
   Nyssaceae
      Nyssa sylvatica var. *biflora* (Walter) Sargent 325

Cyperales
   Cyperaceae
      Scleria pauciflora Muhl. *ex* Willd. 306
   Poaceae
      Andropogon scoparius Michaux 331
      Aristida stricta Michaux 331
      Arundinaria gigantea (Walter) Muhl. 332
      Panicum hemitomon Schultes 303

Ericales
   Cyrillaceae
      Cyrilla racemiflora L. 333
   Ericaceae
      Gaylussacia dumosa (Andrz.) T. & G. 334
      G. frondosa (L.) T. & G. 301
      Lyonia lucida (Lam.) K. Koch 335
      Vaccinium crassifolium Andrews 336

Eriocaulales
   Eriocaulaceae
      Lachnocaulon minus (Chapman) Small 337

Fagales
   Fagaceae
      Quercus virginiana Miller 338
      Q. laevis Walter 339

Juncales
   Juncaceae
      Juncus scirpoides Lam. 340

Laurales
   Lauraceae
      Persea borbonia (L.) Sprengel 341
      Sassafras albidum (Nuttall) Nees 342

Liliales
   Liliaceae
      Smilax auriculata Walter 343

Magnoliales
   Magnoliaceae
      Magnolia virginiana L. 344

Myricales
   Myricaceae
      Myrica cerifera L. 345

Myrtales
   Melastomataceae
      Rhexia mariana L. 312

Polygalales
   Polygalaceae
      Polygala lutea L. 313

---

*Collections of John B. Taggart, Department of Natural Resources and Community Development, Division of Parks and Recreation, Raleigh.

Polygonales
Polygonaceae
Polygonella polygama (Vent.) Engelm. & Gray 316

Sarraceniales
Dionaeaceae
Dionaea muscipula Ellis 315

Scrophulariales
Lentibulariaceae
Utricularia biflora Lam. 318
Scrophulariaceae
Agalinis purpurea (L.) Pennell 317
Seymeria cassioides (J. F. Gmelin) Blake 319

Theales
Hypericaceae
Hypericum reductum P. Adams 346

GYMNOSPERMS

Coniferales
Pinaceae
Pinus palustris Miller 300
P. serotina Michaux 347

D

Ocracoke Island Natural Area*

Hyde County, North Carolina; along N.C. Highway 12, 1.5 miles northeast of Ocracoke, N.C.; 35°07'30"N, 75°55'57"W; Ocracoke, N.C., quad. 7.5 min., 1950
Atlantic Plain; Coastal Plain; Embayed Section
0-21 ft; 0-7 m
ca. 700 acres

Ownership: Cape Hatteras National Seashore.

Administration: Cape Hatteras National Seashore.

Land use: Recreational.

Land classification: Unknown.

Danger to integrity: Unknown.

Publicity Sensitivity: None presently known.

Significance and protection priority: Appears to be state significant. Site is in no apparent jeopardy.

Reasons for priority ratings: A relatively undisturbed area that exhibits excellent community zonation within a limited area; included are most, if not all, the characteristic Community Types that typically are found in North Carolina salt marshes and dune complexes.

Management recommendations: Natural area should be managed for scientific research, as an educational resource, and for preservation of a good example of a barrier island community complex.

Data sources: None.

General scientific references: None.

General documentation and authentication: Data for this area were collected by A. E. Radford's Botany 235 class from the University of North Carolina at Chapel Hill on 1 October 1977. Specimens were deposited in the NCU Herbarium by T. A. Atkinson, Department of Botany, UNC-Chapel Hill in 1977.

---

*Report originally compiled by Lee J. Otte, Department of Geology, UNC at Chapel Hill. Report edited by Deborah K. Strady Otte, Department of Botany, UNC at Chapel Hill.

Discussion

Ocracoke Island, roughly 17 miles (27 km) long and averaging about one-half mile (0.8 km) wide, consists mostly of barren beaches, sand flats, and dunes (Brown, 1959). In 1959, three areas known as the Tarhole Plains, the Great Swash, and The Plains composed the major sand flats on the island. Brown (1959) reported that typical barrier island vegetation covered the Great Swash. Tarhole Plains and The Plains, however, were devoid of vegetation and regularly were covered with sea water to a depth of four to six inches during normal high tides and were covered completely during storm tides.

The Plains, the area analyzed for this report, now supports a vegetation cover that includes most of the typical barrier island herb and shrub Community Types found along the North Carolina coast. This cover has developed in less than 20 years and provides an excellent example of how rapidly plants establish as distinct communities once the proper environmental conditions are met. In the last 20 years, the ocean side of The Plains has built itself up beyond reach of normal tides and of most storm tides. The formation of the foredune on The Plains furnishes one possible explanation. Brown (1959), in a review of a Stratton and Hollowell report (1940), reported that the National Park Service constructed over one-half mile of sand fence along the ocean side of The Plains during the mid-1930s. The fence was constructed in such a fashion that a low dune with a broad base would form to aid in the prevention of erosion by the regular tidal flooding. Possibly it took over 20 years, until after 1959, for this man-made dune to accomplish the task for which it was constructed. A low, broad foredune still exists today, with negligible dune development behind it. Behind the foredune occurs an area of low, poorly developed dunes and depressions indicative of a very young dune system. The whole Plains area still lies very low, as evidenced by the fact that the tidal marsh on the sound side of The Plains still extends more than halfway across the island and by the fact that no point in The Plains rises over five feet (1.5 m) in elevation.

Adjacent relict dunes (the First Hammock Hills) to the northeast of The Plains end abruptly at the edge of The Plains. This suggests that The Plains was at one time higher in elevation and that a large storm-induced washover severely eroded the area.

Whatever the origin of The Plains, it stimulates interest in vegetation because of its young age. The remainder of this discussion is devoted to describing the different vegetational zones on the adjacent relict dunes and in The Plains itself, here divided into the active dune field and the tidal marsh. This discussion includes species and communities found in these zones and a comparison of the zonation pattern with that described in the literature as the typical vegetation of the southeastern Atlantic Coast of North America. A summary statement presents the significance of the area with reasons why it should be set aside and preserved.

A. RELICT DUNES

Well-developed barrier island forests, known as maritime forests, generally are restricted to the strand areas of older landward-lying dunes (Wells, 1928). These areas experience very short hydroperiods and very few perennial winds.

Wells (1928) stated that two consocies of the maritime forest socies exist, both successional to *Quercus virginiana*, the usual dominant of the maritime forest. *Ilex vomitoria* predominates one consocies on exposed seaward sites. The other consocies, growing in more protected situations, usually contains *I. vomitoria* but often forms almost pure stands of *Juniperus virginiana*, *Myrica cerifera*, and *Zanthoxylum clava-herculis*.

The second socies described by Wells (1928) is very similar to the shrub community found on the protected northwest side of the relict dunes, known as the First Hammock Hills. Here *M. cerifera* dominates the community, but *I. vomitoria*, *Baccharis halimifolia*, *J. virginiana*, *Salix caroliniana*, and *Z. clava-herculis* also occur. The Hammock Hills relict dunes, because of their short lateral extent and the angle at which they cross the island, do not provide full protection from salt spray for the plants inhabiting their back side. The front, top, and upper back slopes of the Hammock Hills are dominated by *Uniola paniculata*, a pioneer dune stabilizer that best develops in areas exposed to high atmospheric salt content (Oosting and Billings, 1942). This clearly shows that the ocean still affects the vegetation on the sound side of this portion of Ocracoke Island. The protection in sheltered pockets behind the dunes, however, is sufficient for the establishment and maintenance of the successional shrub stage.

Possibly *Q. virginiana* eventually can become established, intermixed with the present shrubs, but it is unlikely that a forest of any size will develop. A *Q. virginiana* community exists less than half a mile to the northeast of the Hammock Hills in an area known as the Hammock Oaks (see Map D1). Here a relict dune system, parallel to the present shoreline, furnishes sufficient protection from salt-laden sea winds for the oaks to grow on the sound side of the dunes.

B. ACTIVE DUNES

The stabilized terrestrial portion of The Plains consists of the broad, low foredune and a belt of low, poorly developed dunes and depressions extending behind the foredune to a highway (N.C. 12) that runs through the field area. At least six different vegetational zones occupy this dune system. Four of these six can be seen in the foredune environment itself (Table D1).

Two species (*Amaranthus pumilus* and *Cakile edentula*), scattered at the base of the foredune slope, reach about 20 feet (6 m) out onto the upper beach. These two species represent the most salt-tolerant of the local flora--they must withstand not only almost constant salt spray, but also periodic or frequent inundation by storm and exceptionally high tides. These two species extend up onto the foreslope

Table D1. *Foredune environments on The Plains, Ocracoke Island.* 1 = upper beach, 2 = frontslope of foredune, 3 = crest of foredune, 4 = backslope of foredune.

Species	1	2	3	4
Amaranthus pumilus	X	X		
Cakile edentula	X	X		
Ammophila breviligulata		X		
Uniola paniculata		X	X	
Panicum amarum		X	X	X
Strophostyles helvola			X	X
Euphorbia polygonifolia			X	X
Solidago sempervirens				X
Hydrocotyle bonariensis				X
Muhlenbergia capillaris				X
Eragrostis spectabilis				X
Cenchrus tribuloides				X

of the foredune (zone 2) where *Ammophila breviligulata*, *Uniola paniculata*, and *Panicum amarum*, three important salt-tolerant, sand-binding grasses (Chapman, 1974), join them. The third zone, the crest of the foredune, is a transition area between the foreslope and the backslope. Of the three foreslope grasses, *P. amarum* is the only one to become established beyond the top of the dune and down the backslope, although all three inhabit the more open dune field behind the foredune. *Strophostyles helvola* and *Euphorbia polygonifolia* first appear on the dune top and spread down the backslope where *Solidago sempervirens*, *Muhlenbergia capillaris*, *Eragrostis spectabilis*, *Cenchrus tribuloides*, and *Hydrocotyle bonariensis* (the dominant backslope plant) meet them. All of these species in addition to a few others continue into the low dune and depression zone that extends from the foredune back to N.C. Highway 12. The species segregate into different, distinct zones within this area. Unfortunately, no detailed analyses were conducted in this terrain. A sparsely populated yet distinct shrub zone, dominated by *Myrica cerifera*, occurs near the road, but once again, no work was done on this community.

If compared with dune communities on other islands along the southeast coast of the United States, the pattern is roughly the same (Chapman, 1974). It differs primarily in the distribution of various species in the zones. Local differences in environmental factors cause these variations.

The young age of the vegetation may account for the sparseness of the shrub zone. A well-developed shrub community naturally would take longer to develop than an herb-dominated system.

Additional work would aid in the determination of the other segregates within the dune field and the shrub community.

C. TIDAL MARSH

Tidal marshes usually develop in shallow marine environments in the vicinity of rivers whose silt-and-mud-laden waters are discharged into the open marine waters. A barrier, whether an island, a spit, a bar, a beach, or some other positive topographic feature, must be present in order to obstruct the outflow of the muddy river waters and trap the suspended sediments. Ocracoke Island, along with the other islands forming the Outer Banks of North Carolina, acts as a barrier to the water flowing out of Pamlico Sound, a large body of water into which several rivers empty. Muddy sediments accumulate on the quiet, sound side of these islands and form a suitable substrate on which brackish and haline plants can grow. Once these plants become established, they increase the rate of sediment accumulation by functioning as baffles and trapping sediments brought into the marsh system by tidal waters.

The southeastern Atlantic Coast of the United States is a subsiding coastline (Chapman, 1974). In some areas this is due only to the postglacial rise in sea level, in others to a combination of sea level rise and actual subsidence or settling of coastal sediments. In areas such as these, coastal marshes can be maintained only if accretion of sediments within the marsh keeps pace with subsidence. The tidal marsh system along the North Carolina coast appears to be persisting and in most cases is increasing in size. As stated earlier, the marsh system in The Plains, on Ocracoke Island, seems to be less than 20 years old and, based on the 1970 photorevision of the U.S.G.S. Ocracoke 7.5 minute quadrangle, as compared with the 1950 edition, is increasing actively in size.

A tidal salt marsh constitutes a complex ecosystem. A large number of environmental factors, dominated by tidal cycles, sediment influx, and local climate, continually exert influences on the marsh. These major factors control substrate composition, rate and depth of flooding, drainage patterns, rate of drainage, water halinity, soil halinity, moisture availability, aeration of the soil, and height of the marsh in relation to mean sea level. All of these factors, in turn, affect the distribution of different plant species within the marsh. Taxonomically different yet structurally similar species form distinct vegetation zones, controlled by distinct combinations of the above ecologic factors and recognized throughout the tidal salt marshes of the world (Chapman, 1976).

A great deal of work has been completed on salt marsh zonation on the Atlantic Coast of North America. Reasons for the zonation are just beginning to be understood. Environmental factors vary drastically from season to season and even from day to day. Determining which factors, during which season or during which part of the day, actually limit different species presents a difficult task.

Spartina alterniflora, a marginal emergent that becomes established on muddy shorelines (Reed, 1947), acts as the primary colonizer in most southeastern North American salt marshes. The plant develops best at mean sea level and normal marine halinity (Reed, 1947; Chapman, 1974), where it grows taller and forms denser colonies than anywhere else in the marsh. It decreases in size and density above and below mean sea level. Once established, *S. alterniflora* often forms dense, pure colonies or frequently associates with other species. These colonies, with the plants' dense, aerial stalks and matted roots, aid in trapping sand, silt, and clay brought in by tidal waters. This addition of sediment helps build the shoreline outward and the marsh higher (Reed, 1947). The Ocracoke *S. alterniflora* zone composes a major portion of the marsh and extends from the water's edge back into the marsh proper. The populations change in density and height within the marsh. Species associates also vary: *Salicornia europaea* grows with the *Spartina alterniflora* in the lower portions of the marsh and *Salicornia virginica* and *Aster tenuifolius* in the upper portion of the marsh.

A normal successional pattern follows the pioneer *S. alterniflora* zone, dependent on continual accretion and build-up of the marsh above mean sea level. This build-up causes less frequent and shallower flooding and increases freshwater influence. Chapman (1974) proposes the following normal successional pattern based on a continual rise in the level of the marsh for the southeastern North American Atlantic Coast salt marshes: first *S. alterniflora*, then succeeding zones dominated by *Distichlis spicata*, *Spartina patens*, *Juncus roemerianus*, and shrubs or freshwater species. Variations occur because of irregularities in the topography of most marshes and the presence of such features as tidal channels or salt pans.

According to studies undertaken around Norfolk, Virginia (Chapman, 1974), and Beaufort, North Carolina (Reed, 1947), the *S. alterniflora* zone commonly is succeeded by a *Distichlis spicata* zone that, in turn, is succeeded by a *Spartina patens* zone. In Georgia, however, Wharton (1978) recognizes only one zone dominated by *Distichlis spicata* and composed of lesser numbers of *Spartina bakeri*, *Limonium carolinianum*, and other species. In The Plains marsh on Ocracoke, *S. patens* and *D. spicata* form distinct zones, each containing a different group of plants. The *D. spicata* zone, with *Limonium carolinianum*, *Salicornia virginica*, and *S. bigelovii*, experiences a more haline influence than the *Spartina patens* zone, which includes *Mikania scandens*, *Sabatia stellaris*, *Andropogon glomeratus*, *Fimbristylis spadicea*, and *Setaria geniculata*.

The *Juncus roemerianus* community inhabits the headwaters of tidal creeks and the inner, landward edge of the marsh (Wharton, 1978). Chapman (1976) considers this *Juncus* zone the salt marsh sere climax. *Juncus* occurs at the upper limit of direct tidal influence, which is inundated only by storms and spring tides (Wharton, 1978). Because of the infrequent flooding by tidal waters, the *Juncus* community does not receive as much sediment influx as the rest of the marsh, therefore usually making it slightly lower in elevation than adjacent grass marshes

(Cooper and Waits, 1973). Such is the case in the *Juncus* zone in The Plains. A tidal channel borders its northeastern edge and a shrub zone its southwestern limit.

The Norfolk marshes (Chapman, 1974) are bordered by a shrub zone transitional between the marsh and the adjacent terrestrial systems. *Baccharis halimifolia* commonly dominates this shrub zone, although *Kosteletzkya virginica* frequently acts as a co-dominant. *Fimbristylis spadicea*, *Borrichia frutescens*, *Pluchea purpurascens*, and other species also inhabit this zone, which contains the greatest number of species of any of the marsh zones. Wharton (1978) considers the marsh edge shrub zone an ecotone between the salt marsh and the terrestrial shore environments. This area is inundated only during spring tides (Wharton, 1978). The young age of The Plains vegetation probably explains the absence of a dense, well-developed shrub zone around the salt marsh. The shrubs present include *Borrichia frutescens* and *Iva frutescens*, which are scattered throughout a mixture of *Spartina patens* and *Fimbristylis spadicea*.

Changes in halinity govern distribution of the three *Salicornia* species (*S. europaea*, *S. virginica*, and *S. bigelovii*) commonly found in North Carolina tidal salt marshes. In turn, halinity depends on changes in substrate composition, rate of drainage, and distance above mean sea level. *Salicornia europaea* is associated with the *Spartina alterniflora* zone, the wettest, muddiest, and closest to normal marine conditions of all the marsh zones. The other two species are found in areas of higher halinity and very often are associated with salt pans or salt flats.

Salt pans of several types exist in many salt marshes (Chapman, 1976). The one in the Ocracoke marsh system appears to be what Chapman (1976) calls a primary pan. This type develops with the marsh in the early stages of formation and represents bare areas that become cut off and surrounded by vegetation. When this happens, water cannot escape as rapidly as from the rest of the marsh after a flood tide. Thus, standing water aids in prevention of colonization, especially during the hotter months, when the water evaporates and becomes increasingly haline. In the case of The Plains marsh, during the hot autumn months of 1977, the central portion of the salt pan contained a surface coating of halite (NaCl) and the subsurface water (found at a depth of 30 cm) had a halinity of more than 60 parts per thousand (ppt.).

Large salt pans within a marsh system can develop their own vegetational zonation, which can vary from season to season and from year to year. In The Plains salt pan, during the fall of 1977, its central portion contained no plants because of the high halinity. Away from the center of the pan, the first plant was *Salicornia bigelovii*, which at first increased in density, but then decreased as halinity dropped and other plants occurred. On the Pamlico Sound side of the pan *Salicornia virginica* and *Spartina alterniflora* were the next species. On the island side of the pan, because of slightly higher, better-drained ground, a different mixture of species occurred, including *Spartina patens*, *Salicornia virginica*, *Limonium carolinianum*, and *Distichlis spicata*.

SUMMARY

The Plains, comprising two systems (dune and marsh) and including adjacent relict dunes (First Hammock Hills), contains most of the typical vegetational zones found on the North Carolina Outer Banks. The only zone missing is the maritime forest; however, this zone occurs only half a mile (0.8 km) to the northeast. Few of the vegetational zones are well established because of the young age of the area.

This part of Ocracoke Island appears to be significant for North Carolina on the state level. Many miles of barrier island vegetation exist on North Carolina's Outer Banks. All, however, except perhaps that in the Cape Hatteras National Seashore, are in potential danger of being developed. A good example of typical barrier island vegetation needs to be preserved. This area is worthy of protection because, as previously stated, it contains most of the barrier island vegetational zones. It is also important because of its young age--having become vegetated only in the last 20 years, the area is probably not yet stabilized. It would be of great scientific value to be able to observe and analyze as this area matures into a stable, climax situation.

Bibliography

Brown, C. A. 1959. Vegetation of the Outer Banks of North Carolina. Louisiana State University Coastal Studies Series 4. 179 pp.

Chapman, V. J. 1974. Salt marshes and salt deserts of the world. 2nd ed. Verlag von J. Cramer, Germany.

________. 1976. Coastal vegetation. 2nd ed. Pergamon Press, Oxford, England.

Cooper, A. W., and E. D. Waits. 1973. Vegetation types in an irregularly flooded salt marsh on the North Carolina Outer Banks. J. Elisha Mitchell Sci. Soc. 89:78-91.

Oosting, H. J., and W. D. Billings. 1942. Factors effecting vegetation zonation on coastal dunes. Ecology 23:131-42.

Reed, J. F. 1947. The relation of the Spartinetum glabrae near Beaufort, North Carolina to certain edaphic factors. Amer. Midl. Naturalist 38:605-14.

Stratton, A. C., and J. R. Hollowell. 1940. Sand fixation and beach erosion control. U.S. Department of the Interior National Park Service Report. Unpublished.

Wells, B. W. 1928. Plant communities of the Coastal Plain of North Carolina and their successional relations. Ecology 9:230-42.

Wharton, C. H. 1978. The natural environments of Georgia. Special Publication of the Georgia Department of Natural Resources, Atlanta.

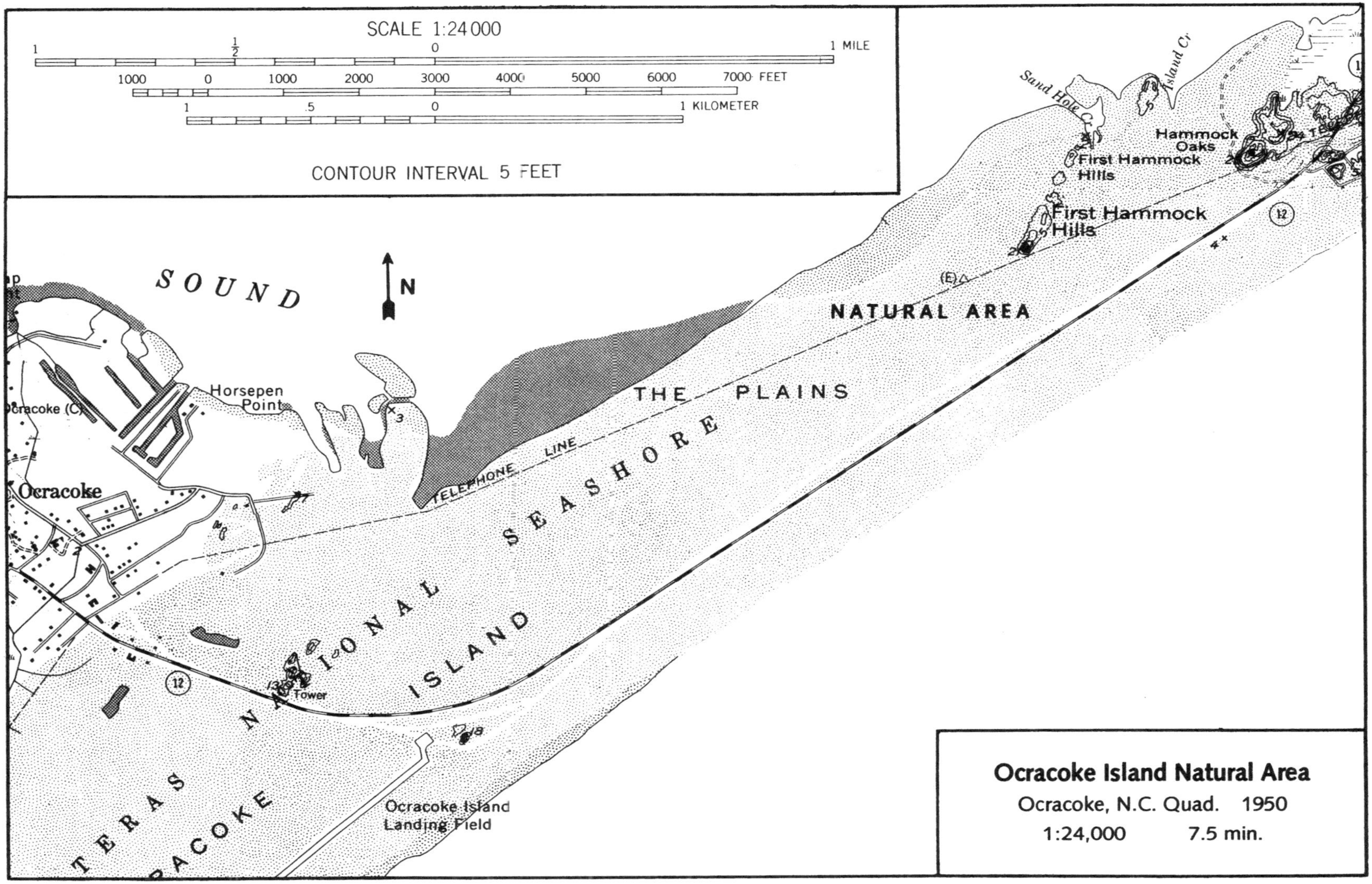

Map D1. *Location of Ocracoke Island Natural Area.*

Natural Area Diversity Summary

CLIMATE: A. Mesothermal; AA. Warm temperate. B. Cool and moist yearly, moderately hot and moderately wet summers, moderately warm and moderately dry winters, with a very long freeze-free period; BB. Warm and moist yearly, moderately hot and moderately wet summers, warm and moderately dry winters, with a very long freeze-free period. C. Similar to local climate; CC. Similar to natural area climate.

SOILS: A. (1-11) Entisol, A. (12) Histosol; AA. (1-4) Psamment, AA. (5-11) Aquent, AA. (12) Saprist. B. (1, 2) Udipsamment, B. (3, 4) Quartzipsamment, B. (5-11) Psammaquent, B. (12) Medisaprist; BB. (1, 2) Typic Udipsamment, BB. (3, 4) Typic Quartzipsamment, BB. (5-11) Typic Psammaquent, BB. (12) Typic Medisaprist. C. (1, 2) Mixed, thermic, C. (3, 4) Sandy, siliceous, thermic, C. (5-11) Mixed, thermic, C. (12) Euic, thermic; CC. (1, 2) Fripp, CC. (3, 4) Duneland Soils, CC. (5-11) Carteret, CC. (12) Hobonny.

GEOLOGY: A. Sedimentary; AA. (1-4) Clastic, AA. (5, 6) Clastic and organic, AA. (7, 8) Clastic and chemical, AA. (9-11) Clastic, AA. (12) Clastic and organic. B. (1-4) Siliceous, B. (5, 6) Siliceous and sulfaceous, B. (7, 8) Siliceous and salinaceous, B. (9-11) Siliceous, B. (12) Sulfaceous and siliceous; BB. (1-4) Eolian dune deposit, BB. (5, 6) Estuarine tidal marsh deposit, BB. (7, 8) Estuarine tidal marsh salt pan deposit, BB. (9-11) Estuarine tidal marsh sand flat deposit, BB. (12) Estuarine tidal marsh deposit. C. Mixed eolian and water-deposited sands and organics; CC. (1-4) Porous, unconsolidated, cross-bedded, quartz sand, CC. (5, 6) Porous, unconsolidated, water-saturated, silty muck, CC. (7) Porous, unconsolidated, salt-encrusted, quartz sand, CC. (8) Porous, unconsolidated, salty, quartz sand, CC. (9-11) Porous, unconsolidated, quartz sand, CC. (12) Porous, unconsolidated, water-saturated, sandy muck.

HYDROLOGY: A. (1-4) Terrestrial, A. (5-12) Estuarine; AA. (1-4) Very dry-xeric, AA. (5-12) Intertidal. B. (1, 3, 4) Fresh and haline, B. (2) Fresh, B. (5-12) Haline; BB. (1-4) Permanently exposed, BB. (5-11) Regularly flooded, BB. (12) Irregularly flooded. C. Mixed estuarine and terrestrial systems ranging from regulary flooded to permanently exposed; CC. (1, 3, 4) Excessively drained system, wetted by fresh rains and haline ocean spray, CC. (2) Excessively drained system, wetted by fresh rains, CC. (5, 6) Poorly drained euhaline system, wetted by haline tidal waters and fresh rains, CC. (7, 8) Somewhat excessively drained hyperhaline system, wetted by haline tidal waters and fresh rain, CC. (9, 10) Somewhat excessively drained polyhaline system, wetted by haline tidal waters and fresh rains, CC. (11) Somewhat excessively drained, euhaline system, wetted by haline tidal waters and fresh rain, CC. (12) Wetted by haline tidal waters and fresh rains.

TOPOGRAPHY: A. Plains; AA. Flat plain. B. Water-deposited barrier island; BB. (1, 2) Wind-deposited relict dune field, BB. (3, 4) Wind-deposited active dune field, BB. (5-12) Water-deposited and organically accumulated estuarine tidal marsh. C. Mixture of tidal flats and active and relict dune fields; CC. (1) Open, variously exposed, smooth to irregular, moderately steep to strongly sloping slopes and flat crest of a relict dune, CC. (2) Open, but protected from salt spray, northwest-facing, smooth to undulatory, gently sloping to sloping, lower back slopes of a relict dune, CC. (3) Open, southeast-facing, smooth, moderately steep to strongly sloping front face of the foredunes, CC. (4) Open, northwest-facing, smooth, moderately steep to strongly sloping slip face of the foredunes, CC. (5, 6) Open, variously exposed, smooth, nearly level mud flats on the lower tidal marsh on the northwest side of the barrier island, CC. (7) Open, variously exposed, nearly level, salt-encrusted, inner portion of a tidal flat salt pan, CC. (8) Open, variously exposed, nearly level, outer portion of a tidal flat salt pan, CC. (9-11) Open, variously exposed, smooth, nearly level, sand flat on the upper portion of the tidal marsh, CC. (12) Open, variously exposed, nearly level, upper margin of the upper tidal flats.

PHYSIOGRAPHY: A. Atlantic Plain; AA. Coastal Plain. B. Embayed Section; BB. Ocracoke Island. C. First Hammock Hills and The Plains; CC. (1) Foreslope and crest of relict dune in First Hammock Hills, underlain by Holocene eolian sands, CC. (2) Backslope of relict dune in First Hammock Hills, underlain by Holocene eolian sands, CC. (3, 4) Foredune on The Plains, underlain by Recent eolian sands, CC. (5, 6) Lower tidal flats on The Plains, underlain by Recent fine clastics and organics, CC. (7, 8) Salt pan on The Plains tidal marsh, underlain by Recent fine clastics and organics, CC. (9-11) Upper tidal flats on The Plains, underlain by Recent fine clastics and organics, CC. (12) Upper margin of tidal marsh on The Plains, underlain by Recent fine clastics and organics.

BIOLOGY:

1. A. TERRESTRIAL GRASS SYSTEM;
   AA. Very tall, rhizomatous, rhizocarpic graminoids. B. Sea Oats relict dune; BB. Poaceae. C. *Uniola paniculata*; CC. *Uniola paniculata*.
2. A. TERRESTRIAL SHRUB SYSTEM;
   AA. Tall, rhizomatous, evergreen shrubs. B. Wax Myrtle dune; BB. Myricales. C. *Myrica cerifera*; CC. *Myrica cerifera*.
3. A. TERRESTRIAL GRASS SYSTEM;
   AA. Very tall, rhizomatous, rhizocarpic graminoids. B. Sea Oats foredune; BB. Poaceae. C. *Uniola paniculata*; CC. *Uniola paniculata*.
4. A. TERRESTRIAL FORB SYSTEM;
   AA. Medium, reptant, deciduous forbs. B. Pennywort foredune; BB. Umbellales. C. *Hydrocotyle bonariensis*; CC. *Hydrocotyle bonariensis*.
5. A. WETLAND GRASS SYSTEM;
   AA. Tall, rhizomatous, rhizocarpic graminoids. B. Cord Grass marsh; BB. Poaceae. C. *Spartina alterniflora*; CC. *Spartina alterniflora*.
6. A. WETLAND GRASS SYSTEM;
   AA. Tall, rhizomatous, rhizocarpic graminoids/Medium, reptant, deciduous forbs. B. Cord Grass marsh; BB. Poaceae/Caryophyllales. C. *Spartina alterniflora*; CC. *Spartina alterniflora/Salicornia virginica*.
7. A. WETLAND FORB SYSTEM;
   AA. Medium, caulescent, summer-green forbs. B. Glasswort flat; BB. Caryophyllales. C. *Salicornia bigelovii*; CC. *Salicornia bigelovii*.
8. A. WETLAND FORB SYSTEM;
   AA. Tall, reptant, deciduous forbs. B. Glasswort flat; BB. Caryophyllales. C. *Salicornia virginica*; CC. *Salicornia virginica*.
9. A. WETLAND GRASS SYSTEM;
   AA. Tall, rhizomatous, rhizocarpic graminoids. B. Salt Grass marsh; BB. Poaceae. C. *Distichlis spicata*; CC. *Distichlis spicata*.
10. A. WETLAND GRASS SYSTEM;
   AA. Very tall, rhizomatous, rhizocarpic graminoids. B. Cord Grass marsh; BB. Poaceae. C. *Spartina patens*; CC. *Spartina patens*.

11. A. WETLAND GRASS SYSTEM;
AA. Tall, rhizomatous, rhizocarpic graminoids. B. Fimbristylis marsh; BB. Cyperaceae. C. *Fimbristylis spadicea*; CC. *Fimbristylis spadicea*.
12. A. WETLAND FORB SYSTEM;
AA. Very tall, cespitose, deciduous forbs. B. Rush marsh; BB. Juncales. C. *Juncus roemerianus*; CC. *Juncus roemerianus*.

Community Diversity Summary 1

Terrestrial grass system
Very tall, rhizomatous, rhizocarpic graminoids

Sea Oats relict dune
Poaceae

UNIOLA PANICULATA
Uniola paniculata

CLIMATE: A. Mesothermal; AA. Warm temperate. B. Cool and moist yearly, moderately hot and moderately wet summers, moderately warm and moderately dry winters, with a very long freeze-free period; BB. Warm and moist yearly, moderately hot and moderately wet summers, warm and moderately dry winters, with a very long freeze-free period. C. Similar to local climate; CC. Similar to natural area climate.
SOILS: A. Entisol; AA. Psamment. B. Udipsamment; BB. Typic Udipsamment. C. Mixed, thermic Typic Udipsamment; CC. Fripp.
GEOLOGY: A. Sedimentary; AA. Clastic. B. Siliceous; BB. Eolian dune deposit. C. Mixed eolian and water-deposited sands and organics; CC. Porous, unconsolidated, cross-bedded, quartz sand.
HYDROLOGY: A. Terrestrial; AA. Very dry-xeric. B. Fresh and haline; BB. Permanently exposed. C. Mixed estuarine and terrestrial systems ranging from regularly flooded to permanently exposed; CC. Excessively drained system, wetted by fresh rains and haline ocean spray.
TOPOGRAPHY: A. Plains; AA. Flat plain. B. Water-deposited barrier island; BB. Wind-deposited relict dune field. C. Mixture of tidal flats and active and relict dune fields; CC. Open, variously exposed, smooth to irregular, moderately steep to strongly sloping slopes and flat crest of relict dune.
PHYSIOGRAPHY: A. Atlantic Plain; AA. Coastal Plain. B. Embayed Section; BB. Ocracoke Island. C. First Hammock Hills and the Plains; CC. Foreslope and crest of relict dune in First Hammock Hills, underlain by Holocene eolian sands.

CANOPY: No canopy species present.

SUBCANOPY: No subcanopy species present.

SHRUB LAYER DOMINANTS: None.
SHRUB LAYER PHYSIOGNOMY: Not applicable.

Shrub Analysis

Species	GF	1 C.S	2 C.S	3 C.S	4 C.S	5 C.S	6 C.S
Tall shrub							
Baccharis halimifolia	H6	-	-	-	+.1	-	-

Number of relevés: 6 Relevé size: 5 m X 5 m

SHRUB SPECIES PRESENT, BUT NOT IN ANALYSIS: TALL SHRUB: *Yucca filamentosa* var. *filamentosa*; TALL DWARF SHRUB: *Opuntia compressa*; VINE: *Smilax bona-nox*.

HERB LAYER DOMINANT: *Uniola paniculata*.
HERB LAYER PHYSIOGNOMY: Very tall, rhizomatous, rhizocarpic graminoids.

Herb Analysis

Species	GF	1 C.S	2 C.S	3 C.S	4 C.S	5 C.S	6 C.S
Tall forbs							
Aster subulata	H11	-	-	-	3.1	+.1	-
Commelina communis	H10	-	-	-	+.1	-	-
Medium forbs							
Croton punctatus	H6	-	-	+.1	+.1	+.1	-
Hydrocotyle bonariensis	H10	-	-	+.1	-	-	-
Small forbs							
Euphorbia polygonifolia	H10	1.1	-	-	-	-	-
Oenothera humifusa	H10	+.1	-	-	-	-	-
Very tall graminoid							
Uniola paniculata	H11	5.4	3.4	5.4	3.4	5.4	5.4
Tall graminoids							
Cenchrus tribuloides	H10	1.1	-	-	-	-	-
Muhlenbergia capillaris	H7	+.1	-	-	-	-	3.2
Triplasis purpurea	H7	-	-	-	1.2	4.2	-

Number of relevés: 6 Relevé size: 1 m X 1 m

HERB SPECIES PRESENT, BUT NOT IN ANALYSIS: TALL FORBS: *Carduus spinosissimus*, *Diodia teres*; VERY TALL GRAMINOID: *Andropogon glomeratus*.

COMMUNITY REFERENCES: None.

COMMUNITY DOCUMENTATION: Site analyzed by A. E. Radford's Botany 235 class, UNC, Chapel Hill, in fall 1977. Specimens deposited in the NCU Herbarium by T. A. Atkinson in 1977.

COMMUNITY ECOLOGICAL CHARACTERIZATION: Vegetationally: Poaceous terrestrial grass system with a sparse shrub layer of widely scattered, mixed shrubs and vines and a closed herb layer dominated by very tall, rhizomatous, rhizocarpic graminoids. Climatically: Warm temperate, mesothermal climate, similar to the local yearly warm and moist climate, which has moderately hot and moderately wet summers, warm and moderately dry winters, and a very long freeze-free period. Pedologically: Mixed, thermic Typic Udipsamment, Fripp soil. Geologically. A clastic, sedimentary, eolian dune deposit, composed of porous, unconsolidated, cross-bedded, siliceous, quartz sands. Hydrologically: An excessively drained, permanently exposed, very dry-xeric, terrestrial system, wetted by fresh rains and haline, wind-blown, ocean spray. Topographically: Open, variously exposed, smooth to irregular, moderately steep to strongly sloping slopes and flat crest of a wind-deposited relict dune field on a water-deposited barrier island located on the seaward edge of a flat plain. Temporally and Spatially: A topoedaphic climax of a psammosere in First Hammock Hills, underlain by Holocene eolian sands on Ocracoke Island in the Embayed Section of the Coastal Plain Province of the Atlantic Plain Region.

Community Diversity Summary 2

Terrestrial shrub system	Wax Myrtle dune
Tall, rhizomatous, evergreen shrubs	Myricales

MYRICA CERIFERA
Myrica cerifera

CLIMATE: A. Mesothermal; AA. Warm temperate. B. Cool and moist yearly, moderately hot and moderately wet summers, moderately warm and moderately dry winters, with a very long freeze-free period; BB. Warm and moist yearly, moderately hot and moderately wet summers, warm and moderately dry winters, with a very long freeze-free period. C. Similar to local climate; CC. Similar to natural area climate.

SOILS: A. Entisol; AA. Psamment. B. Udipsamment; BB. Typic Udipsamment. C. Mixed, thermic Typic Udipsamment; CC. Fripp.

GEOLOGY: A. Sedimentary; AA. Clastic. B. Siliceous; BB. Eolian dune deposit. C. Mixed eolian and water-deposited sands and organics; CC. Porous, unconsolidated, cross-bedded, quartz sand.

HYDROLOGY: A. Terrestrial; AA. Very dry-xeric. B. Fresh; BB. Permanently exposed. C. Mixed estuarine and terrestrial systems ranging from regularly flooded to permanently exposed; CC. Excessively drained system, wetted by fresh rains.

TOPOGRAPHY: A. Plains; AA. Flat plain. B. Water-deposited barrier island; BB. Wind-deposited relict dune field. C. Mixture of tidal flats and active and relict dune fields; CC. Open, but protected from salt spray, northwest-facing, smooth to undulating, gently sloping to sloping, lower backslopes of relict dunes.

PHYSIOGRAPHY: A. Atlantic Plain; AA. Coastal Plain. B. Embayed Section; BB. Ocracoke Island. C. First Hammock Hills; CC. Backslope of relict dune, underlain by Holocene eolian sands.

CANOPY: Height, dbh, and age not determined.
CANOPY DOMINANTS: None.
CANOPY PHYSIOGNOMY: Not applicable.
CANOPY ANALYSIS: Only a species presence list completed. Species list--SMALL TREES: *Juniperus virginiana*, *Salix caroliniana*, *Zanthoxylum clava-herculis*.

SUBCANOPY: No subcanopy species present.

SHRUB LAYER DOMINANT: *Myrica cerifera*.
SHRUB LAYER PHYSIOGNOMY: Tall, rhizomatous, evergreen shrubs.

Shrub Analysis

Species	GF	1 C.S	2 C.S	3 C.S	4 C.S	5 C.S	6 C.S
Tall shrubs							
Baccharis halimifolia	H6	-	5.4	-	-	-	-
Ilex vomitoria	H6	-	-	-	-	+.1	4.4
Myrica cerifera	H11	5.4	5.4	5.4	5.4	5.4	5.4
Vines							
Parthenocissus quinquefolia	H18	-	-	-	-	-	1.1
Smilax auriculata	H18	-	-	-	-	+.1	3.4

Number of relevés: 6 Relevé size: 5 m X 5 m

SHRUB SPECIES PRESENT, BUT NOT IN ANALYSIS: None.

HERB LAYER DOMINANTS: None.
HERB LAYER PHYSIOGNOMY: Not applicable.

Herb Analysis

Species	GF	1 C.S	2 C.S	3 C.S	4 C.S	5 C.S	6 C.S
Tall forbs							
Commelina communis	H10	+.1	-	-	-	-	-
Physalis viscosa	H11	+.1	-	-	-	-	-
Solanum gracile	H6	-	-	-	-	+.1	-
Medium forb							
Hydrocotyle bonariensis	H10	+.1	-	-	+.1	-	-
Very tall graminoids							
Juncus roemerianus	H7	-	-	3.4	-	-	-
Spartina patens	H11	-	-	-	5.4	3.4	-
Tall graminoid							
Muhlenbergia capillaris	H7	-	+.1	-	-	-	-

Number of relevés: 6 Relevé size: 1 m X 1 m

HERB SPECIES PRESENT, BUT NOT IN ANALYSIS: TALL FORB: *Gnaphalium obtusifolium*; SMALL FORB: *Lippia nodiflora*; TALL GRAMINOID: *Juncus validus*.

COMMUNITY REFERENCES: None.

COMMUNITY DOCUMENTATION: Site analyzed by A. E. Radford's Botany 235 class, UNC, Chapel Hill, in fall 1977. Specimens deposited in NCU Herbarium by T. A. Atkinson in 1977.

COMMUNITY ECOLOGICAL CHARACTERIZATION: Vegetationally: Myricalean terrestrial shrub system with a sparse canopy of widely scattered, mixed small trees, a closed shrub layer of tall, rhizomatous, evergreen shrubs, and a sparse herb layer of scattered forbs and graminoids. Climatically: Warm temperate, mesothermal climate, similar to the local yearly warm and moist climate, which has moderately hot and moderately wet summers, warm and moderately dry winters, and a very long freeze-free period. Pedologically: Mixed, thermic Typic Udipsamment, Fripp soil. Geologically: A clastic, sedimentary, eolian dune deposit, composed of porous, unconsolidated, cross-bedded, siliceous, quartz sands. Hydrologically: An excessively drained, permanently exposed, very dry-xeric terrestrial system, wetted by fresh rains. Topographically: Open, but protected from salt spray, northwest-facing, smooth to undulating, gently sloping to sloping, lower backslopes of a wind-deposited relict dune field on a water-deposited barrier island located on the seaward edge of a flat plain. Temporally and Spatially: A topoedaphic climax of a psammosere in First Hammock Hills, underlain by Holocene eolian sands, on Ocracoke Island in the Embayed Section of the Coastal Plain Province of the Atlantic Plain Region.

Community Diversity Summary 3

Terrestrial grass system
Very tall, rhizomatous, rhizocarpic graminoids

Sea Oats foredune
Poaceae

UNIOLA PANICULATA
Uniola paniculata

CLIMATE: A. Mesothermal; AA. Warm temperate. B. Cool and moist yearly, moderately hot and moderately wet summers, moderately warm and moderately dry winters, with a very long freeze-free period; BB. Warm and moist yearly, moderately hot and moderately wet summers, warm and moderately dry winters, with a very long freeze-free period. C. Similar to local climate; CC. Similar to natural area climate.
SOILS: A. Entisol; AA. Psamment. B. Quartzipsamment; BB. Typic Quartzipsamment. C. Sandy, siliceous, thermic Typic Quartzipsamment; CC. Duneland.
GEOLOGY: A. Sedimentary; AA. Clastic. B. Siliceous; BB. Eolian dune deposit.

C. Mixture of eolian sands and water-deposited, fine clastics and organics; CC. Porous, unconsolidated, cross-bedded, quartz sand.

HYDROLOGY: A. Terrestrial; AA. Very dry-xeric. B. Fresh and haline; BB. Permanently exposed. C. Mixed estuarine and terrestrial systems ranging from regularly flooded to permanently exposed; CC. Excessively drained system, wetted by fresh rains and haline ocean spray.

TOPOGRAPHY: A. Plains; AA. Flat plain. B. Water-deposited barrier island; BB. Wind-deposited active dune field. C. Mixture of tidal flats and active and relict dunes; CC. Open, southeast-facing, smooth, moderately steep to strongly sloping front face of the foredunes.

PHYSIOGRAPHY: A. Atlantic Plain; AA. Coastal Plain. B. Embayed Section; BB. Ocracoke Island. C. The Plains dune field; CC. Foredune, underlain by Recent eolian sands.

CANOPY: No canopy species present.

SUBCANOPY: No subcanopy species present.

SHRUB LAYER: No shrub species present.

HERB LAYER DOMINANT: *Uniola paniculata*.

HERB LAYER PHYSIOGNOMY: Very tall, rhizomatous, rhizocarpic graminoids.

HERB ANALYSIS: Only a species presence list completed. Species list--TALL FORBS: *Amaranthus pumilus*, *Cakile edentula*; VERY TALL GRAMINOIDS: *Ammophila breviligulata*, *Uniola paniculata*; TALL GRAMINOID: *Panicum amarum*.

COMMUNITY REFERENCES: None.

COMMUNITY DOCUMENTATION: Site analyzed by A. E. Radford's Botany 235 class, UNC, Chapel Hill, in fall 1977. Specimens deposited in the NCU Herbarium by T. A. Atkinson in 1977.

COMMUNITY ECOLOGICAL CHARACTERIZATION: Vegetationally: Poaceous terrestrial grass system with an open herb layer dominated by very tall, rhizomatous, rhizocarpic graminoids. Climatically: Warm temperate, mesothermal climate, similar to the local yearly warm and moist climate, which has moderately hot and moderately wet summers, warm and moderately dry winters, and a very long freeze-free period. Pedologically: Sandy, siliceous, thermic Typic Quartzipsamment, Duneland soil. Geologically: A clastic, sedimentary, eolian dune deposit, composed of porous, unconsolidated, cross-bedded, siliceous, quartz sands. Hydrologically: Excessively drained, permanently exposed, very dry-xeric terrestrial system, wetted by fresh rains and haline, wind-blown ocean spray. Topographically: Open, southeast-facing, smooth, moderately steep to strongly sloping front face of the foredunes of a wind-deposited, active dune field on a water-deposited barrier island located on the seaward edge of a flat plain. Temporally and Spatially: A topoedaphic climax of a psammosere on the foredune of The Plains dune field, underlain by Recent eolian sands, on Ocracoke Island in the Embayed Section of the Coastal Plain Province of the Atlantic Plain Region.

Community Diversity Summary 4

Terrestrial forb system
Medium, reptant, deciduous forbs

Pennywort foredune
Umbellales

HYDROCOTYLE BONARIENSIS
Hydrocotyle bonariensis

CLIMATE: A. Mesothermal; AA. Warm temperate. B. Cool and moist yearly, moderately hot and moderately wet summers, moderately warm and moderately dry winters, with a very long freeze-free period; BB. Warm and moist yearly, moderately hot and moderately wet summers, warm and moderately dry winters, with a very long freeze-free period. C. Similar to local climate; CC. Similar to natural area climate.
SOILS: A. Entisol; AA. Psamment. B. Quartzipsamment; BB. Typic Quartzipsamment. C. Sandy, siliceous, thermic Typic Quartzipsamment; CC. Duneland.
GEOLOGY: A. Sedimentary; AA. Clastic. B. Siliceous; BB. Eolian dune deposit. C. Mixture of eolian sands and water-deposited, fine clastics and organics; CC. Porous, unconsolidated, cross-bedded, quartz sand.
HYDROLOGY: A. Terrestrial; AA. Very dry-xeric. B. Fresh and haline; BB. Permanently exposed. C. Mixed estuarine and terrestrial systems ranging from regularly flooded to permanently exposed; CC. Excessively drained system, wetted by fresh rains and by occasional haline ocean spray.
TOPOGRAPHY: A. Plains; AA. Flat plain. B. Water-deposited barrier island; BB. Wind-deposited active dune field. C. Mixture of tidal flats and active and relict dune fields; CC. Open, northwest-facing, smooth, moderately steep to strongly sloping slip face of the foredunes.
PHYSIOGRAPHY: A. Atlantic Plain; AA. Coastal Plain. B. Embayed Section; BB. Ocracoke Island. C. The Plains dune field; CC. Foredune, underlain by Recent eolian sands.

CANOPY: No canopy species present.

SUBCANOPY: No subcanopy species present.

SHRUB LAYER DOMINANTS: None.
SHRUB LAYER PHYSIOGNOMY: Not applicable.
SHRUB ANALYSIS: Only a species presence list completed. Species list--VINE: *Strophostyles helvola*.

HERB LAYER DOMINANT: *Hydrocotyle bonariensis*.
HERB LAYER PHYSIOGNOMY: Medium, reptant, deciduous forbs.
HERB ANALYSIS: Only a species presence list completed. Species list--TALL FORB: *Solidago sempervirens*; MEDIUM FORB: *Hydrocotyle bonariensis*; SMALL FORB: *Euphorbia polygonifolia*; TALL GRAMINOIDS: *Cenchrus tribuloides*, *Eragrostis spectabilis*, *Muhlenbergia capillaris*, *Panicum amarum*.

COMMUNITY REFERENCES: None.

COMMUNITY DOCUMENTATION: Site analyzed by A. E. Radford's Botany 235 class, UNC, Chapel Hill, in fall 1977. Specimens deposited in the NCU Herbarium by T. A. Atkinson in 1977.

COMMUNITY ECOLOGICAL CHARACTERIZATION: Vegetationally: Umbellalean terrestrial forb system with sparse, scattered, twining, deciduous vines and an open herb layer dominated by medium, reptant, deciduous forbs. Climatically: Warm temperate, mesothermal climate, similar to the local yearly warm and moist climate, which has moderately hot and moderately wet summers, warm and moderately dry winters, and a very long freeze-free period. Pedologically: Sandy, siliceous, thermic Typic Quartzipsamment, Duneland soil. Geologically: A clastic, sedimentary, eolian dune deposit, composed of porous, unconsolidated, cross-bedded, siliceous, quartz sands. Hydrologically: Excessively drained, permanently exposed, dry, terrestrial system, wetted by fresh rains and occasionally by haline, wind-blown ocean spray. Topographically: Open, northwest-facing, smooth, moderately steep to strongly sloping slip face of the foredunes of a wind-deposited, active dune field on a water-deposited barrier island located on the seaward edge of a flat plain. Temporally and Spatially: A topoedaphic climax of a psammosere on the foredune of The Plains dune field, underlain by Recent eolian sands, on Ocracoke Island in the Embayed Section of the Coastal Plain Province of the Atlantic Plain Region.

Community Diversity Summary 5

Wetland grass system	Cord Grass marsh
Tall, rhizomatous, rhizocarpic graminoids	Poaceae

SPARTINA ALTERNIFLORA
Spartina alterniflora

CLIMATE: A. Mesothermal; AA. Warm temperate. B. Cool and moist yearly, moderately hot and moderately wet summers, moderately warm and moderately dry winters, with a very long freeze-free period; BB. Warm and moist yearly, moderately hot and moderately wet summers, warm and moderately dry winters, with a very long freeze-free period. C. Similar to local climate; CC. Similar to natural area climate.

SOILS: A. Entisol; AA. Aquent. B. Psammaquent; BB. Typic Psammaquent. C. Mixed, thermic Typic Psammaquent; CC. Carteret.

GEOLOGY: A. Sedimentary; AA. Mixed clastic and organic. B. Siliceous and sulfaceous; BB. Estuarine tidal marsh deposit. C. Mixture of eolian sands and water-deposited, fine clastics and organics; CC. Porous, unconsolidated, water-saturated, silty muck.

HYDROLOGY: A. Estuarine; AA. Intertidal. B. Haline; BB. Regularly flooded. C. Mixed estuarine and terrestrial systems ranging from regularly flooded to permanently exposed; CC. Poorly drained euhaline system, wetted by haline tidal waters and fresh rains.

TOPOGRAPHY: A. Plains; AA. Flat plain. B. Water-deposited barrier island; BB. Water-deposited and organically accumulated estuarine tidal marsh. C. Mixture of tidal flats and active and relict dunes; CC. Open, variously exposed, smooth, nearly level mud flats on the lower tidal marsh on the northwest side of the barrier island.

PHYSIOGRAPHY: A. Atlantic Plain; AA. Coastal Plain. B. Embayed Section; BB. Ocracoke Island. C. The Plains tidal marsh; CC. Lower tidal flats, underlain by Recent fine clastics and organics.

CANOPY: No canopy species present.

SUBCANOPY: No subcanopy species present.

SHRUB LAYER: No shrub species present.

HERB LAYER DOMINANT: *Spartina alterniflora.*
HERB LAYER PHYSIOGNOMY: Tall, rhizomatous, rhizocarpic graminoids.

Herb Analysis

Species	GF	1 C.S	2 C.S	3 C.S
Tall forb				
Salicornia europaea	H6	1.1	1.1	1.1
Tall graminoid				
Spartina alterniflora	H11	4.4	3.4	5.4

Number of relevés: 3 Relevé size: 1 m X 1 m

HERB SPECIES PRESENT, BUT NOT IN ANALYSIS: None.

COMMUNITY REFERENCES: None.

COMMUNITY DOCUMENTATION: Site analyzed by A. E. Radford's Botany 235 class, UNC, Chapel Hill, in fall 1977. Specimens deposited in the NCU Herbarium by T. A. Atkinson in 1977.

COMMUNITY ECOLOGICAL CHARACTERIZATION: Vegetationally: Poaceous wetland grass system with an open herb layer dominated by tall, rhizomatous, rhizocarpic graminoids. Climatically: Warm temperate, mesothermal climate, similar to the local yearly warm and moist climate, which has moderately hot and moderately wet summers, warm and moderately dry winters, and a very long freeze-free period. Pedologically: Mixed, thermic Typic Psammaquent, Carteret soil. Geologically: A mixed clastic and organic, sedimentary, estuarine tidal marsh deposit, composed of siliceous and sulfaceous, porous, unconsolidated, water-saturated, silty muck. Hydrologically: Poorly drained, regularly flooded, intertidal, euhaline estuarine system wetted by haline tidal waters and fresh rains. Topographically: Open, variously exposed, smooth, nearly level mud flats on the lower tidal marsh on the northwest side of a water-deposited barrier island located on the seaward edge of a flat plain. Temporally and Spatially: A topoedaphic climax of a hydrosere on The Plains lower tidal flat, underlain by Recent estuarine tidal flat sediments, on Ocracoke Island in the Embayed Section of the Coastal Plain Province of the Atlantic Plain Region.

Community Diversity Summary 6

Wetland grass system
Tall, rhizomatous, rhizocarpic graminoids/
Medium, reptant, deciduous forbs

Cord Grass marsh
Poaceae/Caryophyllales

SPARTINA ALTERNIFLORA/SALICORNIA VIRGINICA
Spartina alterniflora

CLIMATE: A. Mesothermal; AA. Warm temperate. B. Cool and moist yearly, moderately hot and moderately wet summers, moderately warm and moderately dry winters, with a very long freeze-free period; BB. Warm and moist yearly, moderately hot and moderately wet summers, warm and moderately dry winters, with a very long freeze-free period. C. Similar to local climate; CC. Similar to natural area climate.
SOILS: A. Entisol; AA. Aquent. B. Psammaquent; BB. Typic Psammaquent. C. Mixed, thermic Typic Psammaquent; CC. Carteret.
GEOLOGY: A. Sedimentary; AA. Clastic and organic. B. Siliceous and sulfaceous; BB. Estuarine tidal marsh deposit. C. Mixture of eolian sands and water-deposited, fine clastics and organics; CC. Porous, unconsolidated, water-saturated, silty muck.
HYDROLOGY: A. Estuarine; AA. Intertidal. B. Haline; BB. Regularly flooded. C. Mixed estuarine and terrestrial systems ranging from regularly flooded to permanently exposed; CC. Poorly drained euhaline system, wetted by haline tidal waters and fresh rains.
TOPOGRAPHY: A. Plains; AA. Flat plain. B. Water-deposited barrier island; BB. Water-deposited and organically accumulated estuarine tidal marsh. C. Mixture of tidal flats and active and relict dunes; CC. Open, variously exposed, smooth, nearly level mud flats on the lower tidal marsh on the northwest side of the barrier island.
PHYSIOGRAPHY: A. Atlantic Plain; AA. Coastal Plain. B. Embayed Section; BB. Ocracoke Island. C. The Plains tidal marsh; CC. Lower tidal flats, underlain by Recent fine clastics and organics.

CANOPY: No canopy species present.

SUBCANOPY: No subcanopy species present.

SHRUB LAYER: No shrub species present.

HERB LAYER DOMINANTS: *Spartina alterniflora/Salicornia virginica.*
HERB LAYER PHYSIOGNOMY: Tall, rhizomatous, rhizocarpic graminoids/Medium, reptant, deciduous forbs.

Herb Analysis

Species	GF	1 C.S	2 C.S	3 C.S
Tall forbs				
Aster tenuifolius	H11	+.1	+.1	+.1
Salicornia virginica	H10	2.3	2.3	2.3
Tall graminoid				
Spartina alterniflora	H11	5.4	5.4	5.4

Number of relevés: 3 Relevé size: 1 m X 1 m

HERB SPECIES PRESENT, BUT NOT IN ANALYSIS: None.

COMMUNITY REFERENCES: None.

COMMUNITY DOCUMENTATION: Site analyzed by A. E. Radford's Botany 235 class, UNC, Chapel Hill, in fall 1977. Specimens deposited in the NCU Herbarium by T. A. Atkinson in 1977.

COMMUNITY ECOLOGICAL CHARACTERIZATION: Vegetationally: Poaceous/Caryophyllalean wetland grass system with a closed tall herb layer of rhizomatous, rhizocarpic graminoids and an open medium herb layer of reptant, deciduous forbs. Climatically: Warm temperate, mesothermal climate, similar to the local yearly warm and moist climate, which has moderately hot and moderately wet summers, warm and moderately dry winters, and a very long freeze-free period. Pedologically: Mixed, thermic Typic Psammaquent, Carteret soil. Geologically: A mixed clastic and organic, sedimentary, estuarine tidal marsh deposit, composed of siliceous and sulfaceous, porous, unconsolidated, water-saturated, silty muck. Hydrologically: Poorly drained, regularly flooded, intertidal, euhaline estuarine system, wetted by haline tidal waters and fresh rains. Topographically: Open, variously exposed, smooth, nearly level mud flats on the lower tidal marsh on the northwest side of a water-deposited barrier island located on the seaward edge of a flat plain. Temporally and Spatially: A topoedaphic climax of a hydrosere on The Plains lower tidal flat, underlain by Recent estuarine tidal flat sediments, on Ocracoke Island in the Embayed Section of the Coastal Plain Province of the Atlantic Plain Region.

Community Diversity Summary 7

Wetland forb system | Glasswort flat
Medium, caulescent, summer-green forbs | Caryophyllales

<u>*SALICORNIA BIGELOVII*</u>
Salicornia bigelovii

CLIMATE: A. Mesothermal; AA. Warm temperate. B. Cool and moist yearly, moderately hot and moderately wet summers, moderately warm and moderately dry winters, with a very long freeze-free period; BB. Warm and moist yearly, moderately hot and moderately wet summers, warm and moderately dry winters, with a very long freeze-free period. C. Similar to local climate; CC. Similar to natural area climate.
SOILS: A. Entisol; AA. Aquent. B. Psammaquent; BB. Typic Psammaquent. C. Mixed, thermic Typic Psammaquent; CC. Carteret.
GEOLOGY: A. Sedimentary; AA. Clastic and chemical. B. Siliceous and salinaceous; BB. Estuarine tidal marsh salt pan deposit. C. Mixture of eolian sands and water-deposited, fine clastics and organics; CC. Porous, unconsolidated, salt-encrusted, quartz sand.
HYDROLOGY: A. Estuarine; AA. Intertidal. B. Haline; BB. Regularly flooded. C. Mixed estuarine and terrestrial systems ranging from regularly flooded to permanently exposed; CC. Somewhat excessively drained hyperhaline system, wetted by haline tidal waters and fresh rains.
TOPOGRAPHY: A. Plains; AA. Flat plain. B. Water-deposited barrier island; BB. Water-deposited and organically accumulated estuarine tidal marsh. C. Mixture of tidal flats and active and relict dunes; CC. Open, variously exposed, nearly level, salt-encrusted, inner portion of a tidal flat salt pan.
PHYSIOGRAPHY: A. Atlantic Plain; AA. Coastal Plain. B. Embayed Section; BB. Ocracoke Island. C. The Plains tidal marsh; CC. Salt pan, underlain by Recent fine clastics and organics.

CANOPY: No canopy species present.

SUBCANOPY: No subcanopy species present.

SHRUB LAYER: No shrub species present.

HERB LAYER DOMINANT: *Salicornia bigelovii*.
HERB LAYER PHYSIOGNOMY: Medium, caulescent, summer-green forbs.

<u>Herb Analysis</u>

Species	GF	1 C.S	2 C.S	3 C.S
Medium forb				
Salicornia bigelovii	H6	3.4	3.4	3.4

Number of relevés: 3 Relevé size: 1 m X 1 m

HERB SPECIES PRESENT, BUT NOT IN ANALYSIS: TALL FORB: *Salicornia virginica*; TALL GRAMINOID: *Spartina alterniflora*.

COMMUNITY REFERENCES: None.

COMMUNITY DOCUMENTATION: Site analyzed by A. E. Radford's Botany 235 class, UNC, Chapel Hill, in fall 1977. Specimens deposited in the NCU Herbarium by T. A. Atkinson in 1977.

COMMUNITY ECOLOGICAL CHARACTERIZATION: Vegetationally: Caryophyllalean wetland forb system with an open to sparse herb layer dominated by medium, caulescent, summer-green forbs. Climatically: Warm temperate, mesothermal climate, similar to the local yearly warm and moist climate, which has moderately hot and moderately wet summers, warm and moderately dry winters, and a very long freeze-free period. Pedologically: Mixed, thermic Typic Psammaquent, Carteret soil. Geologically: A mixed, clastic and chemical, sedimentary, estuarine tidal marsh salt pan deposit, composed of siliceous and salinaceous, porous, unconsolidated, salt-encrusted, quartz sands. Hydrologically: Somewhat excessively drained, regularly flooded, intertidal, hyperhaline estuarine system, wetted by haline tidal waters and fresh rains. Topographically: Open, variously exposed, nearly level, salt-encrusted, inner portion of a tidal flat salt pan on the northwest side of a water-deposited barrier island located on the seaward edge of a flat plain. Temporally and Spatially: A topoedaphic climax of a hydrosere in a salt pan on The Plains lower tidal flat, underlain by Recent estuarine tidal flat sediments, on Ocracoke Island in the Embayed Section of the Coastal Plain Province of the Atlantic Plain Region.

Community Diversity Summary 8

Wetland forb system
Tall, reptant, deciduous forbs

Glasswort flat
Caryophyllales

SALICORNIA VIRGINICA
Salicornia virginica

CLIMATE: A. Mesothermal; AA. Warm temperate. B. Cool and moist yearly, moderately hot and moderately wet summers, moderately warm and moderately dry winters, with a very long freeze-free period; BB. Warm and moist yearly, moderately hot and moderately wet summers, warm and moderately dry winters, with a very long freeze-free period. C. Similar to local climate; CC. Similar to natural area climate.

SOILS: A. Entisol; AA. Aquent. B. Psammaquent; BB. Typic Psammaquent. C. Mixed, thermic Typic Psammaquent; CC. Carteret.

GEOLOGY: A. Sedimentary; AA. Clastic and chemical. B. Siliceous and salinaceous;

BB. Estuarine tidal marsh salt pan deposit. C. Mixture of eolian sands and water-deposited, fine clastics and organics; CC. Porous, unconsolidated, salty, quartz sand.

HYDROLOGY: A. Estuarine; AA. Intertidal. B. Haline; BB. Regularly flooded. C. Mixed estuarine and terrestrial systems ranging from regularly flooded to permanently exposed; CC. Somewhat excessively drained, hyperhaline system, wetted by haline tidal waters and fresh rains.

TOPOGRAPHY: A. Plains; AA. Flat plain. B. Water-deposited barrier island; BB. Water-deposited and organically accumulated estuarine tidal marsh. C. Mixture of tidal flats and active and relict dunes; CC. Open, variously exposed, nearly level, outer portion of a tidal flat salt pan.

PHYSIOGRAPHY: A. Atlantic Plain; AA. Coastal Plain. B. Embayed Section; BB. Ocracoke Island. C. The Plains tidal marsh; CC. Salt pan, underlain by Recent fine clastics and organics.

CANOPY: No canopy species present.

SUBCANOPY: No subcanopy species present.

SHRUB LAYER: No shrub species present.

HERB LAYER DOMINANT: *Salicornia virginica*.
HERB LAYER PHYSIOGNOMY: Tall, reptant, deciduous forbs.

Herb Analysis

Species	GF	1 C.S	2 C.S	3 C.S
Tall forbs				
Aster tenuifolius	H11	-	-	+.1
Limonium carolinianum	H6	-	+.1	+.1
Salicornia virginica	H10	5.4	5.4	5.4
Tall graminoid				
Spartina alterniflora	H11	1.1	1.1	1.1

Number of relevés: 3 Relevé size: 1 m X 1 m

HERB SPECIES PRESENT, BUT NOT IN ANALYSIS: None.

COMMUNITY REFERENCES: None.

COMMUNITY DOCUMENTATION: Site analyzed by A. E. Radford's Botany 235 class, UNC, Chapel Hill, in fall 1977. Specimens deposited in the NCU Herbarium by T. A. Atkinson in 1977.

COMMUNITY ECOLOGICAL CHARACTERIZATION: Vegetationally: Caryophyllalean wetland forb system with a closed herb layer dominated by tall, reptant, deciduous forbs. Climatically: Warm temperate, mesothermal climate, similar to the local yearly warm and moist climate, which has moderately hot and moderately wet summers, warm and moderately dry winters, and a very long freeze-free period. Pedologically: Mixed, thermic Typic Psammaquent, Carteret soil. Geologically: Mixed clastic and chemical, sedimentary, estuarine tidal marsh salt pan deposit, composed of siliceous and salinaceous, porous, unconsolidated, salty, quartz sands. Hydrologically: Somewhat excessively drained, regularly flooded, intertidal, hyperhaline estuarine system, wetted by haline tidal waters and fresh rains. Topographically: Open, variously exposed, nearly level, outer portion of a tidal flat salt pan on the northwest side of a water-deposited barrier island located on the seaward edge of a flat plain. Temporally and Spatially: A topoedaphic climax of a hydrosere in a salt pan on The Plains lower tidal flat, underlain by Recent estuarine tidal flat sediments, on Ocracoke Island in the Embayed Section of the Coastal Plain Province of the Atlantic Plain Region.

Community Diversity Summary 9

Wetland grass system — Salt Grass marsh
Tall, rhizomatous, rhizocarpic graminoids — Poaceae

DISTICHLIS SPICATA
Distichlis spicata

CLIMATE: A. Mesothermal; AA. Warm temperate. B. Cool and moist yearly, moderately hot and moderately wet summers, moderately warm and moderately dry winters, with a very long freeze-free period; BB. Warm and moist yearly, moderately hot and moderately wet summers, warm and moderately dry winters, with a very long freeze-free period. C. Similar to local climate; CC. Similar to natural area climate.

SOILS: A. Entisol; AA. Aquent. B. Psammaquent; BB. Typic Psammaquent. C. Mixed, thermic Typic Psammaquent; CC. Carteret.

GEOLOGY: A. Sedimentary; AA. Clastic. B. Siliceous; BB. Estuarine tidal marsh sand flat deposit. C. Mixture of eolian sands and water-deposited, fine clastics and organics; CC. Porous, unconsolidated, quartz sand.

HYDROLOGY: A. Estuarine; AA. Intertidal. B. Haline; BB. Regularly flooded. C. Mixed estuarine and terrestrial systems ranging from regularly flooded to permanently exposed; CC. Somewhat excessively drained, polyhaline system, wetted by haline tidal waters and fresh rains.

TOPOGRAPHY: A. Plains; AA. Flat plain. B. Water-deposited barrier island; BB. Water-deposited and organically accumulated estuarine tidal marsh. C. Mixture of tidal flats and active and relict dunes; CC. Open, variously exposed, smooth, nearly level, sand flat on the upper portion of the tidal marsh.

PHYSIOGRAPHY: A. Atlantic Plain; AA. Coastal Plain. B. Embayed Section; BB. Ocracoke Island. C. The Plains tidal marsh; CC. Upper tidal flats, underlain by Recent fine clastics and organics.

CANOPY: No canopy species present.

SUBCANOPY: No subcanopy species present.

SHRUB LAYER: No shrub species present.

HERB LAYER DOMINANT: *Distichlis spicata.*
HERB LAYER PHYSIOGNOMY: Tall, rhizomatous, rhizocarpic graminoids.

Herb Analysis

Species	GF	1 C.S	2 C.S	3 C.S
Tall forbs				
Limonium carolinianum	H6	-	+.1	-
Salicornia virginica	H10	-	+.1	-
Medium forb				
Salicornia bigelovii	H6	1.1	+.1	-
Very tall graminoid				
Spartina patens	H11	-	+.1	1.3
Tall graminoid				
Distichlis spicata	H11	3.4	2.4	2.4

Number of Relevés: 3 Relevé size: 1 m X 1 m

HERB SPECIES PRESENT, BUT NOT IN ANALYSIS: None.

COMMUNITY REFERENCES: None.

COMMUNITY DOCUMENTATION: Site analyzed by A. E. Radford's Botany 235 class, UNC, Chapel Hill, in fall 1977. Specimens deposited in the NCU Herbarium by T. A. Atkinson in 1977.

COMMUNITY ECOLOGICAL CHARACTERIZATION: Vegetationally: Poaceous wetland grass system with an open herb layer dominated by tall, rhizomatous, rhizocarpic graminoids. Climatically: Warm temperate, mesothermal climate, similar to the local yearly warm and moist climate, which has moderately hot and moderately wet summers, warm and moderately dry winters, and a very long freeze-free period. Pedologically: Mixed, thermic Typic Psammaquent, Carteret soil. Geologically: A clastic, sedimentary, estuarine tidal marsh sand flat deposit, composed of siliceous, porous, unconsolidated, quartz sands. Hydrologically: Somewhat excessively drained, regularly flooded, intertidal, polyhaline estuarine system, wetted by haline tidal waters and fresh rains. Topographically: Open, variously exposed, smooth, nearly level, sand flat on the upper portion of a tidal marsh on the northwest side of a water-deposited barrier island located on the seaward edge of a flat plain. Temporally and Spatially: A topoedaphic climax of a hydrosere on the upper portion of The Plains tidal marsh, underlain by Recent tidal flat sediments, on Ocracoke Island in the Embayed Section of the Coastal Plain Province of the Atlantic Plain Region.

Community Diversity Summary 10

Wetland grass system
Very tall, rhizomatous, rhizocarpic graminoids

Cord Grass marsh
Poaceae

SPARTINA PATENS
Spartina patens

CLIMATE: A. Mesothermal; AA. Warm temperate. B. Cool and moist yearly, moderately hot and moderately wet summers, moderately warm and moderately dry winters, with a very long freeze-free period; BB. Warm and moist yearly, moderately hot and moderately wet summers, warm and moderately dry winters, with a very long freeze-free period. C. Similar to local climate; CC. Similar to natural area climate.

SOILS: A. Entisol; AA. Aquent. B. Psammaquent; BB. Typic Psammaquent. C. Mixed, thermic Typic Psammaquent; C. Carteret.

GEOLOGY: A. Sedimentary; AA. Clastic. B. Siliceous; BB. Estuarine tidal marsh sand flat deposit. C. Mixture of eolian sands and water-deposited, fine clastics and organics; CC. Porous, unconsolidated, quartz sand.

HYDROLOGY: A. Estuarine; AA. Intertidal. B. Haline; BB. Regularly flooded. C. Mixed estuarine and terrestrial systems ranging from regularly flooded to permanently exposed; CC. Somewhat excessively drained, polyhaline system, wetted by haline tidal waters and fresh rains.

TOPOGRAPHY: A. Plains; AA. Flat Plain. B. Water-deposited barrier island; BB. Water-deposited and organically accumulated estuarine tidal marsh. C. Mixture of tidal flats and active and relict dunes; CC. Open, variously exposed, smooth, nearly level, sand flat on the upper portion of the tidal marsh.

PHYSIOGRAPHY: A. Atlantic Plain; AA. Coastal Plain. B. Embayed Section; BB. Ocracoke Island. C. The Plains tidal marsh; CC. Upper tidal flats, underlain by Recent fine clastics and organics.

CANOPY: No canopy species present.

SUBCANOPY: No subcanopy species present.

SHRUB LAYER DOMINANTS: None.
SHRUB LAYER PHYSIOGNOMY: Not applicable.

Shrub Analysis

Species	GF	1 C.S	2 C.S	3 C.S
Vine				
Mikania scandens	H19	-	3.2	+.1

Number of relevés: 3 Relevé size: 1 m X 1 m

SHRUB SPECIES PRESENT, BUT NOT IN ANALYSIS: None.

HERB LAYER DOMINANT: *Spartina patens*.
HERB LAYER PHYSIOGNOMY: Very tall, rhizomatous, rhizocarpic graminoids.

Herb Analysis

Species	GF	1 C.S	2 C.S	3 C.S
Tall forb				
Sabatia stellaris	H6	-	-	+.1
Very tall graminoids				
Andropogon glomeratus	H7	3.4	-	1.4
Spartina patens	H11	5.4	5.4	5.4
Tall graminoids				
Fimbristylis spadicea	H11	2.4	4.4	1.4
Setaria geniculata	H7	+.1	3.2	-

Number of relevés: 3 Relevé size: 1 m x 1 m

HERB SPECIES PRESENT, BUT NOT IN ANALYSIS: None.

COMMUNITY REFERENCES: None.

COMMUNITY DOCUMENTATION: Site analyzed by A. E. Radford's Botany 235 class, UNC, Chapel Hill, in fall 1977. Specimens deposited in the NCU Herbarium by T. A. Atkinson in 1977.

COMMUNITY ECOLOGICAL CHARACTERIZATION: Vegetationally: Poaceous wetland grass system with sparse, scattered, twining, deciduous vines and a closed herb layer dominated by very tall, rhizomatous, rhizocarpic graminoids. Climatically: Warm temperate, mesothermal climate, similar to the local yearly warm and moist climate, which has moderately hot and moderately wet summers, warm and moderately dry winters, and a very long freeze-free period. Pedologically: Mixed, thermic Typic Psammaquent, Carteret soil. Geologically: A clastic, sedimentary, estuarine tidal marsh sand flat deposit, composed of siliceous, porous, unconsolidated, quartz sands. Hydrologically: Somewhat excessively drained, regularly flooded, intertidal, polyhaline estuarine system, wetted by haline tidal waters and fresh rains. Topographically: Open, variously exposed, smooth, nearly level, sand flat on the upper portion of a tidal marsh on the northwest side of a water-deposited barrier island located on the seaward edge of a flat plain. Temporally and Spatially: A topoedaphic climax of a hydrosere on the upper portion of The Plains tidal marsh, underlain by Recent tidal flat sediments, on Ocracoke Island in the Embayed Section of the Coastal Plain Province of the Atlantic Plain Region.

Community Diversity Summary 11

Wetland grass system
Tall, rhizomatous, rhizocarpic graminoids

Fimbristylis marsh
Cyperaceae

FIMBRISTYLIS SPADICEA
Fimbristylis spadicea

CLIMATE: A. Mesothermal; AA. Warm temperate. B. Cool and moist yearly, moderately hot and moderately wet summers, moderately warm and moderately dry winters, with a very long freeze-free period; BB. Warm and moist yearly, moderately hot and moderately wet summers, warm and moderately dry winters, with a very long freeze-free period. C. Similar to local climate; CC. Similar to natural area climate.
SOILS: A. Entisol; AA. Aquent. B. Psammaquent; BB. Typic Psammaquent. C. Mixed, thermic Typic Psammaquent; CC. Carteret.
GEOLOGY: A. Sedimentary; AA. Clastic. B. Siliceous; BB. Estuarine tidal marsh sand deposit. C. Mixture of eolian sands and water-deposited, fine clastics and organics; CC. Porous, unconsolidated, quartz sand.
HYDROLOGY: A. Estuarine; AA. Intertidal. B. Haline; BB. Regularly flooded. C. Mixed estuarine and terrestrial systems ranging from regularly flooded to permanently exposed; CC. Somewhat excessively drained, euhaline system, wetted by haline tidal waters and fresh rain waters.
TOPOGRAPHY: A. Plains; AA. Flat plain. B. Water-deposited barrier island; BB. Water-deposited and organically accumulated estuarine tidal marsh. C. Mixture of tidal flats and active and relict dunes; CC. Open, variously exposed, smooth, nearly level, sand flat on the upper portion of the tidal marsh.
PHYSIOGRAPHY: A. Atlantic Plain; AA. Coastal Plain. B. Embayed Section; BB. Ocracoke Island. C. The Plains tidal marsh; CC. Upper tidal flats, underlain by Recent fine clastics and organics.

CANOPY: No canopy species present.

SUBCANOPY: No subcanopy species present.

SHRUB LAYER DOMINANTS: None.

SHRUB LAYER PHYSIOGNOMY: Not applicable.

Shrub Analysis

Species	GF	1 C.S	2 C.S	3 C.S	4 C.S	5 C.S
Tall shrubs						
Borrichia frutescens	H11	4.4	3.4	-	-	-
Iva frutescens	H6	-	4.3	-	-	-

Number of relevés: 5 Relevé size: 3 m X 3 m

SHRUB SPECIES PRESENT, BUT NOT IN ANALYSIS: None.

HERB LAYER DOMINANT: *Fimbristylis spadicea*.
HERB LAYER PHYSIOGNOMY: Tall, rhizomatous, rhizocarpic graminoids.

Herb Analysis

Species	GF	1 C.S	2 C.S	3 C.S	4 C.S	5 C.S
Tall forbs						
Aster tenuifolius	H11	-	+.1	-	1.1	+.1
Limonium carolinianum	H6	-	-	-	1.1	+.1
Pluchea purpurascens	H6	-	-	+.1	-	-
Salicornia virginica	H10	-	-	+.1	1.1	+.1
Very tall graminoid						
Spartina patens	H11	5.4	2.4	5.4	1.4	2.4
Tall graminoid						
Fimbristylis spadicea	H11	3.4	5.4	5.4	5.4	5.4

Number of relevés: 5 Relevé size: 1 m X 1 m

HERB SPECIES PRESENT, BUT NOT IN ANALYSIS: None.

COMMUNITY REFERENCES: None.

COMMUNITY DOCUMENTATION: Site analyzed by A. E. Radford's Botany 235 class, UNC, Chapel Hill, in fall 1977. Specimens deposited in the NCU Herbarium by T. A. Atkinson in 1977.

COMMUNITY ECOLOGICAL CHARACTERIZATION: Vegetationally: Cyperaceous wetland grass system with a sparse shrub layer of tall, rhizomatous shrubs and a closed herb layer dominated by tall, rhizomatous, rhizocarpic graminoids. Climatically: Warm temperate, mesothermal climate, similar to the local yearly warm and moist climate, which has moderately hot and moderately wet summers, warm and moderately dry winters, and a very long freeze-free period. Pedologically: Mixed, thermic Typic Psammaquent, Carteret soil. Geologically: Clastic, sedimentary, estuarine tidal marsh sand deposit, composed of siliceous, porous, unconsolidated, quartz sands. Hydrologically: Somewhat excessively drained, regularly flooded, intertidal, euhaline estuarine system, wetted by haline tidal waters and fresh rains. Topographically: Open, variously exposed, smooth, nearly level, sand flat on the upper portion of the tidal marsh on the northwest side of a water-deposited barrier island located on the seaward edge of a flat plain. Temporally and Spatially: A topoedaphic climax of a hydrosere on the upper portion of The Plains tidal marsh, underlain by Recent tidal flat sediments, on Ocracoke Island in the Embayed Section of the Coastal Plain Province of the Atlantic Plain Region.

Community Diversity Summary 12

Wetland forb system	Rush marsh
Very tall, cespitose, deciduous forbs	Juncales

JUNCUS ROEMERIANUS
Juncus roemerianus

CLIMATE: A. Mesothermal; AA. Warm temperate. B. Cool and moist yearly, moderately hot and moderately wet summers, moderately warm and moderately dry winters, with a very long freeze-free period; BB. Warm and moist yearly, moderately hot and moderately wet summers, warm and moderately dry winters, with a very long freeze-free period. C. Similar to local climate; CC. Similar to natural area climate.

SOILS: A. Histosol; AA. Saprist. B. Medisaprist; BB. Typic Medisaprist. C. Euic, thermic Typic Medisaprist; CC. Hobonny.

GEOLOGY: A. Sedimentary; AA. Clastic and organic. B. Siliceous and sulfaceous; BB. Estuarine tidal marsh deposit. C. Mixture of eolian sands and water-deposited, fine clastics and organics; CC. Porous, unconsolidated, water-saturated, sandy muck.

HYDROLOGY: A. Estuarine; AA. Intertidal. B. Haline; BB. Irregularly flooded. C. Mixed estuarine and terrestrial systems ranging from regularly flooded to permanently exposed; CC. Wetted by haline tidal waters and fresh rains.

TOPOGRAPHY: A. Plains; AA. Flat plain. B. Water-deposited barrier island; BB. Water-deposited and organically accumulated estuarine tidal marsh. C. Mixture of tidal flats and active and relict dunes; CC. Open, variously exposed, nearly level, upper margin of the upper tidal flats.

PHYSIOGRAPHY: A. Atlantic Plain; AA. Coastal Plain. B. Embayed Section; BB. Ocracoke Island. C. The Plains tidal marsh; CC. Upper margin of tidal marsh, underlain by Recent fine clastics and organics.

CANOPY: No canopy species present.

SUBCANOPY: No subcanopy species present.

SHRUB LAYER: No shrub species present.

HERB LAYER DOMINANT: *Juncus roemerianus*.
HERB LAYER PHYSIOGNOMY: Very tall, cespitose, deciduous forbs.

Herb Analysis

Species	GF	1 C.S	2 C.S	3 C.S
Very tall forb				
Juncus roemerianus	H7	5.5	5.5	5.5

Number of relevés: 3 Relevé size: 1 m X 1 m

HERB SPECIES PRESENT, BUT NOT IN ANALYSIS: None.

COMMUNITY REFERENCES: None.

COMMUNITY DOCUMENTATION: Site analyzed by A. E. Radford's Botany 235 class, UNC, Chapel Hill, in fall 1977. Specimens deposited in the NCU Herbarium by T. A. Atkinson in 1977.

COMMUNITY ECOLOGICAL CHARACTERIZATION: Vegetationally: Juncalean wetland forb system with a closed herb layer dominated by very tall, cespitose, deciduous forbs. Climatically: Warm temperate, mesothermal climate, similar to the local yearly warm and moist climate, which has moderately hot and moderately wet summers, warm and moderately dry winters, and a very long freeze-free period. Pedologically: Euic, thermic Typic Medisaprist, Hobonny soil. Geologically: A mixed clastic and organic, estuarine tidal marsh deposit, composed of siliceous and sulfaceous, porous, unconsolidated, water-saturated, sandy muck. Hydrologically: Irregularly flooded, intertidal, haline estuarine system, wetted by haline tidal waters and fresh rains. Topographically: Open, variously exposed, nearly level, upper margin of the upper tidal flats on the northwest side of a water-deposited barrier island located on the seaward edge of a flat plain. Temporally and Spatially: A topoedaphic climax of a hydrosere on the upper portion of The Plains tidal marsh, underlain by Recent tidal flat sediments, on Ocracoke Island in the Embayed Section of the Coastal Plain Province of the Atlantic Plain Region.

Master Species Presence List

ANGIOSPERMS

Asterales

  Asteraceae

    Aster subulata L.
    A. tenuifolius L.
    Baccharis halimifolia L.
    Borrichia frutescens (L.) DC.
    Carduus spinosissimus Walter
    Gnaphalium obtusifolium L.
    Iva frutescens L.
    Mikania scandens (L.) Willd.
    Pluchea purpurascens (Swartz) DC.
    Solidago sempervirens L.

Capparales

  Brassicaceae

    Cakile edentula (Bigelow) Hooker

Caryophyllales

  Amaranthaceae

    Amaranthus pumilus Raf. 2475*

  Cactaceae

    Opuntia compressa (Salisbury) MacBride

  Chenopodiaceae

    Salicornia bigelovii Torrey
    S. europaea L.
    S. virginica L.

Celastrales

  Aquifoliaceae

    Ilex vomitoria Aiton

Commelinales

  Commelinaceae

    Commelina communis L.

Cyperales

  Cyperaceae

    Fimbristylis spadicea (L.) Vahl.

  Poaceae

    Ammophila breviligulata Fernald
    Andropogon glomeratus (Walter) BSP.
    Cenchrus tribuloides L.
    Distichlis spicata (L.) Greene
    Eragrostis spectabilis (Pursh) Steudel
    Muhlenbergia capillaris (Lam.) Greene
    Panicum amarum Ell.
    Setaria geniculata (Lam.) Beauvois
    Spartina alterniflora Loisel.
    S. patens (Aiton) Muhl.
    Triplasis purpurea (Walter) Chapman
    Uniola paniculata L.

Euphorbiales

  Euphorbiaceae

    Croton punctatus Jacquin
    Euphorbia polygonifolia L.

Gentianales

  Gentianaceae

    Sabatia stellaris Pursh

Juncales

  Juncaceae

    Juncus roemerianus Scheele
    J. validus Coville

Lamiales

  Verbenaceae

    Lippia nodiflora (L.) Michaux

Liliales

  Liliaceae

    Smilax auriculata Walter
    S. bona-nox L.
    Yucca filamentosa L. var. *filamentosa*

Myricales

  Myricaceae

    Myrica cerifera L.

Myrtales

  Onagraceae

    Oenothera humifusa Nuttall

---

*Collections of T. A. Atkinson, Department of Botany, University of North Carolina, Chapel Hill.

Plumbaginales
Plumbaginaceae
Limonium carolinianum (Walter) Britton

Polemoniales
Solanaceae
Physalis viscosa L. ssp. *maritima* (M. A. Curtis) Waterfall
Solanum gracile Link

Rhamnales
Vitaceae
Parthenocissus quinquefolia (L.) Planchon

Rosales
Fabaceae
Strophostyles helvola (L.) Ell.

Rubiales
Rubiaceae
Diodia teres Walter

Salicales
Salicaceae
Salix caroliniana Michaux

Sapindales
Rutaceae
Zanthoxylum clava-herculis L. 2474

Umbellales
Apiaceae
Hydrocotyle bonariensis Lam.

GYMNOSPERMS
Pinales
Cupressaceae
Juniperus virginiana L.

E

Significant Site Reconnaissance Report

Roanoke River Bluffs*

Halifax County, North Carolina; along the south bank of the Roanoke River about 0.5 mile NE of intersection of NC 561 and SR 1148.
1.5 miles SE of town of Halifax.
36°18'30" N, 77°34'00" W; Halifax, N.C., quad. 7.5', 1974
Wicomico and Penholoway Terraces; Atlantic Coastal Plain; Coastal Plain
33'-75' elevation
ca. 50 acres

Ownership: Georgia-Pacific Corporation.

Administration: Same as ownership.

Land use: Current usages are primarily hunting and some fishing activities. Extensive logging operations have occurred on the level portions of the tract up to the edges of the slopes and bluff adjacent to the river. The slopes and bluff have been disturbed minimally by some selective cutting.

Land classification: Unknown.

Danger to integrity: The site is in immediate jeopardy. The Georgia-Pacific Corporation acquired the tract several years ago as a building site for a proposed hardwood pulp mill. There are conflicting reports concerning the company's present plans for the site. An adjacent property owner says the company has decided not to build and is looking for a suitable buyer to develop the tract as an industrial site. Local newspapers claim that Georgia-Pacific is waiting for Halifax County to implement a countywide water system before construction will begin. The company planned to dam several embayments near the river to develop wastewater settling ponds. This would inundate much of the mixed hardwood community containing many disjunct species and destroy a portion of the *Trillium sessile* population, unknown elsewhere in North Carolina. This would also annihilate a three-acre natural pond containing a beaver colony and *Wolffia papulifera* and *Wolffia columbiana*, both endangered peripheral species in North Carolina.

Public sensitivity: Halifax Community College, 10 miles away, has expressed an interest in using the site for field trips and as an outdoor classroom for its biology and ecology classes. A recent nature hike organized by Merrill Lynch was attended by over fifty local residents, who expressed much interest in seeing the site and its features preserved in a natural state.

---

*Data compiled and report written by J. Merrill Lynch (539 Henry Street, Roanoke Rapids, N.C. 27870). The format follows that prescribed by Radford (1977). Inventory work done under contract with the North Carolina Natural Heritage Program, North Carolina Department of Natural Resources and Community Development, Raleigh, N.C.

Significance and protection priority: Appears to be nationally significant. The site is in immediate jeopardy.

Reasons for priority rating:

Endangered or threatened species and communities:

1. Presence of *Lutra canadensis* (River Otter), which is a candidate for national threatened status by the Office of Endangered Species, Fish and Wildlife Service.
2. Presence of *Marmota monax* (Woodchuck), disjunct from mountain populations.
3. Presence of *Urtica chamaedryoides* (Stinging Nettle) and *Trillium sessile* (Sessile Trillium), only known populations in North Carolina.
4. Presence of at least four species that are on the North Carolina list of endangered and threatened plants: *Carex jamesii*, *Isopyrum biternatum*, *Wolffia papulifera*, *Wolffia columbiana*, endangered peripheral species.
5. Presence of at least five species that are on the North Carolina list of endangered and threatened birds: *Cathartes aura* (Turkey Vulture), *Coragyps atratus* (Black Vulture), *Buteo lineatus* (Red-shouldered Hawk), *Meleagris gallopava* (Turkey), *Dendroica cerulea* (Cerulean Warbler).
6. Presence of *Dendroica cerulea* (Cerulean Warbler), one of only two known breeding populations in the Coastal Plain south of Maryland, and disjunct by more than two hundred km from the nearest mountain populations.
7. Presence of at least 40 species of vascular plants disjunct from populations in mountains and Piedmont. Several disjunctions of more than two hundred km.
8. Highly unusual mixture of southern, mesic canopy species (*Quercus michauxii*, *Celtis laevigata*) and a northern, montane herbal assemblage (*Dicentra* spp., *Hydrophyllum canadense*, *Isopyrum biternatum*). This intermingling of southern and northern floral elements is probably a result of plant community migrations during the Pleistocene glaciations.
9. Transition area between mixed hardwoods "cove" forest, alluvial floodplain forest, and swamp forest.
10. Excellent example of community type: *Acer negundo/Asimina triloba-Lindera benzoin/Nemophila microcalyx*.
11. Unusually large specimens of *Quercus michauxii* (Swamp Chestnut Oak, dbh 3.5 ft.), *Quercus rubra* (Northern Red Oak, dbh 5.5 ft.), *Acer saccharum* (Sugar Maple, dbh 2.5 ft.), *Aesculus sylvatica* (Yellow Buckeye, dbh 1.0 ft.), *Carya* sp. (hickory, dbh 3.5 ft.), *Platanus occidentalis* (Sycamore, dbh 4.5 ft.).
12. Regionally significant occurrence of Alfisols, Udalfs, Hapludalfs. This soil type has a high base content (pH 6.5-8.0) and is associated with a diverse assemblage of plant species that occur on circumneutral-basic soils. Many of these species are disjunct from populations in the Piedmont and/or mountains. Examples include *Adiantum pedatum* (Maidenhair Fern), *Campanula americana* (Bellflower), *Trillium sessile* (Sessile Trillium), *Urtica chamaedryoides* (Stinging Nettle), *Dicentra cucullaria* (Dutchman's Breeches), *Dicentra canadensis* (Squirrel Corn).
13. Naturally blocked embayment drainage. Siltation by the Roanoke River has blocked the drainage of several creeks resulting in formation of embayments. One is a permanent three-acre pond and the other a temporary pool during periods of heavy rainfall.
14. Bluffs along the Roanoke River are of geological interest.

General diversity: The natural area is an excellent example of a relict, disjunct mixed hardwood community more commonly occurring in the Piedmont and mountains. The high base content of the soil (soil order: Alfisols) is highly unusual for the Coastal Plain. The combination of the rich, basic soil and the primarily north-facing slopes creates a unique microenvironment. This is reflected by the high percentage of disjunct plant species that have distributions primarily in the mountains and/or Piedmont of North Carolina. An alluvial floodplain community is immediately adjacent to the mixed hardwood slope community. The soil of the alluvial community also has a high base content and supports a rich diversity of calcareous plants. Some slopes more distant from the river contain American Beech as the dominant canopy species and also a much different herb assemblage than the slopes mentioned above. Soil is coarser on these slopes, contains a higher percentage of quartzite pebbles, and is therefore more acidic. Some *Kalmia latifolia* (Mountain Laurel) occurs here along with *Galax aphylla* (Galax) and *Hepatica americana* (Hepatica), species that occur on more acidic soils. The natural area is an excellent site for the study of the association of soil pH, soil texture, and slope aspect with various plant communities.

Management recommendations: The natural area should be preserved partly as wilderness and partly managed. Management plans should include consideration of endangered species preservation, scientific research and educational resources, and unique community preservation. Further study is needed of the environmental parameters involved in the distribution of the basic soils and the associated distinctive plant community.

Data sources:

1. Bill Johnson, Halifax County Forest Ranger, Scotland Neck, N.C.
2. Albert E. Radford, Department of Botany, UNC-CH.
3. Wayne Shorter, Soil Conservation Service, Halifax, N.C.
4. Julie H. Moore, North Carolina Natural Heritage Program, Raleigh, N.C.

General scientific references:

Nesom, G. L., and M. Treiber. 1977. Beech-mixed hardwoods communities: A topoedaphic climax on the North Carolina Coastal Plain. Castanea 42: 119-40.

Radford, A. E. 1976. Vegetation-habitats-floras-natural areas in the southeastern United States: Field data and information. University of North Carolina Student Stores, Chapel Hill.

________. 1977. A natural area and diversity classification system. A standardized scheme for basic inventory of species, community, and habitat diversity. University of North Carolina Student Stores, Chapel Hill.

________, H. E. Ahles, and C. R. Bell. 1968. Manual of the vascular flora of the Carolinas. The University of North Carolina Press, Chapel Hill.

Richards, H. G. 1950. Geology of the Coastal Plain of North Carolina. Trans. Amer. Philos. Soc. 40:1-83.

General documentation and authentication: General field reconnaissance of the site was made in May-June and October-December 1978 by Merrill Lynch. Subsequent visits were made in March-May 1979 by Lynch, Julie Moore (staff botanist, North Carolina Natural Heritage Program), and A. E. Radford.

Recorded Flora

Aceraceae
Acer negundo
A. rubrum
A. saccharinum
A. saccharum ssp. *floridanum*

Amaryllidaceae
Zephyranthes atamasco
Hypoxis hirsuta

Anacardiaceae
Rhus radicans

Annonaceae
Asimina triloba

Apiaceae
Hydrocotyle sp.
**Osmorhiza longistylis*
**Chaerophyllum procumbens* var. *procumbens*

Aquifoliaceae
Ilex opaca

Araceae
Peltandra virginica
Arisaema dracontium

Aristolochiaceae
**Asarum canadense*
**Hexastylis virginica*
**H. minor*

Aspidiaceae
Athyrium asplenioides
Cystopteris protrusa
Polystichum acrostichoides

Aspleniaceae
Asplenium platyneuron

Asteraceae
Prenanthes altissima
Hieracium venosum
Krigia dandelion
Cacalia atriplicifolia
Elephantopus tomentosus
E. carolinianus
**Eupatorium rugosum*
E. coelestinum
Antennaria sp.
**Erigeron pulchellus*
Aster sp.
Solidago sp.
Chrysogonum virginianum var. *virginianum*
Verbesina sp.
Arnica acaulis

Balsaminaceae
Impatiens capensis

Berberidaceae
Podophyllum peltatum

Betulaceae
**Corylus americana*
Betula nigra
Carpinus caroliniana
Ostrya virginiana

Bignoniaceae
Campsis radicans
Anisostichus capreolata

Boraginaceae
Myosotis macrosperma

Brassicaceae
**Cardamine concatenata*

Campanulaceae
**Campanula americana*

Caprifoliaceae
Lonicera japonica
L. sempervirens
Sambucus canadensis

Caryophyllaceae
**Stellaria pubera*
S. media
**Silene virginica*

---

*Piedmont and/or mountain disjuncts.

Celastraceae
Euonymus americanus

Commelinaceae
Commelina virginica

Cornaceae
Cornus florida

Cupressaceae
Juniperus virginiana

Cyperaceae
Carex jamesii
C. oligocarpa
C. laxiculmis
C. digitalis
C. albursina

Diapensiaceae
Galax aphylla

Ericaceae
Chimaphila maculata
Rhododendron nudiflorum
Kalmia latifolia
Oxydendrum arboreum
Vaccinium sp.

Fabaceae
Cercis canadensis
Desmodium nudiflorum
Vicia sp.

Fagaceae
Fagus grandifolia
Quercus alba
Q. stellata
Q. michauxii
Q. rubra var. *rubra*
Q. falcata
Q. shumardii

Fumariaceae
Dicentra canadensis
D. cucullaria
Corydalis flavula

Geraniaceae
Geranium maculatum

Hamamelidaceae
Liquidambar styraciflua

Hippocastanaceae
Aesculus sylvatica

Hydrophyllaceae
Nemophila microcalyx
Hydrophyllum canadense

Iridaceae
Sisyrinchium sp.

Juglandaceae
Juglans nigra
Carya cordiformis
C. ovata

Juncaceae
Luzula acuminata

Lamiaceae
Pycnanthemum incanum
Scutellaria ovata
Lamium purpureum
Salvia lyrata

Lauraceae
Lindera benzoin

Lemnaceae
Spirodela oligorrhiza
Lemna perpusilla
Wolffia columbiana
W. papulifera

Liliaceae
Smilax sp.
Trillium sessile
Erythronium americanum
E. umbilicatum
Polygonatum biflorum
Uvularia sessilifolia

Loganiaceae
Gelsemium sempervirens

Loranthaceae
Phoradendron serotinum

Magnoliaceae
Liriodendron tulipifera

---

*Piedmont and/or mountain disjuncts.

Moraceae
Morus rubra

Myricaceae
Myrica cerifera

Nymphaeaceae
Nymphaea odorata

Nyssaceae
Nyssa sylvatica var. *sylvatica*
Nyssa aquatica

Oleaceae
Fraxinus americana var. *americana*
F. pennsylvanica var. *pennsylvanica*
F. caroliniana
Chionanthus virginicus
Ligustrum sinense

Onagraceae
Ludwigia decurrens

Ophioglossaceae
Botrychium virginianum

Orchidaceae
Cypripedium acaule
Goodyera pubescens
**Aplectrum hyemale*
**Tipularia discolor*

Orobanchaceae
**Epifagus virginiana*

Oxalidaceae
Oxalis dillenii
O. violacea

Papaveraceae
**Sanguinaria canadensis*

Phrymaceae
**Phryma leptostachya*

Phytolaccaceae
Phytolacca americana

Pinaceae
Pinus taeda
P. echinata

Platanaceae
Platanus occidentalis

Polypodiaceae
Polypodium polypodioides

Portulacaceae
**Claytonia virginica*

Pteridaceae
**Adiantum pedatum*

Ranunculaceae
**Delphinium tricorne*
**Isopyrum biternatum*
**Thalictrum thalictroides*
Ranunculus abortivus
**Hepatica americana*

Rhamnaceae
Berchemia scandens

Rosaceae
Potentilla sp.
Geum virginianum
Agrimonia sp.
Duchesnea indica
Rubus sp.
Amelanchier arborea

Rubiaceae
Houstonia caerulea
Galium sp.

Salicaceae
Salix nigra
Populus heterophylla

Saururaceae
Saururus cernuus

Saxifragaceae
**Hydrangea arborescens* var. *arborescens*
**Heuchera americana*
**Saxifraga virginiensis*

Scrophulariaceae
Verbascum thapsus
Agalinis purpurea

---

*Piedmont and/or mountain disjuncts.

Simaroubaceae
Ailanthus altissima

Staphyleaceae
**Staphylea trifolia*

Taxodiaceae
Taxodium distichum

Ulmaceae
Ulmus americana
U. rubra
Celtis laevigata

Urticaceae
Laportea canadensis
**Urtica chamaedryoides*

Violaceae
Viola pedata
V. papilionacea

Vitaceae
Parthenocissus quinquefolia
Vitis rotundifolia

Recorded Fauna

Mammals

White-tailed Deer (*Odocoileus virginianus*)	B
Woodchuck (*Marmota monax*)	B
Eastern Cottontail (*Sylvilagus floridanus*)	B
Eastern Gray Squirrel (*Sciurus carolinensis*)	B
Beaver (*Castor canadensis*)	B
Southern Flying Squirrel (*Glaucomys volans*)	B
Opossum (*Didelphis marsupialis*)	B
Eastern Mole (*Scalopus aquaticus*)	B
Raccoon (*Procyon lotor*)	B
River Otter (*Lutra canadensis*)	B?
Red Bat (*Lasiurus borealis*)	B

Birds

Wood Duck (B)
Black Duck (W)
Turkey Vulture (B)
Sharp-shinned Hawk (W)
Red-shouldered Hawk (B)
Turkey (B)
Green Heron (B)
Spotted Sandpiper (T)
American Woodcock (B)
Mourning Dove (B)
Screech Owl (B)
Barred Owl (B)
Chuck-will's-widow (B)
Belted Kingfisher (B)
Mallard (W)
Hooded Merganser (W)
Black Vulture (B)
Red-tailed Hawk (B)
Osprey (T)
Bob White (B)
Great Blue Heron (T)
Solitary Sandpiper (T)
Ring-billed Gull (W)
Yellow-billed Cuckoo (B)
Great Horned Owl (B)
Whip-poor-will (B)
Ruby-throated Hummingbird (B)
Common Flicker (B)

---

*Piedmont and/or mountain disjuncts.

Pileated Woodpecker (B)
Yellow-bellied Sapsucker (W)
Downy Woodpecker (B)
Great Crested Flycatcher (B)
Eastern Wood Pewee (B)
Blue Jay (B)
Fish Crow (T)
Tufted Titmouse (B)
Winter Wren (W)
Catbird (B)
American Robin (B)
Hermit Thrush (W)
Veery (T)
Golden-crowned Kinglet (W)
Cedar Waxwing (W)
White-eyed Vireo (B)
Red-eyed Vireo (B)
Prothonotary Warbler (B)
Black-throated Blue Warbler (T)
Parula Warbler (T)
Blackpoll Warbler (T)
Ovenbird (B)
Kentucky Warbler (B)
American Redstart (B)
Brown-headed Cowbird (B)
Scarlet Tanager (B)
Cardinal (B)
Indigo Bunting (B)
American Goldfinch (B)
Dark-eyed Junco (W)
Swamp Sparrow (W)
Red-bellied Woodpecker (B)
Hairy Woodpecker (B)
Eastern Kingbirg (B)
Acadian Flycatcher (B)
Rough-winged Swallow (T)
Common Crow (B)
Carolina Chickadee (B)
White-breasted Nuthatch (B)
Carolina Wren (B)
Brown Thrasher (B)
Wood Thrush (B)
Swainson's Thrush (T)
Blue-gray Gnatcatcher (B)
Ruby-crowned Kinglet (W)
Starling (T)
Yellow-throated Vireo (B)
Black-and-white Warbler (T)
Cerulean Warbler (B)
Myrtle Warbler (W)
Yellow-throated Warbler (B)
Pine Warbler (B)
Louisiana Waterthrush (B)
Hooded Warbler (B)
Common Grackle (B)
Orchard Oriole (B)
Summer Tanager (B)
Evening Grosbeak (W)
Blue Grosbeak (B)
Rufous-sided Towhee (B)
White-throated Sparrow (W)
Song Sparrow (W)

Key: B = Breeding resident
T = Transient
W = Winter resident

Reptiles and Amphibians

Eastern Mud Turtle (*Kinosternan subrubrum*)
Box Turtle (*Terrapene carolina*)
Yellow-bellied Turtle (*Chrysemys scripta*)
Eastern Painted Turtle (*Chrysemys picta*)
Ground Skink (*Leiolopisma laterale*)
Five-lined Skink (*Eumeces fasciatus*)
Fence Lizard (*Sceloporus undulatus*)
Red-bellied Water Snake (*Natrix erythrogaster*)
Northern Brown Snake (*Storeria dekayi*)
Eastern Garter Snake (*Thamnophis sirtalis*)
Black Rat Snake (*Elaphe obsoleta*)
Copperhead (*Agkistrodon contortrix*)
Fowler's Toad (*Bufo woodhousei fowleri*)
Cricket Frog (*Acris gryllus*)
Spring Peeper (*Hyla crucifer*)
Gray Treefrog (*Hyla versicolor*)
Bullfrog (*Rana catesbeiana*)
Southern Leopard Frog (*Rana utricularia*)

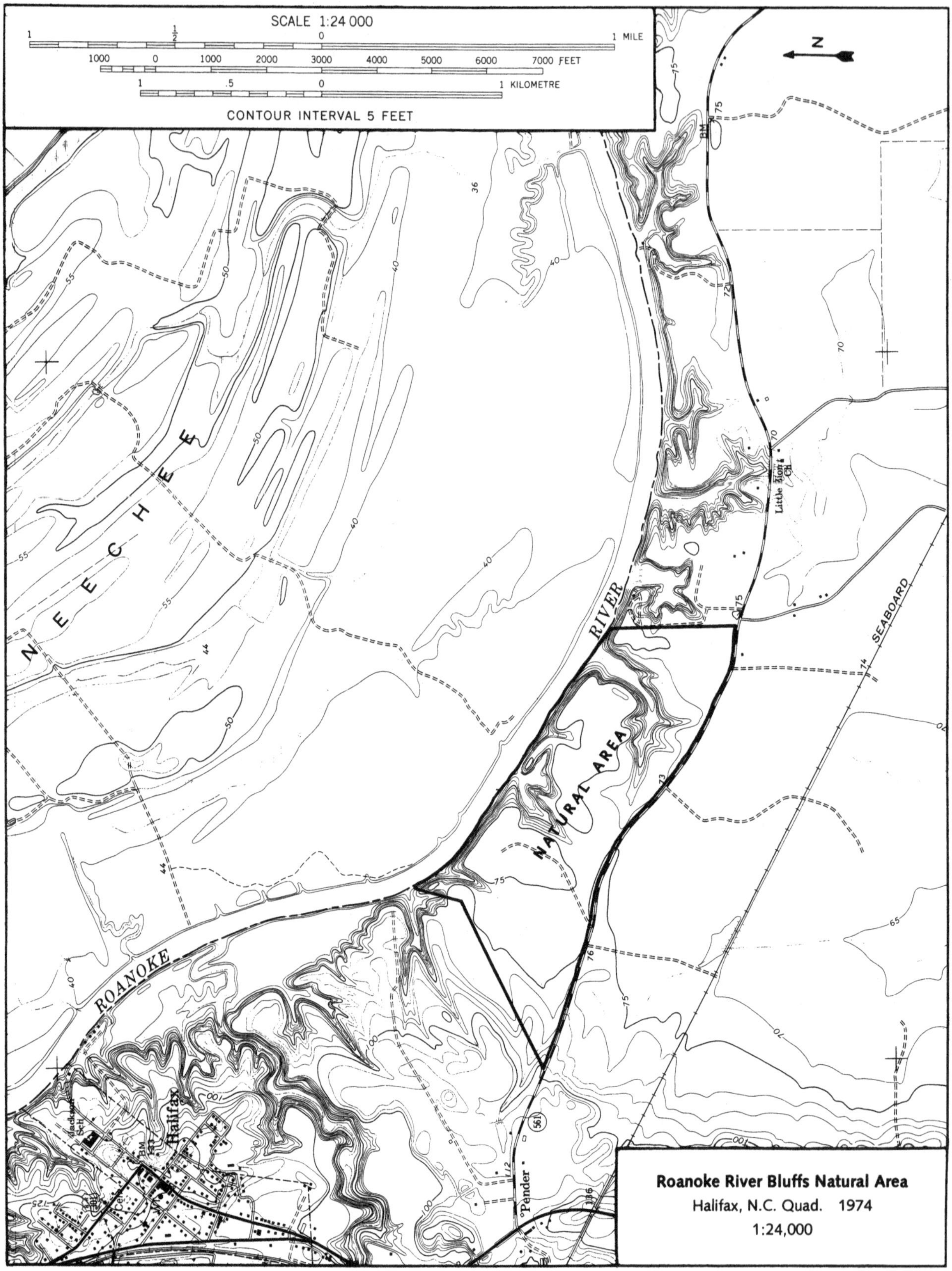

Map E1. *Location of Roanoke River Bluffs Natural Area.*

F

Field Data Reconnaissance Report

South Hyco Creek*

14 June 1978

Location: 12 Person County, N.C. USGS 7.5' Quadrangle: Hurdle Mills, N.C.

Description: South Hyco Creek
Slope: 20°-30° Aspect: Mostly SE Elevation: 360'-500'
Geology: Precambrian or Lower Paleozoic-Carolina Slate Belt, mafic volcanics: chiefly tuffs, breccias, and flows; also felsic fragmental and flow material. In part of sedimentary origin; lenses of bedded slate.
Other: pH 5.0.

CANOPY: BEECH-MIXED HARDWOODS

Acer rubrum
Carya ovata
Carya tomentosa 24", 29"
Fagus grandifolia 24"
Juglans nigra
Liquidambar styraciflua
Liriodendron tulipifera
Nyssa sylvatica
Quercus alba 18", 30"
Quercus rubra 27"

SUBCANOPY: MIXED

Carpinus caroliniana
Cercis canadensis
Cornus florida
Diospyros virginiana
Hamamelis virginiana
Juniperus virginiana
Morus rubra
Pinus virginiana

SHRUBS: NO DOMINANTS

Crataegus sp.
Euonymus americanus
Ilex decidua
Rhododendron nudiflorum
Rhus radicans
Vaccinium vacillans
Viburnum acerifolium
Viburnum rafinesquianum

TRANSGRESSIVES

Acer saccharum
Carya tomentosa
Prunus serotina
Sassafras albidum

VINES

Lonicera japonica
Parthenocissus quinquefolia
Rhus radicans
Vitis rotundifolia

*Report compiled by Alan S. Weakley, Department of Botany, UNC-Chapel Hill, and Jimmy R. Dickerson, North Carolina State University, Raleigh.

HERBS: MIXED

Agrostis sp.
Botrychium virginianum
Chimaphila maculata
Chrysogonum virginianum
Cimicifuga racemosa
Dianthus armeria
Epifagus virginiana
Euphorbia corollata
Galium circaezans
Goodyera pubescens
Hepatica americana
Heuchera americana
Panicum sp.
Polystichum acrostichoides
Silene virginica
Smilacina racemosa
Thalictrum thalictroides
Thelypteris hexagonoptera
Uvularia perfoliata

G

Field Data Reconnaissance Report

North Mayo River*

3 July 1978

Location: 29 Henry County, Va. USGS 7.5' Quadrangle: Spencer, Va.

Description: Slope on N. Mayo River
Slope: 40°-60° Aspect: NNW Elevation: 820'-900'
Geology: Precambrian or Paleozoic--Leatherwood granite: biotite, muscovite granite, locally porphyritic. Includes in N.C., Paleozoic--Granite: massive to weakly foliated, even-grained to porphyritic granitic rocks.
Other: pH 6.0; many fox burrows, cut over.

CANOPY: *FAGUS*-MIXED, locally NO LAYER
Acer rubrum 14", 15", 24"
Carya glabra 10"
Carya sp.
Fagus grandifolia 12"
Fraxinus pennsylvanica 10"
Liriodendron tulipifera 12"
Magnolia acuminata 11"
Oxydendrum arboreum 8", 10"
Quercus rubra 15"
Robinia pseudo-acacia 18", 17"
Sassafras albidum 20"--Ht. 80'
Tilia heterophylla 12"--Ht. 13'

SUBCANOPY: *CARPINUS*
Asimina triloba
Carpinus caroliniana 10"
Cercis canadensis
Cornus florida
Hamamelis virginiana 4"--Ht. 25'
Lindera benzoin 1"--Ht. 20'
Liriodendron tulipifera
Magnolia acuminata

SHRUBS: *LINDERA*
Kalmia latifolia
Lindera benzoin (very large-20')
Rhododendron maximum
Viburnum acerifolium
Viburnum prunifolium

TRANSGRESSIVES
Acer rubrum
Cercis canadensis
Fraxinus sp.

VINES
Lonicera japonica
Parthenocissus quinquefolia
Rhus radicans

---

*Report compiled by Alan S. Weakley, Department of Botany, UNC-Chapel Hill, and Jimmy R. Dickerson, North Carolina State University, Raleigh.

HERBS: MIXED

Adiantum pedatum
Arisaema triphyllum
Asarum canadense
Aster sp.
Athyrium thelypteroides
Botrychium virginianum
Cimicifuga racemosa
Heuchera americana
Impatiens capensis
Iris verna
Laportea canadensis
Podophyllum peltatum
Polystichum acrostichoides
Sanguinaria canadensis
Smilacina racemosa
Thalictrum thalictroides
Thelypteris hexagonoptera
Thelypteris noveboracensis
Viola sp.

Part III

Natural Heritage Resource Information

The treatments in Part III are presented to make the text more useful to academic and heritage workers. A specialized terminology has evolved in relation to conservation and heritage work, so a glossary has been included. The major effort of conservation today is through state natural heritage programs. An effective protection and preservation device for our national natural heritage is the registry of natural areas. A major computer standardization scheme is through the element file system. The major element files are those listed in the rare species classifications. The determination of natural boundaries, the grading of quality of a natural community, and the detection of past disturbances in natural areas are procedures basic to decisions on acquisition of property for protection and preservation.

A

Natural Heritage Glossary*

Conservation: The wise or rational use of resources (especially natural resources) emphasizing protection from loss or depletion.

Conservation Easements: Rights acquired over private property for the purpose of using or preserving certain identified resources or features.

Criterion (pl. criteria): A standard or guideline against which a resource can be evaluated.

Significance Criteria: Standards or guidelines developed for listing resources on the National Register of Natural Areas (proposed) at the national, state, or local significance level.

Critical Area: An area where uncontrolled or incompatible development could result in significant damage to the environment, life, or property, or the long-term public interest, which is of more than local significance (adapted from the proposed 1973 National Land Use Policy and Planning Assistance Act).

Critical Habitat: (1) The specific areas within the geographical area occupied by a listed threatened or endangered species on which are found those physical or biological features essential to the conservation of the species and which may require special management considerations or protection; and (2) specific areas outside the geographical area occupied by a listed threatened or endangered species, upon a determination by the secretary of the interior that such areas are essential for the conservation of the species (adapted from the Endangered Species Act Amendments of 1978, P.L. 95–632, 92 Stat 3751).

Diversity (Biological): The variety of plant and animal species located within a given geographical area; areas of high diversity are characterized by a great variety of species.

Natural Diversity: Plant and animal species that have evolved over the last 600 million years and the discrete types of terrestrial and aquatic communities and ecosystems in which these species live.

Habitat: The natural environment that sustains a plant or animal organism, population, or community.

Heritage:

National: That collection of resources important to Americans because they are significant aspects of our history and culture and/or significant elements of our natural environment.

Historic: The collection of districts, sites, buildings, structures, and objects significant in American history, architecture, archeology, or culture (adapted from the 1966 National Historic Preservation Act).

Natural: Representative examples of the full array of discrete types of terrestrial and aquatic communities, geologic features, landforms, and habitats of native plant and animal species that may be eliminated without deliberate protection (adapted from the proposed National Heritage Policy Act).

Inventory: A systematic collection of resource information based on a classification system.

Natural Heritage Inventory: A systematic classification and identification of the full array of terrestrial and aquatic communities, landforms, geological features, and habitats of native plant and animal species.

Landmark:

National Natural: A representative example of the nation's natural history, including terrestrial and aquatic communities, landforms, geological features, and habitats of native plant and animal species, judged by the secretary of the interior to possess national significance in illustrating or interpreting the nation's natural history.

Registry: The National Registry of Natural Landmarks, now administered by the Heritage Conservation and Recreation Service, which lists areas that have been designated by the secretary of the interior as national natural landmarks.

Landscape: The natural features, such as hills, forests, fields, and bodies of water, that distinguish one part of the earth's surface from another part; usually that portion of land that can be comprehended in a single view.

Cultural: Natural settings or geographic features that are historically associated with and significant to an understanding of other cultural resources because of

*Prepared by the Heritage Conservation and Recreation Service, U.S. Department of the Interior.

their proximity or relationship, such as Harper's Ferry, West Virginia; Mount Vernon, Virginia; or Walden Pond in Massachusetts.

Natural Area: An area of land or water that either retains or has reestablished its natural character and values, although it need not be completely undisturbed, which provides scientific, educational, recreational, or inspirational benefits.

Classification: A system that defines, identifies, and catalogs discrete terrestrial and aquatic communities, landforms, geological features, and habitats of native plant and animal species and provides the basis for the natural heritage inventory.

Ecological Reserve: A natural area that has been afforded some form of legal protection.

Experimental Ecological Reserves: A system of research areas surveyed by The Institute of Ecology that is dedicated to experimental research with manipulation and modification permitted.

Research Natural Areas: A system of natural areas administered by the Federal Committee on Ecological Reserves that seeks to establish a representative array of natural ecosystems and their inherent processes as baseline areas for educational and scientific research purposes.

Wilderness: A specific wilderness established and protected under the provisions of the Wilderness Act of 1964, as amended.

Natural Heritage Act: Refers to the proposed National Heritage Policy Act, which will provide the basis for a national program to identify, select, and protect natural areas and historic places of significance to the nation's heritage.

Natural Heritage Program: The natural component of the National Heritage Program, which will consist of a network of voluntary state natural heritage programs that will systematically classify, identify, and assess the range of terrestrial and aquatic communities, landforms, geological features, and habitats of native plant and animal species located within their boundaries.

Natural Heritage Registry: Refers to the proposed National Register of Natural Areas that would be established by the proposed National Heritage Policy Act. The proposed register will include natural areas of national, state, and local significance, with nationally significant natural areas designated as national natural landmarks, and will be administered by the Heritage Conservation and Recreation Service. The existing National Registry of Natural Landmarks will be incorporated into the proposed National Register of Natural Areas.

Preservation: A subset of the larger concept of conservation that refers to activities that encourage the maintenance, rehabilitation, or restoration of historic resources.

Natural Resource: The universe of natural entities that are useful or essential to mankind.

River:

Free-Flowing: Rivers or portions of rivers that are unmodified by the works of man, or, if modified, still retain their natural scenic qualities or recreation opportunities.

Recreational: Rivers or sections of rivers readily accessible by road or railroad that may have some development along their shoreline and may have undergone some impoundment or diversion in the past.

Scenic: Rivers or sections of rivers free of impoundments, with shorelines or watersheds still largely undeveloped but accessible in places by roads.

Wild: Rivers or sections of rivers free of impoundments and generally inaccessible except by trails, with watersheds or shorelines essentially primitive and waters unpolluted.

Species: A group of organisms living in similar surroundings that reproduce exclusively with each other.

Endangered: Any species that is in danger of extinction throughout all or a significant portion of its range (from the Endangered Species Act of 1973).

Endemic: Any species that is confined naturally to a certain limited area or region; native to a restricted locality.

Rare: Any species that has small or scattered populations throughout its range but has not been designated as endangered or threatened.

Threatened: Any species that is likely to become an endangered species within the foreseeable future throughout all or a significant portion of its range (from the Endangered Species Act of 1973).

Wetland: Land where the water table is at or near the surface for most of the year.

B

State Natural Heritage Programs:
Background Information*

The state natural heritage programs, undertaken in cooperation with state governments, are aimed at creating a continuing process for identifying significant natural areas and setting protection priorities within a given state. The innovative heritage inventory focuses first on the components, or elements, of natural diversity. For the purposes of the inventory, an "element" is defined as a natural feature of particular interest because it is exemplary, unique, or endangered on a statewide or national basis. For example, an element can be a Bald Eagle, a Ginseng plant, or a virgin stand of Longleaf Pine trees.

A classification of element "types," such as special animal species, is the cornerstone of the heritage inventory. Researchers catalog a state's vulnerable plant and animal species, plant communities, aquatic types, and critical habitats, as well as outstanding geologic features. Information on the existence, numbers, condition, status, and location of all significant examples can then be determined.

The element-based inventory approach is an advance in assuring a comprehensive inventory of a state's natural diversity. Most previous inventories have been conducted on a site-by-site basis, which means that researchers inventory a site that has already been judged an important area. Site inventories can mean that subjective judgment will substitute for intensive, objective research. Element-based inventories indicate important sites because critical elements are pinpointed on them. The heritage inventories can help assure that little-known areas rich in ecological diversity will not escape attention. Also, the ongoing, dynamic inventories continually update and refine data, so that their practical value is constantly enhanced.

A typical heritage program is conducted under an agreement set forth in a contract between the appropriate arm of a state government and The Nature Conservancy. Funding may be provided by one or more foundations and/or state agencies pooling their resources, matched by the U.S. Department of the Interior's Land and Water Conservation Fund. At the end of the contract period, a heritage program is usually integrated into state government. Both funding and administrative arrangements are flexible.

Since 1974, heritage programs have begun in seventeen states—South Carolina, Tennessee, West Virginia, Mississippi, North Carolina, Ohio, New Mexico, Oregon, Oklahoma, Washington, Indiana, Kentucky, Rhode Island, Wyoming, Massachusetts, Arkansas, and Minnesota and there is a Tennessee Valley Authority Region Heritage Program. Plans call for heritage programs in 25 states by 1982, which, in conjunction with TNC's other efforts in natural areas inventory, will enable it to form a national system.

The Conservancy, as well as other organizations, will use the information generated by the inventories to guide land protection efforts. In addition to actual acquisition, the information yielded by the inventories can assist planners and developers in decision making so that needless conflicts between conservation and development interests can be avoided. In fact, the heritage programs have already proved to be of practical value to the agencies and organizations that initially supported them.

---

*Prepared by Joy Davis, The Nature Conservancy, 1800 N. Kent St., Arlington, Virginia 22209.

C

Registry of Natural Areas

North Carolina*

SECTION .0200—
REGISTRY OF NATURAL HERITAGE AREAS

.0201 OBJECTIVES OF REGISTRY

The Registry of Natural Heritage Areas is a recognition program based upon an official list of significant natural areas derived from the Natural Heritage Program's inventory of elements of natural diversity. The registry is a voluntary, non-regulatory, non-binding recognition program. Objectives of the Registry of Natural Heritage Areas are to:

(1) protect significant examples of the total diversity of natural features occurring in the state;
(2) establish reserves for breeding stocks of endangered, threatened, or otherwise unique species of plants and animals;
(3) encourage educational activities and scientific research;
(4) preserve unique and unusual natural features;
(5) protect natural areas against uses which would destroy their natural conditions.

History Note: Statutory Authority G.S. 113–3; 113–8; Eff.

.0202 CRITERIA FOR ELIGIBILITY

(a) For an area to qualify as a Natural Heritage Area and thus be eligible for registration, it must possess one or more of the following natural values:

(1) a habitat for individual species of plants or animals that are in danger of or threatened by extirpation;
(2) an exemplary terrestrial plant community;
(3) an exemplary aquatic community type;
(4) an outstanding geologic or geomorphic feature that illustrates geologic processes or the history of the earth;
(5) a unique or unusual natural feature such as virgin conditions or unusual vegetation types;
(6) other biological or ecological phenomena of significance, such as a major bird rookery or bat colony.

(b) Furthermore, an area shall be evaluated with respect to the following factors:

(1) presence of natural values not adequately represented in previously registered Natural Heritage Areas;
(2) diversity of natural types of flora and fauna;
(3) quality and viability of the natural features (*i.e.*, self-sufficiency of the natural ecosystem when properly managed; degree of vulnerability to disturbances and intrusions);
(4) absence of damaging land uses, logging, grazing, erosion, intrusion by exotic species, etc., or extent to which past disturbances have altered natural features. Considering that nearly all areas of the state have been altered by human intrusions to some extent and considering that certain natural elements require manipulative management, an area should not be denied recognition solely because of past disturbances;
(5) capability of being managed so as to protect and maintain natural features in a natural condition; a buffer zone is desirable to assure protection (a buffer zone, where possible, should follow naturally defensible boundaries and should help protect the site against adverse effects from the use and development of adjacent land; the buffer zone may be included in the designated area but need not itself possess special natural values);
(6) compatibility of protective management prac-

*North Carolina Department of Natural Resources and Community Development, North Carolina Administrative Code 15 12H.

tices with current use practices on adjacent lands;

(7) scientific and educational value.

History Note: Statutory Authority G.S. 113–3; 113–8;
Eff.

.0203 DESIGNATION PROCESS

(a) Nomination.

(1) Nominations for the Registry of Natural Heritage Areas may be made by the Natural Heritage Program staff, by other public agencies, by members of the natural areas advisory committee, and by any other resident or property owner of the state. Nominations shall be submitted to the North Carolina Natural Heritage Program.

(2) The Natural Heritage Areas Nomination Form, or equivalent information, shall be submitted in order to provide the Natural Heritage Program with general information on location, owner or administering agency, current use, and natural significance of proposed areas. The Natural Heritage Program staff shall, if necessary, conduct an on-site evaluation of a nominated area in order to gather additional information on which to determine that the area meets eligibility criteria. The nomination form can be obtained at the address of the Natural Heritage Program.

(3) After reviewing sufficient information on a nominated area, the Natural Heritage Program staff shall determine if an area qualifies for the registry and shall document its findings in an evaluation report with recommendations for action. For each nominated site worthy of registration, the Natural Heritage Program staff coordinator shall prepare a Statement of Significance and shall sign a Statement of Recommendation that the area is found eligible for the registry. The Natural Heritage Program staff coordinator, in cases when a site is found not to meet eligibility criteria, shall prepare a negative report explaining the decision. The natural areas advisory committee shall review both positive and negative recommendations.

(4) Nominations initiated by the public or other agencies shall be accepted or rejected (with a full explanation accorded upon rejection) by the department within six months of receipt. The nominator may petition for consideration again if significant new information is forwarded to the Natural Heritage Program.

(b) Notification of Landowner or Administrator. Once an area is nominated and appears potentially eligible for registration, the Natural Heritage Program staff shall notify the owner or administering agency. This notification may not be necessary at this point if the owner nominated or knew about the nomination of the property. The owner may request that the property be or not be considered further for registration.

(c) Review Process. All nominations and recommendation statements shall be submitted by the Natural Heritage Program to the natural areas advisory committee for its review and approval. The natural areas advisory committee shall receive and review nominations at its regularly scheduled meetings. The chairman or acting chairman of the advisory committee, upon committee approval of the nomination, shall sign the Statement of Recommendation before approval by the director of the division of parks and recreation and submission to the assistant secretary of the department. Before making his recommendation to the secretary, the assistant secretary of the department shall solicit reviews and comment upon the nomination from all appropriate divisions of the department and other appropriate agencies. Recommendation statements and a report on the owner's willingness to accept registration shall then be forwarded to the secretary for final decision on eligibility.

(d) Designation. Upon receipt of the recommendations from the Natural Heritage Program and natural areas advisory committee and reviews by appropriate divisions and agencies, the secretary shall have the option of approving or not approving an area as eligible for the registry. The registration of a site is ultimately the voluntary decision of the landowner or administering agency.

History Note: Statutory Authority G.S. 113–3; 113–8;
Eff.

.0204 REGISTRATION

(a) A natural area shall become officially registered when a voluntary agreement to protect and manage the site for its specified natural values has been signed by the owner and the secretary, according to requirements of this Rule. The owner shall be given a certificate signifying the inclusion of the area on the registry.

(b) After the secretary approves an area as eligible for registration, the Natural Heritage Program shall offer the owner or administering agency the opportunity of placing the designated part of the property on the registry in return for signing a non-binding agreement (promise of intent) to manage the site for the protection of the significant natural elements. Natural Heritage Area Registry status for an area shall become effective upon the signing of the letter of agreement by the secretary and the landowner or administering agency.

(c) In cases when an area recommended to the registry is administered by the department, the secretary shall have the decision of registering or not registering the area, upon receiving the recommendation of the management agency. In cases when another public agency other than the department is the administrator or owner of an area which is recommended to the registry, the registration will become effective upon the signing of the agreement by the secretary and the responsible executive of the administering agency.

(d) Upon signing of the agreement, the department shall present the owner or administering agency with a certificate which indicates the area is a registered Natural Heritage Area. The owner or a competent volunteer shall be requested to report to the Natural Heritage Program at least once a year on the condition of the area. The Natural Heritage Program shall maintain a file on each registered area that contains complete documentation, annual status reports, and management reports.

(e) The owner must be advised that it is his option to publicize the registration.

History Note: Statutory Authority G.S. 113–3; 113–8;
Eff.

.0205 RECISION

(a) The registration agreement may be terminated after 30 days from notification by either party. Such termination removes the area from the registry.

(b) The secretary may rescind recognition if the owner fails to carry out the promised protection practices. The written agreement between the landowner and the department shall request 30 days notification by either party before the agreement is terminated. The secretary has the authority to rescind registry status for any area on department administered property. Such action should occur only after it has been clearly shown that there is a higher, better and more important use for an area. Recision shall not affect existing statutory protection for an area.

(c) Anyone may petition the department to remove an area from the registry if he believes the site no longer deserves recognition. The petition for removal must explain the changes that have occurred since the area was registered. After considering the petition, the secretary, upon recommendation of the Natural Heritage Program staff and natural area advisory committee may order removal by signing a recision order.

(d) Recision shall remove the area from the registry, and the owner or administering agency shall be requested to return the certificate. Anyone who believes he is disadvantaged by any of the steps in the process described in this Rule may seek an administrative hearing as provided by the departmental administrative hearing rules as located in 15 NCAC 1B. 0200.

History Note: Statutory Authority G.S. 113–3; 113–8;
Eff.

.0206 PUBLIC ACCESS

Registration of a natural area does not provide or require rights of public access to a registered area. Visitors must obtain the permission of the landowner or managing agency before entering the property. The landowner or managing agency retains the option to restrict publicity and access to the property.

History Note: Statutory Authority G.S. 113–120.6;
113–3; 113–8;
Eff.

.0207 MANAGEMENT OF REGISTERED NATURAL HERITAGE AREAS

(a) The guiding standards for managing and using registered Natural Heritage Areas are to protect their natural values and to maintain the areas in as nearly a natural condition as possible. Because each area is likely to be dissimilar to all others in the natural elements present and in certain other aspects, the department administering registered natural areas should be encouraged to develop management plans for the sites. Management of an area shall be in a manner intended to protect or enhance its natural value. The department shall design a boundary sign for registered natural areas for the optional use by the owner or administering agency.

(b) Any owner of a registered Natural Heritage Area may request management advice from the department.

History Note: Statutory Authority G.S. 113–3; 113–8;
Eff.

D

Lowest Common Denominator Element File*

Abstract

The Nature Conservancy's Natural Heritage Program furnishes its users with the technical skills and methodologies they need to develop a comprehensive system for identifying and preserving ecologically significant natural features. The Lowest Common Denominator (LCD) approach to ecological inventory focuses on the elements of natural diversity. We feel confident that our element-by-element methodology provides the best means yet developed for identifying preservation and land use priorities.

The LCD Element File system (LCD-ELF) is an automated data base management system designed to facilitate the collection, management and use of ecological data. It was designed to act as a vehicle for efficient flow of data from the field to the user. The system consists of a data base and six program modules. The program modules offer the capability to create and manage the data base, permit modifications to individual records, provide retrieval and certain statistical and reporting capabilities, supply routines for plotting the ecogeographical data, and include a backup system for the data base.

The LCD record format is designed to incorporate the minimum amount of data necessary for analysis of the existence, quantity, status, and distribution of both individual elements and element ensembles. The format promotes rapid achievement of comprehensive data coverage by allowing an incremental approach to ecological inventory. It thus eliminates bottlenecks caused by voluminous record transcription and storage.

It is our belief that the element-by-element approach to ecological inventory augmented by the power and flexibility of the computer system represents a breakthrough in conservation methodology. This manual will introduce the user to our philosophy and provide all the instructions necessary for productive use of the LCD-ELF system.

A. Classification of the Elements of Natural Diversity

The Program begins by developing a classification of the elements of natural diversity that exist throughout the state. This classification serves as a skeletal structure for organizing natural landscape information so that the dissimilarity of certain elements is recognized and like elements are grouped to facilitate direct comparisons based upon objective data rather than subjective judgment.

The units of classification may be thought of as "targets" for both data collection and protection activities; they help to insure that nothing is overlooked in inventory process. The data category types include plant communities, aquatic habitats, geologic features, and endangered or otherwise significant special species.

As a working taxonomy, this classification retains the flexibility to be changed as the state's priorities dictate; individual categories may be added as they are identified, existing categories may be subdivided, and whole new classes of elements may be added, such as soil types or historic and cultural elements.

B. An Element-by-Element Approach to Ecological Inventory

Most natural areas inventories have been designed to operate on a *site* basis, which presents several difficulties not present in the *element* basis used by the Heritage Program. One difficulty lies in establishing ecological

*Excerpts from H. P. Moyseenko, R. E. Jenkins, S. Woodall, and L. A. Miller. 1977. Lowest common denominator element file: An information management system. 3rd ed. The Nature Conservancy, Arlington, Virginia.

boundaries; this is no simple matter, given the complexity of the natural environment. Another is that inventories usually define and then rate sites on simple "niftiness" basis rather than as a result of any systematic consideration of ecological needs. Most serious of all, a rigorous comparison of one site to another is impossible because each site is unique, by virtue of either its unduplicated characteristics or its particular juxtaposition of elements.

The Heritage approach circumvents these difficulties by focusing primarily on individual features or elements defined by the classification. Since data are collected on an element-by-element basis, we can easily compare elements of the same type wherever they occur. Herein lies the strength of our methodology: by comparing like things we can make more precise estimates of their significance, rarity, and of the relative quality of each element occurrence as a representative of a species or biotic community.

We feel that keying the inventory process to elements of natural diversity is the best approach advanced so far for identifying preservation priorities. We also believe that as the Heritage data base becomes more complete, the potential uses of the system will increase. The comparative methodology which helps us to identify and document the relative critical importance of various natural lands will be of great utility to land use planners. By mapping the collected data, we will provide a great deal of "ground level" information that can be used in conjunction with satellite-provided remote sensing data to find signatures—unique patterns—for particular elements. In turn, the remote sensing data will help us gain a more comprehensive appreciation of the ecological condition of the total land cover.

Yet another application of the Heritage methodology will be in the field of environmental impact assessment, which has long been hindered by a lack of state, regional or national perspective. It is impossible to judge the criticality or significance of any individual site (or alterations to that site) by reference to the site alone. By collecting standardized Heritage data in the environmental impact survey process, decision makers will be able to use the Heritage information system to gain the perspective necessary for estimating relative site significance. Our system is not intended to compete with broad-base state land use computer systems. Rather, it is designed to complement the land use planning process by refining the ecological input.

Through the implementation of the automated Lowest Common Denominator Element File (LCD-ELF) system, we can compile and organize a great deal of critical ecological data. We can analyze those data to set priorities for natural areas preservation and we can help effectively utilize limited financial resources for their actual protection. We have the advantages that we are dealing with real and definable entities in biological species and habitats and that we can collect data incrementally as financial resources allow. The LCD record format is relatively short, and standardized. It provides a framework for recording, storing, and retrieving the minimum amount of data necessary for analysis of element characteristics and distribution. Thus, it allows broad but efficient data coverage and reduces the data management bottlenecks frequently caused by extensive record compilation and transcription.

We collect data in distinct "quantum layers." In the first inventory swath, we compile the smallest amount of meaningful correlative data on as many significant elements as possible. After we store a comprehensive list of elements and their locations in the data base, we focus specifically on sites whose element ensembles are worthy of closer inspection. Only then do we conduct intensive field surveys and other supplemental data collection. We incorporate data which are not recorded in the LCD-ELF format into manual files.

C. Handling the LCD Element File: A General Discussion

When we speak of an element's existence relative to a geographical place, we call it an "element occurrence." This term clarifies the distinction between an element that has been written up in a study as existing *somewhere* in a state or region and the reported occurrence of an element on the ground. Each LCD record in the data base represents a body of information on a separate element occurrence.

The automated LCD-ELF system allows you to retrieve information on one element occurrence, or to retrieve information on all of the reported occurrences of a particular type of element. You can also obtain specific information about all element occurrences sharing some common attribute, such as location.

An example of the kind of search and retrieval request you could make would be for all information about occurrences of *Dendrocopos borealis* on highly protected areas in Hoke County. The computer program would select all element occurrence records that meet these criteria, and print in usable form the information it finds. You could also have an outline of Hoke County drawn by a CalComp plotter and have all reported occurrences of *Dendrocopos borealis* plotted within it. This drawing could be overlaid onto the U.S.G.S. 7.5 minute topographic map (or maps) which contains Hoke County, allowing you to easily determine where the species is reported to exist on the real landscape.

Computer programs for the LCD-ELF offer many other combinations of search and retrieval capabilities, which you will find discussed later in this document. You can see from the example given above that the LCD can make information accessible in a variety of ways that can be very useful for determining how rare an element is, how it is distributed on the real landscape, and how it relates to other features of the landscape.

We have designed the automated LCD-ELF system with the user in mind. Someone with little or no computer experience can easily manage and retrieve information from the LCD Element File by applying the techniques documented in this manual. Once you have studied this text, you should find it easy to input new data and to update, retrieve and display information that is currently in the system.

D. Lowest Common Denominator Element File: A Computerized Information Handler

Let us examine now several concepts about the uses of an information handler. Generally, an information handler should have provisions for managing, locating, protecting, and displaying data. First, we must create a data base by deciding what kinds of data to collect and then by conducting an inventory to obtain the data.

Second, we record data according to a prescribed format, and file it in a medium—an information handler. The medium could be index cards, or 8½ x 11 paper which is manually placed into file drawers or stored in separate shoe boxes, or an automated information management system. Regardless of application, any new data must be merged and catalogued with the current collection already assembled in existing files. Normally, new information is added according to a standard format. Third, the desired data should be easily retrievable from the storage medium.

In a manual system, the process of searching is fundamentally simple; however, because of time or cost limitations, the inclination to carry out a full search is often repressed. Unfortunately, these limitations determine what data will be used. Manually searching and locating data is tedious, for not only must the user search multiple files by hand, but the selected information must also be correlated with other general inputs. The ability to permute, correlate, cross-reference and interrelate stored data is important for analyzing the contents of a storage medium.

Let us now consider the many tasks to which a computer-based system might be applied. For example, a computer system could locate all the elements of ecological significance, recreational entities, etc. within a geographical base area. It could then determine which of these entities are relatively rare or abundant so that comparisons with other like entities could be made. Such tasks cannot be assigned to manual filing systems, simply because they are time-prohibitive. By using the computer system approach we can build a versatile and accessible ecological data base. Whenever additional data are obtained, they can be added to the computer files for permanent storage. The ecological data can be continually augmented and refined in response to additional data obtained from field surveys and evaluations. As the quantity of ecological data increases, the search and reporting capabilities offered by the computer program allows users to efficiently cross-reference, compare and interrelate the information.

The LCD-ELF system allows the user to retrieve data relevant to a type of element, and data on element occurrences. It is possible to determine which areas share common characteristics such as protectedness versus non-protectedness, and to pinpoint all land holdings that are owned by the federal or state government. Thus, the system is an invaluable aid in setting ecological priorities for site protection. The LCD-ELF provides an index of relative rarity by showing which elements have few or no reported occurrences. The relative measure of rarity or abundance of data types becomes more accurate as the system gains data.

The LCD-ELF system allows the user to generate maps and map overlays which identify and illustrate areas which have particular significance because of their element concentrations. For example, one might want to see the distribution of endangered species in relation to protected sites. The computer program can retrieve data from both categories and produce one map showing both the location of all the endangered species and the location of all the protected sites.

The above discussion was intended to provide the reader with an appreciation of automated environmental data management capabilities. However, one caveat: data bases are not intended to provide a full remedy for the indecisions, oversights and inadequacies of past resource management strategies; rather, the central theme of computerized data management systems is that decisions regarding natural ecological diversity can be quickly approached. However, penalties for errors in judgment will still remain. Computer systems cannot provide decision makers with all the facts they need. Computers may on occasion provide decision makers with a spurious impression of certainty. Computers should not be regarded as a panacea—even if machines may eventually compete with men in matters of decision making. The output of a system is only as refined as the input.

```
            INDEX KEY: SP_AQUILA_CHRYSAETOS_001      (V)
                STATE: TN                                          COORDINATES: 362700N874300W
               COUNTY: 161                                         USGS QUAD MAP NAME: CUMBERLAND CITY
           INDEX CODE: 11.507.001.3608746                          SPECIAL STATUS:
    NAME OF SITE/AREA: CROSS CREEKS NATIONAL WILDLIFE REFUGE       PROTECTION STATUS: 2
     NUMBER OF OWNERS: 01                                          DATE OF INFORMATION:  741123
      PRINCIPAL OWNER: US FWS                                      SIZE:
       SOURCE OF LEAD: JACK SITES TN HERITAGE PROGRAM
  GENERAL DESCRIPTION: GOLDEN EAGLE, SEEN ON BLUFF ON RIGHT SIDE OF GRAVEL RD ABOUT 2 MI UPSTREAM
                       FROM HEADQUARTERS, CCNWR NEAR DOVER, TN

CONTENTS OF MANUAL FILE:     DIRECTIONS     BOUNDARY ON TOPO     AERIALS     FIELD SURVEY     OWNER INFO     REFERENCES
                                                    B                                             O

            INDEX KEY: SP_AQUILA_CHRYSAETOS_002     (IV)
                STATE: TN                                          COORDINATES: 361800N87080W      (N)
           INDEX CODE: 11.507.002.3608732                          USGS QUAD MAP NAME: CHEATAM DAM
    NAME OF SITE/AREA: CHEATAM LAKE                                SPECIAL STATUS:
     NUMBER OF OWNERS: 01                                          PROTECTION STATUS: 2
      PRINCIPAL OWNER: CORPS                                       DATE OF INFORMATION:  751221
       SOURCE OF LEAD: MIKE BIERLY TN ORNITHOLOGICAL SOCIETY NASHVILLE
  GENERAL DESCRIPTION: GOLDEN EAGLE  1 SPECIMEN COUNTED ON TOS BIRD COUNT.

CONTENTS OF MANUAL FILE:     DIRECTIONS     BOUNDARY ON TOPO     AERIALS     FIELD SURVEY     OWNER INFO     REFERENCES
                                                    B                                             O              R

            INDEX KEY: SP_ASTRAGALUS_TENNESSEENSIS_006     (V)
                STATE: TN                                          COORDINATES: 360400N861800W     (N)
               COUNTY: 189                                         USGS QUAD MAP NAME: VINE
           INDEX CODE: 02.202.006.3608613                          SPECIAL STATUS:
    NAME OF SITE/AREA: CEDARS OF LEBANON ST PK                     PROTECTION STATUS: 2
     NUMBER OF OWNERS: 01                                          DATE OF INFORMATION:  660000
      PRINCIPAL OWNER: TN DOC                                      SIZE:
       SOURCE OF LEAD: CAROL F CAUDLE 1968 PHD DISSERTATION VANDERBILT UNIVERSITY
  GENERAL DESCRIPTION: TENNESSEE MILK VETCH IN CEDARS OF LEBANON ST PK

CONTENTS OF MANUAL FILE:     DIRECTIONS     BOUNDARY ON TOPO     AERIALS     FIELD SURVEY     OWNER INFO     REFERENCES
```

Sample LCD-ELF Printout

E

Classification of Rare Species*

We hope that this account of the endangered and threatened vascular plants will aid legislators and conservationists in the formulation of new regulations or laws for increased protection of the native North Carolina flora and the habitats in which they exist. It should also alert the general public to the precarious status of numerous plant species and unique environments in the state. Such a report should be useful to individuals and organizations interested in designating "natural areas," and maintaining wild flower sanctuaries or botanical gardens. The information contained here should be valuable in the preparation of "environmental impact statements" and publicity or educational programs regarding the endangered plants of the state. It is also hoped that this report will give the impetus to an ongoing and vigorous research program leading to a greater understanding of the taxonomy, ecology, and reproductive biology of these organisms.

We sincerely trust that the report will not encourage people to dig up the few remaining specimens from their natural environment and transplant them into private gardens, where they rarely survive. Such selfish, private exploitation must not occur. Commercial interests should turn to nursery propagation rather than digging plants from their native habitats.

A. Status

The committee is recognizing 409 *Rare* species of native vascular plants of North Carolina, which represents 12% of the total vascular flora. Because of their rarity, relatively few of these are known by the public, or by amateur and professional botanists. Nevertheless, there is deep concern for the preservation of all of these, and their status in the state should be monitored constantly. Omitted from this list are chance hybrids which may occur rarely between closely related species, and also taxa at the varietal level unless they are considered quite distinct and occupy unique habitats in the state.

A *Rare* plant is difficult to define precisely since the concept of rareness involves two variables: first, the overall distribution; and second, the relative density or frequency of individual plants within that distribution. This difficulty is compounded by the fact that the limits of both variables are entirely subjective. Figure E1 illustrates this relationship. The extreme in rareness, point A, would be a species like the Bladen buttercup, represented by very few individual plants and also restricted to an extremely limited geographical area. The other two extremes are also considered "rare," but this status may not be so obvious until the total picture of both range and density is considered. Point B is represented by a plant such as Lewis' heart leaf, which occurs over a fairly broad range but in very low density. Point C is represented by a plant like Oconee bells, which is locally quite abundant but still rare since its total distribution is very limited. The concept of rareness includes all possible intermediates between these three extremes. The dashed line on the graph represents the subjective limit of the concept of rareness, and any species too common and widely distributed to be included within this limit would not be considered rare.

Plant species are rare in North Carolina for various reasons. They may be at the periphery of their range, or they may be long-range disjuncts or endemics. They are considered as either *Endangered* or *Threatened* by the very fact that they are rare. *Peripheral* species may be fairly common either north, west, or south of North Carolina, but rare at the terminus of their distribution here. We argue that these native, peripheral species represent an integral part of the North Carolina flora and a probably significant element of the total genetic diversity of the species, and that this state is the "keeper" of this one important segment of the entire range. Their elimination here may represent a considerable loss in

*Excerpt from J. W. Hardin and Committee. 1977. Vascular Plants. Pages 56–138 in J. E. Cooper, S. S. Robinson, and J. B. Funderburg, eds. Endangered and threatened plants and animals of North Carolina. North Carolina State Museum of Natural History, Raleigh.

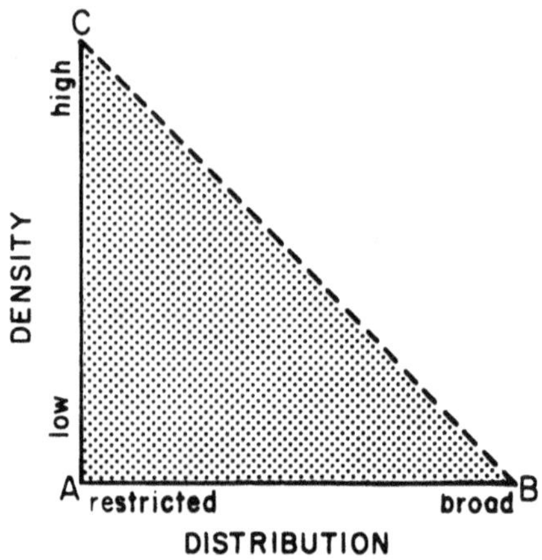

Figure E1. *Relative rareness of plants as a function of distribution and density.*

terms of the gene pool, and also a great loss in terms of the natural history of North Carolina. An example is the Palmetto palm which, although common farther south, is a rare and unique element in the forests of Smith Island and the southern coast (Figure E2A). The 318 *Endangered* and *Threatened* peripheral species are listed in Table E3. [Only an excerpt from Hardin's Table 3 appears in this book.]

The endemics, long-range disjuncts, and those species that are rare throughout are of greatest concern. Conservatively, there are 91 of these, listed in Table E2 [only an excerpt, as with Table 3, appears here] and also described individually. We are very fortunate that these represent but 2.7% of the vascular flora of North Carolina, in contrast to the estimated 10% endangered and threatened species for the entire United States. Preservation must start with these 91 species, for they are extremely vulnerable to alteration in their habitats and are either immediately endangered or likely to become endangered if not provided with protection.

A long-range *disjunct* in North Carolina is a rare segment of a species population which is significantly separated, by a few hundred miles or more, from the main area of distribution where it may be quite common. For example, Wright's cliff-brake fern occurs in the North Carolina Piedmont, nearly 1,000 miles east of its normal range in the southwestern United States (Figure E2B). The Dwarf polypody fern, which occurs in North America only in Macon County, is a disjunct from the American tropics. Such populations often represent a significant relict of past geological times, and usually mark a unique and interesting habitat within the state which should be preserved.

An *endemic*, for our purposes, is a species which has its native area totally confined to a small area of North Carolina and possibly adjacent parts of neighboring states, existing nowhere else in the world except possibly under cultivation. Examples of North Carolina endemics are Cooley's meadowrue and Mountain hudsonia, known only from the southeast and Burke County, respectively (Figure E2C). The range of certain other endemics may extend over state lines. Examples would be Venus' fly-trap, which occurs in three counties of South Carolina (Figure E2D); Lewis' heartleaf, which extends into southeastern Virginia (Figure E2E); and the southern Appalachian endemic, Spreading avens, which is also in eastern Tennessee (Figure E2F).

An *extinct* species is one which was endemic and known to exist naturally in North Carolina during earlier times, but is no longer found. An example is Bigleaf Scurfpea, collected only once in 1897 in Polk County. An *extirpated* species is a disjunct or peripheral species which is no longer found in North Carolina but still occurs elsewhere. For example, Sweet gale, which normally ranges from Pennsylvania northward, once occurred as a disjunct in Henderson County. If these apparently extinct or extirpated species are ever found again in North Carolina, they should automatically be considered *Endangered*.

Exploitation, either private or commercial, is a potential threat. Such plants as Ginseng and Goldenseal are collected for the crude drug trade, and extensive digging of the roots can be a definite threat to continued survival. When Chinese ginseng was eliminated long ago from its native area by over-exploitation, the multimillion dollar trade turned to our eastern North American species. It, too, is now threatened, even though widespread. Venus' fly-trap has become a popular novelty and house plant in recent years. Although locally common, it is threatened with over-exploitation by commercial interests, who should turn to greenhouse or nursery grown plants entirely. Rareness is also an unfortunate enticement for many enthusiasts to dig such plants for their private gardens. Certain dealers even advertise "rare" plants. Backyard gardeners will have to show their appreciation of these plants by leaving them in their native habitats, and should transfer their efforts to the preservation of "natural areas" or other means of protection and management. "Management" in such cases may not always mean leaving the habitats alone, for many of these species exist in early successional stages of the vegetation. Periodic cutting or burning may be a necessary management practice for them to persist.

The list of 91 species presented here supersedes the preliminary list published by the North Carolina Department of Natural and Economic Resources in 1973. The NER list was hastily drawn up and completed prior to basic decisions regarding the categories to be included. These 91 are also very different from the 144 species

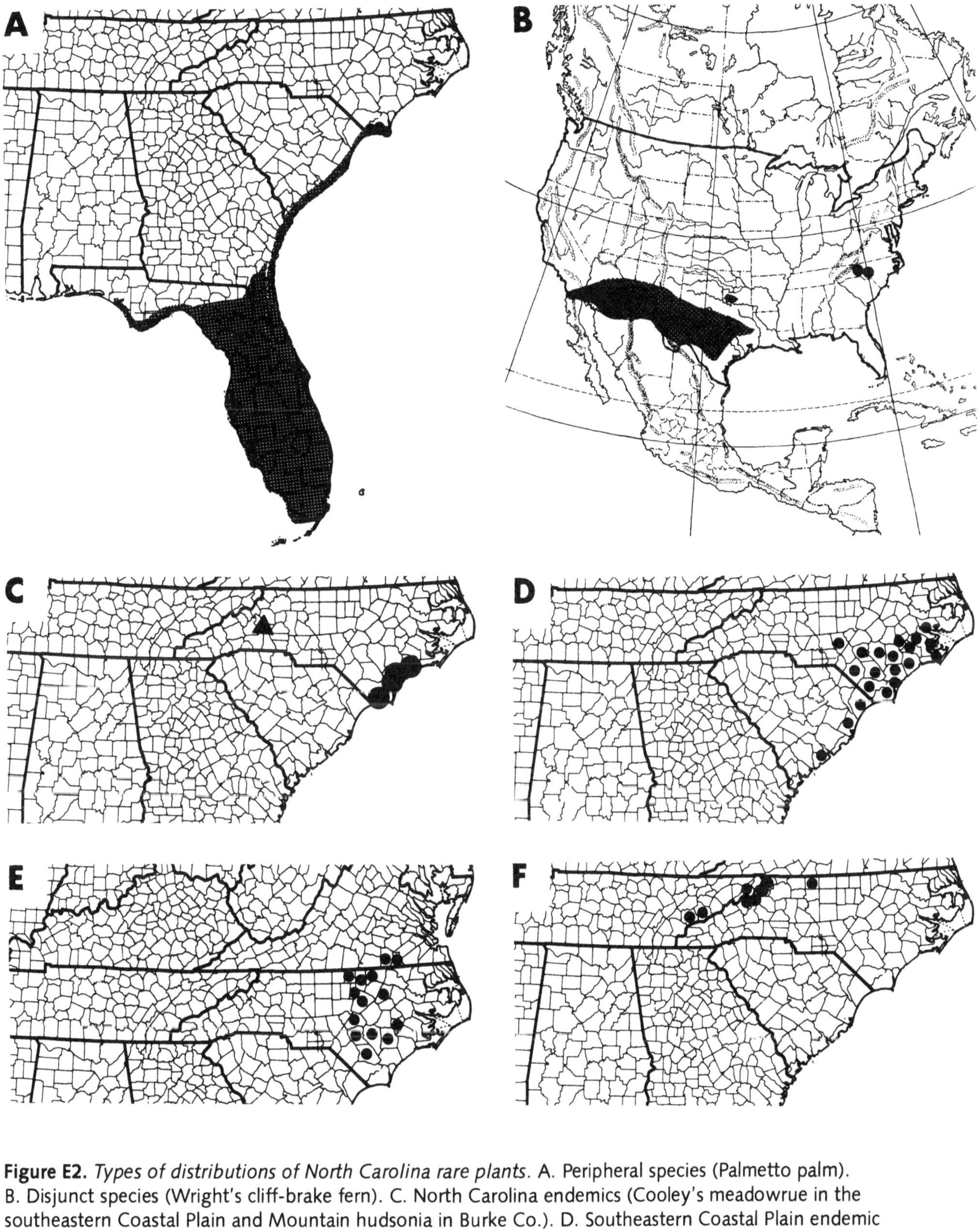

Figure E2. *Types of distributions of North Carolina rare plants*. A. Peripheral species (Palmetto palm). B. Disjunct species (Wright's cliff-brake fern). C. North Carolina endemics (Cooley's meadowrue in the southeastern Coastal Plain and Mountain hudsonia in Burke Co.). D. Southeastern Coastal Plain endemic (Venus' fly-trap). E. Piedmont-Coastal Plain endemic (Lewis' heart leaf). F. Southern Appalachian endemic (Spreading avens).

previously placed on a "conservation list" prepared by the North Carolina Garden Clubs and the North Carolina Wild Flower Preservation Society. The plants on their list are, for the most part, the showy or popular species in our flora such as bloodroot, mayapple, flowering dogwood, and mountain laurel. These are all important, and we need to guard against their possible over-exploitation by the "plant diggers," but they are not rare and cannot legitimately be considered of great concern. This present list also differs from the federal one published in the *Federal Register* (*40*[127], pt. V, 1975), principally by the inclusion of long-range disjuncts. Those species which are extinct, endemic, or endangered or threatened throughout their range appear also on the federal list.

The county distributions of the 91 extinct, endangered, or threatened species are shown on a composite map (Figure E3) and in Table E1 [once again, only an excerpt appears here], and indicate concentrations in certain areas. These cluster areas reflect unique habitats where endemics or disjuncts persist. Our preservation efforts will have to concentrate on habitats and unique plant communities, for their protection is the obvious means of conserving the individual species.

B. Explanation of Lists and Species Accounts

Table E2 is an alphabetical list of the 91 species [here selected examples] of primary concern. The status of each, and the account reference numbers, are given after the names. Every effort should be made to preserve these species as they exist in North Carolina. We prefer to think of them as constituting a single group of primary concern, with little distinction between *Endangered* and *Threatened* categories. If any order of priority is necessary, it should be: (1) endemics, (2) endangered and threatened throughout, and (3) disjuncts.

Most nomenclature, and the order of the families in the species accounts, follows Radford et al. (1968). Reference to the more complete descriptions in the "Manual" is given at the end of the general description of each species, e.g. (Man. p. 6).

Table E3 lists the native species [here selected examples] which are rare, either endangered or threatened in North Carolina, but peripheral in distribution. They appear here in the order in which they occur in the "Manual." Although found in adjacent states, care should be taken that these species do not become extirpated here.

These lists and the information presented in this report will remain open for additions as well as deletions, changes of status, and additional county distributions as our knowledge about each species in North Carolina increases. Anyone having recommendations is urged to contact the Chairman of the Plant Committee. Information will be maintained at the Herbarium, Department of Botany, North Carolina State University, Raleigh, NC 27607.

The preparation and writing of this entire report was a cooperative effort by R. L. Kologiski, J. R. Massey, J. F. Matthews, J. D. Pittillo, A. E. Radford, and J. W. Hardin, Chairman.

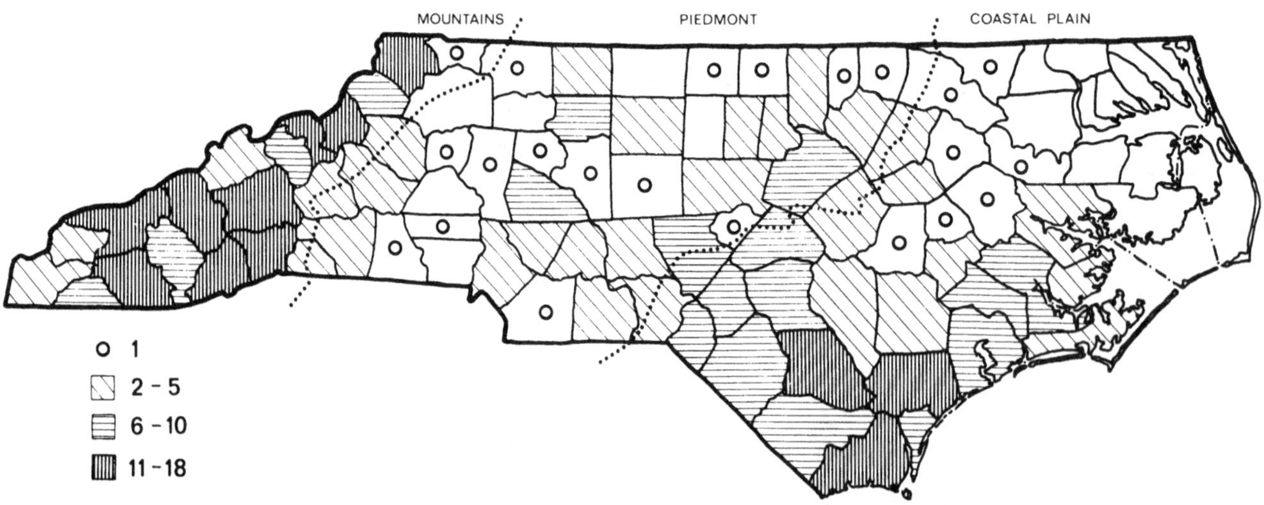

Figure E3. *Distributions of the 91 extinct, endangered, and threatened plant species, as indicated by the number of species in each county.*

Table E1. *County distributions of vascular plants of primary concern.* Numbers refer to species accounts. Counties without such species are omitted. [This is an excerpt from the original table.]

Alexander 2
Alleghany 55
Anson 42, 49, 50
Ashe 18, 19, 20, 25, 54, 55, 58, 65, 69, 70, 85
Avery 10, 18, 20, 25, 30, 32, 51
Beaufort 17, 52, 73
Bladen 4, 13, 17, 38, 44, 47, 49, 52, 68, 82, 91
Brunswick 9, 13, 29, 45, 49, 52, 62, 64, 66, 73, 76, 86, 89, 91
Buncombe 7, 16, 18, 23, 25, 28, 30, 34, 35, 37, 49, 54, 80, 85, 87, 88
Burke 30, 63, 65, 72, 84, 85
Cabarrus 36, 83
Caldwell 20, 30, 84
Carteret 28, 52, 76, 89
Caswell 1
Chatham 36, 67
Cherokee 29, 59, 75
Clay 8, 11, 12, 19, 20, 28, 30, 59
Henderson 7, 22, 24, 26, 29, 39, 31, 33, 37, 49, 54, 65, 80, 87
Hoke 49, 61, 62, 68, 82, 91
Iredell 36
Jackson 20, 24, 28, 30, 43, 55, 65, 70, 87, 88
Johnston 9, 28, 49, 91
Jones 3, 4, 13, 28, 30, 41, 52, 91
Lee 65
Lenoir 13, 38, 52
Lincoln 39
McDowell 25, 54, 72, 85
Macon 6, 16, 18, 19, 20, 28, 30, 34, 35, 42, 54, 55, 59, 65, 69, 70, 72, 79, 87
Madison 35, 43, 54, 59, 65
Martin 46
Mecklenburg 28, 36, 65, 83
Mitchell 14, 15, 18, 20, 25, 51, 53, 54, 55, 57, 58, 65, 70, 78, 84, 85, 90

Table E2. *Vascular plants of primary concern.* Letter following the common name denotes status; see key below. Number following letter refers to species account number in text. [This is an excerpt from the original table.]

KEY:
A = Extinct or extirpated
B = Endangered endemic
C = Endangered throughout
D = Endangered disjunct
E = Threatened endemic
F = Threatened throughout
G = Threatened disjunct
H = Exploited

Species	Common Name
Arenaria godfreyi	Godfrey's sandwort C 41
Asplenium monanthes	Single-sorus spleenwort fern D 5
Astilbe crenatiloba	Roan false goat's-beard A 53
Betula papyrifera var. *cordifolia*	Mountain paper birch D 34
Cacalia rugelia	Rugel's ragwort B 79
Calamagrostis porteri	Porter's reedgrass G 8
Calamovilfa brevipilis	Riverbank sandreed F 9
Cladrastis lutea	Yellowwood FH 59
Dionaea muscipula	Venus' fly-trap EH 52
Eriocaulon lineare	Linear pipewort G 22
Lindernia saxicola	Rock false pimpernel A 75

Table E3. *Peripheral, native and rare species which are endangered (E), threatened (T), extirpated (Ex), or exploited (Ep) in North Carolina.* The column headed S is Status. [This is an excerpt from the original table.]

Species	S	Area of occurrence in N.C.
Lycopodium porophilum	T	Southern Mountains
Botrychium lunarioides	T	Southern Piedmont
Trichomanes petersii	E	Southern Mountains
Cheilanthes alabamensis	E or Ex	Mountains
Potamogeton natans	Ex	Mountains
Carex jamesii	E	Inner Coastal Plain
Habenaria orbiculata	Ep, T	Northern Mountains
Geum laciniatum	E	Mountains
Lonicera canadensis	E	Mountains
Chrysoma pauciflosculosa	E	Southwestern Coastal Plain

F

Categories and Boundaries of Natural Areas and Features

by John White*

A. Natural Area Categories and Significant Features

A natural area may be categorized according to the type of significant feature it contains. A **significant feature** is a feature that allows a site to qualify as a significant natural area. By making a list of potential significant features and then seeking occurrences of these features, one avoids the problems that arise by first deciding that a site is a "natural area" and then subjectively rationalizing the site's significance.

The following classification of natural area types was developed for the Illinois Natural Areas Inventory. The seven categories are listed in Table F1 and are described in the following paragraphs.

1. Summary of Categories

Category I: High-quality terrestrial or wetland natural communities. These areas have natural communities that are relatively undisturbed, so that they reflect as nearly as possible the natural condition at the time of settlement. Areas in this category are chosen because they have high-quality natural communities, as described in Section G, which begins on page 371.

Category II: Habitats with endangered species. These sites have vertebrate animals or vascular plants that are in danger of extirpation from the state.

Category III: Habitats with relict species. Sites are recognized as natural areas if they have outstanding assemblages of plants that are relicts of a past climatic period.

Category IV: Outstanding geologic features. Localities that are outstanding representatives of the state's geologic diversity are listed as geologic areas. In addition to natural features, artificial sites such as abandoned quarries could qualify as geologic areas.

Category V: Nature preserves or lands that are managed and used for natural science studies. Lands that are specially managed and used as natural areas for teaching and research or as nature preserves are included in this category, even though the natural communities might have been disturbed. This category includes areas such as dedicated state nature preserves and federal research natural areas, as well as areas maintained by schools.

Category VI: Unique natural features. A few significant natural areas do not fit into any of the above categories. These are sites of unique natural features, which are often small areas with unusual floristic, faunistic, or ecological features. Examples include: (1) a cave with an outstanding invertebrate fauna, (2) a large bat hibernaculum, and (3) a cliff habitat that has an unusual assemblage of plants that cannot be considered endangered or relict species.

Category VII: Outstanding aquatic features. Some streams and lakes are listed as natural areas because they are relatively unpolluted and natural habitats for native aquatic life. These areas are distinct from Category I wetlands because the water quality and the fauna are the bases for recognizing the area; in contrast, the undisturbed character of the vegetation is the primary consideration in Category I wetlands.

Exceptional features. A natural area must have at least one significant feature. In addition, it may have **exceptional features** which add to the preservation value of the area but are not important enough to be the reason for identifying a natural area. For example, an exceptional feature may be a stand of undisturbed forest that is smaller than the minimum acreage required for a significant feature. Another exceptional feature may be an animal that is rare but is not threatened or endangered in the state. After a site has been recognized as a natural area because of the occurrence of a significant

*Natural Quality Service, Urbana, Illinois.

Table F1. *Natural Area Categories and Significant Features*

Natural Area Category		Significant Feature
I.	Ecological area	High-quality terrestrial or wetland natural community
II.	Endangered species habitat	Habitat with endangered species
III.	Relict species habitat	Habitat with relict species
IV.	Geologic area	Outstanding geologic feature
V.	Natural science study area	Nature preserve or land that is managed and used for natural science studies
VI.	Unique natural area	Unique natural feature
VII.	Aquatic area	Outstanding aquatic feature

feature, then the presence of exceptional features should be noted.

2. Land Condition Classes

Natural land. The part of a natural area that is relatively undisturbed is termed **natural land**. It is defined according to natural quality, and it includes Grade A, B, and C natural communities. (See Section G for a description of grades.) Although the boundaries of natural land and the significant feature may coincide, natural land often extends beyond the significant feature. When natural land does extend beyond the significant feature, it consists of Grade C natural communities or Grade A and B stands that are too small to qualify as significant features.

Buffer land. Land within the natural area boundaries that is not natural land is designated **buffer land**. Buffer land consists of Grade D or E natural communities that are included in the natural area to ensure potential for protection of the significant feature.

B. Natural Area Boundaries

1. Category I natural areas. The boundaries of a Category I natural area are determined by the boundaries of the significant feature, natural land, and buffer land as described later. Following are the general guidelines for determining natural area boundaries.

a. Natural area boundaries should follow the boundaries of natural features, which may or may not coincide with artificial lines or boundaries. Examples of the boundaries of natural features include the edges of woods and the limits of watersheds. Artificial lines or boundaries include roads, fences, and property lines. The boundaries of a natural feature often coincide with artificial lines because of changes in land use along these lines. For instance, the edge of a forest may follow a property line because the forest on the opposite side of the property line has been cleared. If the edge of the natural land does not follow a straight artificial boundary such as a property line, the natural area boundary should not be extended to an artificial boundary unless the additional area is needed as buffer land.

b. Natural area boundaries should not be drawn arbitrarily. Three guidelines, listed in order of priority, should be used to determine boundaries. (1) Sharp changes in natural quality should be used as first choice. (2) If there is no sharp break (for example, if a Grade C forest extends a great distance from the significant feature), then watershed boundaries should be used as natural area boundaries. (3) Occasionally even watershed boundaries are unclear or unsuitable; in such a case, some other boundary may be used, provided that the reason for choosing the boundary is recorded.

c. Natural area boundaries should be conservative. The natural area should have the minimum area needed to (1) include the significant feature, (2) include enough additional natural land to represent adequately the diversity of the area, and (3) ensure potential for the area's protection and management. Unnecessary land should not be included because this often introduces more management problems, causes extra work during field surveys, and may lead to questions about the validity of the boundaries.

d. Acquisition factors should not be considered when determining the boundaries of a natural area. These practical factors include (1) ownership boundaries, (2) monetary value of the land, (3) availability of the land, (4) access, and (5) the potential of adjacent land for uses such as nature interpretation and maintenance facilities. These considerations go beyond the scope of simply identifying natural areas.

The significant feature is the most important part of a natural area; additional natural land should be included only to the extent that it complements and protects the significant feature. For example, if the significant feature is a band of old growth dry-mesic forest midslope on a forested bluff, the natural area boundaries should extend vertically to include the dry blufftop and the mesic lower slopes, but the boundaries should not extend laterally along the bluff far from the significant feature. If the significant feature occurs on one side of a deep, narrow ravine, then both sides of the ravine should be in the natural area; but if the significant feature is on the side of a broad valley, then the natural area should be limited to one side.

Designation of buffer land should be similarly restricted. It may be argued that no area is safe from outside influences, even if the entire watershed is managed to preserve the area; however, buffer land should be restricted to the immediate area needed to protect the natural land from direct influences and to provide manageable boundaries. The need to include the natural diversity of an area must be tempered with the need to have a manageable area with defensible boundaries.

2. *Other categories of natural areas.* The above guidelines also apply to natural areas other than Category I. The boundaries of other kinds of natural areas usually extend little if any beyond the significant feature boundaries, though, because in these cases, land is rarely added to the significant feature to add diversity. Category II and III natural areas are usually small, and the boundaries include only the habitat with the endangered or relict species, plus any land needed to protect that habitat. For natural areas that have only Category IV, V, VI, or VII significant features, the natural area boundary usually coincides with the significant feature boundary.

C. Boundaries of Significant Features

1. *Category I significant features.* A Category I significant feature is an area of land or water that is relatively undisturbed. Two points of view are useful for determining the presence or boundaries of a Category I significant feature in the field. First, ask, "If all the land is the same natural quality as the part I am viewing, will the tract qualify as a natural area?" If the answer is yes, then the land is part of the significant feature, provided that the acreage is large enough. Second, ask, "Could the land I am viewing qualify alone as a natural area if (a) all the natural land around it were cleared, but (b) sufficient buffer land were provided?" If the answer is yes, then the land is part of the significant feature, provided that the acreage is large enough.

The boundaries of a significant feature may coincide with the boundaries of a natural area, but the natural area often extends beyond the significant feature. For example, a tract may consist of a steep hillside with a mature second growth forest (not a significant feature) and a series of one-quarter to two-acre Grade A hill prairies (the significant features). The mature second growth forest is designated as natural land; but, in the absence of the hill prairies, the hillside would not be outstanding and the tract would not qualify as a natural area.

2. *Other categories of significant features.* The boundaries for significant features other than Category I are drawn to include the feature that is the reason for recognizing the natural area. The boundaries of a Category II significant feature include the population of the endangered plant or the critical habitat of the endangered animal. A Category III significant feature coincides with the habitat of the relict plants. The boundaries of a Category V significant feature are the same as the boundaries of the area specifically dedicated to nature studies or nature preservation. The boundaries of a Category IV, VI, or VII significant feature include the outstanding geologic feature, unique natural feature, or aquatic feature.

D. Boundaries of Natural Land and Buffer Land

1. *Natural land.* After the significant feature is delimited, one must decide how much more natural land

should be in the natural area boundaries. Two factors should be considered: sufficient natural land should be added to (1) represent the natural diversity of the area and (2) help protect the significant feature.

a. For Category I natural areas, the focus is on the significant feature, but the boundaries should include enough adjacent land to include the full diversity of natural communities directly associated with the significant feature. For other categories of natural areas, the boundaries should be more conservative.

b. Natural land also serves to protect the significant feature, but including excessive natural land, far removed from the significant feature, can cause management problems. A Grade C natural community should not be included in the natural area boundaries if it (1) is distant from the significant feature, (2) does not serve to protect the significant feature, and (3) does not help provide manageable boundaries for the natural area.

2. *Buffer land.* Buffer land is disturbed land that would be excluded from the natural area, except for (1) the need to identify land with uses or conditions that must be controlled to avoid damage to the natural land and (2) the need for manageable boundaries. These two reasons are explained below.

a. If the current use or condition of land adjoining the natural land is damaging to the natural land, then the land must be included as buffer. For instance, if runoff from a garbage dump is damaging the natural land, then the dump must be included in the area boundaries as buffer because its use must be changed to protect the natural land. If the pastured heads of ravines above a natural forest are eroding and causing siltation in the natural land, then the heads of the ravines must be included as buffer land.

The adjoining land must have a use or condition that has a significant impact on the natural land in order to be designated buffer. To determine whether the impact is significant, one should decide whether continuance of the incompatible use or condition will change the natural quality of the natural land. If the land use or condition is not causing rapid changes and has been long-standing, but the natural land still qualifies for inclusion in a natural area, then one should seriously question whether the adjoining land should be designated as buffer land.

Only the current condition of adjoining land should be considered when designating buffer land; changes in land use should not be anticipated. For example, a woodlot may be surrounded by cropland, and if it is not significantly affected by agricultural chemicals and unnatural runoff, then it does not require buffer land. One should not designate a strip of cropland surrounding the woodlot as buffer land simply because the construction of apartments (a change in land use) next to the forest would have a significant impact on the area.

b. Natural land often surrounds or adjoins Grade D or E land that does not have current uses or conditions detrimental to the natural land, but should be included as buffer land to provide manageable boundaries to the natural area. Three examples are: (1) Grade E and D land (such as a cultivated or abandoned field) wholly within a complex of wetland and prairie should be included in the natural area to avoid an "inholding," but its disturbed quality should be recognized by designating it as buffer land. (2) A series of narrow, cleared ridgetops that extend into a Grade B forest should be included in the natural area boundaries as buffer land to give the area manageable and defensible boundaries, even though the fields have no direct damaging effect on the forest. (3) The natural area boundaries should be drawn around an entire woodlot, even though the edges and corners of the woodlot may be Grade D (regrowth). These regrowth stands may not be critical to the continued maintenance of the area as a natural area, but they should be included as buffer land to give the natural area manageable boundaries.

G

Grading Natural Quality of a Natural Community

by John White*

Natural quality is defined as a measure of the effects of disturbance to a natural community. Natural quality is expressed by a system of grades, which are affected by both artificial and natural disturbances. Some procedures for detecting disturbances are given in Section H, which begins on page 374.

A. Grading Artificial Disturbances

The grading system provides terms for describing the relative amount of successional instability or change in a community's natural diversity, species composition, and structure resulting from disturbance. The grades are summarized as follows:

Grade A: Relatively stable or undisturbed communities
Grade B: Late successional or lightly disturbed communities
Grade C: Mid-successional or moderately to heavily disturbed communities
Grade D: Early successional or severely disturbed communities
Grade E: Very early successional or very severely disturbed communities

1. *Grade A. Relatively stable or undisturbed communities.* Ideally, a Grade A community has a structure and composition that has reached stability and does not show the effects of disturbance by humans. This grade includes a range of conditions: the community may be gradually changing or it may have been lightly disturbed. Examples are (1) old growth, ungrazed forest, (2) prairie with undisturbed soil and natural plant species composition, and (3) wetland with unpolluted water, unaltered water level, and natural vegetation.

2. *Grade B. Late successional or lightly disturbed communities.* A Grade B community is a former Grade A community that either (1) has recently been lightly disturbed, or (2) has been moderately to heavily disturbed in the past, but has recovered significantly. If the community was recently disturbed, it was not disturbed so heavily that the original structure and composition were destroyed. If the community was disturbed in the past, it has reverted so that it is reaching stability and no longer is rapidly changing. Examples are (1) old growth forest that was selectively logged five years ago, (2) old second growth forest that has been moderately affected by grazing but now is in the late recovery stage, (3) prairie with somewhat weedy composition because the soil was graded fifteen years ago, and (4) wetland in which the original water level has been altered, changing the species composition locally but not destroying the structure and natural diversity of the community.

3. *Grade C. Mid-successional or moderately to heavily disturbed communities.* A Grade C community either (1) has been moderately to heavily disturbed (and may or may not be reverting), or (2) has been severely disturbed and has reverted significantly. The disturbance to a Grade C community has been so great that the original structure has been destroyed, and often the composition has been changed significantly. This grade includes a broad range of degrees of disturbance and of recovery. Examples are (1) heavily grazed, old growth forest, (2) young to mature second growth forest, (3) prairie that has been grazed for so long that many native species have been replaced by weeds, and (4) wetland with artificial water level that has changed the structure and composition of the vegetation.

4. *Grade D. Early successional or severely disturbed communities.* A Grade D community either (1) has been severely disturbed and has not recovered significantly, or (2) has been very severely disturbed but has begun to recover. A Grade D community has been so heavily

*Natural Quality Service, Urbana, Illinois.

disturbed that its structure (and usually composition) has been severely altered and is rapidly undergoing succession. (If the disturbance is constant, such as with continual grazing, the community may be stable and may not be succeeding.) Examples are (1) recently clear-cut forest, (2) mature second growth, severely grazed forest, (3) railroad prairie remnant with graded soil, dominated by weeds, with many native species missing, and (4) wetland that has been artificially flooded or drained, greatly changing the vegetation.

5. *Grade E. Very early successional or very severely disturbed communities.* A Grade E community has been so severely disturbed that the original community has been removed, and either (1) the site is going through the first stages of secondary succession, or (2) the natural biota is nearly gone. A Grade E community has very few or no higher plants or animals of the original community, and the land surface often is altered. Examples are (1) newly cleared land, (2) cropland, (3) improved pastureland, (4) railroad embankment, and (5) paved parking lot.

B. Grading Natural Disturbances

The concept of grading natural quality originally evolved to describe the relative amount of change in a community caused by direct, artificial disturbance by humans. Natural disturbances often affect a community in a manner similar to unnatural disturbances. Two examples are: (1) An old growth, ungrazed upland forest is Grade A. If this forest is severely disturbed by a tornado, it may appear to have been clear-cut by humans. Even though the forest had never been disturbed by people, an "A" grade is misleading because the trees are felled. A "D" grade is also misleading, however, because the forest was not affected by human disturbance. (2) An open community of pioneer herbs and invertebrates on a river bluff of eroding glacial drift may be completely untouched by people because it is inaccessible and because erosion would remove any evidence of human disturbance. It would be misleading, however, to give a naturally disturbed community such as this the same grade as a stable, undisturbed community.

The following classification of successional types provides a framework for grading naturally disturbed communities and distinguishing them from artificially disturbed areas: (1) relatively stable or undisturbed communities; (2) secondary successional communities caused by artificial disturbance; (3) secondary successional communities caused or maintained by natural disturbance; (4) primary successional communities caused or maintained by natural disturbance; (5) formerly stable communities now undergoing further primary succession.

1. *Relatively stable or undisturbed communities.* These are typical Grade A communities, described at the beginning of this section. The type may be undergoing gradual changes. The community may be maintained by fire or unusual soil, but it is stable under natural conditions. Although this type is described as "undisturbed," it includes communities that are disturbed lightly.

2. *Secondary successional communities caused by artificial disturbance.* A community may be disturbed and undergo secondary succession because of natural disturbances such as windstorms, diseases, insect outbreaks, severe fires, and a temporary, unusually high water table. These disturbances are not continual: often they are one-time, catastrophic events, and they do not always maintain the community at an earlier stage of succession. An old growth forest damaged by wind would be graded as follows:

B*n*: Enough canopy trees removed to approximate heavy selective logging

C*n*: Most overstory trees downed, leaving only young to mature trees

D*n*: Forest leveled by windstorm, leaving only saplings and shrubs

Communities affected by other natural disturbances are graded in a manner similar to the preceding example. For instance, the grade of a forest damaged by death of American Elms from Dutch Elm disease would depend on the percentage of the trees removed from the stand.

3. *Primary successional communities caused or maintained by natural disturbance.* Some communities either are (1) created and undergo succession because of natural disturbance, or are (2) kept perpetually youthful by natural disturbance. In the first group are riverbank forests, and in the second group are young tree and shrub communities on river bluffs of eroding glacial drift. The grade of these communities depends on the stage of succession. A riverbank forest would be graded as in the following example:

B*n*: Old Silver Maple (*Acer saccharinum*) forest: overstory trees are old to mature. The stand has reached stability because the very old trees are removed by floodwaters.

C*n*: Mature to young Silver Maple forest: overstory trees are mature to young. The stand usually is not stable: it may become older in time or floodwaters may keep the stand relatively young.

D*n*: Very young riverbank forest: trees are very

young. Common trees are Black Willow (*Salix nigra*), Cottonwood (*Populus deltoides*), Silver Maple, and Sycamore (*Platanus occidentalis*).

E*n*: River bar: exposed alluvium with no vegetation or only herbaceous vegetation.

4. *Formerly stable communities now undergoing further succession.* Many prairie and savannah communities are undergoing succession that is changing their structure and composition or even eliminating them. These communities were originally maintained by fire. The absence of fires is a natural disturbance caused by humans. The natural quality of these communities depends on the degree of change from the condition that could have been expected if the pre-European settlement condition of periodic fires had continued. For example, a prairie would be graded as follows:

B*n*: Sparse or scattered woody invasion, which probably has not eliminated any prairie species.

C*n*: Heavy woody invasion, which probably has eliminated some prairie species and soon threatens to eliminate the prairie unless management is started.

D*n*: Former prairie, covered by brush or young trees, with only scattered prairie plants.

The presence in low numbers (or in small patches) of trees and shrubs that are normally found in prairies does not affect the grade. Small stands of woody invasion are natural features that do not lower the grade. Savannahs that are becoming closed forests, and other communities that are undergoing similar succession started by human activities, are graded according to the degree of change from presettlement conditions.

C. Application of the Grading System

This classification of natural quality was developed for the Illinois Natural Areas Inventory as a basis for selecting high-quality natural areas and for describing the disturbances in all types of natural areas. The Illinois inventory project recognized as natural areas all sites with Grade A or B land or water that met minimum acreage criteria. In addition, the degree of disturbance was described and mapped according to grades for all types of natural areas, including sites chosen for features other than high-quality natural communities.

The grading system is based on degree of disturbance, alone; other factors such as acreage and the presence of endangered species were not considered when determining natural quality. Although these other factors are important for determining the overall preservation value of a natural area, they were kept separate so that natural quality described only the degree of disturbance.

The natural quality grades comprise a simple classification with standardized terms. Its simplicity is a limitation because such a wide variety of disturbance and successional attributes is reduced to a set of letter grades. Consequently, in addition to classifying natural quality in terms of a grade, it is necessary for natural heritage workers to describe the degree of disturbance and the successional stage in more detail, explaining the reasons for assigning the grade. Interpretation and use of the grading system were very uniform among fieldworkers in Illinois, and the natural quality classification has been adopted by workers in other states.

H

Some Procedures for Detecting Disturbances

by John White*

Knowledge of past disturbances is important in selecting natural areas and understanding the history of use in identified natural areas. Artificial disturbances may be detected by examining aerial photos, by aerial surveys, and by ground surveys. This discussion is based on midwestern natural communities, and only forests and prairies are discussed. Many of the principles that apply to prairies can be used with wetlands and savannahs.

A. Examining Aerial Photos

This discussion applies to black-and-white panchromatic aerial photos. The most commonly available photos of this type are from the U.S. Department of Agriculture's Agricultural Stabilization and Conservation Service (ASCS), which has photographed most parts of the country on a roughly seven-year cycle. Photos are available as 1:20,000 or 1:40,000 scale contact prints or as enlargements. A commonly used scale is 1:7,920 (eight inches per mile).

1. Forests. Woodlots that have been protected from disturbances often have straight boundaries and square corners that follow property lines. A disturbed stand usually has an irregular boundary that follows steep slopes and wet areas. Such boundaries indicate that the landowner has cleared as much timber as is economical and that the remainder has very seldom been protected from logging or grazing.

An old growth, undisturbed forest on soil that is not especially limiting to tree growth has a continuous, relatively even canopy with large-crowned trees. A canopy that is open, uneven, or composed of small-crowned trees usually indicates logging or grazing disturbances. A young to mature second growth stand has a dense, even canopy of small-crowned trees. A stand that has had recent, selective logging exhibits distinct, small gaps in the canopy. A stand that has recently been heavily logged has a ragged appearance. If such a stand has recovered for several years, it has a pebbly appearance because the older surviving trees cast rounded shadows on the dense new canopy of young trees.

Small trees and openings in the forest canopy are not necessarily results of disturbance when they coincide with soil types and topographic situations that might support flatwoods, xeric upland forests, wet floodplain forests, and similar communities.

Rectilinear or sharp changes in the canopy within a stand usually are the boundaries of timber cutting. A stereoscope is useful for determining the relative height of trees and irregularities in the canopy. Stereoscopes generally are not needed except when concentrating on a specific site, and they may be cumbersome to use with aerial photo enlargements.

Grazing can be detected with the following clues and techniques:

a. Look for livestock trails (thin, whitish lines) that extend from a pasture or barnlot into a woodlot. Care must be taken to distinguish intermittent streams and unimproved roads from stock trails.

b. If the pasture immediately adjoining a woodlot is trampled bare, then the livestock in the pasture are being fenced out of the forest.

c. A pond in a woodlot is usually for watering livestock.

d. If there is an indistinct, uneven boundary between a woodlot and a pasture, then the forest most likely is grazed.

e. If a woodlot is fenced, but trees do not extend to the fence, then it probably is grazed. Otherwise trees and shrubs would have invaded the open area between the fence and the forest edge.

*Natural Quality Service, Urbana, Illinois.

f. A forested slope between an upland pasture and a bottomland pasture is almost invariably grazed.

g. A forest with large gaps in its canopy and broad areas in which the ground is evident probably is grazed.

2. Prairies. In contrast to forests, natural prairie remnants often have irregular boundaries. Straight boundaries usually indicate that the prairie is fenced and grazed. Undisturbed prairies often persist in irregularly shaped patches that coincide with soil that cannot be cultivated or grazed. If a prairie remnant (or woodlot) is isolated in cropland, with no water source, and there is no livestock lane leading to it, there is a relatively high probability that it is not being grazed. Such areas may be grazed in the winter after the crops have been harvested.

Prairies have been nearly eliminated from the Midwest, so remnants that can be detected on aerial photos merit examination in the field unless the photo shows that the prairie has been severely disturbed. Even though prairies must be examined from an airplane or on the ground, aerial photos are useful for learning about past disturbances.

Parallel lines in a prairie may indicate past cultivation or mowing. The lines are caused by differences in vegetation along furrows or along ruts or gouges caused by machinery. The lines may persist for many years after the disturbance, and they do not necessarily indicate that the prairie is too disturbed to qualify as a natural area. Mowing may have little lasting effect on a prairie, but ruts from mowing in wet soil may persist for years. A prairie may recover from cultivation, especially if the soil is sand and there is an adjacent prairie remnant to serve as a source for colonizing the disturbed area.

On aerial photos, currently cultivated or mowed areas in a prairie are clear, bright, whitish areas, usually with sharp, straight sides. The mowed or plowed areas have a light tone because they reflect sunlight better than undisturbed grass.

Closely grazed grassland has an even gray tone. In contrast, an area with tall grasses or forbs has a coarser and more textured appearance, with greater variability in tones. Heavily grazed areas have whitish patches of bare soil or thin, light lines from livestock trails. Trampled areas are most prominent along fences, at gates, in fence corners, and around areas that provide feed, water, or shelter.

B. Examining Old Aerial Photos

Examining old aerial photos of specific sites is important for several reasons. Past disturbances such as clearing, cultivation, and timber harvests may be detected even though they are no longer apparent in newer photos. The contrast in vegetation between different communities is sometimes greater in old photos. This is particularly true with flatwoods, hill prairies, and glades, which in general were once more open but are blending with the surrounding forest because of woody plant succession. Some of the earlier photos have finer resolution and greater contrast than some of the newer prints.

Examining a series of aerial photos from different years increases the chance that an especially clear or useful photo will be found. For example, an Illinois county was photographed one year when the leaves on the rare Chestnut Oaks (*Quercus prinus*) had their autumn coloration but the surrounding forest was still green. The stands of Chestnut Oak appear on these photos as prominent light-toned patches in the otherwise dark gray forest.

C. Aerial Surveys

The aerial survey should be based on preparation provided by examining maps and aerial photos. Flying provides a closer and more recent look than is possible by examining aerial photos, and it allows fieldworkers to detect disturbances quickly, without conducting time-consuming ground surveys.

1. Forests. Direct evidence of logging includes the presence of stumps, tops from felled trees, logging roads, and an open, uneven canopy from removal of trees. The length of time since logging may be determined by noting (1) the degree of decay of the stumps, which varies according to tree species and soil conditions, and (2) the size and approximate age of trees that are replacing ones that were removed. Many other clues, described below, can be used to evaluate logging disturbances from an airplane.

With experience, it is possible to judge the overall age of a stand by studying the size and form of individual trees and the structure of the stand. Young second growth trees are relatively small and slender and have an immature growth form, with many ascending branches. Old growth trees are large, with relatively few, large, spreading or ascending limbs. Trees that have sprouted from the stumps of cut trees are often multiple-trunked. (Grazing, burning, and clearing of the understory can also injure saplings and result in multiple-stemmed trees.) Often the largest trees in a forest are poorly formed individuals that were never cut because they were unsuitable for lumber; such trees are readily recognized from an airplane. If all of the largest trees are poorly formed, then the stand is probably second growth. Young oaks in cut-over stands often retain their leaves throughout the winter and are highly visible indicators of disturbance. Recently logged floodplain forest characteristically displays dense stands of yellow grass in winter.

Old growth stands of certain forest types have exceptionally large, well-formed trees that are conspicuous from the air. For instance, outstanding floodplain forest along one midwestern river invariably has huge Bur Oaks (*Quercus macrocarpa*); and dry upland forest in southern Illinois has massive Post Oaks (*Quercus stellata*) if it is old growth.

Grazing disturbances can be evaluated from an airplane by looking for damage to the understory. This is difficult unless the damage is severe, because the branches of the overstory trees obscure and confuse the condition of the understory. It is possible to view the understory from an airplane by looking into the edge of the forest at a low angle; binoculars are useful for this. A heavily grazed forest has a relatively clean appearance because the understory has been removed. This is especially apparent if the damage stops at a fence inside the woods and the grazed part can be compared with the rest of the forest. A stand with a long history of grazing has an open canopy and overstory trees with unnaturally low, spreading limbs that have replaced understory trees. Even after grazing has ceased, the lower limbs persist for years until gradually replaced by new understory growth.

Other evidence of grazing includes livestock and stock trails. Trails are especially prominent on steep hillsides and radiating from gates. They are also prominent on the inside corner of a fenced, L-shaped woodlot, because the livestock take the shortest path between the arms of the fenced area. Livestock tend to form paths on the shortest or easiest route, so the paths often follow ridgetops or barriers such as creeks. If there are fences, at least one side of the fence probably has been grazed at some time. Hog sheds, stock ponds, and watering tanks are other indications. Ponds are rarely constructed in forests unless they are for watering stock. Trampled, bare earth, particularly along fences, in fence corners, and at gates, is a sign of grazing.

Thorny or unpalatable trees and shrubs increase with grazing. Most have a characteristic growth form, phenology, or color that allows quick recognition from an airplane. Common midwestern species include Multiflora Rose (*Rosa multiflora*), Osage Orange (*Maclura pomifera*), Honey Locust (*Gleditsia triacanthos*), Hawthorns (*Crataegus*), Gooseberries (*Ribes*), and Red Cedar (*Juniperus virginiana*).

There may be indirect evidence of grazing. If a woods adjoins a pasture and is not fenced from the pasture, then it is grazed to some extent. If a wooded slope has cleared pastureland above or below it, then the woods is almost invariably grazed. If a woodlot is isolated in cropland, then it probably is not grazed during the growing season, unless a fenced lane connects the woodlot with a barnlot or pasture.

2. Prairies. The second group of clues listed above for detecting grazing disturbances to forests can also be applied to prairies. Livestock trails are especially prominent at gates, salt blocks, and feed bunks. On hill prairies, the trails form parallel terraces on the slope.

In the fall and winter, most prairie grasses are yellow, orange, or reddish-brown. Exotic cool-season grasses may be green, light yellow, brown, or whitish in winter.

Thorny and unpalatable shrubs such as Red Cedar, Multiflora Rose, and Hawthorns indicate grazing. Absence of lower branches on shrubs usually indicates that livestock are still grazing the prairie.

Only heavily disturbed prairies can be rejected during the aerial survey, because species composition can be examined in detail only by ground surveys.

Some apparent indicators of disturbance actually may be signs that a prairie has not been grazed recently and may have recovered to high natural quality. For example, a sand prairie that has a pine plantation probably has not been grazed since the trees were planted. A gravel prairie or dolomite prairie that has a quarry probably has been protected from grazing during the mining operation.

Seasonal Considerations

The dormant season, from autumn to early spring, is the most advantageous time for aerial surveys. At this time, the interior of the forest is not obscured by a leafy canopy, and native prairie grasses are brightly colored and highly conspicuous. Clear, cold days allow the smoothest flights.

The low angle of the sun in midwinter has advantages and disadvantages. The crowns of trees are highlighted, so it is easier to determine the size and age of trees, but long, dark shadows obscure the understory and forest floor. Especially in the afternoon, the sun's rays are reddish, and the distinctive colors of Bald Cypresses and prairie grasses are enhanced. The observer has more problems with glare from the low angle of the sun unless the sky is overcast.

A brief period in early spring when leaves and flowers are appearing is the most advantageous for surveys of forests. The expanding leaves form a thin veil that accents the crowns of individual trees and shrubs, but the canopy is not so dense that the ground is obscured.

A person familiar with the phenology of flowering plants can use this knowledge to much advantage. Many individual species can be identified by the color of their flowers. Shrubs often produce leaves before overstory trees, so the condition of the understory is more apparent in early spring.

Aerial searches for prairies during the summer are not effective because prairie grasses cannot be distinguished from other grasses. There is little advantage to flying while forbs are in flower, because few species can be identified with certainty unless they characteristically form large, dense colonies. Weedy forbs are easily mistaken for conservative prairie species from an airplane.

Late summer and early autumn are best for aerial surveys of wet prairies and wetlands with herbaceous vegetation. Many forbs that are indicators of disturbance are in flower in late summer, and they can be identified by an experienced observer. The contrasts among species of graminoids is greatest in early autumn, when each dominant species has a different color and texture. Killing frosts occur early in lowlands because of cold air drainage, and the color of individual species changes from week to week after the first frosts, so the surveyor should become familiar with the condition of the vegetation immediately before flying.

Some ways that snow can aid aerial surveys are listed below to indicate the wide variety of techniques available to the aerial surveyor.

a. A snow cover provides a white background against which forests can be examined. Without a snow cover, the form of individual trees is obscured by the surrounding trees, and the trees blend with the forest floor. With a snow cover, trees stand out as individuals, and the branching patterns are more apparent.

b. Stumps are more easily seen in the snow, unless they are covered by a deep snow. As the snow melts, stumps are especially apparent because a ring of snow melts away from each stump before the rest of the ground is exposed.

c. Large, old growth trees can be spotted more easily if snow clings to their large limbs after it has fallen from the branches of smaller trees.

d. The contrast between grazed and ungrazed parts of a forest is increased by snow, especially if the forest is observed along a fence that separates the two parts of the forest. The snow-covered ground on the grazed side of the fence has a clearer appearance because the snow is obscured less by herbaceous vegetation and shrubs.

e. A light, powdery snowfall causes grazing trails to stand out as white lines in a background of brown leaf litter. In deeper snow, active stock trails stand out as dark lines.

f. Removal of the groundcover and understory by grazing can sometimes be detected by studying snowdrifts. If the drifts extend very far into the forest from the edge of the woodlot, then there probably is not enough understory remaining to blunt the force of the wind.

g. Livestock, fences, and livestock shelters can be detected more readily in the snow.

h. Snow can enhance the visibility of hill prairies. Snow-covered hill prairies stand out as white patches among dark trees. As the snow melts, hill prairies are accented again because the snow melts from the prairies first, causing them to be brown patches against a background of snow.

i. Snow persists on protected slopes that face north and east after it has melted elsewhere. The persisting snow makes it possible to see every small ravine, which is an important aid in marking the location of small hill prairies in rugged topography.

j. Surveying for seep springs is ideal after a fresh snowfall, because the springs and spring runs appear dark against the snow.

k. Snow assists in the detection of small ponds and sloughs in forested land. If they are frozen, they appear as white patches, which makes them prominent in timber stands. If they are not frozen, they appear as blackish patches among the snow.

l. Air escaping from the upper entrances of caves melts snow, which sometimes makes the mouths of caves more prominent. However, the white background from snow makes it impossible to detect caves from the air by looking for fog streaming from the entrances.

D. Ground Surveys

Determining the degree of disturbance to forests and prairies by on-site inspections relies mainly on an evaluation of the natural quality of the vegetation. Although all components of a natural community are important, plant communities are the best indicators of the history and present condition of a community. Disturbances are reflected in the vegetation's structure (age, distribution, size, and other characteristics of individuals) and species composition. Although the types and abundance of animals are good indicators of an area's natural quality, detailed surveys of animals are not needed to decide how a terrestrial or wetland natural community has been disturbed. The fauna depends on the plant communities, and an analysis of the vegetation provides information for predicting the kinds of animals that probably are present but are not apparent during brief surveys.

Vegetation sampling and plant species lists should not be used by themselves to determine the natural quality of an area. Rather, the degree of disturbance should be analyzed by studying the structure and composition of the vegetation, along with other indicators of disturbance such as unnatural changes in the soil or water. Some of the indicators that apply to forests and prairies are discussed in the remainder of this section.

1. Forests. A primary indicator of the history of a stand is the age of the overstory trees. The following classification provides uniform terms for describing the age of a forest:

Age of stand	*Age of overstory trees*	
Old growth	Very old	120 years +
Old second growth	Old	90–120 years
Mature second growth	Mature	40–90 years
Young second growth	Young	20–40 years
Regrowth	Very young	10–20 years

The age classes for the forest stand correspond with age classes for individual trees. For example, an old growth stand has a predominance of very old trees (120 years or older) in the overstory. Sometimes a stand consists of a mixture of different ages and must be described, for example, as "all-aged second growth" or "young to mature second growth with scattered very old trees."

Assessment of damage from grazing is complicated by the different rates at which forests recover from grazing, but the following basic terminology may be used to describe current grazing damage:

None—There is no evidence, or almost no evidence, of grazing.

Light—Some evidence of grazing damage is present. A browse line is developing, and natural understory reproduction has stopped. There is a small gap in the age of the understory and an increase in thorny species.

Moderate—Evidence of grazing is obvious. There is a definite gap in the natural understory, and thorny species are well established. Grazing trails are well established.

Heavy—Understory has been replaced by thorny shrubs. Gaps are beginning to develop in the overstory.

Severe—Large gaps have developed in the overstory, and thorny species are entering the canopy. The edges of the woods (along fences and particularly in fence corners) may be without trees or shrubs, because of continual, prolonged trampling.

Very severe—Understory has been eliminated or is dying. Overstory trees are being killed. Soil is bare and eroding.

Examples of clues for determining the history of disturbance of a forest are listed below:

a. In a typical, undisturbed, old growth forest, the overstory trees have tall, straight, clear trunks, with relatively few, large, ascending and spreading limbs in the canopy. On dry soil, the trees are smaller and the crowns may be more widely spreading. An experienced person can judge the age of a tree with close accuracy by observing the site conditions and the size and form of the tree. Increment borings may be used when necessary to determine the age of individual trees.

b. Forked trunks result from injuries such as fire and grazing damage. Multiple-trunked trees may be stump sprouts, the result of logging.

c. Trees with large lower limbs and broad crowns usually indicate a history of grazing. Some stands with broad-crowned trees may be former savannahs that have succeeded to forest.

d. An unusually high or low density of trees within a size class indicates past disturbance, usually logging or grazing. The disturbances have either removed a size class or have suppressed growth or reproduction. After the disturbance, an unusually high density of young trees may result as the stand recovers.

e. An abundance of certain species indicates grazing. These include thorny and unpalatable species, such as Multiflora Rose, Honey Locust, Osage Orange, Poison Ivy (*Rhus radicans*), Hawthorns, Prickly Ash (*Xanthoxylum americanum*), Gooseberries, and Blackberries (*Rubus*). Some understory species such as Musclewood (*Carpinus caroliniana*) and most *Viburnum* species are sensitive and decrease with grazing. Other species such as Elms (*Ulmus*) and Sugar Maples (*Acer saccharum*) decrease under grazing pressure but may become abundant after the livestock are removed.

f. The age of trees in a stand may indicate the period of grazing. For example, if the oldest thorny invaders are forty years old, then grazing probably began about forty years ago. If a formerly grazed upland forest has ten-year-old Hackberries (*Celtis occidentalis*), Elms, and Sycamores (*Platanus occidentalis*), then grazing probably ceased ten years ago.

g. Information from owners or local residents can help reconstruct the history of disturbance of a woodlot, but these sources often do not reliably recall past disturbances.

2. Prairies. Vegetation characteristics that may indicate disturbances include the following:

a. Low densities of either forbs or grasses.

b. Clones and dense colonies of single species, particularly plants that spread by rhizomes or readily germinate on bare soil, such as Canada Goldenrod (*Solidago canadensis*), Tall Thoroughwort (*Eupatorium altissimum*), and Prairie Dock (*Silphium terebinthinaceum*).

c. Abundance of species that generally increase with disturbance, such as Yellow Coneflower (*Ratibida pinnata*), Heath Aster (*Aster pilosus*), and Canada Ticktrefoil (*Desmodium canadense*).

d. Absence of species that decrease with disturbance, such as Cream False Indigo (*Baptisia leucophaea*), Purple Prairie Clover (*Petalostemum purpureum*), and Compass Plant (*Silphium laciniatum*).

Soil characteristics that indicate disturbance include:

a. A plow layer of well-mixed soil (often with a loss of natural soil structure and a compacted plow sole at the lower boundary of the plow layer).

b. Removal of the A horizon by erosion or mechanical means (determined by comparing the depth of the A horizon in similar soil on adjacent land).

Part IV

Field Inventory and Research Procedures

A. Basic Inventory
B. Reconnaissance Inventory
C. Population Status Inventory
D. Class Field Organization
E. Archeology and Ecological Diversity Classification
F. A Species Biology Research Proposal

The notes and worksheets in Sections A–D of Part IV are included as suggestions for efficient and cost-effective inventory. A classification for archeology is introduced and an example of a basic inventory of an archeological site using the ecological diversity classification is presented. The species biology proposal, based on the methods and principles described in Part I, Chapter 4, can be used as a guide or model for research on species of special concern.

Introduction to Part IV

The data forms (worksheets) in Tables A1–C9, designed for field use in the inventory of natural areas and populations, have been found to be effective for class, team, or individual use in basic, reconnaissance, and population status inventory. The final inventory report can be made directly from these completed data sheets, with the exception of the discussion section for basic inventory, which requires literature perspective. The data and information requested on these forms represents the desired information—hierarchical element entries, items of significance, characterizations, and other information—for the types of inventories. The expertise or time seldom will be available on any one field day or period to procure all of the data requested. Subsequent visits by appropriately trained individuals may be required. The necessary information on seasonal flora and fauna, for example, will have to be obtained at the proper time. Inventory progress for any natural area or population can be determined very easily by consulting a manual file of completed data forms prepared for that particular site.

This part is presented as an aid for more effective and efficient use of the Ecological Diversity Classification and the Species Biology Information Systems. The notes presented in Section A are the result of the experience gained from the use of these worksheets by students and heritage workers. Many of these suggestions have come from members of ecosystematics classes. These aids are provided for problem areas in the system, for additional information not given in Chapter 2, for more detailed explanation of format, such as arrangement of species in tables and lists, and for the use of field equipment. References and lists of equipment are included for each theme, or component. The worksheets in Section B have been used in the North Carolina Natural Heritage Program and in a floristics course at the University of North Carolina at Chapel Hill. The inventory sheets in Section C were used by Massey and Whitson during the summer of 1979 for obtaining population data on threatened and endangered species in the southern Blue Ridge Mountains.

The format to be followed for the final typed report of the Community Diversity Summary and Ecological Characterization is found in Table 10, page 100; for the Species Population Summary and Ecological Characterization in Table 11, page 101; and for the entire Natural Area Report in Table 9, page 97. Significance and priority are treated in Tables 12–15, pages 103 through 108. References cited in the notes are listed in the bibliography at the end of the book.

To thoroughly understand the relationships between the field inventory worksheets and the final natural area reports, compare Table 1 (p. 7) with page 192, Table 2 (p. 37) with page 194, Table 3 (p. 48) with page 194, Table 4 (p. 55) with page 194, and so forth.

Section D is an example of team organization for basic inventory. Section E includes a classification and example of archeological resource inventory. Section F is presented as an example of a sound research proposal for the study of species biology.

The basic advice to any class, team, or individual doing this type of work is "do your homework," that is, perform an information or pilot inventory and make a study of the species expected in the natural area prior to the on-site visit. The more one knows about an area prior to a visit, the greater the amount of information one may acquire during the actual field study. Finally, in the field, identify the hierarchical elements of diversity correctly, observe ecological relationships intelligently, describe the natural features precisely, and record the data and information accurately and completely. Follow format instructions meticulously, especially for characterizations. Use available maps of every type pertinent to the work and area. Maps are basic tools and guides, but always remember that the fundamental purpose of any on-site study is to discover "ground truth."

A

Basic Inventory

I. General Information and Summary

(Tables A1–A2)

A. Table A1 (General Information, p. 395)

Before proceeding with this section, read and study Section III of Chapter 3 and Sections C, E, F, G, and H of Part III.

ADMINISTRATION: For notes concerning this section, see page 96. Refer to Part II for examples. Most of this section is completed before going into the field.

PRIORITIES AND RECOMMENDATIONS: For notes concerning this section, see page 102, where significance and priority are discussed in detail. Also refer to Table 12 (p. 103), Table 13 (p. 104), Table 14 (p. 106), and Table 15 (p. 108). See pages 229 to 230 for a well-written example.

PROTECTION SUGGESTIONS AND MANAGEMENT RECOMMENDATIONS: See page 98 for discussion and pages 148 and 230 for examples.

B. Table A2 (Natural Area Diversity Summary and Documentation, p. 396)

GENERAL DOCUMENTATION: See page 96 and reports in Part II.

DATA SOURCE(S): Individual(s) knowledgeable of site and respective addresses (professional address if available).

GENERAL REFERENCE(S): References that specifically treat the study site or natural area under investigation. Format for references follows that of the Bibliography in this book, which in turn follows that presented in the *CBE Style Manual* (Ayars et al., 1972). Abbreviations for journals should follow an accepted reference; for instance, in botany we use *Botanico-Periodicum-Huntianum* (B-P-H) (Lawrence et al., 1968). All references for reports must be cited in the correct manner.

GENERAL DOCUMENTATION AND AUTHENTICATION: The following statement always occurs in this category:

Area analyzed by ___*(individual, class, etc.)*___ from ___*(agency, school, etc.)*___ on ___*(date)*___. Specimens deposited in ___*(herbarium)*___ by ___*(collector)*___ on ___*(date)*___.

The category also includes the names and respective professional addresses of those individuals contacted regarding the analysis, for example, a professional geologist for rock identifications or pedologist for soils determinations.

REPORT AUTHOR(S): The individual who compiles the report and writes the discussion. This is denoted by a footnote on the first page of the report.

REPORT EDITOR(S): The individual who edits the report before publication. This also is included in the footnote mentioned under the category report author.

NATURAL AREA DIVERSITY SUMMARY (See p. 99 and reports in Part II): This represents a collection of all entries at each hierarchical level (A.–CC.) for each component (I.–VII.) in the site and provides an easily accessible listing of abiotic and biotic features. The worksheet shows the format to be followed in the summary, including punctuation, spacing, and so forth. Usually this requires a separate sheet of paper.

Special problems occur in a large area that contains several Community Types. In this case, often several hierarchical element entries exist for some or all of the hierarchical levels (A.–CC.) in an abiotic component (II.–VII.). A specific format is used. The following soils excerpt from the Iron Mine Hill natural area report in this book exemplifies the format:

A. (1–6, 8, 9)* Inceptisol, A. (7) Ultisol; AA. (1–6, 8, 9) Umbrept, AA. (7) Udult. B. (1–6, 8, 9) Haplumbrept, B. (7) Hapludult; BB. (1–6, 8, 9) Typic Haplumbrept, BB. (7) Humic Hapludult. C. (1–6, 8, 9) Coarse-loamy, mixed, mesic, acidic Typic Haplumbrept, C. (7) Fine-loamy, mixed, mesic, acidic Humic Hapludult; CC. (1–6, 8, 9) Burton stony loam, CC. (7) Porters.

II. Biotic Component

(Tables A3–A9, pp. 395–403)

A. General Comments Regarding Tables A3, A6, A7, A8, A9

1. DOMINANTS (See p. 29): Names of dominant species.

2. PHYSIOGNOMY (See p. 35): A physiognomic description of the dominant species based on I.AA. Biotic Subsystem. For a list of descriptors, see the list of hierarchical elements, pages 37–38. Definitions for most of these may be found in Mueller-Dombois and Ellenberg (1974).

Growth form (see p. 29 for definition) is determined by field observations along with consulting a flora or manual (such as Radford et al., 1968). Size is also based on field observations. Note that if a plant is herbaceous and scapose, the inflorescence height is used. **Duration** (length of time that a plant or any of its component parts exists—Radford et al., 1974) can often be determined in the field; however, for certainty refer to a flora or manual.

Two growth form terms most commonly utilized for trees are excurrent and deliquescent. These terms do not appear in Mueller-Dombois and Ellenberg (1974) and are defined here. An **excurrent** tree possesses a single, undivided trunk with lateral branches, as in most gymnosperm trees. A **deliquescent** tree branches out into numerous trunks and lacks a main axis, as in most dicot trees.

3. VEGETATION REFERENCE(S): These are references that specifically deal with the Community Type or locality under consideration in the site analysis. Format of references follows that of Bibliography in this book (see notes on p. 383).

4. VEGETATION DOCUMENTATION: The following statement should be included under this heading:

Area analyzed by *(individual, class, etc.)* from *(agency, school, etc.)* on *(date)*. Specimens deposited in *(herbarium)* by *(collector)* on *(date)*.

In addition, if an expert identifies plant specimens or authenticates identifications, give the name and address of the individual.

5. SPECIES PRESENT, BUT NOT IN ANALYSIS: These are species that occur at the site but do not appear in the field sampling analysis. The canopy team notes the canopy species, the subcanopy and shrub team notes the subcanopy and shrub species, and the herb team notes the herb species.

B. Table A3 (Vegetation—Tree Stratum [Canopy], p. 397) (See pp. 25–26.)

1. HEIGHT, DBH, AGE: Select the largest specimens of the canopy dominants and record heights, diameters at breast height (dbh's), and ages. Heights can be calculated by using an optical reading clinometer. Instructions for its use in computing heights are generally included in a pamphlet accompanying the instrument. Diameters can be taken with a dbh tape or a standard measuring tape (in this case dbh is obtained by dividing by pi–3.14). A nonmetallic tape yields the best results. Ages of trees can be determined by counting annual growth rings from wood cores extracted with an increment borer. Cores may be stored in plastic drinking straws, labeled in some manner (perhaps masking tape), and brought back to the laboratory. If annual rings cannot be clearly distinguished, a phloroglucinol solution can be used. As a cleaner, WD-40 should be sprayed on and inside the corer after each use. Pruning paint should be applied to the wound in the tree. For detailed instructions on the care and use of the increment borer, consult the 1978 catalog of Forestry Suppliers, Inc.

*Numbers refer to Community Types.

2. SIZE CLASSES: Observe and record the various size and age classes of canopy species, for example, seedlings, saplings, transgressives, mature trees. Also note various dbh classes. This indicates whether or not the forest is reproducing itself.

3. CANOPY ANALYSIS: Data forms for collecting and for calculating field data are provided on pages 398 and 399, respectively. See this page for notes and suggestions regarding these forms.

The table provided on the worksheet should be filled in only when the data are in final form, ready to be typed. Record the species in order of importance values, starting with the highest value and ending with the lowest value. See canopy analysis tables in Part II for examples.

The "*d*" directly under the canopy analysis table symbolizes the calculation "**average distance**," which can be determined with the following formulas:

total distance = sum of all distances measured in all sampling points
average distance (d) = total distance/total number of trees in all sampling points

The **number of individuals per acre** can be calculated in the following manner:

number of individuals/acre = 43,560 ft./(average distance)^2

4. CANOPY SPECIES PRESENT, BUT NOT IN ANALYSIS: Arrange the species in a list according to the following rules: (1) divide into size classes; (2) alphabetize by genus; and (3) within genus alphabetize by specific epithets. See Part II for examples. In the final typed copy size classes appear in all capital letters. Species names are underlined. See examples in Part II for punctuation format.

C. Table A4 (Quarter-Point Field Data Sheet, p. 398)

This worksheet is for collecting canopy and subcanopy quarter-point data in the field. The species in each quarter of the sample point is recorded along with respective distances and diameters. Diameters are converted into basal areas in the laboratory. For details regarding field procedures, see page 27.

SIZE CLASSES: Note size classes (for example, seedlings, saplings, transgressives) and dbh classes for woody species.

SEEDLINGS: Alphabetically list woody species present as seedlings.

TRANSGRESSIVES: Alphabetically list woody species present as transgressives.

D. Table A5 (Quarter-Point Data Chart, p. 399)

This worksheet is for calculating canopy and subcanopy quarter-point data. Refer to pages 27 to 28 for methods of data computation.

E. Tables A6 and A7 (Vegetation—Tree Stratum [Subcanopy], pp. 400 and 401) (See p. 26.)

Two different worksheets are provided for the subcanopy because it may be analyzed in one of two ways —by the quarter-point method used for canopy (Table A6) or by the relevé method used for shrubs and herbs (Table A7). Which method is employed depends upon the particular site—if the subcanopy species are fairly large and the layer is well defined and well developed, use the quarter-point method. Otherwise, the relevé method (see p. 29) will suffice. Sometimes a subcanopy may not be present or at least not well developed, in which case no analysis may be necessary.

1. Table A6

See above comments regarding Table A3 and introductory comments (p. 384) regarding all five vegetation tables.

2. Table A7

For all categories except the analysis table, see above comments regarding Table A3 and introductory comments (p. 384) regarding all five vegetation tables.

SUBCANOPY ANALYSIS TABLE:

1. GROWTH FORM: This is defined on page 29. A list of growth forms is on pages 37 to 38. These terms are defined in Mueller-Dombois and Ellenberg (1974).

2. COVER: This is defined on page 27. A table (from Radford et al., 1974) showing cover values appears below. It is suggested that a copy of the table be taped to a field clipboard.

R - one specimen in area, few if any nearby
+ - Cover less than 1%
Cover 1 - Cover 1%–5%
Cover 2 - Cover 5%–12.5%
Cover 3 - Cover 12.5%–25%
Cover 4 - Cover 25%–50%
Cover 5 - Cover 50%–100%

3. SOCIABILITY: This is defined on page 29. A table (from Radford et al., 1974) showing sociability values appears below. Once again, it is suggested that a copy of the table be taped to a field clipboard.

Sociability 5 - Plants occurring in great crowds (pure populations)
Sociability 4 - Plants in small colonies, in extensive patches, or in carpets
Sociability 3 - Plants in small patches or cushions
Sociability 2 - Plants grouped or tufted
Sociability 1 - Plants growing singly

4. ARRANGEMENT OF SPECIES IN TABLE: Arrange the species in the analysis table according to the following rules (see pp. 209 and 265 for an example):
   a. Divide the subcanopy species into size classes (commonly only one size class is present).
   b. List the species by beginning with the tallest trees and working down to the shortest.
   c. Within each size class list the trees alphabetically by genus and within each genus alphabetically by specific epithets.

5. SUBCANOPY SPECIES PRESENT, BUT NOT IN ANALYSIS: Follow the same rules for arrangement outlined above for the analysis table. See page 219 for an example. In the final typed copy size classes appear in all capital letters. Species names are underlined. See the example for punctuation format.

F. Table A8 (Vegetation—Shrub Stratum, p. 402) (See p. 26.)

1. The relevé method is used to analyze the shrub stratum (see p. 29). For all categories except the analysis table, refer to introductory comments (p. 384) regarding all five vegetation tables.

2. Woody vines and transgressives are included with the shrub analysis. A **woody vine** is a plant with elongate, nonself-supporting, climbing or clambering, woody stems. A **transgressive** is an immature individual of a woody species.

3. SHRUB ANALYSIS TABLE:
   a. Growth form, cover, and sociability are discussed above with reference to Table A7.
   b. Arrange the species in the analysis table according to the following rules (see pages 204–205 for an example):
      (1) Keep transgressives as a separate category. Always list these first (before the shrubs).
      (2) Keep woody vines as a separate category. Always list these last (after the shrubs).
      (3) Divide the shrubs into size classes (see p. 37).
      (4) List the shrubs by beginning with the tallest (giant shrubs) and working down to the shortest (very low dwarf shrubs).
      (5) Within the transgressives, the woody vines, and each size class list the species alphabetically by genus and within each genus alphabetically by specific epithets.

4. SHRUB SPECIES PRESENT, BUT NOT IN ANALYSIS: Follow the same rules for arrangement outlined above for the analysis table. See page 205 for an example. In the final typed copy size classes appear in all capital letters. Species names are underlined. See the example for punctuation format.

G. Table A9 (Vegetation—Herb Stratum, p. 403) (See p. 26.)

1. The relevé method is used to analyze the herb stratum (see p. 29). For all categories except the analysis table, refer to introductory comments (p. 384) regarding all five vegetation tables.

2. Herbaceous vines are included with the herb analysis.

3. HERB ANALYSIS TABLE:
   a. Growth form, cover, and sociability are discussed above with reference to Table A7.
   b. Arrange the species in the analysis table according to the following rules (see pp. 205–206 for an example):
      (1) If mosses and/or lichens form a conspicuous part of the community, keep these listed separately and place them at the end (after all herbs) of the analysis table.
      (2) Separate the herbs into herbaceous vines, forbs, graminoids, ferns, and fern allies.
      herbaceous vine—an elongate, weak-stemmed, often climbing plant
      forb—any herbaceous plant other than grasses, sedges, or rushes

graminoid—any herbaceous plant that is a grass (Poaceae), a sedge (Cyperaceae), or a rush (Juncaceae)
fern—any herbaceous plant that is a flowerless, seedless vascular plant of the class Filicopsida, characteristically reproduces by means of spores, and has fronds (megaphylls)
fern ally—any herbaceous plant that is a flowerless, seedless vascular plant of the classes Lycopsida (clubmosses, spikemosses, quillworts) and Sphenopsida (horsetails), characteristically reproduces by means of spores, and has very small, needlelike or scalelike leaves (or long, linear-subulate, grasslike leaves in quillworts) which are microphylls

(3) Within each type of plant group listed in (2) separate into size classes (see. p. 37).
(4) List in the table in the following order: from the tallest forb (extremely tall) to the smallest forb (very small), from the tallest graminoid to the smallest graminoid, from the tallest fern to the smallest fern, from the tallest fern ally to the smallest fern ally.
(5) Within each of these subdivisions based on type of herb and size, list the species alphabetically by genus and within each genus alphabetically by specific epithets.

4. HERB SPECIES PRESENT, BUT NOT IN ANALYSIS: Follow the same rules for arrangement outlined above for the analysis table. See page 206 for an example. In the final typed copy size and herb classes appear in all capital letters. Species names are underlined. See the example for punctuation format.

EQUIPMENT FOR VEGETATION ANALYSIS:

AGE—increment borer, pruning paint, WD-40, plastic straws, masking tape for labels

DBH—dbh tape or standard measuring tape

HEIGHT—clinometer, standard measuring tape

QUARTER-POINT ANALYSIS
clipboard
#4 pencils
data sheets (Table A4)
worksheet (Table A10)
blank sheets for presence list
standard measuring tape
dbh tape
identification manual
hand lens
collecting bag for unknown specimens

RELEVÉ ANALYSIS
clipboard
#4 pencils
data sheets (Tables A7, A8, A9)
blank sheets for presence list
data sheets (Tables A7, A8, A9)
blank sheets for presence list
device for outlining 5 m × 5 m and 1 m × 1 m for relevés
cover and sociability tables (taped to clipboard)
identification manual
hand lens
collecting bag for unknown specimens

WORKSHEET INFORMATION SOURCES:
Küchler (1964)—vegetation map
Regional forest survey maps
State vegetation maps
Pertinent local flora, fauna, and vegetation studies and reports
Local or regional identification manual
Quarter-point and relevé data
Species lists
Cronquist (1968)
Curtis and Cottam (1962)
Foster and Gifford (1974)
Mueller-Dombois and Ellenberg (1974)
Radford et al. (1974)

H. Table A10 (I. Vegetation—Biotic Communities, p. 404)

See examples in the Natural Area Diversity Summaries and Community Diversity Summaries in the reports presented in Part II.

A. Choose from hierarchical elements listed on page 37. See discussion on page 35.

AA. Choose from hierarchical elements listed on pages 37 to 38 and compose the physiognomic description in the following manner—size class, growth form, duration modifying plant habit. Definitions are found in Mueller-Dombois and Ellenberg (1974). See discussion on page 35.

B. Composed by investigator, although many standard terms are already in use. Consult list of selected examples on pages 38 to 39. See discussion on page 34.

BB. Species Associates list composed by investigator

after analysis. Refer to list of selected examples on page 39. For Taxonomic Order, choose from hierarchical elements listed on pages 39–40. See discussion on pages 32–34.

C. & CC. Both of these are delimited by analysis and nomenclature in each community site by the investigator. For selected examples, see pages 40 to 45. I.C. is discussed on page 32 and I.CC. on pages 30–32. Also read pages 25–27 and pages 29–30. For nomenclature format of I.CC., see page 30.

VEGETATION ECOLOGICAL CHARACTERIZATION: Complete a description according to the format at the bottom of page 100. See examples on page 36. Note that physiognomic descriptions of all vegetational strata are included in the characterization, even if the layer has no true dominants or co-dominants. A density modifier (sparse, open, closed) precedes the physiognomic description. See page 27 for definitions of these density terms.

WORKSHEET INFORMATION SOURCES (For specific references used at each hierarchical level, refer to the discussion on pages indicated above in I.A. through I.CC.):

Küchler (1964) vegetation map
Regional forest survey maps
State vegetation maps
Pertinent local flora, fauna, and vegetation studies and reports
Local or regional identification manual
Quarter-point and relevé data
Species lists
Cronquist (1968)
Curtis and Cottam (1962)
Foster and Gifford (1974)
Mueller-Dombois and Ellenberg (1974)
Radford et al. (1974)

III. Abiotic Components

(Tables A11–A16, pp. 405–410)

A. General Comments Regarding Tables A11–A16 (pp. 405–410)

Several features consistently appear on each of these tables. To avoid needless repetition these features are discussed here.

REFERENCE(S): Cite specific references pertinent to the community or site under investigation. Do not include general references (for example, *The National Atlas*, Cowardin et al. [1972]), which are always consulted. Format for references should follow that outlined on page 383 under the category General References.

DOCUMENTATION: Most often this consists of the names and addresses of those individuals contacted regarding the analysis, for example, a professional geologist, pedologist, or hydrologist. Include their professional addresses.

NOTES ON MINOR FEATURES: These represent notes on minor abiotic features (usually form a microhabitat) present in the site that exert some control over occurrences and distributions of Community Types and species, particularly threatened and endangered species.

MICROHABITAT INDICATORS: Indicate the species inhabiting the specific microhabitat feature(s) noted in the previous category.

ECOLOGICAL CHARACTERIZATION: Use the format indicated at the bottom of page 100. Examples appear in Part II in the Community Diversity Summaries and also at the end of each section in Chapter 2.

B. Table A11 (II. Climate, p. 405)

(An example of II.A.–II.CC. appears on p. 47.)

A. Choose from hierarchical elements listed on page 48. See discussion on page 45. Complete before going into the field by using Goode and Espenshade (1950) and Trewartha (1954).

AA. Same procedure and page references as A.

B. & BB. Data are collected and a description is written by using the modifiers provided on pages 48 to 49. On the worksheet an area in which to record data from the literature is provided. Below this a space for the climatic description is provided. Note that this description (not the data) is the entry to be used in the system for the Community and Natural Area Diversity Summaries.

B. See discussion on page 46. Complete before going into the field by using *The National Atlas* (U.S. Department of the Interior, 1970).

BB. See discussion on page 46. Complete before going into the field by using the climatic supplement for the state (U.S. Department of Commerce, 1965). Select as a reference point the weather station that is nearest to and most closely resembles the study site.

C. & CC. These are completed while in the field by using observation and common sense and by comparing the study site with the information gathered in II.B. and BB.

C. Choose from hierarchical elements on page 49. See discussion on page 46.

CC. Choose from hierarchical elements on page 49. See discussion on page 46.

CLIMATIC REFERENCE(S): Cite the title of the climatic supplement used in II.BB.

WEATHER STATION(S): The station used for data gathering in II.BB.

MAP REFERENCE(S): Local climatic maps consulted.

EQUIPMENT:
worksheet (Table A11)

CLIMATIC INFORMATION SOURCES:
Climatic data
Goode and Espenshade (1950)
Trewartha (1954)
U.S. Department of the Interior (1970), *The National Atlas*
U.S. Department of Commerce (1965), Climatic Supplements

C. Table A12 (III. Soils, p. 406)

(An example of III.A.–III.CC. appears on p. 54.)

Before going into the field, study soils and drainage maps and consult the literature, for example, county soil surveys and local soil studies and reports. If unfamiliar with soil taxonomy, characteristics, or methods, refer to the following texts: Soil Survey Staff (1951, 1975), Black et al. (1965), Thompson and Troeh (1973), Brady (1974), Steila (1976), and Buol et al. (1980).

The field process that obtains best results for determining system levels A. through CC. is outlined here. References required for these determinations appear under equipment at the end of this section.

A pedon (at its smallest one square meter) represents the typical unit of sampling in a soil survey. Therefore, the ideal technique for examining soil is to dig a soil pit, one square meter in size and deep enough to record all soil horizons, in an area that appears typical of the site under investigation. See the list of equipment needed at the end of this section. Generally in large study sites some variation exists in soil depth, horizon development, pH, and other factors. In many cases time does not permit the excavation of a soil pit wherever a variation (suggested either by change in vegetation or by some other indicator) occurs. In other situations, such as in a small population of a threatened or endangered species, a standard soil pit would be impractical because it possibly could injure or even kill members of the population by disrupting the local drainage rate and direction. In these situations, a twist earth auger, a bucket type soil auger, or a tube sampler (soil probe) proves very useful for spot checking soil variation. Fairly accurate soil horizon data can be collected using augers.

For detailed chemical analyses of soils, samples need to be collected and saved. These can be sent to soil testing laboratories at county or state agricultural offices. Either horizon samples or a homogenized sample consisting of equal portions of each horizon can be collected. The sampling technique depends on the type of information sought.

Study sites underlain by Histosols (organic soils) often grade downward into peat, the thickness of which should be determined. This type of work requires a peat sampler.

A. Choose from hierarchical elements listed on page 55. See discussion on pages 51 to 52.

AA. Choose from hierarchical elements listed on page 55. See discussion on page 52.

B. Determined in the field and/or in the laboratory. For selected examples, see page 55. See discussion on pages 52 to 53.

BB. Determined in the field and/or in the laboratory. For selected examples, see page 55. See discussion on page 53.

C. Determined in the field and/or in the laboratory. For selected examples, see page 55. See discussion on page 53.

CC. Determined in the field and/or in the laboratory. For selected examples, see page 55. See discussion on pages 53 to 54.

SOIL ANALYSIS TABLE: Used to record data collected from soil pit.

Texture—Refer to Radford (1976) or Thompson and Troeh (1973). It is recommended that these texture classes be photocopied and taken into the field.

Color—Use the Munsell Soil Color Chart (1975).

pH—Refer to Radford (1976) or Soil Survey Staff (1951). It is recommended that these pH classes be photocopied and taken into the field.

SOIL MOISTURE CLASS(-ES): Refer to Radford (1976) or Soil Survey Staff (1951). Once again, it is recommended that these classes be photocopied and taken into the field.

SOIL DOCUMENTATION: Complete this category only if a professional was consulted.

EQUIPMENT (If unfamiliar with equipment, consult a Forestry Suppliers, Inc., catalog):

shovel
twist earth auger
bucket type soil auger
tube sampler (soil probe)
peat sampler
pH kit
Munsell soil color chart
distilled water
Kimwipes
standard measuring tape or meter stick
sample bags
reference slide for particle sizes
field references—
  photocopies of texture, pH, and soil moisture classes
  soil series identification device (for example, *The Soils of North Carolina* by Lee, 1955)
  Soil Survey Staff (1972, 1975). These are too bulky to carry in the field but should be kept in the field vehicle.
  Steila (1976)
worksheet (Table A12)

SOILS INFORMATION SOURCES:

Soils data
National soil maps (such as *The National Atlas*)
State soil maps
County soil maps and surveys
Local soil maps, surveys, studies, and reports
Local soil agent
Local agricultural agent
Identification device for series (for example, Lee, 1955)
Black et al. (1965)
Brady (1974)
Munsell Soil Color Chart (1975)
Radford (1976)
Soil Survey Staff (1951, 1972, 1975)
Steila (1976)
Thompson and Troeh (1973)

D. Table A13 (IV. Geology, p. 407)

(An example of IV.A.–IV.CC. appears on p. 62.)

Most of North America has been investigated geologically. Before attempting any geologic field work for any type of field study the available literature and maps should be consulted. Many good sources from which to obtain pertinent references exist. The Geological Society of America publishes *The Bibliography and Index of Geology* monthly and lists all geologic publications by date, author, subject, and location. Many state geological surveys supply bibliographies of work completed in each state. Local colleges, universities, and museums frequently keep reference lists and/or collections of publications dealing with the surrounding area. Geologists associated with these and other geologically oriented concerns frequently are familiar with local geology. Publications that deal generally or specifically with a given study site provide information that aids a geologist or any field scientist. Most of the geologic hierarchical elements can be determined before going into the field with a geologic publication pertinent to the study site. This geologic information can provide clues as to the presence of soil types and type of vegetation, the landscape pattern based on rock erosion, and the presence of certain rock outcrops that might serve as microhabitats for certain species. Always, however, base the final hierarchical element determinations on field work. Be sure to collect a sample for closer examination and/or expert determination.

A. Choose from hierarchical elements listed on page 63. See discussion on pages 56 to 57.

AA. Choose from hierarchical elements listed on page 63. See discussion on page 57.

B. For selected examples see page 63. See discussion on pages 57 to 58.

BB. For selected examples see pages 63 to 64. See discussion on page 58.

C. & CC. A description is written for these two levels and is composed of modifiers selected from the following categories. Compose in a clear, sensible, readable form.

Rock-sediment type—Selected examples occur on pages 64 to 65 and discussion on pages 59 to 60.

Deformational history—Selected examples occur on page 65 and discussion on page 60.

Percentage of area that consists of exposed rock—Choose from hierarchical elements on page 65. See discussion on page 60.

Manner in which rock is exposed—Selected examples occur on page 65 and discussion on page 60.

Degree of weathering—Choose from hierarchical elements on page 66. See discussion on pages 60 to 61.

Rock-sediment fabric (see p. 61).

  Structure—Selected examples occur on page 66 and discussion on page 61.

  Strike and dip—These are measured in the field with a Brunton compass. For methodology, consult Lahee (1961) and Compton (1962). See the selected example on page 66 and discussion on page 61.

Porosity—Choose from hierarchical elements on page 66. See discussion on page 61.

Permeability—Choose from hierarchical elements on page 66. See discussion on page 61.

Degree of consolidation—Choose from hierarchical elements on page 66. See discussion on page 61.

Particle size—Choose from hierarchical elements on page 66. See discussion on page 61.

EQUIPMENT:
rock hammer
hand lens
Brunton compass
maps
sample bags
worksheet (Table A13)

GEOLOGIC INFORMATION SOURCES:
Geologic data
Geologic map of the United States
State geologic maps (7.5 and 15 minute maps available for some states)
County geologic maps and surveys
Local geologic maps, surveys, studies, and reports
General geology textbooks:
Bayly (1968)
Billings (1972)
Blatt et al. (1972)
Compton (1962)
Gary et al. (1972)
Hamblin and Howard (1975)
Kemp (1940)
Lahee (1961)
Pettijohn (1975)
Trowbridge (1962)

E. Table A14 (V. Hydrology, p. 408)

(An example of V.A.–V.CC. appears on p. 70.)

Once again, before entering the field, study drainage and hydrologic maps and consult the literature, such as local hydrologic studies and reports.

Many of the parameters in Natural Area and Community Hydrologic Site Types often will not be attainable because of lack of time, equipment, and expertise. Features that can be noted fairly easily and should be completed include water chemistry, drainage, and water depth. A general water color should always be indicated even if equipment is not available to detect true and apparent colors. The other factors should be measured when undertaking a detailed site analysis over an extended period of time, for example, in heritage inventory.

A. Choose from hierarchical elements on page 71. See discussion on pages 67 to 68.

AA. Choose from hierarchical elements on page 71. See discussion on pages 68 to 69.

B. Choose from hierarchical elements on page 71. See discussion on page 69.

BB. Choose from hierarchical elements on page 71. See discussion on page 69.

C. & CC. A description is written and is composed of modifiers selected from the categories listed on the worksheet. Most often, only water chemistry (see p. 72 for hierarchical elements and p. 69 for discussion) and drainage (see p. 72 for hierarchical elements and p. 69 for discussion) will be determined.

EQUIPMENT:
water pH kit
salinometer
worksheet (Table A14)

HYDROLOGIC INFORMATION SOURCES:
Hydrologic data
Regional runoff, aquifer, and water balance maps
Drainage maps (state and local)
County and local hydrologic surveys, studies, and reports
Cowardin et al. (1977)
Reid (1961)
Soil Survey Staff (1951)
Welch (1948)

F. Table A15 (VI. Topography, p. 409)

(An example of VI.A.–VI.CC. appears on p. 78.)

The topographic classification scheme requires familiarity with two aspects of landform analysis: map and field interpretation. As in the previous abiotic component worksheets, topographic maps should be studied and the literature consulted before attempting any field work. The final analysis always should be based upon careful field observations, which often necessitate a tremendous amount of reconnoitering. Several references concerning map interpretation and general methodology might prove helpful: Greitzer (1944), Lobeck and Tellington (1944), Lahee (1961), Compton (1962), Way (1973), and Hamblin and Howard (1975).

A. Choose from hierarchical elements on page 79 by consulting Hammond (1964) or *The National Atlas* (U.S. Department of the Interior, 1970). See discussion on pages 74 to 75.

AA. Choose from hierarchical elements on page 79 by consulting Hammond (1964) or *The National Atlas* (U.S. Department of the Interior, 1970). See discussion on page 75.

B. Determined in the field. For selected examples, see pages 79 to 80. See discussion on pages 75 to 76.

BB. Determined in the field. Please note that agent and process are indicated only if they differ from those in VI.B. For selected examples, see pages 80 to 81. See discussion on page 76.

C. & CC. A description is written from field observations and data for these two levels and is composed of modifiers selected from the following categories. Compose in a clear, sensible, readable form.

Landform—Selected examples occur on pages 80 and 81 and discussion on page 76.

Shelter—Choose from hierarchical elements on page 81. See discussion on page 77.

Aspect—Determine with a compass and choose from hierarchical elements on page 81. See discussion on page 77.

Slope angle—Determine with a clinometer and choose from hierarchical elements on page 82. See discussion on page 77.

Profile—Choose from hierarchical elements on page 82. See discussion on page 77.

Surface patterns—Extremely careful scrutiny of the study site is required. Selected examples occur on page 82 and discussion on page 77.

Position—Note that this feature is utilized only when the study site or community occupies a slope or a portion thereof. Choose from hierarchical elements on page 82. See discussion on page 77.

EQUIPMENT:
topographic maps
geologic maps often help
orthophotoquad maps (if available)
clinometer
compass
altimeter

TOPOGRAPHIC INFORMATION SOURCES:
Topographic data
Regional topographic maps (1:250,000)
Local topographic maps (1:62,500 and 1:24,000)
Geomorphological studies and reports
Regional and local hydrographic charts
State drainage maps
Orthophotoquad maps
Introductory geomorphology texts:
- Fairbridge (1968)
- Garner (1974)
- Hinds (1943)
- Thornbury (1969)

Compton (1962)
Gary et al. (1972)
Greitzer (1944)
Hamblin and Howard (1975)
Hammond (1964)
Lahee (1961)
Lobeck and Tellington (1944)
U.S. Department of the Interior (1970)
Way (1973)

G. Table A16 (VII. Physiography, p. 410)

(An example of VII.A.–VII.CC. appears on p. 85.)

The first three levels (A., AA., B.) of physiography are determined previous to any field work by consulting Fenneman (1931, 1938) or *The National Atlas* (U.S. Department of the Interior, 1970). A study of topographic and geologic maps also should be made before entering the field. The last three levels as well as the geography section are completed in the field.

A. Choose from hierarchical elements on page 87 by consulting Fenneman (1931, 1938) or *The National Atlas* (U.S. Department of the Interior, 1970). See discussion on page 83.

AA. Choose from hierarchical elements on page 87 by consulting Fenneman (1931, 1938) or *The National Atlas* (U.S. Department of the Interior, 1970). See discussion on pages 83–84.

B. Choose from hierarchical elements on pages 87 to 88 by consulting Fenneman (1931, 1938) or *The National Atlas* (U.S. Department of the Interior, 1970). See discussion on page 84.

BB. Determine in the field by observation and use of topographic maps. See selected examples on page 89 and discussion on page 84.

C. Determine in the field by observation and use of topographic maps. See selected examples on page 89 and discussion on page 84.

CC. Community position—Determine in the field by observation and use of topographic maps. See selected examples on page 89 and discussion on page 85.

Geologic formation and age—Obtained from

official geologic maps and publications. See selected examples on page 89 and discussion on page 85.

GEOGRAPHY: See page 85 for discussion. Latitude and longitude readings are obtained from maps. Elevation readings are taken with an altimeter.

SIZE OF AREA: This is calculated by placing a plastic Modified Acreage Grid over the study site outline delimited on a topographic map. This grid can be obtained from Forestry Supplies, Inc.

SUCCESSIONAL STAGE: This pertains to the state of habitat development: pioneer (beginning or early successional) stage, seral (transient or middle successional) stage, or climax (stable or late successional) stage (see p. 19). References that contain general discussions of succession include Clements (1916, 1928, 1936), Odum (1971), and Mueller-Dombois and Ellenberg (1974).

SERE TYPE: **Sere** is defined as the entire sequence of communities that successively replace one another in a given area from the pioneer to the climax stage. Several types of seres exist, for example, hydrosere, halosere, psammosere, pelosere, lithosere, subsere. See pages 18 to 19 and Radford et al. (1974) for definitions of these sere types and the general references mentioned under successional stage for general discussion.

EQUIPMENT:
geologic maps
topographic maps
worksheet (Table A16)

PHYSIOGRAPHIC INFORMATION SOURCES:
Physiographic data
Local topographic maps
Geologic maps
Local geographic studies and reports
Drainage maps
Fenneman (1931, 1938)
U.S. Department of the Interior (1970), *The National Atlas*

IV. Community Documentation and Ecological Characterization

(Table A17, p. 411)

Table A17 (Community Documentation and Ecological Characterization, p. 411)

COMMUNITY REFERENCE(S): List only references that deal specifically with the community under investigation (see p. 96). For notes regarding format for references, see page 383.

COMMUNITY DOCUMENTATION: State who collected the data, the field work date, who documented the species with voucher specimens, where these specimens were deposited, and the date of deposition. Use the following format:

Area analyzed by *(individual, class, etc.)* from *(agency, school, etc.)* on *(date)*. Specimens deposited in *(herbarium)* by *(collector)* on *(date)*.

Indicate individuals responsible for professional identifications, analyses, or authentications of community data. Include their professional addresses. See Community Diversity Summaries in Part II for examples.

COMMUNITY ECOLOGICAL CHARACTERIZATION: Use the format indicated at the bottom of page 100. This category is discussed on page 388. Examples appear in Part II in the Community Diversity Summaries. Common sense must be employed when composing the ecological characterization, that is, the components must make sense and sound relatively readable.

V. Population Level Inventory

(Table A18, pp. 411–412)

Table A18 (D. Population Level Inventory: Endangered and Threatened Species, pp. 411–412)

The population level features are discussed briefly on pages 90 to 92 and extensively in Chapter 4. Because categories II.D. through VII.D. follow the format for II.CC. through VII.CC., consult the discussions and hierarchical element lists that pertain to the CC. level.

The pages where these appear are listed below:

	Discussions	*Lists*
II.CC.	46	49
III.CC.	53–54	55
IV.CC.	58–62	64–67
V.CC.	69–70	72–73
VI.CC.	76–77	81–82
VII.CC.	85	89

VI. Notes Regarding Collection

EQUIPMENT:
collecting bags
newspaper
water (to wet down plants)
rubber bands
trowel
blank sheets for presence list
tree clippers
field notebook
plastic bags (for aquatics)
hand lens
identification manual

REFERENCES:
identification floras or manuals

Radford et al. (1974) contains an excellent chapter concerning the collection and preparation of plant specimens

VII. Notes Regarding Master Species Presence List

See discussion on pages 35–36. Note that this list requires time and effort, hence, it is not prepared in the field.

The following information should be included:

a master list of species with respective authors

Community Type membership for each species if several Community Types occur in the site

a collection number for each documented species

a reference to the collector and his/her professional address

a reference to depository for voucher specimens

microhabitat preferences for species

the reference base for nomenclature in each major group

For format used, see page 35. Also refer to pages 225 to 228 for an example.

REFERENCES:
identification floras and manuals
Cronquist (1968)
Foster and Gifford (1974)
Radford et al. (1974)

Table A1. *General Information*

Natural Area ______________________ Size ______________________

Team ______________________ Field Work Date ______________________

Administration

Ownership:

Administration:

Land use:

Land classification:

Danger to integrity:

Publicity sensitivity:

Priorities and recommendations

Significance:

Endangered and threatened species and/or communities:

Diversity:

Condition:

Distribution:

Humanistic feature:

Productivity:

Protection suggestions:

Management recommendations:

Table A2. *Natural Area Diversity Summary and Documentation*

Natural Area ____________________ Community Type ____________________

Team ____________________ Field Work Date ____________________

General documentation

Data source(s):

General reference(s):

General documentation and authentication:

Report author(s) and professional address(es):

Report editor(s) and professional address(es):

Natural area diversity summary:

CLIMATE: A. Climatic Regime; AA. Climatic Subregime. B. Sectional Climate; BB. Local Climate. C. Natural Area Climatic Site Type; CC. Community Climatic Site Type.

SOILS: A. Soil Order; AA. Soil Suborder. B. Soil Great Group; BB. Soil Subgroup. C. Soil Family; CC. Soil Type.

GEOLOGY: A. Rock System; AA. Rock Subsystem. B. Rock-Sediment Chemistry; BB. Rock-Sediment Occurrence. C. Natural Area Geologic Site Type; CC. Community Geologic Site Type.

HYDROLOGY: A. Hydrologic System; AA. Hydrologic Subsystem. B. Water Chemistry; BB. Water Regime. C. Natural Area Hydrologic Site Type; CC. Community Hydrologic Site Type.

TOPOGRAPHY: A. Topographic System; AA. Topographic Subsystem. B. Landscape; BB. Landform. C. Natural Area Topographic Site Type; CC. Community Topographic Site Type.

PHYSIOGRAPHY: A. Physiographic Region; AA. Physiographic Province. B. Physiographic Section; BB. Local Landform. C. Natural Area Physiographic Site Type; CC. Community Physiographic Site Type.

BIOLOGY: A. Biotic System; AA. Biotic Subsystem. B. Community Cover Class; BB. Community Class. C. Community Cover Type; CC. Community Type.

Table A3. *I. Vegetation—Tree Stratum(-a)*

Natural Area ______________________ Community Type ________________

Team ____________________________ Field Work Date ________________

CANOPY

Species height dbh age

Species height dbh age

Size classes:

Canopy dominant(s):

Canopy physiognomy: Size, growth form, duration:

CANOPY ANALYSIS

Species	I.V.	Rel. Den.	Rel. Dom.	Rel. Freq.

Number of points: ____ d: ____ Number of individuals/acre: ____

Canopy species present, but not in analysis:

Canopy species associates:

Vegetation references:

Vegetation documentation:

Table A4. *Quarter-Point Field Data Sheet*

Point Number	Species	Distance	DBH	Basal Area
1				
2				
3				
4				
5				

Point Number	Species	Distance	DBH	Basal Area
6				
7				
8				
9				
10				

Size classes:

Seedlings:

Transgressives:

Table A5. *Quarter-Point Data Chart*

Species	No. of Quarter-Points in which Species Occurred	No. of Trees	Total Basal Area	Relative Density	Relative Dominance	Relative Frequency	Importance Value

Table A6. *I. Vegetation—Tree Stratum(-a)*

Natural Area ______________________ Community Type ______________________

Team ______________________ Field Work Date ______________________

SUBCANOPY

Species: height dbh age

Species: height dbh age

Size classes:

Subcanopy dominant(s):

Subcanopy physiognomy: Size, growth form, duration:

SUBCANOPY ANALYSIS

Species	I.V.	Rel. Den.	Rel. Dom.	Rel. Freq.

Number of points: _____ d: _____ Number of individuals/acre: _____

Subcanopy species present, but not in analysis:

Subcanopy species associates:

Vegetation references:

Vegetation documentation:

Table A7. *I. Vegetation—Tree Stratum(-a)*

Natural Area ______________________ Community Type __________________

Team ______________________________ Field Work Date _________________

SUBCANOPY

Species: height dbh age

Species: height dbh age

Size classes:

Subcanopy dominant(s):

Subcanopy physiognomy: Size, growth form, duration:

SUBCANOPY ANALYSIS

Species	GF	1 C.S	2 C.S	3 C.S	4 C.S	5 C.S	6 C.S	7 C.S	8 C.S	9 C.S	10 C.S

Number of relevés: ____ Relevé size: ____

Subcanopy species present, but not in analysis:

Subcanopy species associates:

Vegetation references:

Vegetation documentation:

Table A8. *I. Vegetation—Shrub Stratum(-a)*

Natural Area ______________________ Community Type ______________________

Team ______________________ Field Work Date ______________________

Shrub layer dominant(s):

Shrub layer physiognomy: Size, growth form, duration:

SHRUB ANALYSIS

Species	GF	1 C.S	2 C.S	3 C.S	4 C.S	5 C.S	6 C.S	7 C.S	8 C.S	9 C.S	10 C.S

Number of relevés: _____ Relevé size: _____

Shrub species present, but not in analysis:

Shrub species associates:

Vegetation references:

Vegetation documentation:

Table A9. *I. Vegetation—Herb Stratum(-a)*

Natural Area ______________________ Community Type ______________________

Team ______________________ Field Work Date ______________________

Herb layer dominant(s):

Herb layer physiognomy: Size, growth form, duration:

HERB ANALYSIS

Species	GF	1 C.S	2 C.S	3 C.S	4 C.S	5 C.S	6 C.S	7 C.S	8 C.S	9 C.S	10 C.S

Number of relevés: _____ Relevé size: _____

Herb species present, but not in analysis:

Herb species associates:

Vegetation references:

Vegetation documentation:

Table A10. *I. Vegetation—Biotic Communities*

Natural Area ____________________ Community Type ____________________

Team ____________________ Field Work Date ____________________

A. Vegetation System:

Biotic System:

Major Habitat:

AA. Vegetation Subsystem:

B. Community Cover Class:

Generic Class:

General Habitat Feature:

BB. Community Class:

Species Associates:

Taxonomic Order:

C. Community Cover Type:

CC. Community Type:

Vegetation ecological characterization:

Table A11. *II. Climate*

Natural Area ____________________ Community Type ____________________

Team ____________________ Field Work Date ____________________

A. Climatic Regime:

AA. Climatic Subregime:

B. Sectional Climate:

DATA:

Mean annual sunshine:
Average annual precipitation:
Freeze-free period:

Avg. mean max. temp. of hottest month:
Avg. mean min. temp. of coldest month:
Avg. mean prec. of driest month:
Avg. mean prec. of wettest month:

WRITTEN DESCRIPTION:

BB. Local Climate:

DATA:

Mean annual sunshine:
Average annual precipitation:
Freeze-free period:

Avg. mean max. temp. of hottest month:
Avg. mean min. temp. of coldest month:
Avg. mean prec. of driest month:
Avg. mean prec. of wettest month:

WRITTEN DESCRIPTION:

C. Natural Area Climatic Site Type:

CC. Community Climatic Site Type:

Climatic reference(s):
Weather station(s):
Map reference(s):

Climatic documentation:

Notes on minor climatic features:
Thermal belts:
Frost pockets:
Air drainage patterns:

Climatic microhabitat indicators (Indicate specific feature for each species):

Climatic ecological characterization:

Table A12. *III. Soils*

Natural Area:
Community Type:
Date:
Description by:
County, State:

Moisture class:
Drainage class:
Erosion class:
pH class:
Parent material:

Landscape position:
Slope:
Aspect:
Other:

Soil documentation:

Map reference:

SOIL ANALYSIS

Horizon	Depth	Matrix Colors	Mottles	Texture	Structure Grade	Structure Class	Structure Type	Consis-tency	Horizon Boundary	pH	Remarks

A. Soil Order:
B. Soil Great Group:
C. Soil Family:

AA. Soil Suborder:
BB. Soil Subgroup:
CC. Soil Series:

Notes on minor soil features (such as, moisture, texture, pH, series):
Soil microhabitat indicators (Indicate specific feature for each species.):

Soil ecological characterization:

Table A13. *IV. Geology*

Natural Area ____________________ Community Type ____________________

Team ____________________ Field Work Date ____________________

A. Rock System:

AA. Rock Subsystem:

B. Rock-Sediment Chemistry:

BB. Rock-Sediment Occurrence:

C. Natural Area Geologic Site Type:
- Rock-sediment type:
- Deformational history:
- Exposed rock cover %:
- Manner in which rock is exposed:
- Weathering of exposed rock:
- Rock-sediment fabric:
  - Structure:
  - Strike:
  - Dip:
  - Porosity:
  - Permeability:
  - Degree of consolidation:
  - Particle size:

CC. Community Geologic Site Type:
- Rock-sediment type:
- Deformational history:
- Exposed rock cover %:
- Manner in which rock is exposed:
- Weathering of exposed rock:
- Rock-sediment fabric:
  - Structure:
  - Strike:
  - Dip:
  - Porosity:
  - Permeability:
  - Degree of consolidation:
  - Particle size:

Geologic reference(s):

Map reference(s):

Geologic documentation:
- Rock determiner:
- Rock authenticator:

Notes on minor geologic features:
- Rock-sediment type:
- Minor rock occurrences:
- Exposure:

Geologic microhabitat indicators (Indicate specific feature for each species):

Geologic ecological characterization:

Table A14. *V. Hydrology*

Natural Area ______________________ Community Type ______________________

Team ______________________ Field Work Date ______________________

A. Hydrologic System:

AA. Hydrologic Subsystem:

B. Water Chemistry:

BB. Water Regime:

C. Natural Area Hydrologic Site Type:

- Water chemistry:
- True color:
- Apparent color:
- Turbidity:
- Drainage:
- Water depth:
- Water temperature:
- Water source:
- Water velocity:
- Light penetration:
- Volume of water moving past a given point/given time:
- Turnover rate:
- Ice cover:
- Dissolved and colloidal substances:
- Substrate:

CC. Community Hydrologic Site Type:

- Water chemistry:
- True color:
- Apparent color:
- Turbidity:
- Drainage:
- Water depth:
- Water temperature:
- Water source:
- Water velocity:
- Light penetration:
- Volume of water moving past a given point/given time:
- Turnover rate:
- Ice cover:
- Dissolved and colloidal substances:
- Substrate:

Hydrologic reference(s):

Map reference(s):

Hydrologic documentation:

Notes on minor hydrologic features:

- Drainage:
- Water color:
- Water chemistry:
- Turbidity:
- Substrate:

Hydrologic microhabitat indicators (Indicate specific feature for each species):

Hydrologic ecological characterization:

Table A15. *VI. Topography*

Natural Area ______________________ Community Type ____________________

Team ______________________________ Field Work Date ___________________

A. Topographic System:

AA. Topographic Subsystem:

B. Landscape:
   Landscape:
   Landscape-forming agent:

BB. Landform:
   Landform:
   Landform-forming agent:

C. Natural Area Topographic Site Type:
   Natural area landform:
   Natural area shelter:
   Natural area aspect:
   Natural area slope angle:
   Natural area profile:
   Natural area surface patterns:
   Natural area position:

CC. Community Topographic Site Type:
   Community landform:
   Community shelter:
   Community aspect:
   Community slope angle:
   Community profile:
   Community surface patterns:
   Community position:

Topographic reference(s):

Map reference(s):

Topographic documentation:

Notes on minor topographic features:
   Landscape-forming processes:
   Surface relief:
   Slope characteristics:

Topographic microhabitat indicators (Indicate specific feature for each species):

Topographic ecological characterization:

Table A16. *VII. Physiography*

Natural Area ______________________ Community Type ____________________

Team ______________________________ Field Work Date ___________________

A. Physiographic Region:

AA. Physiographic Province:

B. Physiographic Section:

BB. Local Landform:

C. Natural Area Physiographic Site Type:

CC. Community Physiographic Site Type:
Community position:
Geologic formation:
Geologic formation age:

Physiographic reference(s):

Map reference(s):

Physiographic documentation:

Geography:
Distance and direction from permanent landmark:

County: State:

Latitude: Longitude:

Elevation:

Size of area:

Successional stage:

Sere type:

Physiographic ecological characterization:

Table A17. *Community Documentation and Ecological Characterization*

Community references:

Community documentation:

Community ecological characterization:

Table A18. *D. Population Level Inventory—Endangered and Threatened Species*

Natural Area ______________________ Community Type ____________________

Team _____________________________ Field Work Date ___________________

Species:

Species legal status
- International:
- National:
- State:
- Local:

References:

I.D. Biotic Associates:

II.D. Population Climatic Site Type:

III.D. Population Soil Site Type:

IV.D. Population Geologic Site Type:

V.D. Population Hydrologic Site Type:

VI.D. Population Topographic Site Type:

Table A18. *(continued)*

VII.D. Population Physiographic Site Type:

Successional stage:

Population inventory
- Population numerical relations
  - Population size
    - Definition of units:
    - Number of units:
    - Approximate number of subunits per unit:
  - Abundance:
  - Density:
  - Frequency:
- Distributional relations:
- Areal relations:

Disturbance type(s) and agent(s):

Population vulnerability or threat:

Size classes:

Phenological classes:

Reproductive classes:

Population vigor:

Microhabitat feature:

Species reference(s):

Species documentation:

Species ecological characterization:

B

Reconnaissance Inventory

Review Chapter 3, IC, and see Part II, E–G. Table B1 is a self-explanatory general report form developed by the North Carolina Heritage Program. Table B2 is based on appropriate tables in Part IV, A. Basic Inventory. Tables B3–B5 are expanded tables for actual recording of data suggested in Table B2.

Table B1. *North Carolina Heritage Program Information and Reconnaissance Report*

ESSENTIAL INFORMATION TO BE INCLUDED IN ALL REPORTS

Name of area:

County:

Location description: (Indicate distance and direction from nearest town, road, intersection, or well-known landmark.)

Topographic quad map reference:
USGS quad name and date:

Attach a map and information indicating the approximate boundaries of the site.

Ownership information: (Include name and addresses, if known.)

Report prepared by: Your name: ______________________________

Other persons knowledgeable about site:

Name | Address

PROVIDE AS MUCH DETAILED INFORMATION ABOUT THE FOLLOWING ITEMS AS POSSIBLE.

Current Use and Protection Status

How is the area being used or managed? Indicate any apparent threats to the viability of the natural area.

Table B1. *(continued)*

Vegetation and Plant Communities

Describe the vegetation or plant communities present, indicating dominant cover species of the major community types. Note any special features of quality or diversity.

Physical Features

Describe topographic and physical features of special interest or significance, such as waterfalls, bluffs, rock outcroppings, natural ponds, fossil deposits, and relict sand dunes.

Rare Plants and Animals

List any rare or unusual plants and animals known to occur on the site. Note status as rare, endemic, disjunct, endangered, or threatened. Indicate population size(s) if known. For animals, indicate whether the species is breeding, wintering, or transient.

Common Name	Scientific Name	Status

Other Significant Features

List any other reasons for considering the site to be an important natural area.

Publications and Scientific References

Author(s)	Source	Date

Flora Species List

List plant occurrences in your field survey or other available lists. List by community, then by strata--canopy, subcanopy, shrubs, vines, herbaceous layer.

Scientific Name	Common Name

Fauna Species List

List animal species known to occur on the site. Indicate status as breeding, wintering, or transient if known.

Common Name	Scientific Name	Status

Table B1. *(continued)*

Conclude with an evaluation of the site's ecological significance.

Provide management recommendations.

Include a site map with boundary lines and demarcation of locations of important natural features.

Table B2. *Natural Area Field Data Reconnaissance Report* (See Chapters 2 and 3.)

NATURAL AREA SUMMARY

Site name: Use topographic name, if possible; if not, coin a name with outstanding feature as part of name.
Locality: Distance and direction from nearest permanent landmark; County, State.
Coordinates: Latitude and longitude; Map reference (Quadrangle name, scale, publication date, and agency); Elevation.
Data collector(s): Name, institution affiliation, home or institutional address, phone number of home and/or business.
Collection date(s):
Owner/Administrator: Include names and addresses. Size of Area:
Data sources: For entire report, major parts of report, and individual items within report.
List threatened and endangered species and indicate status of each.
Indicate significant features of natural area, always including condition of community.
Indicate current use and protection status.
Make management and protection recommendation.
Include topographic map showing site with distinctly marked boundaries.

BIOTIC AND ABIOTIC INVENTORY
(For each Community Type)
BIOLOGY

Biotic System: Biotic Cover Type:
Biotic Community Type:
Dominant size classes: Species with dbh.
Largest specimens: By species with dbh, height, age.

HABITAT

CLIMATE:

SOIL FAMILY:	pH class	Texture class	Munsell color
A1			
A2			
B			

GEOLOGY:
Formation: Rock-sediment chemistry:
Rock type:

Table B2. *(continued)*

HYDROLOGY
- Water chemistry class: — Drainage:
- Water regime:

TOPOGRAPHY:
- Landform or water body: — Community profile:
- Community shelter: — Community surface pattern:
- Community aspect: — Community position:
- Community angle:

PHYSIOGRAPHY:
- Region: — Province: — Section:
- Physiographic feature (for example, drainage system, mountain range, plain, or area):

NATURAL AREA SPECIES LISTS

List vascular plant species by strata: canopy, subcanopy, shrub, transgressives, vines, epiphytes, and herbs.

List animal species by groups: mammals, birds, reptiles, amphibians, and so forth. Indicate residence status, relative abundance, and breeding status, where possible.

Table B3. *Natural Area Floristics General Information Report*

Site name:

Locality:

Coordinates: — Map reference:

Physiographic region: — Province: — Section:

Physiographic feature:

Natural area site description:

Size: — Elevation:

Boundaries (map and survey lines):

Ownership:

Administration:

Table B3. *(continued)*

Land use:

Land classification:

Danger to integrity:

Publicity sensitivity:

Endangered and threatened species and/or communities:

Diversity significance:

Distribution significance:

Protection suggestions:

Management recommendations:

Data collectors:

Data collection date:

Data documentation:

Data source(s):

General references:

Table B4. *Floristics Habitat Report*

Site name: | Data collectors:
State: | County: | Data collection date:

HABITAT

Biotic System: | Biotic Cover Type:

Biotic Community Type:

Dominant size classes:

Largest specimens:

Dominant life form:

Climatic Subregime:

Soil Family: | Soil Series:

	pH class	Texture class	Munsell color
A			
A			
B			
B			
C			

Geologic formation:

Rock-Sediment Chemistry: | Rock type:

Water Chemistry: | Drainage:

Water Regime:

Community Landform or Water Body:

Community shelter: | Community aspect:

Community angle: | Community profile:

Community surface pattern: | Community position:

Succession stage:

Significant habitat features:

Profile showing community distribution in relation to major habitat features:

Table B5. *Natural Area Flora*

Site name:		Data collector(s):
State:	County:	Data collection date:

C

Population Status Inventory

Tables C1–C9 are field data forms used in population inventory. Consult Chapter 4, IA, for instructions on updating Table CI and Chapter 4, IIA, for instructions on conducting population, habitat, and threat inventory.

Table C1. *Species General Information Update Form*

Species ______________________________ Date ______________________

Population Name ________________________ Population Number __________

Phenology:
- Month ______
- Vegetative ______
- Flowering ______
- Fruiting ______

Predominant phase:
- ______ Vegetative
- ______ Flowering
- ______ Fruiting/Sporulating

Historical distribution:
- ______ Verification
- ______ New Station
- ______ Uncertain

Reference(s) used: ______________________
____
____

Habitat preference:

1. Habitat types
    1. Bog
    2. Swamp
    3. Marsh
    4. Grassland
    5. Savannah
    6. Chaparral--shrub
    7. Desert
    8. Tundra
    9. Ice
    10. Forest
        1. Gymnosperm
        2. Angiosperm
        3. Mixed
    11. Other
2. Topographic types
    1. Mountains, hills, and ridges
    2. Scarps, bluffs, cliffs, and escarpments
    3. Benches and terraces
    4. Basins
    5. Lakes, bays, ponds, and pools
    6. Valleys, gorges, and channels
    7. Plains and flats
    8. Beaches
    9. Other
3. Substrate types
    1. Water
        1. Fresh
        2. Saline
        3. Haline
    2. Soil
        1. General type
        2. Texture
    3. Rock
        1. Origin type
            1. Igneous
            2. Sedimentary
            3. Metamorphic
            4. Uncertain
        2. Specific type
            1. Limestone
            2. Sandstone
            3. Shale
            4. Granite
            5. Other
    3. Humus or organic layers

Table C1. *(continued)*

Habitat development state:

_____ Early/Pioneer _____ Mid/Transient _____ Late/Climax _____ Uncertain

Comments:

Author:

Table C2. *Locality Reconnaissance Form*

Species ______________________ Date ______________

Locality Name ______________________ Population Number ________

State ________ County _______ Duration of search _____ Person hours ( )

USGS quad map ______________________________

Distance/Direction (description and attached map):

Locality information source(s) ______________________
____

Species present ________ Authentication method ______________
Species absent ________ ______________

Comments:

ADDITIONAL LOCALITY AND INVENTORY INFORMATION FORM

Vouchers: Specimen # ______________ Photograph # ______________

Lat. ___° ___' ___" N; Long. ___° ___' ___" W Range/Township/Section ________

Table C2. *(continued)*

Land inventory:

Owner

Contact
Name

Address

Phone ( )

Designation

Current Use

Past Use
Recent past (<10 years)
Historic past (>10 years)

Comments:

Table C3. *Population Inventory Form*

Scaled population map (attach)

Unit characteristics
Unit definition:

Subunit definition (size and number per unit):

Population characteristics

Class Definition	Origin			Phenology			Vigor			Distribution		Observations
	S	A	U	V	FL	FR	VV	V	NV	U	C	

Comments:

Table C3. *(continued)*

Biotic interrelations

	LF	STM	FL	FR	OTHER	FAMILY/ORDER/SPECIES
Browse, graze						
Suck						
Parasite (Plt/Anim)						
Infestation						
Visitors						
Other ______						

Comments (such as degree of interrelations, location, proportion, weather conditions):

Unit/Subunit census summary (include sample areas on map)

	AREA SIZE		AREA COUNT	
Total Population		(A/E)		(A/E)
Subsample ( )		(A/E)		(A/E)
Subsample ( )		(A/E)		(A/E)
Subsample ( )		(A/E)		(A/E)
Subsample ( )		(A/E)		(A/E)
Subsample ( )		(A/E)		(A/E)
Subsample ( )		(A/E)		(A/E)
Subsample ( )		(A/E)		(A/E)

Census detail

Population/	Class	Number			Percentage			
Subsample	Symbol	V	FL	FR	V	FL	FR	Total

Table C3. *(continued)*

Census detail

Population/ Subsample	Class Symbol	Number V	Number FL	Number FR	Percentage V	Percentage FL	Percentage FR	Total
Total								

Comments (proportion, location of flowering, vigor, uniformity):

Table C4. *Vegetation Inventory Form*

Species	Population Habitat 1	2	3	4	5	Extrapopulation Habitat 1	2	3	4	5
Canopy (___m x ___m samples)										
Subcanopy (___m x ___m samples)										

Table C4. *(continued)*

Species	Population Habitat 1	2	3	4	5	Extrapopulation Habitat 1	2	3	4	5
Shrubs (___m x ___m samples)										
Herbs (___m x ___m samples)										
Cryptogams (___m x ___m samples)										

Table C5. *Vegetation Summary*

	Population Habitat	Extrapopulation Habitat
Vegetation Uniformity:	(Yes/No)	(Yes/No)
Community Type:		

Table C6. *Biotic Associate Inventory Form*

Species	Population Hab.						Extrapopulation	
	C	c	S	H	K		Present	Absent

Table C7. *Abiotic Factor Inventory Form*

	Population Habitat	Extrapopulation Habitat
CLIMATE		
Uniformity of:	(Yes/No)	(Yes/No)
Temperature Relations		
Precipitation Relations		
Moisture Effectiveness		
SOIL		
Uniformity of:	(Yes/No)	(Yes/No)
Family/Series		
Depth		
pH		
Texture		
GEOLOGY		
Uniformity of:	(Yes/No)	(Yes/No)
Rock System		
Rock Exposure, %		
Manner of Exposure		
HYDROLOGY		
Uniformity of:	(Yes/No)	(Yes/No)
Chemistry		
Drainage		
Depth		
TOPOGRAPHY		
Uniformity of:	(Yes/No)	(Yes/No)
Landform		
Shelter		
Aspect		
Angle		
Profile		
Position		
Surface		
PHYSIOGRAPHY		
Uniformity of:	(Yes/No)	(Yes/No)
Position		
Altitude		

Table C8. *Habitat Disturbance Inventory Form*

(Qualitative/Quantitative/Activity/Agent/Objective)

	Population Habitat	Extrapopulation Habitat
VEGETATION		
Removed	______	______
Uprooted	______	______
Trampled	______	______
Dead, dying	______	______
Introductions	______	______
LITTER		
Removed	______	______
Upheaved	______	______
Trampled	______	______
Added	______	______
SOIL		
Removed	______	______
Upheaved	______	______
Compacted	______	______
Overburdened	______	______
Exchanged	______	______
WATER		
Removed	______	______
Increased	______	______
Enriched	______	______

Potential disturbance/threat (Activity/Agent/Objective/Extent):

Other Comments:

Table C9. *Author Information Form*

Author(s)
Name(s) ______ ______ ______
Address(es) ______ ______ ______
______ ______ ______
______ ______ ______
Phone numbers ( ) ______ ( ) ______ ( ) ______
Signature ______ ______ ______
Date ______ ______ ______
Permanent plots established: Yes _____ No _____; Monitoring plans ______
Voucher deposition

D

Class Field Organization

(Botany 235, Fall 1978)

	PROVINCE									PROVINCE		
	CP	BR	BR	BR	PP	RV	CU	CP	CP	CP	CP	CP
Day	1	2	3	4	5	6	7	8	9	10	11	12
TEAMS \ DUTIES												
A.	I	II	III	IV	V	VI	I	II	III	IV	V	VI
B.	II	III	IV	V	VI	I	II	III	IV	V	VI	I
C.	III	IV	V	VI	I	II	III	IV	V	VI	I	II
D.	IV	V	VI	I	II	III	IV	V	VI	I	II	III
E.	V	VI	I	II	III	IV	V	VI	I	II	III	IV
F.	VI	I	II	III	IV	V	VI	I	II	III	IV	V

DUTIES (See Chapter 3, Section IV)

I. Cerebration—Natural area summary, discussion, and climatology

II. Geology—Geology, topography, and physiography

III. Soils—Soils and hydrology

IV. Canopy—Canopy and subcanopy analysis

V. Shrub-Herb—Shrub and herb layer analysis

VI. Collecting—Master presence list and minor habitat indicators

TEAMS

A. Students 1, 2

B. Students 3, 4

C. Students 5, 6

D. Students 7, 8

E. Students 9, 10

F. Students 11, 12

PROVINCE

BR = Blue Ridge

RV = Ridge and Valley

CU = Cumberland Plateau

PP = Piedmont

CP = Coastal Plain

Notes: Assignments are arranged to give each student equal inventory and analysis training time in each component of diversity. Usually two sites and two community types per natural area are inventoried and analyzed per class field day. The duties in a twelve-field-day course with twelve students are rotated so that each student will make one discussion presentation and literature study (cerebration) for one natural area in the Coastal Plain and another in the Appalachian Highlands.

E

Archeology and Ecological Diversity Classification

*Editorial Comments**

Man's past can be considered a natural as well as a cultural heritage. If one assumes that man is an integral part of nature, then the history of man's interaction with the elements of nature can be recorded and studied in the same manner as other natural sciences. The Archeological Resource Inventory presented here represents an attempt to arrange the multitude of man's "cultural phenomena by their similarities and differences in time and space" (Schneider and Dittmar, this publication, this section). The format of this inventory classification scheme follows the same general style presented in Chapter 2. In this particular resource inventory, chronology is considered the first controlling principle, resulting in the higher levels of the inventory hierarchy being devoted to the temporal arrangement of cultural affiliations. The lower levels of the system deal with specific spatial phenomena and typologies, as determined by on-site inventories. As will be noticed, most examples used in the hierarchical elements of the resource inventory are concentrated in the eastern United States, an area with which the authors are most familiar. As with the other inventory classifications presented in this volume, however, this system is easily adaptable to any other part of North America.

Because the major goal of the Archeological Resource Inventory is to present an organized and consistent technique to record and classify the history of man, it becomes important to realize that the history of man, as reflected in archeological sites, does not exist "*in vacuo*; rather, as expressions of human nature." These sites "reflect man's dependency upon and utilization of the natural environment" (Schneider and Dittmar, this publication, this section). One must therefore understand that the environment in which man lives influences man's culture and, after the passing of man, the environment influences and, in many ways, controls the manner in which the record of man's passing is preserved. In order to better understand this influence, a method must be developed to recognize and record all possible natural processes and products that might in some manner affect an archeological site. The Natural Heritage Classification Scheme presented in Chapter 2 should prove to be very helpful in archeological inventories. The Dropzone Archaic Archeological Site report (p. 442) illustrates the quantity of essential, yet nonarcheological, data that can be systematically collected, recorded, and eventually used in the interpretation of individual sites. These data can also serve an important role when the history of development of a number of sites is compared. The search for additional sites of particular affiliations could become more successful after common elements that are seen to be major controlling factors for original occupation or for site preservation are recognized and understood.

Archeological Resource Inventory†

Overview

If we compare the scope and character of archeology today versus 16 years ago, we would quickly conclude that our nation has assumed a vested interest in its cultural heritage. Foremost among the signals heralding such interest are four major pieces of federal legislation requiring survey, assessment, evaluation, and preserva-

*Prepared by A. E. Radford, D. K. S. Otte, and L. J. Otte.
†Kent Schneider, Regional Archeologist, U.S.D.A. Forest Service, 1720 Peachtree Street, N.W., Atlanta, Georgia. Ed Dittmar, Department of Geology, University of North Carolina, Chapel Hill.

tion of cultural resources that could be influenced by federally sanctioned or supported undertakings. Most states have now enacted similar authorities, thereby making cultural (archeological) resource awareness a truly public concern.

A certain amount of confusion about archeological resource management has resulted from unfamiliarity with environmental assessment and archeological process prescribed and guided by law. At the present time, almost all archeological work proceeds through three interrelated phases in assessing the impact of an undertaking. The undertaking proposed is surveyed (examined) for the presence of archeological sites. Typically, an archeologist walks over an area collecting and sampling evidence of former human occupation or activities. The information gathered often can reveal when and how long sites were used and may indicate relative archeological values. If no sites are found, or if those encountered are not significant (as measured by federal guidelines), the undertaking may proceed. Surface or shallow subsurface archeological surveys do not always return evidence sufficient to assess a site's significance. The site must be tested to determine whether surface indications truly reflect its potential. When a significant site occurs within the boundaries of a proposed undertaking, the effect of the undertaking upon the site must be mitigated. The most favorable form of mitigation is preservation, but preservation may not be prudent or feasible. Mitigation may involve total site excavation or intensive excavation of the more important portions of the site.

The great demands for archeological surveys and assessments are being met by the archeological profession through an emerging speciality that we, here, term "archeological technician." The technician, using tools borrowed principally from the mathematical and physical sciences, is trained to conduct comprehensive on-the-ground surveys to locate sites and to assess sites by their contents, contexts, and integrity. The archeological profession, however, has been unable to provide the technician with a data collection format that provides uniformity and reliability in information recording and, hence, comparability among and between classes of data.

The classification scheme presented here addresses directly the need for a uniform data collection format. It is not truly hierarchical in nature; rather, it is organized by priorities with data gathering and organization assuming the first order. The assumption here is that one must know archeological data in its broader contexts (environment) before he can propose alternatives for site preservation or mitigation (that is, management).

Introduction to the Archeological Resource Inventory

As is true with natural processes, human behavior and the products thereof appear in seemingly endless varieties. Human behavior is socially generated: systems of belief and the resulting products transcend the life of a given community and are transmitted with modifications from one generation to the next. Languages, religions, political and economic systems, and their associated accoutrements are but few examples of man's diverse response to himself and the natural environment. There are, however, commonalities identifiable within these expressions that make man a unique life form, but one whose worldwide presence is clearly recognizable. Man is very dependent upon the natural environment; the cultural record is replete with the tools and trappings through which man has adapted to and exploited the natural systems with which he interrelates.

Archeology, one of the subdisciplines of anthropology (the study of man), strives to understand or interpret human behavior from its products, not from observations of behavior in process. The causal agents (people) are gone; their ideas of the cosmos and world and their role in these, however, are recorded in the physical evidences (artifacts, food remains, features, structures) associated with former human activities.

Archeological surveys and excavations provide the raw data from which former human activities are interpreted. The survey allows site discovery and limited interpretation from observed or recovered cultural and environmental phenomena. Through excavation, archeological sites literally are taken apart; the vertical and horizontal contexts of site contents are recorded meticulously and later analyzed and reported. Archeological sites, however, do not exist *in vacuo*; rather, as expressions of human nature, they reflect man's dependency upon and utilization of the natural environment. With the increasing pressure to discover sites, to preserve few and mitigate many, archeologists must develop uniform conceptual tools that permit rapid and complete field recovery of cultural and relevant natural data.

To understand past human events, activities, and processes, we can order cultural phenomena by their similarities and differences. The system can be arranged hierarchically on the basis of temporal occurrence. The higher levels (A. and AA.) consist of broad temporal categories, and each can be determined from data in the next lower level. The levels B., BB., and C. can be determined from information in one or more of the respective lower levels. The most detailed lower levels (C. and CC.) consist of value elements—cultural phenomena such as artifacts retrieved from the field that were once meaningful to the maker or user. Vertical and horizontal context must be recorded. It is here that the system exhibits characteristics unadaptable to true hierarchical arrangement. The system described below can be employed in gathering a broad range of cultural and environmental observations that will contribute to an organized understanding of our cultural heritage. An outline of components of an archeological site report, including site designation and management status, appears in Table E1.

Explanation of Archeological Hierarchies

Cultural and environmental conditions and processes adduced from physical aspects and spatial relationships of archeological phenomena may be placed in a relative time frame to permit analysis of cultural manifestations. Generally, in this hierarchical arrangement, spatial phenomena and typologies are considered relative to chronology; that is, chronology is the first controlling principle. This approach was found to be most useful by Jennings (1968, p. 25) in his synthesis on North American prehistory.

It is important to keep in mind that the system presented here is intended for use as an inventory and serves as a means to logically organize, summarize, and compare archeological site-natural site information. The archeological and cultural-chronological elements have been arranged in a hierarchical system (as much as possible) to facilitate archeological site comparisons at initial levels of investigation and to meet cultural resource management needs. The cultural resource inventory system should not be construed as a theoretical paradigm for cross-cultural comparisons or as a comprehensive archeological synthesis. No attempt is made to arbitrarily classify levels of cultural manifestations; rather, a descriptive base that employs common archeological definitions is outlined. The degree to which data elements are incorporated within the base depends upon the specific needs of the investigator. The classificatory examples apply predominantly to areas of the eastern United States, but the system has been structured generally so that regional or local variations in chronologies, features, or typologies may be substituted or incorporated. Suggestions and critiques are solicited so that the inventory can be modified and refined for the widest possible applicability.

Major Period (A.). The highest level in the archeological inventory recognizes the major divisions of human history. The **prehistoric period** is considered to be the geographic area occupied before the introduction of writing or written records. The **historic period** is the part of the past that has been recorded in writing. The **protohistoric period** is a locally variable interval between the prehistoric and historic periods: "Protohistory is the study of peoples who were living after history began but who themselves did not have writing" (Hole and Heizer, 1973, p. 6).

At this level, the major period or periods of site occupation should be identified by an archeologist based on the level of investigation being undertaken. For example, a surface reconnaissance of an archeological site may involve mapping of standing historic structures, but prehistoric levels of the site may go undetected unless subsurface testing is done. The amounts and kinds of detailed archeological information gathered will be influenced by site management needs.

Cultural Affiliation (AA.). The three major chronological periods can be subdivided based on cultural traits. As noted previously, prehistoric and often protohistoric identities are based solely on interpretations or artifactual (including food remains and the like) and/or architectural remains. In historic site analyses, the physical evidence is used to analyze aspects not available in written records and to verify and add detail to historic accounts. See Table E2, page 436.

The prehistoric cultural traditions (such as Paleo-Indian and Archaic) listed here apply generally to areas of the eastern United States and Canada referred to as the Eastern Woodlands Culture Area by Willey (1966, p. 248). For a review of these subdivisions, with general appraisals of absolute dates and approximate time ranges, the reader is referred to Griffin (1978, pp. 51–56). We acknowledge that other chronological divisions and classifications exist (see, for example, Caldwell, 1958; Willey and Phillips, 1963).

Cultural Complexes and Components (B.). The cultural period divisions recognized in level AA. can be further subdivided into complexes or component parts. After Jennings (1968, p. 7), a **complex** is considered as a cluster of associated physical objects and their relationships at a particular site or level within a site. Component is a closely related term commonly used in eastern North American archeology. A **component** is a single level at a single site, represented by the associated objects and trails recovered in that level (Jennings, 1968, p. 24). Stratified (multicomponent) sites may contain several different components or culturally distinct complexes. Components and complexes, which are often useful as relative chronological entities, are identified by stratigraphically controlled inventories of elements in lower levels (BB., C., and CC.).

Most of the complexes and components listed in Table E2 (p. 436) are examples selected from recent archeological reports. Other regional and local complexes can be incorporated as needed.

Table E1. *Archeological Site Designation and Management Status*

I. Site Number and Location

- State site number
- Site name
- County
- UTM zone
  - Easting
  - Northing
- District
  - Section
    - Land lot
- Map source
  - USGS quad
  - County
  - Other
    - Map data
    - Map scale
- Site elevation (above mean sea level)

II. Site Investigation and Documentation

- Principal investigator
  - Site supervisor
  - Agency
  - Dates of investigation
- Investigation
  - Documentary
  - Survey
    - Surface (percent of site surveyed)
    - Subsurface testing (percent of site tested)
  - Excavation or mitigation
    - Partial (area and percent)
    - Complete
- Curation
  - Collection location
- Documentation
  - Document locations
- Informants

III. Site Status and Management

- Nature of site
  - Site type
  - Standing architecture
  - Intact features
    - Surface
    - Subsurface
  - Midden
- Site size
  - Length in meters
  - Width in meters
  - Orientation of length

IV. Preservation state

- Undisturbed
- Percent disturbed and nature of disturbance

Table E1. *(continued)*

- Eroded
- Cultivated
- Flooded
- Pot hunted
- Vandalized
- Destroyed

Dates of observation

V. Preservation Prospect

- Safe
- Endangered or threatened
  - Natural processes
    - Mechanical
    - Chemical
    - Specific
  - Human
    - Mechanical
    - Chemical
    - Specific

Owner

- Private
- Municipal
- County
- State
- Federal

VI. Federal or State Register Status

- National Heritage Trust
- National Natural Landmark
- National Historic Landmark
- National Register
  - Significance
    - National
    - Regional
    - Local
  - Status
    - Registered
    - Nominated
    - Eligible
    - Ineligible
    - Deleted
- State or Local Heritage Trust
- Other

VII. Management Recommendation

- Preservation intact
  - Implementation
- Further testing (investigative survey)
- Partial excavation
- Restoration and reconstruction
  - Implementation
- "Complete" excavation
- Mitigation

Site Type (BB.). As defined by Deetz (1967, p. 11), an **archeological site** is "a spatial concentration of material evidence of human activity." This is, of course, one of many operational definitions (Schiffer and Gummerman, 1977, p. 183). A wide variety of archeological site types of varying proportions have been recognized. As can be seen in the selected examples (BB.), some site type designations give more specific indications of use-activity than others. Individual features, such as those listed in (C.), may be considered sites if they occur in isolation (Chang, 1972, p. 9).

Generally, site type designations relate primarily to spatial aspects and secondarily to time. In many cases, the site type is obviously indicative of approximate age and, therefore, aids in chronological determinations found in the higher hierarchical levels. For this reason, the lists of site types (selected examples) have been divided into major period categories (prehistoric, protohistoric, historic). When types cannot be listed for each known component, the types listed are for the predominant components. See Table E2, page 437.

Artifact assemblages and organic remains are basic elements that characterize cultural complexes and make possible their differentiation. In level (C.), specific site features are described in greater detail than in level (B.). For all features and many types of artifacts, size and orientation measurements should be included in the descriptions. Selected prehistoric and protohistoric examples are provided in Table E2.

Diagnostic Value Elements (C.). These are the "nuts and bolts" of the archeological trade and include a broad range of artifact kinds as well as environmental materials (faunal, botanic). One must be careful in entering appropriate data because form and function often overlap terminologically. See Table E2, page 438.

Measurements and Proportions (CC.). Included in this level are measurements of dominant components and diagnostic value elements. Site maps and profile/feature drawings, which are efficient means of displaying spatial relationships, should be included at this level. See Table E2, page 441.

Bibliography

Caldwell, J. R. 1958. Trend and tradition in the prehistory of the Eastern United States. American Anthropological Association, Memoir 88.

Chang, K. C., ed. 1968. Settlement archaeology. National Press, Palo Alto, California.

———. 1972. Settlement patterns in archaeology. Addison-Wesley modules in anthropology, 24. Addison-Wesley, Reading, Mass.

Coe, J. L. 1964. The formative cultures of the Carolina Piedmont. Trans. Amer. Philos. Soc. Number 54.

Deetz, J. 1967. Invitation to archaeology. Natural History Press, New York.

Fitting, J., ed. 1973. The development of North American archaeology: essays in the history of regional traditions. Anchor Press, Garden City, N.Y.

Griffin, J. B., ed. 1952. Archeology of the eastern United States. University of Chicago Press, Chicago.

———. 1978. Eastern United States. Pp. 51–70 *in* R. E. Taylor and G. W. Meighan, eds. Chronologies in New World archaeology. Academic Press, New York.

Hodder, I., and C. Orton. 1976. Spatial analysis in archaeology. Cambridge University Press, Cambridge.

Hole, F., and R. F. Heizer. 1973. An introduction to prehistoric archeology. Holt, Rinehart and Winston, New York.

Jennings, J. D. 1968. Prehistory of North America. McGraw-Hill Book Company, New York.

King, T. F. 1978. The archeological survey: Methods and uses. Cultural Resource Management Studies, Office of Archeology and Historic Preservation, U.S. Department of the Interior, Washington, D.C.

———, P. P. Hickman, and G. Berg. 1977. Anthropology in historic preservation: Caring for culture's clutter. Academic Press, New York.

McGimsey, C. R. 1972. Public archeology. Seminar Press, New York.

McKern, W. C. 1939. The midwestern taxonomic method as an aid to archaeological culture study. Amer. Antiquity 4(4):301–13.

Oakley, C. B., and E. Futato. 1975. Archeological investigations in the Little Bear Creek Reservoir. Research Series No. 1. Office of Archeological Research, University of Alabama, University.

Schiffer, M. B., and G. J. Gummerman, eds. 1977. Conservation archeology: A guide for cultural resource management studies. Academic Press, New York.

Willey, G. R. 1966. An introduction to American archaeology. Vol. 1. North and Middle America. Prentice-Hall, Inc., Englewood Cliffs, N.J.

———, and P. Phillips. 1963. Method and theory in American archaeology. University of Chicago Press, Chicago.

Table E2. *Archeological Hierarchical Elements*

A. Major Period

Prehistoric	Protohistoric	Historic

AA. Cultural Affiliation

Prehistoric

Paleo	Archaic	Woodland
Mississippian	Early	Early
Early	Middle	Middle
Mature	Late	Late
Late		

Protohistoric

Historic

Indian	Nonaboriginal

Unknown

B. Cultural Complexes and Components

(The examples given are selected from the Eastern Woodlands Culture Area and are variously determined from diagnostic lithic types, pottery types, or other diagnostic features or combinations of features.)

Prehistoric

Paleo-Indian		
Preclovis	Clovis	Enterline
Williamson	Quad	Cumberland
Early Archaic		
Dalton	Hardaway	Palmer
Kirk		
Middle Archaic		
Stanly	Eva	Morrow Mountain
Guilford	Three Mile	Stallings Island
Savannah River	Mt. Taylor	Lauderdale
Late Archaic		
Stallings Island	Savannah River	Orange
Lauderdale		
Early Woodland		
Watts Bar	Badin	Dunlap
Kellog		
Middle Woodland		
Badin	Copena	Swift Creek
Hamilton		

Table E2. *(continued)*

Late Woodland		
Napier	Late Swift Creek	
Early Mississippian		
Hopewell	Yadkin	
Mature and Late Mississippian		
Moundville	Fort Walton	Appalachee
Etowah	Lamar	Uwharrie
Pee Dee	Caraway	Duck River
Dallas	Mouse Creek	Early Qualla
	Protohistoric	
Late Qualla	Overhill	Boyd
Tugalo	Dallas	
	Historic	
Indian		
Cherokee	Yuchi	Creek
Chickasaw	Choctaw	Timucua
Catawba	Virginia Algonquians	Tutelo
Delaware	Susquehanna	
Nonaboriginal		
Hispano-Colonial	Franco-Colonial	Anglo-Colonial
Revolutionary War	American Pioneer Expansion	Early Federal Period
Pioneer Industrial	Civil War	
Unknown		
BB. Site Type		
Prehistoric		
Selected examples:		
procurement	camp	village
town	house	farmstead
lithic station	quarry	burial mound
effigy mound	pictograph	petroglyph
rock shelter	cave	fish trap
kill	artifact cache	artifact scatter
shell midden	rock alignment	
Protohistoric		
Selected examples:		
camp site	village	town
house	farmstead	

Table E2. *(continued)*

Historic

Selected examples:

Agricultural		
field	terrace	
barn	granary	stock pen
Commercial-Industrial		
store-office	factory	warehouse storage building
gristmill	sawmill	
mine	quarry	furnace
kiln	still	
Domestic-Public		
house	outhouse	cellar
cave	church	school
tavern	cemetery	trash dump
Military		
fort	breastworks	camp
battleground	rifle pit	entrenchment
Transportation		
trail	road	railroad
tunnel	railroad station	
Water-related		
bridge	dam	levee
canal	ditch	pier landing
ferry	mill pond	well
sewer	water tank	ship or boat

C. Diagnostic Value Elements

Architecture and Features

Surface

Selected examples:

Platform mound		
size	subplatforms	ramps
other		
Effigy mound		
earth	rock	
	zoomorphic	
	anthropomorphic	
Rock mound		

Subsurface

Selected prehistoric and protohistoric examples:

Domestic house	Pit	Hearth
rectangle	cooking	rock
square	storage	clay
oval	burial	raised
round	quarry	basin
other	other	other
wall trench		
individual post		
wattle and daub		
bark		
other		

Table E2. *(continued)*

Burial	Organic Remains	Preparation
primary		mortar
secondary		bedrock
bundle		fieldstone
flexed		
cremation		
urn		
other		
Weir	Trap	Windbreak
Use area	Palisade	
Artifacts by component		
Ceramics		
Selected examples:		
Type		
Prehistoric		
Deptford	Etowah	Lamar
Napier	Swift Creek	Cartersville
Other	Unknown prehistoric	
Historic		
Mocha	Pearlware	Staffordshire
Delftware	Other	Unknown historic
Kind		
Pipe--smoking	Pipe--terracotta	Plate
Cup	Bowl	Jar
Pitcher	Figurine	Brick
Insulator	Other	
Part		
Body shard	Rim shard	Base shard
Handle	Tetrapod	Node
Adorno type	Other	
Paste/Temper		
Grit	Sand	Shell
Grog	Limestone	
Decoration		
Prehistoric		
Plain	Simple stamped	Complicated stamped
Checked stamped	Linear check stamped	Bold incised
Punctate	Fibric impressed	Unidentified stamped
Historic		
Plain	Transfer print	Hand painted
Finger painted (polychrome)	Willow transfer pattern	Polychrome
	Underglazed polychrome	Luster decorated
Brown ironware	Blue/gray ironware	White saltglazed
Black		
Amount		

Table E2. *(continued)*

Lithics		
Selected examples:		
Type		
Standing Boy	Kirk	Stanly
Palmer	Morrow Mountain	Hamilton
Madison	Unidentified	
Kind		
Chipped		
Flake	Projectile point	Biface
Percussion	Complete	Scraper
Bifacially retouched	Tip	Drill
	Base	Knife
Unifacially retouched	Other	Other
		Unidentified
Uniface	Core	Debitage
Scraper		
Drill		
Knife		
Other		
Unidentified		
Ground/Polished		
Axe	Celt	Discoidal
Atlatl weight	Gorget	Dead
Pipe	Hoe	Stone disc
Stone bowl	Grooved stone	Netsinker
Pendant	Pestle	Mortar
Shaped hammerstone	Anvil	Edge-ground implement
Other	Unidentified	
	Ground stone	
	Polished stone	
	Perforated stone	
Other lithics		
Hematite/limonite (ochre "pencil")	Cut mica	Mica plate or crystal
	Quartz crystal	
Glass		
Color		
Green	Clear	Blue
Purple	Milk	
Kind		
Marble	Window	Jar
Bottle	Plate	Globe
Chandelier		
Part		
Lip	Neck	Base
Handle		
Amount		

Table E2. *(continued)*

Metal		
Material		
Iron	Nickel	Copper
Silver	Gold	Alloy
Kind		
Breastplate	Earspool	Nail
Bolt	Nut	square
Button	Buckle	wire
Hook/pin	Eyes	Minié ball
Coin	Cannon	Other
Amount		
Faunal Materials		
Shell		
Gorget	Bead	Hook
Bone		
Gorget	Beads	Awl
Pin	Needle	Hook
Teeth/Claws		
Shark	Bear	Other
Floral Materials		
Basket	Sandal	Digging stick
Projectile shaft	Cloth	Net
Other		
Absolute Dates		
C14 (calibrated; uncalibrated)		
Dendrochronology		
Palynology		
Other		
CC. Measurements and Proportions		
Site Area Dimensions		
Length	Width	Orientation of length
Outline shape	Other	
Profiles		
Occupation layer characteristics	Midden depths	Overlaps
	Intrusions	Other
Artifact Distributions-Densities		

Dropzone Archaic Archeological Site

Union County, Georgia; accessible via U.S. 76 approximately 9 miles west of Blairsville, turn south (left) onto Mulky Gap Road to the point where the paved road ceases and a gravel road begins. Site is S 45° W approximately 300 meters.
Land lots 102, 115, District 10
Mulky Gap, Ga., quad. 7.5 min., 1965
Appalachian Highlands; Blue Ridge; Southern Section
ca. 5 acres

Ownership: Chattahoochee-Oconee National Forests.
Administration: Chattahoochee-Oconee National Forests.
Land use: Clear-cut ca. 1975; hunting.
Land classification: Unknown.
Dangers to integrity: Off-road vehicles; a timber road bisects a portion of the site. Also surface collection by amateurs.
Publicity sensitivity: This is a significant site, but security cannot be assured. Low-key publicity recommended.
Significance and protection priority: The site has local and regional cultural resource significance.
Reasons for priority rating: The site dates approximately B.C. 7000–2000 and contains innumerable tools, blanks, preforms, and lithic debitage. The site has not been subjected to agricultural practices and, hence, has not been plowed. One possible hearth has been uncovered during limited archeological testing. These and other features are anticipated in situ.
Management recommendations: Preservation of site; seedlings have been planted.
Data sources: "Archeological Investigation of an Archaic Site on Proposed Army Dropzone" (report number 77Br401s-1).
General references: None.
General documentation and authentication: Site intensively surface collected and tested (limited) by Kent A. Schneider and staff during the spring and summer of 1977. Specimens analyzed by lithic specialists at the University of Georgia, where specimens are stored. Data sheets on each artifact are available.

Discussion

The site lies on Land Lots 102 and 115, District 10, Union County, Georgia. The site occupies a knoll and is protected by a cove formed by Hicks Ridge. According to a "Clearing and Erosion Plan" prepared in part by Kenneth Day (forest hydrologist), the site locale has a mean elevation of approximately 2,178 feet, a maximum relief of 155 feet, and a mean land slope of 16 percent (Day, 1975). The dominant soils are Bradson and Tate soils, which developed from weathered granite gneiss and mica schist. The internal drainage of these soils is medium. Alluvial soils occur near streams adjacent to the site. Internal drainage is poor here. Prior to a timber sale in 1975, dominant tree species included Northern Red Oak, White Oak, and Hickory, with some Yellow Poplar. At the time of the cultural resource survey (1977), Yellow Poplar seedlings and saplings predominated. The primary understory was Mountain Laurel.

In formulating an assessment strategy for the site, recourse to records, documents, and discussions with local informants was considered essential. Archaic sites that exhibit both vertical and horizontal integrity (stratigraphically intact) are infrequent yet vital to studies dealing with lithic procurement, distribution, and manufacturing processes. Thus, to gain insight into past land management activities, as well as the presence of other cultural resources in the general area of the site, relevant deeds on file at the Forest Supervisor's Office, Gainesville, Georgia, were investigated; local informants Frank Thompson and Paul Mull (who assisted in the survey and whose ancestors received original land grants from the state of Georgia) were questioned; site files derived from Georgia State University and the University of Georgia were examined. Excluding information obtained from Thompson and Mull, little information on past ground-disturbing activities or other site loci was available. Forest Service records suggested that the lands adjacent to the site had been intermittently thinned of timber for decades prior to purchase of the land by the federal government in 1936. These early logging operations in high elevations often relied upon mule and horse for timber skidding and removal by cart; presumably, ground impact was considerably less when contrasted with modern machine removal techniques.

Brief Survey Description

An on-site inspection, commencing 4 March 1977, revealed the presence of lithic debris scattered over a 52,000-square-foot area cleared in November 1976. The impacted area was confined to a knoll that measured approximately 100,000 square feet and represented one-thirty-fifth of a proposed parachute dropzone scheduled to be similarly cleared. The materials observed included quartz and chert flakes exhibiting bifacial thinning (some heat-treated), blades, bifaces, projectile points, a possible spokeshave, and other utilized debitage. Reconnaissance of the area demonstrated the presence of parent outcrops suitable for tool manufacture. Reconnaissance also suggested that the lithic materials occurred in clusters as a result of (or despite) mechanical clearing.

Controlled surface collection of site materials in 400-square-foot grid units followed by minimal subsurface testing guided by an artifact frequency analysis was selected as the most expedient initial assessment strategy. The most salient factors guiding such decision were the presence of large flakes detached from parent outcrops as well as blanks and tools resulting from the reduction of these flakes. In its exposed state, loss or dispersion of these materials from the site through erosion and/or amateur artifact collection appeared inevitable. If left untouched, artifact frequency analyses as well as determination of any in situ cultural phenomena would be impaired seriously.

Brief Description of Survey Results

The intensive surface survey, confined to 52,000 square feet of the site, was conducted on a uniform level of intensity (that is, all collection units were treated as equally as possible). From the 132 collection units, 678 specimens were recovered. Of these, 37 percent exhibited bifacial thinning; 4 percent were classed as bifaces; 3 percent were projectile points; and 6 percent were considered utilized debitage including scrapers (side, end, turtleback), gravers, and possible drills/perforators. The balance of the assemblage was classed as miscellaneous and block flakes. Ninety-eight percent of the assemblage was quartz; 2 percent was chert.

Specimen distribution, grouped into two classes in Figure E1, requires explanation, which in the present preliminary report can only be suggestive. Classes *Flakes* and *Projectile Points* exhibit a major cluster in the area 140W/140N, yet minor flake/point clusters occur in pockets. A cluster trend in a southeast-northwest mode is suggested by the paucity of materials in the northeast portion of the site. An on-the-ground inspection revealed that machine clearing originated in the area of the temporary bench mark and tended to fan outward toward the northwest portion of the site. Hence, one is tempted to infer that artifact distribution and concentration reflect machine blade movement over the ground surface. On the other hand, ground disturbance by blade was not uniform; areas most heavily disturbed occur north, northwest, and northeast of the noted concentration.

Subsurface testing on the machine-cleared area sheds little additional light. Of the seven auger tests conducted, one bifacially thinned flake was found at a depth of 16 inches (160W/140N); although native quartz occurred in abundance, no other recognizably utilized materials were recovered. The subsurface tests were conducted on less than 1 percent of the grid units sampled.

Summary and Recommendations

Archaic lithic materials exposed by the machine clearing of stumps and logging slash have been collected systematically. Preliminary analysis of artifact type and frequency suggests in situ cultural relationships expressed by a cluster as well as pockets of tools in association with manufacturing debitage. Subsurface tests, though inconclusive, suggest that the site may have a depth of 24 cm in some locations and may extend south of the machine-cleared area.

Geographically, the site is suited ideally to seasonal exploitation of the flora with its attendant fauna by early inhabitants of the area. Protected from high winds by a cove formed by Hicks Ridge, the site overlooks a valley still rich in game and vegetation. East of and adjacent to the site is an active stream, the bed of which has exposed workable parent materials native to the area.

From a behavioral standpoint, the site warrants further investigation before additional ground-disturbing activities occur. Little is known of the Archaic stage in the east, even less in north Georgia. The data collected suggest that the site is multicomponent and that the full range of tool manufacture occurred here. Should massive ground disturbance such as cultivation indeed be absent, then a high probability exists that in situ relationships in tool manufacture and subsurface features will be discovered. Moreover, behavioral patterns might be inferred from these activity areas. Investigation aimed at securing information sufficient to arrive at a determination of the site's eligibility for National Register status is essential in planning for future land use.

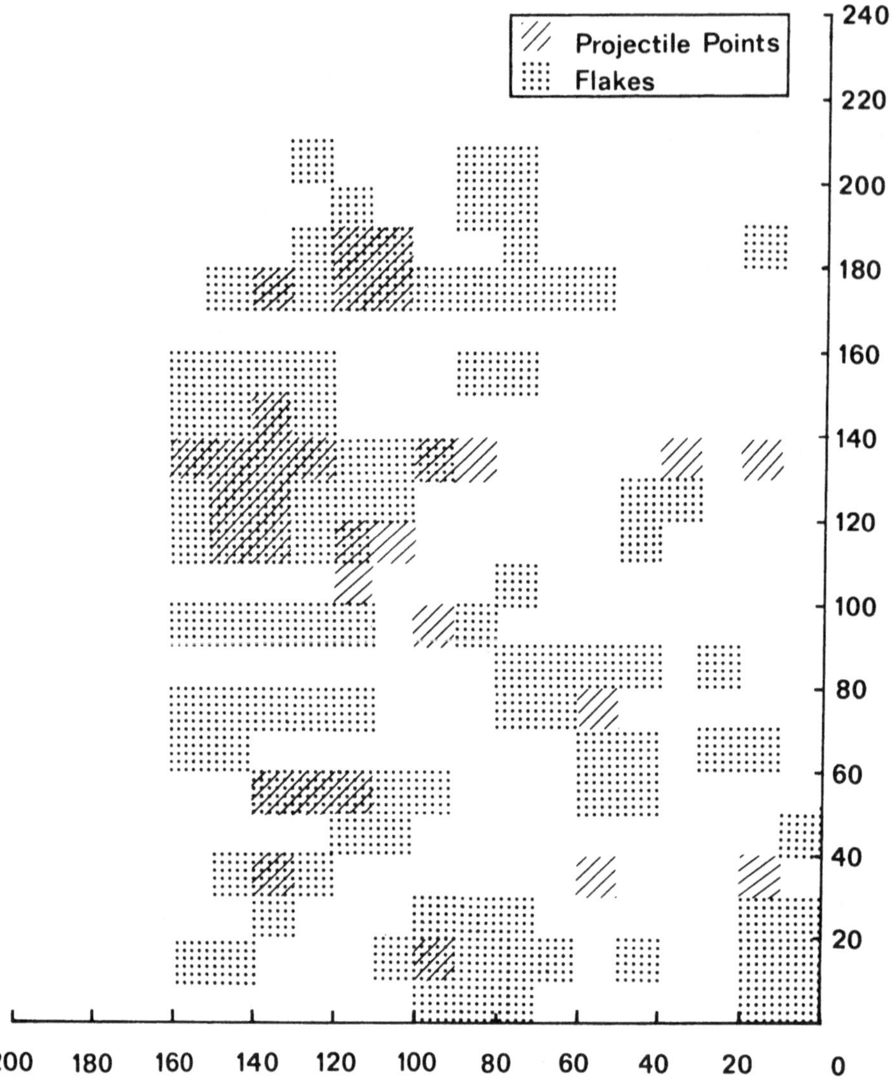

Figure E1. *Artifact distribution.*

Management Summary

1. Impact of proposed projects
   a. The proposed U.S. Army Dropzone encompasses 40 acres of land that will be machine-cleared. The exposed portion of the site lies within the southwest limits of the dropzone. Site limits have not been determined; additional ground clearance may destroy unexposed portions of the site.
   b. A proposed U.S. Forest Service road providing access to Compartment 402 will bisect the exposed portion of the site. Site limits have not been determined; construction of roads and turnarounds may destroy unexposed portions of the site.
2. Proposal for site management. A portion of the site lies exposed and subject to data loss through erosion and amateur collection. Although the surface of the exposed area has been systematically collected, the depth of the site and the presence of in situ cultural relationships have not been fully assessed. Nor have site limits been defined. To determine National Register eligibility, the following work outline is recommended. Should the site not warrant nomination, then the work conducted will provide clearance for dropzone and road construction activities and contribute to knowledge of the Archaic stage.
   a. Thorough literature search of regional Archaic sites;
   b. Test excavations to subsoil on exposed areas exhibiting high and low artifact frequencies;
   c. Test excavations outside exposed area to determine

presence of in situ cultural relationships and site limits.
Estimated time: three weeks
Estimated cost: $6,000

3. Alternatives for site management
   a. Avoidance. Total avoidance seems difficult to justify because a portion of the site has been exposed and is subject to data loss through erosion and amateur collection.
   b. Preservation. Determination of National Register eligibility could be delayed indefinitely, but the exposed portion of the site should be stabilized to prevent data attrition. Stabilization should include covering the exposed portion of the site with plastic or other suitable membrane and covering the membrane with fill. The fill should be seeded.
4. Mitigation. Should the site qualify for National Register status and mitigation is sought, then the following tasks should be accomplished:
   a. Thorough literature search of Archaic sites, regional and national;
   b. Hand removal of understory and expansion of extant grid into unexposed portions of the site;
   c. Removal of the forest litter by fire rake and collection of surface materials per grid unit;
   d. Excavation of high and low artifact frequency areas to subsoil.
   Estimated time: six weeks
   Estimated cost: $12,000

Bibliography

Day, K. 1975. U.S. Army dropzone. Manuscript on file, Forest Supervisor's Office, Gainesville, Georgia.

Natural Area Diversity Summary

CLIMATE: A. Mesothermal; AA. Warm temperate. B. Cool and moist yearly, moderately hot and moderately wet summers, moderately warm and moderately wet winters, with an average freeze-free period; BB. Cool and moist yearly, warm and moderately dry summers, moderately warm and moderately wet winters, with a short freeze-free period. C. Cooler and wetter than local climate; CC. Similar to natural area climate.

SOILS: A. Ultisol; AA. Udult. B. Hapludult; BB. Typic Hapludult. C. (1) Clayey, oxidic, mesic Typic Hapludult, C. (2) Fine-loamy, mixed, mesic Typic Hapludult; CC. (1) Bradson, CC. (2) Similar to Tate.

GEOLOGY: A. (1) Sedimentary, A. (2) Metamorphic; AA. (1) Clastic, AA. (2) Metasedimentary and metaplutonic. B. (1) Siliceous, B. (2) Acidic to intermediate and siliceous; BB. (1) Stream terrace deposit, BB. (2) Regional and dynamic metamorphism. C. (1) Mixed terrace gravels, sands, and muds, C. (2) Biotite gneiss ("granite gneiss") injected with pegmatite dikes, and quartzite which is locally tightly folded and faulted; CC. (1) On-site surface composed of loosely consolidated stream terrace gravel consisting of coarse, angular quartzite pebbles in a clayey-sandy matrix, CC. (2) Bedrock is weakly foliated, fine- to medium-grained biotite gneiss (moderately to highly weathered in nearby roadcut exposures) injected with small pegmatite dikes interlayered with medium- to very coarse-grained quartzite. Quartzite occurs as subangular gravel, cobbles, and boulders (float) densely scattered across the site as components of the stream terrace deposit. Biotite gneiss occurs as scattered cobbles and boulders on the site.

HYDROLOGY: A. Terrestrial; AA. Dry-mesic to wet. B. Fresh; BB. Permanently exposed to seasonally saturated. C. Moderately well-drained to poorly drained slopes and stream terrace; CC. Moderately well-drained lower slopes and poorly drained, seasonally saturated stream terraces and seepages. The site is crossed by approximately 5,300 ft of perennial streams with a stream gradient of 0.07 foot $foot^{-1}$ (7%).

TOPOGRAPHY: A. Hills and mountains; AA. Low mountains. B. Stream valleys between water-eroded ridges-low mountains; BB. Lower ridge slopes and cove near the head of a perennial stream valley. C. Same as local landform; CC. Gently sloping, elongate, northeast-trending knoll on a lower ridge slope extension and adjacent stream terrace in the southwest area of a partly sheltered to sheltered, rolling to irregular, gently sloping to sloping, east-facing cove with a concavoconstant north-south profile.

PHYSIOGRAPHY: A. Appalachian Highlands; AA. Blue Ridge. B. Southern Section; BB. Chattahoochee National Forest, Nottely River drainage basin. C. Valley of north- to east-flowing unnamed tributary of Mulky Gap Branch Creek, located between Hicks Gap Ridge and Hicks Ridge; CC. Stream terrace composed of sands and gravels and adjacent to an elongate northeast-trending knoll extension and underlain by Precambrian biotite gneiss and quartzite.

BIOLOGY: A. Terrestrial broadleaf forest system; AA. Large, deliquescent, deciduous trees/Normal, rhizomatous, evergreen shrubs. B. Oak-hickory forest; BB. Fagales-Juglandales/Ericales. C. Mixed oaks–hickories; CC. Mixed oaks–hickories/*Kalmia latifolia*.

ARCHEOLOGY: A. Prehistoric; AA. Archaic. B. Early Archaic Kirk and Middle Archaic Savannah River, Stanley,

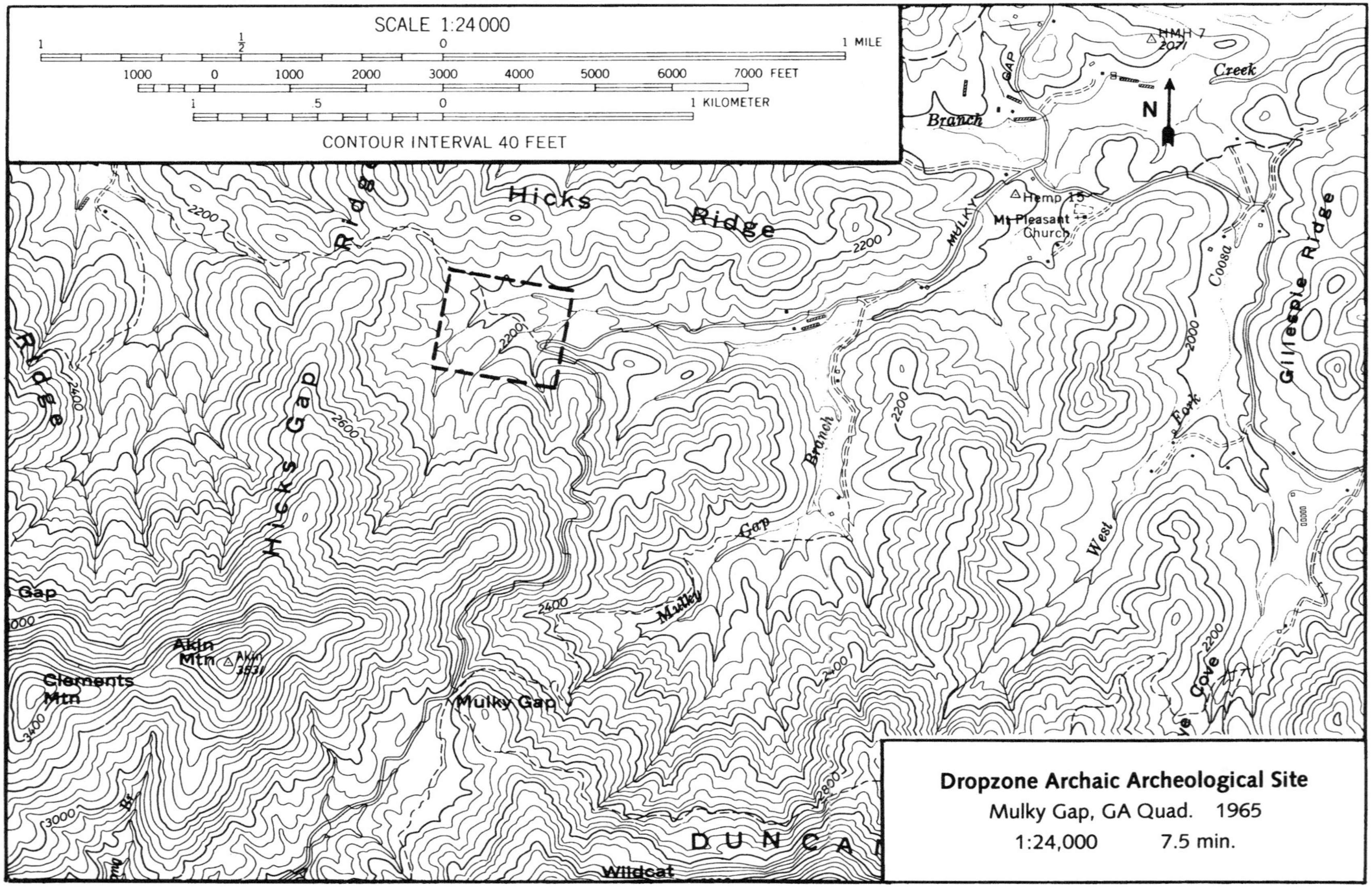

Map E1. *Location of Dropzone Archaic Archeological Site.*

and Morrow Mountain; BB. Lithic station, containing campsite, quarry, artifact scatter, and artifact cache. C. Surface scatter includes rock hearth possibly used for cooking, storage pits, quarry, burial pits, and numerous artifacts; CC. Site approximately five acres in size. Its western limits have not been determined, but it measures approximately 100 meters north to south and lies in a small cove with intermittent streams marking its northern and southern limits.

F

A Species Biology Research Proposal:

The Maintenance, Reproductive, Dispersion, and Establishment Processes in Two Closely Related Caprifoliaceae Species: *Diervilla rivularis* Gattinger and *Diervilla sessilifolia* Buck.*

Three species of *Diervilla* (Bush Honeysuckle) occur in the southeastern United States (Hardin, 1968). *Diervilla lonicera* extends from Newfoundland to Saskatchewan and southward in the Appalachians to North Carolina and Tennessee. *Diervilla sessilifolia*, an endemic of the upper elevations of the southern Appalachians, is widely distributed in the states of Georgia, Alabama, South Carolina, Tennessee, and North Carolina. *Diervilla rivularis*, of much more limited distribution, was first described by Gattinger (1888). It occurs in mountain woods habitats in a total of twelve counties in Alabama, Georgia, North Carolina, and Tennessee (Whetstone, 1978). Cooper et al. (1977) consider both the limited distribution and human development of suitable mountain habitats as the major threats to the continued survival of *D. rivularis*.

Both *D. rivularis* and *D. sessilifolia* are erect shrubs up to two meters tall with opposite, somewhat lanceolate, nearly sessile leaves five to eighteen centimeters long (Cooper et al., 1977). Both species have yellow, unequally five-lobed corollas that turn reddish as the flowers age.

Some confusion, still unresolved, exists over the taxonomic status of these two species (Whetstone, personal communication, 1978). Ahles (1964) considers *D. rivularis* a variety of *D. sessilifolia*, whereas Hardin (1968) considers the two species distinct based primarily on pubescence of the branchlet, leaf, pedicel, and calyx in *D. rivularis*, compared to the glabrous condition of these parts in *D. sessilifolia*. This study hopes to confirm the validity of the taxonomic distinction between these two species.

Hardin (1968) states that he has never observed the two species growing sympatrically, except under artificial cultivation. Whetstone (personal communication, 1978) has observed the two species growing sympatrically in at least one site in Alabama.

Hardin (1968) addresses inconclusively the evolutionary relationship between three species. He suggests that there are two possibilities. *Diervilla rivularis* and *D. sessilifolia* could be most closely related and a product of divergent evolution; or *D. rivularis* could be a disjunct and divergent part of *D. lonicera*, assuming that *D. lonicera* was once distributed throughout the Appalachians before portions of its range were truncated during the Pleistocene. A similar possibility is that *D. rivularis* could be a product of reticulate evolution when *D. lonicera* became sympatric with *D. sessilifolia* because of habitat pressure from glaciation. Hardin concludes that *D. rivularis* and *D. sessilifolia* are the most closely related species but chooses to treat them as distinct species. He suggests that more work on their habitat preference and population structure needs to be done.

The limited distribution of *D. rivularis* compared to the

*Prepared by Mary Sue Henifin, The New York Botanical Garden, Bronx, New York, in partial fulfillment of the requirements for Species Biology–Methods and Techniques (Botany 236) sponsored by the Highlands Biological Station and the University of North Carolina. This proposal incorporates the basic concepts of the species biology approach applied to the study of two closely related species, one relatively common and widely distributed, and the other rare, localized, and proposed as threatened.

widespread distribution of *D. sessilifolia* raises questions about the relative degrees of success in maintenance, reproductive, dispersion, and establishment processes in the life cycles of these two species. Working with a closely related common species such as *D. sessilifolia* allows one to perform manipulative experiments and remove plant parts for further studies, activities that may not be possible for study of the rarer species. Hopefully, conclusions that apply to *D. rivularis* may be drawn from such manipulative work with *D. sessilifolia* populations. Working with two closely related species with different distribution patterns can also help provide a clearer explanation of the selective effect of habitat pressures on the evolutionary history and resulting life cycle processes of these two species.

This research proposal is divided into four parts. Part I describes the proposed research on the maintenance processes and habitat preferences in these two species. Part II outlines the research to determine the reproductive biology. Part III continues with the dispersion and establishment section of this proposal. Part IV presents the conclusions. Parts I–III are self-contained units, each capable of generating a complete research product. The combined time sequence of the experiments and field studies described in this investigation are presented in a summary work plan.

I. Maintenance Processes and Habitat Preference

A. Objectives of Investigation

This work begins with two related major hypotheses to test. A series of questions is developed for each hypothesis to determine its validity (adapted from Massey and Whitson, 1978).

1. Hypothesis I

There is a distinct habitat difference between the two species. One reason for the relative rareness of *D. rivularis* is that its habitat preference type is more limited and/or of a type more subject to human disturbance.

a. How can the general habitat type of *D. sessilifolia* be characterized?

b. How can the general habitat type of *D. rivularis* be characterized?

c. In what habitat type(s) do these species grow sympatrically?

d. Are there intermediate habitats where either species occurs?

e. What types of habitat types support intermediate phenotypes or hybrids between the two species (if such phenotypes or hybrids exist)?

2. Hypothesis II

There is a critical phase in the life cycle of *D. rivularis* that is more vulnerable to biological or environmental disturbance, resulting in more limited population numbers and a negative trend in existing populations.

a. What is the area size of the populations of these two species?

b. Is there a range of classes in the two species?

c. What are the origins of the classes?

d. What are the percentages of each class in the population?

e. What are the spatial relations of the classes?

f. What is the survivorship of each class progressing to the next class?

g. What is the relative vigor of the populations?

h. Is there a difference in gene frequency (both in terms of polymorphism and of relative degree of heterozygosity) in these populations?

In answering these questions, several theoretical parameters can also be determined. What are the life history consequences of natural selection (Gadgil and Bossert, 1970)? Is there a pattern in the life history of these two species and how does it relate to the environment (Murphy, 1968; Schaffer and Gadgil, 1975)?

B. Methods and Techniques

In gathering data to answer the above questions and in confirming or negating the above hypotheses the work has been divided into the following self-contained units.

1. Field Work—Season I

a. A population of *D. rivularis* will be surveyed in each county where it occurs (a total of 12 counties in the states of Alabama, Georgia, North Carolina, and Tennessee). The Radford system of habitat classification will be employed (Radford et al., 1978) to gather information on hydrologic, biotic, climatic, pedologic, geologic, topographic, and physiographic factors. Twelve populations of *D. sessilifolia* in the states of Alabama, Georgia, North Carolina, and Tennessee will be classified in the same manner. In each of these populations approximate numbers of individuals will be determined and voucher specimens will be taken, where not detrimental to the population, and rooted for laboratory work.

b. Three populations of each species will be designated for more precise field work. All individuals in these populations will be marked, and the information indicated in the population maintenance monitoring sheet (Figure F1) will be gathered. A careful map of the population showing locations of individuals will be made. Photographs will be taken. Seeds will be gathered, where not detrimental to the population.

2. Laboratory Work—Winter of Season I

a. The habitat data will be carefully analyzed to determine habitat preference and biotic associates of *D. rivularis* and *D. sessilifolia*.

b. The cuttings from these two species will be placed in different growth chamber regimes varying the following parameters: moisture, light intensity, soil pH and texture, and other factors determined crucial in the habitat analysis.

c. Preliminary gel electrophoresis runs will be made to work out 12 enzyme systems to use in assaying degree of polymorphism in the six intensive-study populations and for determination of vegetatively reproduced individuals versus sexually reproduced individuals (Davis, 1964; Shaw and Prasad, 1970; Johnson, 1975).

3. Field Work—Season II

a. Each population will be monitored once a month using the data sheet in Figure F1. Mortality of individuals and recruitment of seedlings into the adult population will be carefully monitored.

b. Material will be taken from marked individuals for electrophoretic analysis. This material will be placed in liquid nitrogen in the field for quick freezing, then placed on dry ice. It will be assayed within a week of collection.

c. Reciprocal transplant studies will be undertaken to see if *D. sessilifolia* and *D. rivularis* can survive in the habitat preference of the other (Cavers and Harper, 1967).

d. Several plants in the intensive-study populations will be sacrificed, if possible, to determine the extent of vegetative reproduction (rhizome initiation).

C. Results

From the methods and techniques described in the preceding section the following results will be obtained.

1. For Hypothesis I:

a. The habitat preferences of *D. sessilifolia* and *D. rivularis*.

b. Whether or not each species can survive in the habitat type of the other.

2. For Hypothesis II:

a. The area size of the populations.

b. The range of classes in the species analyzed by population and by species.

c. The origins of the classes (either sexual or asexual).

d. The percentages of each class by population and by species.

e. The spatial distribution of classes in each population.

f. The survivorship of each class progressing to the next class analyzed by population and by species.

g. The vigor of the populations and of the species.

h. Differences in gene frequencies analyzed by population and by species.

Population Number: ______________

Date: ____________________

Unit #	Size (height in meters)	# stems	# Primary branches	# flowers	# fruits	$\pm$ cotyledons

Figure F1. *Population Maintenance Monitoring Sheet*

D. Discussion

Demographic parameters are very important in determining plant population dynamics, but very little demographic work has been done on plant populations when compared to the study of animal populations (Harper and White, 1974). The work outlined above will hopefully contribute important information on detailed life histories that can be used to explore several theoretical models (Gadgil and Bossert, 1970; Schaffer and Gadgil, 1975). The analysis of the electrophoretic results will also aid in the estimation of fitness from population data (Prout, 1969). These data will also be used to test the hypothesis that habitat specialists display significantly lower genic variation than habitat generalists (Nevo and Cleve, 1978).

II. The Reproductive Processes

A. Literature Relationship to Topic of Investigation

Diervilla rivularis and *D. sessilifolia* are shrubs that appear to reproduce both sexually and asexually. Both species have a zygomorphic corolla that is dilated upward with five exerted stamens and a capitate stigma. The corolla is light yellow with a darker yellow middle that turns red as the corolla fades. Early work by Lovell (1900) on *D. diervilla* indicated that the color change had no effect on number of insect visitors. Lovell also found that both the yellow and red flowers produced nectar. He indicated that the stigma was no longer receptive when the flower turned red.

A significant amount of work has been done on the development and evolution of the floral structure in the Caprifoliaceae, although this work does not specifically deal with the genus *Diervilla*. Artiushenko (1951) described the development of the blossom and fruit in the Honeysuckle family. Hewins (1900) gave an early description of the organogeny of the flower as well as the embryology of the Caprifoliaceae. Horne (1914) studied evolution in relationship to floral syndromes in Hamamelidaceae, Caprifoliaceae, and Cornaceae. Bailey (1929) discussed reproductive biology in honeysuckles and weigelas of interest to the cultivator. Two early papers subjectively discussed information of interest to pollination studies. Bartlett (1907) discussed flower color in the American *Diervillas*, and Lovell (1900) observed visitors to various species of American Caprifoliaceae.

This literature survey has identified information of interest in planning reproductive biology studies, but none of the literature reports quantitative studies similar to the ones proposed here. Hopefully, by relating information on reproductive biology to habitat preference and population structure of these two species, the evolutionary relationship of these two species can be more fully determined. Theory leading to a more complete model of natural selection and the evolution of reproductive effort (Hirshfield and Tinkle, 1975) will also be a result of this study.

B. Objectives of Investigation

The following questions (adapted from Massey and Whitson, 1978) are designed to test the two related, major hypotheses presented.

1. Hypothesis I

Both *D. rivularis* and *D. sessilifolia* have developed distinct reproductive patterns related to their habitat preferences and evolutionary relationship.

a. Is reproduction occurring
   (1.) in populations of *D. sessilifolia*?
   (2.) in populations of *D. rivularis*?

b. What types of reproduction are occurring
   (1.) in populations of *D. sessilifolia*?
   (2.) in populations of *D. rivularis*?

c. What breeding systems are operative
   (1.) in populations of *D. sessilifolia*?
   (2.) in populations of *D. rivularis*?

d. What pollination systems are operative
   (1.) in populations of *D. sessilifolia*?
   (2.) in populations of *D. rivularis*?

2. Hypothesis II

There is a particularly vulnerable process (or processes) in the reproductive biology of *D. rivularis* that, in part, accounts for its more limited distribution.

a. What is the reproductive capacity or status of populations of *D. sessilifolia*?

b. What is the reproductive capacity or status of populations of *D. rivularis*?

c. What comparisons can be drawn between the reproductive statuses of the two species?

d. What relationships can be drawn between reproductive statuses and the evolutionary positions of these two species?

In developing answers to the above questions, information will be generated that can contribute to the construction of more theoretical parameters. How does evolution shape the development of optimal reproductive strategies (Schaffer, 1974)? By what mechanisms does natural selection determine evolution of reproductive effort (Hirshfield and Tinkel, 1975)?

C. Methods and Techniques

In gathering data to answer the above questions and to confirm or negate the above hypotheses, the work has been divided into self-contained units, each of which can generate a research product.

1. Field Work—Season I

For intensive study of the reproductive biology of *D. rivularis* and *D. sessilifolia* ten inflorescences on each of five plants in each of two populations of each species (out of the three populations for each species used for intensive habitat study) will be frequently monitored. Each of these inflorescences and flowers will be marked with colored string. The information described below will be collected (see Figure F2, reproductive biology data sheet).

a. Number of reproductive units per total population number was determined in the habitat section of this proposal (see Figure F1).

b. Dates of opening of flower buds, stigma receptivity, pollen release, change in flower color, and development of fruit to maturity.

c. Number of flowers/inflorescence.

d. Number of inflorescences/plant (total number will be counted).

e. Number of fruits/flower.

f. Number of seeds/fruit (these seeds will be collected for viability studies).

g. Where not detrimental to the survival of the population, ten inflorescences from five plants in each population will be taken and placed in preservative to allow determination in the laboratory of number of pollen grains/anther and number of ovules/flower. Ten cuttings from each population will also be established in growth chambers, and photoperiods for flowering will be determined.

h. Where not detrimental to the survival of the population, two one-meter-square areas will be carefully excavated, and the pattern of vegetative reproduction will be mapped.

2. Laboratory and Greenhouse Work—Winter of Season I

a. From the preserved material, number of anthers/flower, number of pollen grains/anther, and number of ovules/flower will be counted. Fifty flowers from each population and one anther from each flower (using anthers from different floral positions) will be used to make these counts (see Figure F3, reproductive biology data sheet).

b. Pollen viability, using the aceto-carmine in glycerine jelly technique (Radford et al., 1974), will be determined. Size range and mean pollen size will be calculated. SEM pictures of the sculpturing of the pollen grains will be taken to determine if these sculpturings play a significant role.

c. Mean seed viability will be determined using tetrazolium.

d. The 10 cuttings from each of the four experimental populations will be rooted in growth chambers. One individual from each of the four populations will be dug up at the end of each month to make determinations of vegetative reproduction patterns. These data will be compared with data from the field excavations.

3. Field Work—Season II

a. Pollination studies.

(1.) Five inflorescences on each of five plants in the four experimental populations will be labeled as

	Date bud appears to open	Date stigma appears receptive	Date of pollen release	Date of change in flower color	Date fruit appears mature	No. fruits/ flower	No. seeds/ fruit
Inflorescence Number: _____ Flower No.: ____ ____ ____ ____ ____ ____ ____ ____ ____ ____ ____ ____ ____ Inflorescence Number: _____ Flower No.: ____ ____ ____ ____ ____ ____ ____ ____ ____ ____ ____ ____ ____							

Figure F2. *Reproductive Biology Data Sheet—Field Work*

Plant #	Flower #	# Anthers	# Ovules	# Pollen grains/anther Position of anther	% Pollen Viability

Figure F3. *Reproductive Biology Data Sheet–Laboratory and Greenhouse Work*

controls. Information on seed set will be gathered.

(2.) Five inflorescences on each of five plants in the four experimental populations will be caged and emasculated to test for apomixis.

(3.) Five inflorescences on each of five plants in the four experimental populations will be caged and then self-pollinated to test for self-compatibility.

(4.) Five inflorescences on each of five plants in the four experimental populations will be caged and left alone to test for self-pollination.

(5.) Flowers will also be caged and emasculated (if self-fertile) and then crossed with pollen obtained from other units of the same population, pollen from different populations of the same species, and pollen from the other *Diervilla* species under study.

(6.) The functional details of the flower structure, including ultraviolet patterns, will be studied in the field.

(7.) The availability of pollen and nectar in *D. rivularis* and *D. sessilifolia* will be determined. Time of day at which this food is available will be noted. Quantity and quality of food will be noted (including using a pocket spectrometer to measure sugar concentration in nectar).

(8.) Each population will be studied for insect and bird visitor behavior. Two hours in the early morning, midmorning, early afternoon, and late afternoon will be spent collecting visitors to the *Diervilla* flowers. Notes will be kept on weather as correlated to vector activity. Note of the position of the pollen on the visitor will be made as well as the types of pollen present, where possible. Careful attention will be given to the method by which vectors visit the plant and their position.

(9.) Five inflorescences on each of five plants in the four experimental populations will be caged. The cages will be removed and then recaged after a vector visits (the flower visited will be marked) to test for pollinator efficiency. Species of vector will be determined.

(10.) Information about vector behavior will be correlated with the phenology of the flower.

(11.) Observations on other species that attract *Diervilla* vectors will be noted.

(12.) If time permits, floral, pollen, and nectar manipulation will be used to determine what attracts the vectors.

4. Laboratory and Greenhouse Work—Winter of Season II

a. Plants established in the greenhouse will be manipulated to promote flowering.

b. The crossing experiments performed in the field will be repeated to ensure their accuracy.

D. Results

From the methods and techniques described in the preceding section, the following results will be obtained.

1. For Hypothesis I:

a. The reproductive biology of *D. sessilifolia* and *D. rivularis*, including sexual and asexual systems, and pollinator behavior.

b. The correlations between the reproductive biology and the habitat preferences of these two species.

2. For Hypothesis II:

a. The reproductive effectiveness of these two species:

(1.) Pollen/ovule ratio
(2.) Ovule/seed ratio
(3.) Average fruit production/plant
(4.) Average seed production/fruit
(5.) Percent seed viability
(6.) Average viable seed production/plant

b. The asexual potential of populations of these two species.

c. The sexual potential of populations of these two species:

(number sexual individuals/population) ×
(mean number fruits/plant) ×
(mean number seeds/fruit) ×
(percent viable seeds) =

sexual reproductive potential

E. Discussion

Effective reproduction is essential to the immediate and evolutionary success of a plant population. If only maintenance of individual units occurs without successful reproduction, a chance catastrophe may destroy a significant portion of a species' gene pool. This study will help contribute to knowledge of the importance of sexual versus asexual reproduction (Thomas and Dale, 1975). It will also contribute to our understanding of the various reproductive strategies in plant populations and their effectiveness (Harper and Ogden, 1970; Sarukhan, 1974).

III. Dispersion and Establishment Processes

A. Objectives of Investigation

This proposal describes both laboratory and field work designed to determine the dispersion and establishment processes of *D. rivularis* and *D. sessilifolia*. There are two major hypotheses with questions to test their validity (adapted from Massey and Whitson, 1978).

1. Hypothesis I

There are critical factors in the dispersion and establishment processes of *D. rivularis* that make it a less successful colonizer than *D. sessilifolia*.

a. What types of propagules are present in these two species and what percentages of the propagules are viable?

b. What types of dispersal systems are operative in these two species?

c. What are the dispersal units and/or agents in these two species?

d. What is the dispersal effectiveness of populations of these two species?

e. Are there new individuals present in these populations and what are their origins?

f. What establishment processes are operative in these populations?

2. Hypothesis II

The dispersion and establishment processes of these two species are related to their habitat preferences.

a. What are the spatial relations of establishment processes?

b. What are the establishment effectivenesses based on origin?

c. What relationship do dispersal and establishment processes have to the habitat preferences of these two species?

d. How have dispersal and establishment processes influenced the evolutionary success and relationship of these two species?

In answering these questions, information will also be used to develop important theoretical parameters in plant life history modeling (Mertz, 1970). Results will be tested against Sarukhan and Gadgil's (1974) mathematical model incorporating multiple modes of reproduction.

B. Methods and Techniques

The methods and techniques for determining the number and type of asexual and sexual propagules and their viability under laboratory conditions were described in Part II, the reproductive biology section of this proposal.

In gathering data to answer the questions on dispersion and establishment and to confirm or negate the hypotheses stated in the above section, the work has been divided into self-contained units, each of which can generate a research product.

The following experiments will be conducted in the four intensive-study populations, two each of *D. sessilifolia* and *D. rivularis*, described in Part II, Reproductive Biology.

1. Field Work—Season I

a. Break-off mechanisms—sexual propagules.

(1.) Infructescences on the 10 plants in each of the

four populations used for intensive field study will be observed for the break-off mechanisms of the fruits.

(2.) One hundred fruits on two plants will be painted with radioactive dye, and clear contact paper will be placed below the plants (sticky side up) to determine how many fruits fall directly below the parent plants.

(3.) Several days will be spent in each population making preliminary determination of other biotic and abiotic dispersal mechanisms.

(4.) If biotic dispersal vectors are observed, piles of a certain number of seeds will be placed where accessible to the dispersal vector and qualitative observations on their removal will be made (Beattie and Lyons, 1975).

b. Break-off mechanism—asexual reproduction.

(1.) In the two one-meter-square plots used to determine vegetative reproduction processes (see Part II) rhizomes will be preserved for laboratory study. They will be examined under the microscope and cultured in petri dishes to determine if fungi and/or bacteria play a role in severing connections between vegetatively connected units.

(2.) Radioactive C^{14} experiments will be done on vegetatively connected units to determine if there is nutrient transport between connected units.

(3.) Observations will be made to determine if biotic or abiotic vectors are used to transport asexual propagules (rhizomes or broken branches).

c. Soil samples will be taken in the four intensive-study populations to a depth of one meter. These soil samples will be divided according to soil depth and sifted to determine number of seeds in the soil bank. The seeds will be tested with tetrazolium for viability.

d. In the two permanent plots (two meters square) set up in each of the four intensive-study populations, seedling germination will be carefully monitored throughout the growing season. The information will be correlated with weather data, soil moisture data, and soil pH to get a preliminary assessment of germination requirements.

e. Seeds will be collected, where not detrimental to the population, for laboratory work.

2. Laboratory and Greenhouse Work—Winter of Season I

Determination of preestablishment processes:

a. Various laboratory manipulations of the seeds will be performed to determine under which conditions they germinate best. Treatments tried will be:

(1.) Seeds will be cold treated (samples will be placed at 0°C and groups of seeds removed at monthly intervals to test for germination).

(2.) Seeds will be scarified both mechanically (with a razor blade) and chemically (with dilute sulfuric acid).

(3.) Seeds will be germinated on a temperature gradient bar to determine optimum germination temperature.

(4.) Seeds will be washed in water for varying lengths of time to determine if leaching may be an important factor.

(5.) Various light/dark regimes for germination will be tested to determine photoperiod requirements.

From these manipulations preliminary conclusions will be made about germination requirements.

b. Careful observations will be made on the growth and morphology of germinating seedlings.

c. Ten plants will be established in the greenhouse from each of the four experimental populations. One plant from each population will be dug up at the end of each month to determine vegetative reproduction patterns. These plants will be placed under different temperature and moisture regimes to determine if these factors affect asexual reproduction.

d. Sections will be made to determine the morphology of root initiation and shoot initiation in asexual reproduction.

3. Field Work—Season II

a. Two one-meter-square plots will be placed in each of the four intensive-study populations in areas of high seedling density. Each seedling will be mapped and monitored frequently. Mortality records will be kept. Soil pH, moisture, and temperature range readings will be made on a weekly basis. Amount of predation will be observed.

b. Presence and absence observations on seedlings In other parts of the population will be noted and records kept on their mortality.

c. Two one-meter-square plots in each of four populations will be denuded of all vegetation. The soil will be sifted with large-gauge sifts to remove any vegetative material to a depth of two meters. Observations on the effects of disturbance on vegetative establishment or seedling establishment will be made from these manipulations.

d. A map will be made of individuals appearing to arise vegetatively in the permanent plots described in (1) and (3).

e. At the end of the growing season each seedling and new vegetative unit will be carefully marked so that it can be identified after the winter.

4. Greenhouse and Laboratory Work—Winter of Season II

Seedlings will be grown in 6″ × 12″ greenhouse flats. Various manipulations will be made to determine their mortality patterns. The following factors will be varied:

a. Temperature

b. Moisture

c. Amount of light

d. Degree of predation (simulated by using a paper punch on the seedling leaves to resemble seedling predation observed in the field)

5. Field Work—Season III

a. The seedlings and new vegetative units marked the previous growing season will be monitored for winter mortality. These units will be monitored throughout the growing season.

b. Attempts will be made to correlate mortality with experimental data from the greenhouse studies.

C. Results

From the methods and techniques described in the previous section the following results will be obtained:

1. For Hypothesis I:

a. The types of propagules present and their viability for *D. rivularis* and *D. sessilifolia*.

b. The dispersal systems of these two species.

c. The dispersal units and agents in these two species.

d. The dispersal effectiveness in these two species. (This can be determined only subjectively, not objectively, from the methods described. It would take at least several more seasons of field work to determine this information objectively.)

e. The establishment processes in these two species.

2. For Hypothesis II:

a. The relationship between habitat and establishment processes.

b. The effectiveness of sexual and asexual establishment.

D. Discussion

Dispersal and establishment patterns in these two species can help explain habitat pressure and colonization success. These studies will help explain important phases in the life history (Mertz, 1970; Wilbur et al.,1974) of these two species.

IV. Work Plan Summary

Field Work—Season I

Radford habitat survey of 12 populations each of *D. rivularis* and *D. sessilifolia*
Careful mapping of three populations of each species for demographic analysis
Monitoring of flowering/fruiting
Determination of break-off mechanisms and seed distribution patterns
C^{14} experiments to determine nutrient flow between vegetatively connected units
Determination of seed soil bank and seed viability
Setting up of permanent plots to monitor seed germination in field

Laboratory and Greenhouse Work—Winter of Season I

Manipulation of habitat parameters on plants established in greenhouse
Preliminary gel electrophoresis
Determination of pollen ovule/ratio
Pollen viability testing and morphology studies
Seed viability
Determination of rhizome initiation patterns
Seed germination tests

Field Work—Season II

Monthly demographic monitoring of two populations of each species
Electrophoretic gene frequency analysis for 12 populations of each species
Reciprocal transplant studies
Determination of extent of rhizome initiation
Pollination studies
Seedling monitoring
Establishment observations on denuded plots

Laboratory and Greenhouse Work—Winter of Season II

Repeat of crossing experiments performed in the field
Seedling establishment experiments

Field Work—Season III

Monitoring of seedlings and new vegetative units

V. Conclusions

From the results, the habitat preference and maintenance, reproductive biology, dispersion, and establishment processes of these two species will be determined. Rarity of preferred habitat may be an element in the rareness of *D. rivularis*. Critical phases in the life cycle of these two species may also explain the relative rareness of *D. rivularis* by pointing out where such processes may have broken down. Selective pressures, as they relate to the life cycle, will be described (Murphy, 1968; Hirshfield and Tinkle, 1975), and the evolutionary relationship between the two species will be explained (Jain and Bradshaw, 1966). Management recommendations for the conservation of *D. rivularis* will be made.

Epilogue

Natural Heritage: Classification, Inventory, and Information is a comprehensive treatment of natural diversity from a heritage perspective. The basic rationale and principles involved in the development of the classification system for ecological diversity are presented in the introductory sections of this publication. The relevance of the classification of diversity to identification, inventory, habitat, and site-species relationships is described in the text. The explanations of the classification and information systems and their application to inventory, analysis, and characterization of diversity form the core of the book.

Concepts, definitions, and principles are integrated with classification and identification of ecological diversity, habitat analysis, determination of site-species relationships, and development of information systems. The inclusive classification and nomenclature systems for communities, habitats, and populations embrace biotic and abiotic components and elements. Systematic methods of inventorying and documenting natural areas, critical habitats, and endangered species are presented for thoroughness and efficiency in collecting, comparing, and evaluating information relative to natural heritage programs.

Philosophically, this work is a holistic approach to understanding and preserving our natural diversity and heritage. The classification systems, methods, and selected examples are based on inventories of hundreds of natural areas, communities, and species populations in the United States.

We feel that this book provides a sound basis for the continued development of the classification of natural diversity, for the improvement of inventory procedures, and for the refinement of information systems. Our primary goals are the fostering of the conservation of our natural heritage; the advancing of field training in ecology, systematics, and evolution; and the determining of site-population relationships for species of special concern. A general aim is the recruitment, enlightenment, and training of the many people who will be needed to successfully identify, inventory, evaluate, manage, and protect the significant features of ecologic and genetic diversity in our natural heritage.

The following sections on a natural heritage program (I) and field training in natural diversity (II) are included as suggestions for the accomplishment of the goals for the fostering of conservation and the advancing of field training, respectively. The significant features presented in the third section (III) are considered necessary for the realization of all three primary goals.

I. A Natural Heritage Program for a Region

A successful natural heritage program for any region requires knowledgeable leadership and many enthusiastic participants interested in conservation who have had training in a wide variety of fields such as natural diversity, history, biology, climatology, hydrology, geology, geomorphology, soils, ecology, taxonomy, geography, conservation, environmental management, law, business, politics, editing, promotion, and property acquisition.

An effective natural heritage program has to integrate the talents of conservation-oriented enthusiasts with the strengths of sponsoring agencies and institutions* for the preservation of the natural diversity of an area. The duties and responsibilities of participating individuals and organizations have to be carefully defined and delimited for an efficient heritage operation. Guidelines for resolving overlapping duties and responsibilities resulting from personal interests and institutional mandates must be established early to develop a program with maximum productivity and minimum friction and misunderstanding. Decisions should be reached on desired conservation products (see B. 3) in the developmental

*Sponsoring agencies and institutions: conservation agencies; heritage trusts; regional museums, herbaria, botanical gardens, garden clubs, field stations; wildlife commissions; federal and state agencies.

stages of any program. Careful planning, the selection of a broadly experienced and active policymaking body,* the hiring of properly trained leadership and heritage staff, and the wise use of the expertise of consultants and advisors will ensure the preservation and protection of the natural heritage of any region.

A suggested outline for a natural heritage program for a local region, province, or state is presented here to: (1) aid in the initial organization of a program; (2) provide a means of gaining perspective in an ongoing program; (3) suggest a system for evaluating the progress of an established program; and (4) present a potential plan for prospective products of the natural heritage program for the region.

Suggested Outline for a Natural Heritage Program

A. Our natural heritage
  1. Introduction to the natural heritage of the region
  2. History of land use and classification
B. Organization, sponsorship, and support for a national heritage program
  1. Organization
    a. Board of governors
    b. Heritage staff
    c. Special consultants
  2. Sponsorship and support
    a. Governmental agencies
    b. Private agencies
    c. Individuals
  3. Major processes and products in a natural heritage program
    a. Property acquisition and protection
    b. Registry of natural areas
    c. Heritage publications
    d. Data bank
    e. Management plans
    f. Basic inventories
    g. Environmental education
    h. Promotions
C. Components and elements of the natural heritage of the region
  1. Biology
  2. Climatology
  3. Soils
  4. Geology
  5. Hydrology
  6. Topography
  7. Physiography
D. Natural communities of the region (selected)
  1. Balds
  2. Spruce-fir forests
  3. Montane and piedmont pine forests
  4. Northern and cove hardwood forests
  5. Appalachian oak forests
  6. Flatrock communities
  7. Aquatic communities
  8. Marshes
  9. Bogs
  10. Swamp forests
  11. Savannahs
  12. Maritime forests
E. Natural landmarks of the region
  1. Landmarks of the region
  2. Natural area system for the region
  3. Land and water preserve system for the region
F. Significant features in the natural heritage of the region
  1. Endangered or threatened species and habitats
  2. Biotic and abiotic diversity
  3. Natural features condition
  4. Distribution features
  5. Humanistic features
G. Protection of our natural heritage
  1. Acquisition and management of natural landmarks and critical habitats
  2. Registry of natural areas
  3. Legal aspects of natural heritage work
H. Catalog of species in the natural heritage of the region
  1. Animals
  2. Plants
  3. Minerals
I. The future of the natural heritage in the region
  1. Present status
  2. The future

*Governing board: natural heritage coordinator, conservationist, legal counsel, real estate specialist, management-planner specialist, botanist, zoologist, geologist.

II. Field Training in Natural Diversity

Interdisciplinary field training in ecological diversity is necessary for the implementation and continuation of a natural heritage program in most regions. A course proposal and syllabus is included as a suggestion for initiation and development of natural diversity study at field and/or conservation-oriented institutions.

FIELD TRAINING IN NATURAL DIVERSITY (6 credit hours). Prerequisites: ecology, physical geology, local flora. An interdisciplinary course in field analysis of habitat, inventory of ecological diversity, and determination of site-species relationships. Six lecture and 24 laboratory and field hours per week. Summer. Staff.

This interdisciplinary field experience in natural diversity has been designed for advanced undergraduates, beginning graduates, and professional workers in ecology, systematics, evolution, conservation, environmental management, environmental law, forestry, wildlife, and natural heritage programs. The course has been developed for those interested in field analysis of habitat, inventory of ecological diversity, and the determination of site-species relationships. The field training involves application of diversity classification and information systems that provides (1) facts basic to decisions on habitat management, resource use and protection, environmental impact, natural area acquisition, and land classifications and (2) information fundamental to an understanding of the distribution, origin, migration, and evolution of species, floras, faunas, communities, and ecosystems.

In addition to the prerequisites, background experience in soils classification, climatology, hydrology, aerial photo and map interpretation, and field zoology is desirable. The prerequisites, background experience, and field training in natural diversity are requisites for the study of ecosystematics, the climactic course in a natural diversity-conservation program.

*Lecture Schedule**

Day	Topic	Reading Assignment
M	Natural diversity: Concepts, definitions, and principles	Pt.1, Ch.1
T	Classification of ecological diversity: Organization and rationale for system	Pt.1, Ch.1
Th	Biological diversity: Concepts, classification, and nomenclature	Pt.1, Ch.2, Sec.1
M	Geologic diversity: Concepts, classification, and nomenclature	Pt.1, Ch.2, Sec.4
W	Pedologic diversity: Concepts, classification, and nomenclature	Pt.1, Ch.2, Sec.3
F	Hydrologic diversity: Concepts, classification, and nomenclature	Pt.1, Ch.2, Sec.5
M	Climatic diversity: Concepts, classification, and nomenclature	Pt.1, Ch.2, Sec.2
W	Physiographic diversity: Concepts, classification, and nomenclature	Pt.1, Ch.2, Sec.7
F	Topographic diversity: Concepts, classification, and nomenclature	Pt.1, Ch.2, Sec.6

*See Part IV for basic references.

Lecture Schedule (continued)

Day	Topic	Reading Assignment
M	Site-species relationships and habitat analysis	Pt.1, Ch.1
W	Inventory: Concepts, types, and procedures	Pt.1, Ch.3, Sec.1
F	Significance: Criteria, evaluation, and priority	Pt.1, Ch.3, Sec.3
M	Endangered plant information program	Pt.1, Ch.4
W	Significance of field training in natural diversity	Pt.1, Ch.1 & Epilogue
F	Review of course	Pt.1, Ch.1

*Field and Laboratory Schedule**

Day	Topic
M	Literature research and ecological characterization problem (laboratory)
T	Literature and characterization problem continued (laboratory)
W	Field techniques for analysis of biological diversity (field)
Th	Off-day
F	Field techniques for analysis of biological diversity (field)
M	Biological data analysis; preparation for field studies in geology (laboratory)
T	Field techniques for analysis of geologic diversity (field)
W	Geologic data analysis; preparation for field studies in pedology (laboratory)
Th	Field techniques for analysis of pedologic diversity (field)
F	Off-day
M	Pedologic data analysis; preparation for field studies in hydrology and climate (laboratory)
T	Field techniques for analysis of hydrologic and climatic diversity (field)
W	Hydrologic and climatic data analysis; preparation for field studies in topography and physiography (laboratory)
Th	Field techniques for analysis of topographic and physiographic diversity (field)
F	Off-day
M	Topographic and physiographic data analysis; preparation for natural area field studies (laboratory)
T	Natural area study (field)
W	Natural area data analysis and synthesis; preparation for natural area field studies (laboratory)
Th	Natural area study (field)
F	Off-day
M	Natural area data analysis and synthesis; preparation for endangered species field studies (laboratory)
T	Endangered species population and status inventory (field)
W	Endangered species population and status reports; preparation for natural area field studies (laboratory)
Th	Natural area study (field)
F	Off-day

*See detailed instructions in Part IV.

III. Significant Features for Inventory, Analysis, and Characterization of Ecological Diversity

The features listed below, in essence, represent the fundamental elements found in a prospectus for a book on natural heritage. In retrospect these special features are the significant aspects of any heritage text that provides a sound basis for natural heritage work—inventory, analysis, and characterization. These elements are also the basic requirements for promoting conservation, developing field training in diversity, and increasing the understanding of the biology and habitat of the species and communities of special concern.

Significant Features

1. Classification system for ecological diversity.
2. Ecological component classifications and descriptions.
3. System of inventory procedures for natural diversity of local areas, regions, and states.
4. Information system for study of endangered species.
5. Documentation systems for species biology and communities.
6. Ecological characterization system for populations, communities, and habitats.
7. Site-species relationship system for determination of critical habitat and management of populations.
8. Habitat analysis procedures for populations and communities.
9. Major ecosystem (aquatic, wetland, terrestrial) circumscriptions.
10. Master set of diversity elements files for state heritage programs.
11. Standardized format for reports for species, communities, and natural areas.
12. Natural diversity summary system for data banking and heritage analysis.
13. Species status summary for threatened and endangered populations.
14. Set of criteria for determination of significance of communities and natural areas.
15. Compilation of natural heritage resource information: archeological resource inventory and classification, registry of natural areas, state heritage programs, least common denominator element files, classification of rare species, grading natural quality of a natural community, some procedures for detecting disturbances, and categories and boundaries of natural areas.
16. Treatment of basic concepts and definitions pertinent to natural heritage work.
17. Set of guiding principles for development of ecological diversity classification and species biology information systems.
18. Selected examples illustrating application of classification, inventory, information, and documentation systems.
19. Many examples of systems application giving information basic to decisions on land classification, habitat management, natural area acquisition, resource use and protection, environmental impact, and stewardship of natural areas.
20. Detailed Instructions for field analysis of habitats and inventory of ecological diversity for populations and communities.

A Concluding Statement

A viable, dynamic, and efficient natural heritage program requires dedicated leadership and the cooperative, enlightened, and sustained efforts of many amateur and professional workers for the preservation, management, and protection of natural diversity. Each national, state, and regional heritage staff must have at least one specialist trained in natural diversity, who has a real enthusiasm for stewardship.

Effective stewardship of our natural heritage demands the procurement of the most qualified individuals with the best possible training in ecological diversity—those who appreciate the relationships of the elements of diversity to time and the effects of man, wind, fire, flood, and disease—those who have a zeal for the preservation of our genetic and ecological heritage.

A. E. Radford
J. R. Massey
D. K. S. Otte
L. J. Otte
P. D. Whitson

Bibliography

Ahles, H. E. 1964. New combinations for some vascular plants of the southeastern United States. J. Elisha Mitchell Sci. Soc. 80:172–73.

Artiushenko, Z. T. 1951. Development of blossom and fruit in the honeysuckle family (Caprifoliaceae). Amorphol. Anat. Pl. 2:131–69.

Ashe, W. W. 1897. Forests of North Carolina. Pp. 139–224 in G. Pinchot and W. W. Ashe. Timber trees and forests of North Carolina. North Carolina Geological Survey Bulletin 6. Raleigh.

Ayars, J. S., P. L. Altman, M. Broadbent, E. J. Huth, D. R. Lincicome, and R. V. Ormes (Committee on Form and Style). 1972. CBE style manual. 3rd ed. American Institute of Biological Sciences, Washington, D.C.

Ayensu, E. S., and R. A. DeFilipps. 1978. Endangered and threatened plants of the United States. Smithsonian Institution and World Wildlife Fund, Inc., Washington, D.C.

Bailey, L. H. 1929. The case of *Diervilla* and *Weigela*: A discussion of bush honeysuckles and weigelas in the interest of the cultivator. Gentes Herb. 2:39–54.

Bailey, R. G. 1976. Ecoregions of the United States. Fish and Wildlife Service, U.S. Forest Service, Ogden, Utah.

Barker, E. D. 1922. A note from Okefinokee. Torreya 22(6):104–6.

Bartlett, H. H. 1907. Flower color in the American *Diervillas*. Rhodora 9:147–48.

Bayly, B. 1968. Introduction to petrology. Prentice-Hall, Inc., Englewood Cliffs, N.J.

Beard, J. S. 1933. The savannah vegetation of northern tropical America. Ecol. Monogr. 23:195–215.

Beattie, A. J., and N. Lyons. 1975. Seed dispersal in *Viola* (Violaceae): Adaptations and strategy. Amer. J. Bot. 62:712–22.

Beaven, G. F., and H. J. Oosting. 1939. Pocomoke Swamp: A study of a cypress swamp on the eastern shore of Maryland. Bull. Torrey Bot. Club 66:367–89.

Bennett, S. 1977. The virgin forest—how can you tell? Yankee, December 1977, pp. 102–7.

Bergman, H. F. 1920. The relation of aeration to the growth and activity of roots and its influence on the ecesis of plants in swamps. Ann. Bot. (London) 34:13–33.

Billings, M. P. 1972. Structural geology. 3d ed. Prentice-Hall, Inc., Englewood Cliffs, N.J.

Black, C. A., D. D. Evans, J. L. White, L. E. Ensminger, and F. E. Clark, eds. 1965. Methods of soil analysis. Part 1. Physical and mineralogical properties, including statistics of measurement and sampling. American Society of Agronomy, Monograph 9. Madison, Wisconsin.

Blatt, H., G. Middleton, and R. Murray. 1972. The origin of sedimentary rocks. Prentice-Hall, Inc., Englewood Cliffs, N.J.

Blumenstock, D. I., and C. W. Thornthwaite. 1941. Climate and man. Pp. 98–127 in U.S. Department of Agriculture. Climate and man. U.S. Government Printing Office, Washington, D.C.

Boufford, D., and E. Wood. 1976. Panther Creek Natural Area. Pp. 270–81 in A. E. Radford. Vegetation—habitats—floras—natural areas in the southeastern United States: Field data and information. University of North Carolina Student Stores, Chapel Hill.

Brady, N. C. 1974. The nature and properties of soils. 8th ed. Macmillan Co., New York.

Braun, E. L. 1950. Deciduous forests of eastern North America. Hafner Press, New York.

Braun-Blanquet, J. 1932. Plant sociology: The study of plant communities. Translated by G. D. Fuller and H. S. Conard. McGraw-Hill Book Co., Inc., New York.

Brown, C. A. 1943. Vegetation and lake level correlations at Catahoula Lake, Louisiana. Geogr. Rev. (New York) 33:435–45.

______. 1959. Vegetation of the Outer Banks of North Carolina. Louisiana State University Coastal Studies Series 4. Baton Rouge.

Brown, D. M. 1941. Vegetation of Roan Mountain: A phytosociological and successional study. Ecol. Monogr. 11:61–97.

Buckman, R. E., and R. L. Quintus. 1972. Natural areas of the Society of American Foresters. Washington, D.C.

Buell, M., and W. A. Wistendahl. 1955. Flood plain forests of the Raritan River. Bull. Torrey Bot. Club 82:463 –72.

Buol, S. W., F. D. Hole, and R. J. McCracken. 1980. Soil genesis and classification. 2d ed. The Iowa State University Press, Ames.

Butler, J. R., E. S. Custer, and W. A. White. 1975. Potential geological natural landmarks: Piedmont region, eastern United States. Prepared for the National Park Service. Unpublished report.

Cajander, A. K. 1926. The theory of forest types. Acta Forest. Fenn. 29.

Caldwell, J. R. 1958. Trend and tradition in the prehistory of the eastern United States. American Anthropological Association, Memoir 88. Menasha, Wisconsin.

Carter, J. 1977. Weymouth Woods management plan—Section I. Division of Parks and Recreation, Raleigh, N.C.

Cavers, P. B., and J. L. Harper. 1967. Studies in the dynamics of plant populations. 1. The fate of seed and transplants introduced into various habitats. J. Ecol. 55:59 –71.

Chang, K. C., ed. 1968. Settlement archaeology. National Press, Palo Alto, Calif.

______. 1972. Settlement patterns in archaeology. Addison-Wesley module in anthropology, 24. Addison-Wesley, Reading, Mass.

Chapman, H. S. 1932. Is the longleaf type a climax? Ecology 13:328 –34.

Chapman, V. J. 1974. Salt marshes and salt deserts of the world. 2d ed. Verlag von J. Cramer, Lehre, Germany.

———. 1976. Coastal vegetation. 2d ed. Pergamon Press, Oxford, England.

Christensen, N. L. 1977. Fire and soil –plant nutrient relations in a pine-wiregrass savannah of the Coastal Plain of North Carolina. Oecologia 31:27 –44.

Clements, F. E. 1916. Plant succession: An analysis of the development of vegetation. Publ. Carnegie Inst. Wash. 242:1 –512.

———. 1928. Plant succession and indicators. Wilson, New York.

———. 1936. Nature and structure of the climax. J. Ecol. 24:252 –84.

———, and V. E. Shelford. 1939. Bioecology. John Wiley & Sons, Inc., New York.

Coe, J. L. 1964. The formative cultures of the Carolina Piedmont. Trans. Amer. Philos. Soc., Vol. 54, Part 5. Philadelphia, Pa.

Coker, W. C., and H. R. Totten. 1945. Trees of the southeastern states. University of North Carolina Press, Chapel Hill.

Committee for Tennessee Rare Plants. n.d. The rare vascular plants of Tennessee. B. E. Wofford, Rapporteur for the Committee, Department of Botany, University of Tennessee, Knoxville, Tenn., 37916. Manuscript.

Committee on Vascular Plants. 1977. Vascular plants. Reprinted from J. E. Cooper, S. S. Robinson, and J. B. Funderburg, eds. Endangered and threatened plants and animals of North Carolina. Bookstore, University of North Carolina, Charlotte.

Compton, R. R. 1962. Manual of field geology. John Wiley & Sons, Inc., New York.

Conant, R. 1975. A field guide to reptiles and amphibians of eastern and central North America. Houghton Mifflin, Boston.

Cooke, R. V., and A. Warren. 1973. Geomorphology in deserts. University of California Press, Berkeley.

Cooper, A. W., and E. D. Waits. 1973. Vegetation types in an irregularly flooded salt marsh on the North Carolina Outer Banks. J. Elisha Mitchell Sci. Soc. 89:78 –91.

Cooper, J. E., S. S. Robinson, and J. B. Funderburg, eds. 1977. Endangered and threatened plants and animals of North Carolina. North Carolina State Museum of Natural History, Raleigh.

Cowardin, L. M., V. Carter, F. C. Golet, and E. T. LaRoe. 1976. Interim classification of wetlands and aquatic habitats of the United States. Fish and Wildlife Service, U.S. Department of the Interior. U.S. Government Printing Office, Washington, D.C.

———. 1977. Classification of wetlands and deep-water habitats of the United States (an operational draft). Fish and Wildlife Service, U.S. Department of the Interior. U.S. Government Printing Office, Washington, D.C.

Cronquist, A. 1968. The evolution and classification of flowering plants. Houghton Mifflin, Boston.

———. n.d. Asteraceae. In A. E. Radford, ed. Vascular flora of the southeastern United States. Manuscript.

Curtis, J. T., and G. Cottam. 1962. Plant ecology workbook. Burgess Publishing Co., Minneapolis, Minn.

Cypert, E. 1961. The effects of fires in the Okefinokee Swamp in 1954 and 1955. Amer. Midl. Naturalist 66:485–503.

Dansereau, P. 1951. Description and recording of vegetation upon a structural basis. Ecology 32:172–229.

———. 1957. Biogeography: An ecological perspective. Ronald Press, New York.

Daubenmire, R. 1952. Forest vegetation of northern Idaho and adjacent Washington, and its bearing on concepts of vegetation classification. Ecol. Monogr. 22:301–30.

———. 1966. Vegetation: Identification of typal communities. Science 151:291–98.

———. 1968. Ecology of fire in grasslands. Advances Ecol. Res. 5:209–66.

———. 1976. The use of vegetation in assessing the productivity of forest lands. Bot. Rev. (Lancaster) 42:115–43.

Davis, B. J. 1964. Disc electrophoresis. Ann. New York Acad. Sci. 121:404–37.

Day, K. 1975. U.S. Army dropzone. Manuscript on file, Forest Supervisor's Office, Gainesville, Ga.

Day, R. K., and D. DenUyl. 1932. The natural regeneration of farm woods following the exclusion of livestock. Purdue University Agricultural Experiment Station Bulletin 368. Lafayette, Indiana.

Dayton, B. R. 1966. The relationship of vegetation to Iredell and other Piedmont soils in Granville County, North Carolina. J. Elisha Mitchell Sci. Soc. 82:108–18.

Dean, G. W. 1969. Forests and forestry in the Dismal Swamp. Virginia J. Sci. 20:166–73.

Deetz, J. 1967. Invitation to archaeology. Natural History Press, New York.

Demaree, D. 1932. Submersing experiments with *Taxodium*. Ecology 13:258–62.

DenUyl, D., O. D. Diller, and R. K. Day. 1938. The development of natural reproduction in previously grazed farmwoods. Purdue University Agricultural Experiment Station Bulletin 431. Lafayette, Indiana.

Dunn, J. 1977. Soil survey of Orange County, North Carolina. Soil Conservation Service, U.S. Department of Agriculture. U.S. Government Printing Office, Washington, D.C.

Dyksterhuis, E. J. 1957. The savannah concept and its use. Ecology 38:435–42.

Egler, F. E. 1977. The nature of vegetation: Its management and mismanagement. An introduction to vegetation science. Aton Forest, Norfolk, Conn.

Endangered Species Act of 1973. Public Law 93–205. 87 Statute 884–903. 93d Congress, S. 1983. In United States statutes at large. 1974. Vol. 87. Published under direction of the Administrator of General Services by the Office of the Federal Register, National Archives and Records Service. U.S. Government Printing Office, Washington, D.C.

Endangered Species Act Amendments of 1978. Public Law 95–632. 92 Statute 3751–67. 95th Congress, S. 2899. Presently unbound.

Fairbridge, R. W., ed. 1968. The encyclopedia of geomorphology. Reinhold Book Corporation, New York.

Federal Information Processing Standards Publication. 1973. Counties and county equivalents of the states of the United States. Federal General Data Standard, Representations and Codes (FIPS 6–2). U.S. Department of Commerce, National Bureau of Standards, Office of ADP Standards Management. U.S. Government Printing Office, Washington, D.C.

Fenneman, N. M. 1931. Physiography of western United States. McGraw-Hill Book Co., Inc., New York.

______. 1938. Physiography of eastern United States. McGraw-Hill Book Co., Inc., New York.

Fernald, M. L. 1950. Gray's manual of botany. 8th ed. D. Van Nostrand, New York.

Fish and Wildlife Service. 1977. International trade in endangered species of wild fauna and flora. Federal Register. February 22. U.S. Department of the Interior. U.S. Government Printing Office, Washington, D.C.

Fitting, J., ed. 1973. The development of North American archaeology; essays in the history of regional traditions. Anchor Press, Garden City, N.Y.

Forestry Suppliers, Inc. 1978. Catalogue Number 28. Spring 1978–Spring 1979. Jackson, Miss.

Fosberg, F. R. 1961. A classification of vegetation for general purposes. Trop. Ecol. 2:1–28.

Foster, A. S., and E. M. Gifford, Jr. 1974. Comparative morphology of vascular plants. W. H. Freeman & Co., San Francisco.

Gadgil, M., and W. H. Bossert. 1970. Life history consequences of natural selection. Amer. Naturalist 104:1024.

Garner, H. F. 1974. The origin of landscapes: A synthesis of geomorphology. Oxford University Press, New York.

Gary, M., R. McAfee, Jr., and C. L. Wolf, eds. 1972. Glossary of geology. American Geological Institute, Washington, D.C.

Gattinger, A. 1888. *Diervilla rivularis*, n. sp. Bot. Gaz. (Crawfordsville) 13:191.

Gemborys, S. R., and E. J. Hodgkins. 1971. Forests of small stream bottoms in the Coastal Plain of southwestern Alabama. Ecology 52:70–84.

General Services Administration, Office of Finance. 1976. Worldwide geographic location codes. U.S. Government Printing Office, Washington, D.C.

Gleason, H. A. 1952. The new Britton and Brown illustrated flora of the northeastern United States and adjacent Canada. Vols. 1–3. New York Botanical Garden, New York.

Goldthwait, R. P., ed. 1975. Glacial deposits. Dowden, Hutchinson, & Ross, Inc., Stroudsburg, Pa.

Goode, J. P., and E. B. Espenshade, Jr. 1950. Goode's school atlas. Rand McNally & Co., New York.

Gray, A. 1842. Notes of a botanical excursion to the mountains of North Carolina . . . Amer. J. Sci. 42:1–49. Published also in London J. Bot. (1842+). Vol. 1.

Gregory, K. J., and D. E. Walling. 1973. Drainage basin form and process: A geomorphological approach. John Wiley & Sons, Inc., New York.

Greitzer, S. L. 1944. Elementary topography and map reading. McGraw-Hill Book Co., Inc., New York.

Griffin, J. B., ed. 1952. Archeology of the eastern United States. University of Chicago Press, Chicago.

______. 1978. Eastern United States. Pp. 51–70 in R. E. Taylor and C. W. Meighan, eds. Chronologies in New World archeology. Academic Press, New York.

Hall, T. F., and W. T. Penfound. 1939. A phytosociological analysis of a cypress-gum swamp in southeastern Louisiana. Amer. Midl. Naturalist 21:378–95.

______. 1943. Cypress-gum communities in the Blue Girth Swamp near Selma, Alabama. Ecology 24:208–17.

Hamblin, W. K., and J. D. Howard. 1975. Exercises in physical geology. 4th ed. Burgess Publishing Co., Minneapolis, Minn.

Hammond, E. H. 1964. Classes of land surface form in the forty-eight states, U.S.A. Annals of the Association of American Geographers. Map Supplement Number 4. U.S. Government Printing Office, Washington, D.C.

Hardin, J. W. 1968. *Diervilla* (Caprifoliaceae) of the southeastern United States. Castanea 33:31–36.

______, and Committee. 1977. Vascular plants. Pp. 56–138 in J. E. Cooper, S. S. Robinson, and J. B. Funderburg, eds. Endangered and threatened plants and animals of North Carolina. North Carolina State Museum of Natural History, Raleigh.

Harper, J. L., and J. Ogden. 1970. The reproductive strategy of higher plants. 1. The concept of strategy with reference to *Senecio vulgaris* L. J. Ecol. 58:681–98.

Harper, J. L., and J. White. 1974. The demography of plants. Pp. 419–36 in R. F. Johnston, P. W. Frank, and C. D. Michener, eds. Annual review of ecology and systematics. Vol. 5. Annual Reviews Inc., Palo Alto, Calif.

Harper, R. M. 1905. Further observations on *Taxodium*. Bull. Torrey Bot. Club 32:105–15.

Henifin, M. S., L. E. Morse, J. L. Reveal, B. MacBryde, and J. I. Lawyer. 1980. Guidelines for the preparation of status reports on rare or endangered plant species. In L. E. Morse and M. S. Henifin, eds. Geographical data organization for rare plant conservation. New York Botanical Garden, New York. In press.

Henry, E. F. 1970. Soils of the Dismal Swamp of Virginia. Virginia J. Sci. 21:41.

Heritage Conservation and Recreation Service (H.C.R.S.). 1978. Ecological classification in the National Heritage Program. Symposium, Washington, D.C., 20–24 March 1978. Unpublished draft.

Hewins, N. P. 1900. A contribution to a knowledge of the organogeny of the flower and of the embryology of the Caprifoliaceae. Proc. Amer. Acad. Arts 49:280–81.

Hinds, N. E. A. 1943. Geomorphology: The evolution of landscape. Prentice-Hall, Inc., New York.

Hirshfield, M. F., and D. W. Tinkle. 1975. Natural selection and the evolution of reproductive effort. Proc. Acad. Nat. Sci. Philadelphia 72:2227–31.

Hitchcock, A. S. 1950. Manual of the grasses of the United States. 2d ed. Revised by A. Chase. U.S. Department of

Agriculture Misc. Pub. No. 200. U.S. Government Printing Office, Washington, D.C.

Hodder, I., and C. Orton. 1976. Spatial analysis in archaeology. Cambridge University Press, Cambridge.

Hole, F., and R. F. Heizer. 1973. An introduction to prehistoric archeology. Holt, Rinehart, and Winston, New York.

Holmgren, P. K., and W. Keuken. 1974. Index herbariorum. Part 1. The herbaria of the world. 6th ed. Oosthoek, Scheltema & Holkema, Utrecht, Netherlands.

Horne, A. S. 1914. A contribution to the study of the evolution of the flower, with special reference to the Hamamelidaceae, Caprifoliaceae, and Cornaceae. Trans. Linn. Soc. London, Bot. 8:239–309.

Hotchkiss, N., and R. E. Stewart. 1947. Vegetation of Patuxent Refuge, Maryland. Amer. Midl. Naturalist 38:1–75.

Jackson, H. S. 1978. Superintendent of Carolina Beach State Park. Personal communication of 17 October to J. Taggart.

Jain, S. K., and A. D. Bradshaw. 1966. Evolutionary divergence among adjacent plant populations. 1. Evidence and its theoretical analysis. Heredity 21:407–41.

Jennings, J. D. 1968. Prehistory of North America. McGraw-Hill Book Co., Inc., New York.

Johnson, G. B. 1975. The use of internal standards in electrophoretic surveys of enzyme polymorphism. Biochem. Genetics 13:833–47.

Kartesz, J. T., and R. Kartesz. 1977. The biota of North America. Part 1, Vascular plants. Vol. 1, Rare plants. B.O.N.A.C., Pittsburgh. Published by authors.

Kemp, J. F. 1940. A handbook of rocks, for use without the petrographic microscope. 6th ed. D. Van Nostrand Co., Inc., New York.

King, C. C., and L. E. Elfner, eds. 1975. Organisms and biological communities as indicators of environmental quality—a symposium. Ohio Biological Survey Informative Circular No. 8. Columbus, Ohio.

King, T. F. 1978. The archeological survey: Methods and uses. Cultural Resource Management Studies, Office of Archeology and Historic Preservation, U.S. Department of the Interior, Washington, D.C.

———, P. P. Hickman, and G. Berg. 1977. Anthropology in historic preservation: Caring for culture's clutter. Academic Press, New York.

Köppen, W. 1931. Grundriss der Klimakunde. Walter de Gruyter Co., Berlin.

Kologiski, R. L. 1977a. Phytosociology of the Green Swamp, North Carolina. Ph.D. dissertation. North Carolina State University, Raleigh.

———. 1977b. The phytosociology of the Green Swamp, North Carolina. Agricultural Experiment Station Tech. Bull. No. 250. Raleigh, N.C.

Komarek, E. V. 1974. Effects of fire on temperate forests and related ecosystems. Pp. 251–77 in T. T. Kozlowski and C. E. Ahlgren, eds. Fire and ecosystems. Academic Press, New York.

Küchler, A. W. 1947. A geographic system of vegetation. Geogr. Rev. (New York) 37:233–40.

———. 1949. A physiognomic classification of vegetation. Ann. Assoc. Amer. Geogr. 39:201–10.

———. 1951. The relation between classifying and mapping vegetation. Ecology 32:275–83.

———. 1964. Potential natural vegetation of the conterminous United States. American-Geographical Society, Special Research Publication No. 36. New York.

———. 1967. Vegetation mapping. Ronald Press, New York.

Kuntze, O. 1891. Revisio Generum Plantarum . . . Pars. 1. Arthur Felix, Leipzig.

Kurz, H., and D. Demaree. 1934. Cypress buttresses and knees in relation to water and air. Ecology 15:36–41.

Lahee, F. H. 1961. Field geology. 6th ed. McGraw-Hill Book Co., Inc., New York.

Lawrence, G. H. M., A. F. G. Buchheim, G. S. Daniels, and H. Dolezal, eds. 1968. B-P-H: Botanico-Periodicum-Huntianum. Hunt Botanical Library, Pittsburgh, Pa.

Lee, W. D. 1955. The soils of North Carolina: Their formation, identification, and use. Tech. Bull. No. 115. North Carolina Agricultural Experiment Station, Raleigh.

Leopold, L. B., and J. P. Miller. 1956. Ephemeral streams—hydraulic factors and their relation to the drainage net. Geological Survey Professional Paper 282-A. U.S. Government Printing Office, Washington, D.C.

Lobeck, A. K., and W. J. Tellington. 1944. Military maps and air photographs: Their use and interpretation. McGraw-Hill Book Co., Inc., New York.

Lovell, J. H. 1900. The visitors of the Caprifoliaceae. Amer. Naturalist 34:37–51.

Lowe, E. N. 1921. Plants of Mississippi: A list of flowering plants and ferns. Mississippi State Geological Survey, Bull. No. 17. Hederman Bros., Jackson, Miss.

Lynch, J. M. 1979. Personal communication to A. E. Radford.

Mackenzie, K. K. 1940. North American Cariceae. New York Botanical Garden, New York.

Mann, V. I., T. G. Clarke, L. D. Hayes, and D. S. Kirstein. 1965. Geology of the Chapel Hill Quadrangle, North Carolina. North Carolina Geological Survey Special Publication 1. Raleigh.

Massey, J. R. 1974. Plant identification. Pp. 522–36 in A. E. Radford, W. C. Dickison, J. R. Massey, and C. R. Bell. Vascular plant systematics. Harper & Row, New York.

———, and P. D. Whitson. 1978. Species information system. Unpublished manuscript. This may be obtained from J. R. Massey, Department of Botany, Herbarium, University of North Carolina, Chapel Hill, N.C. 27514.

———. 1980. Species biology, the key to plant preservation. Rhodora 82: 97–103.

McGimsey, C. R. 1972. Public archeology. Seminar Press, New York.

McKern, W. C. 1939. The midwestern taxonomic method as an aid to archaeological culture study. Amer. Antiquity 4(4):301–13.

McVaugh, R. 1943. The vegetation of the granitic flat-rocks of the southeastern United States. Ecol. Monogr. 13:119–66.

Merriam, C. G., and Committee. 1963. Webster's seventh new collegiate dictionary. H. O. Houghton & Co., Cambridge, Mass.

Mertz, D. B. 1970. Notes on methods used in life-history

studies. Pp. 4–17 in J. H. Connell, D. B. Mertz, and W. W. Murdoch, eds. Readings in ecology and ecological genetics. Harper & Row, New York.

Mohlenbrock, R. H. 1967. The illustrated flora of Illinois: Ferns. Southern Illinois University Press, Carbondale, Ill.

Morris, W., ed. 1969. The American heritage dictionary of the English language. American Heritage Publishing Co., Inc., and Houghton Mifflin Company, New York.

Moyseenko, H. P., R. E. Jenkins, S. Woodall, and L. A. Miller. 1977. Lowest common denominator element file: An information management system. 3d ed. The Nature Conservancy, Arlington, Va.

Mueller-Dombois, D., and H. Ellenberg. 1974. Aims and methods of vegetation ecology. John Wiley & Sons, New York.

Muenscher, W. C. 1944. Aquatic plants of the United States. Comstock Publishing Co., Ithaca, N.Y.

Munsell Soil Color Chart. 1975. Macbeth Division of Kollmorgen Corp., Baltimore, Md.

Murphy, G. 1968. Pattern in life history and the environment. Amer. Naturalist 97:409–13.

National Historic Preservation Act. 1966. Public Law 89–665. 80 Statute 915–19. 89th Congress, S. 3035. In United States statutes at large. 1967. Vol. 80, Part 1. Published under direction of the Administrator of General Services by the Office of the Federal Register, National Archives and Records Service. U.S. Government Printing Office, Washington, D.C.

National Land Use Policy and Planning Assistance Act. 1973. S. Rept. No. 93–197. 93d Congress, S. 268. In Senate Reports. Miscellaneous reports on public bills. Vol. 3. U.S. Government Printing Office, Washington, D.C.

Nemeth, J. C. 1968. The hardwood vegetation and soils of Hill Demonstration Forest, Durham County, North Carolina. J. Elisha Mitchell Sci. Soc. 84:482–91.

Nesom, G. L., and M. Treiber. 1977. Beech–mixed hardwoods communities: A topoedaphic climax on the North Carolina Coastal Plain. Castanea 42:119–40.

Nevo, E., and H. Cleve. 1978. Genetic differentiation during speciation. Nature 275:125–26.

North Carolina Department of Conservation and Development. 1958. Geologic map of North Carolina. Raleigh.

North Carolina State Highway Commission. 1970. Municipal, state, primary, and interstate highway systems (map). [Raleigh].

Oakley, C. B., and E. Futato. 1975. Archeological investigations in the Little Bear Creek Reservoir. Research Series No. 1. Office of Archeological Research, University of Alabama, University.

Odum, E. P. 1971. Fundamentals of ecology. W. B. Saunders Co., Philadelphia.

Oosting, H. J. 1942. An ecological analysis of the plant communities of the Piedmont, North Carolina. Amer. Midl. Naturalist 28:1–126.

———. 1956. The study of plant communities. W. H. Freeman & Co., San Francisco.

———, and W. D. Billings. 1942. Factors effecting vegetation zonation on coastal dunes. Ecology 23:131–42.

Otte, D. K. Strady. 1978. Classification of vegetation in relation to the Radford natural diversity classification system. Department of Botany, University of North Carolina, Chapel Hill. Unpublished report.

Otte, L. 1977. Roan Mountain bald communities. Class report for Botany 235 (A.E. Radford, professor). Department of Botany, University of North Carolina, Chapel Hill.

Penfound, W. T. 1952. Southern swamps and marshes. Bot. Rev. (Lancaster) 18:413–46.

———, and T. F. Hall. 1939. A phytosociological analysis of a tupelo gum forest near Huntsville, Alabama. Ecology 20:358–64.

Penfound, W. T., and E. S. Hathaway. 1938. Plant communities of the marshlands of southeastern Louisiana. Ecol. Monogr. 8:1–56.

Pessin, L. S. 1933. Forest associations in the uplands of the lower Gulf Coastal Plain. Ecology 14:1–14.

Pettijohn, F. J. 1975. Sedimentary rocks. 3d ed. Harper & Row, New York.

Phillips, E. A. 1959. Methods of vegetation study. Holt, Rinehart, and Winston, Inc., New York.

Pittillo, J. D. 1976. Potential natural landmarks of the Southern Blue Ridge Portion of the Appalachian Ranges Natural Region. Department of Biology, Western Carolina University, Cullowhee, N.C.

Prout, T. 1969. The estimation of fitness from population data. Genetics 63:949–67.

Putnam, J. A., G. M. Furnival, and J. S. McKnight. 1960. Management and inventory of southern hardwoods. Handbook 181. U.S. Department of Agriculture, Washington, D.C.

Radford, A. E. 1976. Vegetation—habitats—floras—natural areas in the southeastern United States: Field data and information. University of North Carolina Student Stores, Chapel Hill.

———. 1977. A natural area and diversity classification system. A standardized scheme for basic inventory of species, community, and habitat diversity. University of North Carolina Student Stores, Chapel Hill.

———. 1979. A report on the Prevot Tract for the Environmental Protection Agency. Unpublished report.

———, H. E. Ahles, and C. R. Bell. 1968. Manual of the vascular flora of the Carolinas. University of North Carolina Press, Chapel Hill.

Radford, A. E., W. C. Dickison, J. R. Massey, and C. R. Bell. 1974. Vascular plant systematics. Harper & Row, New York.

Radford, A. E., and D. L. Martin. 1975. Potential ecological natural landmarks: Piedmont Region, eastern United States. Prepared and privately published for the National Park Service.

Radford, A. E., L. J. Otte, and D. K. Strady Otte. 1978. Natural heritage classification and information systems: Ecological diversity classification and inventory. University of North Carolina Student Stores, Chapel Hill.

Raunkiaer, C. 1937. Plant life forms. Clarendon, Oxford.

Reed, J. F. 1947. The relation of the Spartinetum glabrae near Beaufort, North Carolina to certain edaphic factors. Amer.

Midl. Naturalist 38:605–14.

Reiche, P. 1943. Graphic representation of chemical weathering. J. Sed. Pet. 13:58–68.

Reid, G. K. 1961. Ecology of inland waters and estuaries. Reinhold Publishing Corp., New York.

Richards, H. G. 1950. Geology of the Coastal Plain of North America. Trans. Amer. Philos. Soc. 40:1–83.

Sarukhan, J. 1974. Studies on plant demography: *Ranunculus repens* L., *R. bulbosa* L., and *R. acris* L. 2. Reproductive strategies and seed population dynamics. J. Ecol. 62:151–78.

———, and M. Gadgil. 1974. Studies on plant demography: *Ranunculus repens* L., *R. bulbosa* L., and *R. acris* L. 3. A mathematical model incorporating multiple modes of reproduction. J. Ecol. 62:921–36.

Schaffer, W. M. 1974. The evolution of optimal reproductive strategies: The effect of age structure. Ecology 55:291–303.

———, and M. D. Gadgil. 1975. Selection for optimal life histories in plants. Pp. 142–57 in M. L. Cody and J. M. Diamond, eds. Ecology and evolution of communities. Belknap Press, Cambridge, Mass.

Schefler, W. C. 1969. Statistics for the biological sciences. Addison-Wesley, Reading, Mass.

Schiffer, M. B., and G. J. Gummerman, eds. 1977. Conservation archeology: A guide for cultural resource management studies. Academic Press, New York.

Sculthorpe, C. D. 1967. The biology of aquatic vascular plants. Edward Arnold (Publishers) Ltd., London.

Sechrest, C. G., and A. W. Cooper. 1970. An analysis of the vegetation and soils of upland hardwood stands in the Piedmont and Coastal Plain of Moore County, North Carolina. Castanea 35:26–57.

Shaw, C., and R. Prasad. 1970. Starch gel electrophoresis of enzymes: A compilation of recipes. Biochem. Genetics 4:297–320.

Shunk, I. V. 1939. Oxygen requirements for germination of seeds of *Nyssa aquatica*, tupelo gum. Science 90:565–66.

Small, J. K. 1933. An Everglade cypress swamp. J. New York Bot. Gard. 34:261–67.

Society of American Foresters. 1954. Forest cover types of North America (exclusive of Mexico). Report of Committee on Forest Types. Washington, D.C.

Soil Conservation Service. 1977. Soil survey of New Hanover County, North Carolina. U.S. Department of Agriculture, Washington, D.C.

———. In press. Soil survey of Edgecombe County, North Carolina. U.S. Department of Agriculture, Washington, D.C.

Soil Survey Staff. 1951. Soil survey manual. Handbook No. 18. U.S. Department of Agriculture, Washington, D.C.

———. 1972. Soil series of the United States, Puerto Rico, and the Virgin Islands: Their taxonomic classification. Soil Conservation Service, U.S. Department of Agriculture, Washington, D.C.

———. 1975. Soil taxonomy: A basic system of soil classification for making and interpreting soil surveys. Agriculture Handbook No. 436. Soil Conservation Service, U.S. Department of Agriculture, Washington, D.C.

Spector, W. S., ed. 1956. Handbook of biological data. W. B. Saunders Co., Philadelphia.

Steila, D. 1976. The geography of soils. Prentice-Hall, Inc., Englewood Cliffs, N.J.

Stratton, A. C., and J. R. Hollowell. 1940. Sand fixation and beach erosion control. U.S. Department of the Interior National Park Service Report. Unpublished.

Stuckey, J. L. 1965. North Carolina: Its geology and mineral resources. North Carolina State University Print Shop, Raleigh.

Sweeting, M. M. 1973. Karst landforms. Columbia University Press, New York.

Teskey, R. O., and T. M. Hinckley. 1977. Impact of water level changes on woody riparian wetland communities. Vol. 2. The southern forest region. Fish and Wildlife Service, U.S. department of the Interior, Washington, D.C.

Thieret, J. W. 1971. Quadrat study of a bottomland forest in St. Martin Parish, Louisiana. Castanea 36:174–81.

Thomas, A. G., and H. M. Dale. 1975. The role of seed reproduction in the dynamics of established populations of *Hieracium floribundum* and a comparison with that of vegetative reproduction. Canad. J. Bot. 53:3022–31.

Thomas, W. 1976. Southeastern Coastal Plain savannah. Pp. 178–79 in A. E. Radford. Vegetation—habitats—floras—natural areas in the southeastern United States: Field data and information. University of North Carolina Student Stores, Chapel Hill.

Thompson, L. M., and F. R. Troeh. 1973. Soils and soil fertility. McGraw-Hill Book Co., Inc., New York.

Thornbury, W. D. 1969. Principles of geomorphology. John Wiley & Sons, Inc., New York.

Torrey, J., and A. Gray. 1838–40. A flora of North America. Wiley & Putnam, New York.

Trewartha, G. T. 1954. An introduction to climate. McGraw-Hill Book Co., Inc., New York.

Tripp, J. T. B. (Counsel). 1979. Report to the Environmental Protection Agency and the U.S. Army Corps of Engineers regarding the Lake Ophelia, Avoyelles Parish, Louisiana wetland determination. Environmental Defense Fund, Inc. Unpublished.

Trowbridge, A. C., ed. 1962. Dictionary of geological terms. Dolphin Books, Doubleday & Co., Inc., Garden City, N.Y. Prepared under the direction of the American Geological Institute.

Twidale, C. R. 1976. Analysis of landforms. John Wiley & Sons Australasia Pty., Ltd., Sydney, Australia.

U.S. Department of Commerce. 1965. Decennial census of United States climate—climatic summary of the United States—supplement for 1951 through 1960. (Each state has a supplement.) U.S. Government Printing Office, Washington, D.C.

U.S. Department of the Interior. 1970. The national atlas of the United States of America. U.S. Government Printing Office, Washington, D.C.

Viosca, P. 1928. Louisiana wetlands and the value of their

wildlife and fishery resources. Ecology 9:216–30.

Vogl, R. J. 1972. Fire in the southeastern grasslands. Proceedings, Tall Timbers Fire Ecology Conference 12:175–98.

Walter, H. 1973. Vegetation of the earth in relation to climate and the eco-physiological conditions. The English Universities Press Ltd., New York.

Way, D. S. 1973. Terrain analysis: A guide to site selection using aerial photographic interpretation. Dowden, Hutchinson, & Ross, Inc., Stroudsburg, Pa.

Weaver, J. E., and F. E. Clements. 1938. Plant ecology. McGraw-Hill Book Co., Inc., New York.

Webb, D. A. 1954. Is the classification of plant communities either possible or desirable? Bot. Tidsskr. 51:362–70.

Weber, M. 1978. Unpublished data for a master's thesis on *Chrysosplenium iowense*. Department of Biology, University of Northern Iowa, Cedar Falls.

Welch, P. S. 1948. Limnological methods. Blakiston Co., New York.

Wells, B. W. 1928a. A southern upland grass-sedge bog: An ecological study. Agricultural Experiment Station Tech. Bull. No. 32. Raleigh, N.C.

______. 1928b. Plant communities of the Coastal Plain of North Carolina and their successional relations. Ecology 9:230–42.

______. 1942. Ecological problems of the southeastern United States Coastal Plain. Bot. Rev. (Lancaster) 8:533–61.

Westhoff, V., and E. van der Maarel. 1973. The Braun-Blanquet approach. Pp. 618–726 in R. H. Whittaker, ed. Handbook of vegetation science. Part 5. Ordination and classification of communities. Dr. W. Junk b.v., Publisher, The Hague.

Wharton, C. H. 1978. The natural environments of Georgia. Special Publication of the Georgia Department of Natural Resources, Atlanta.

Whelan, T., III. 1979. A preliminary assessment of soil characteristics and wetlands: Prevot Tract, Avoyelles Parish, Louisiana. Prepared for Environmental Protection Agency, Region VI. Unpublished report.

Whetstone, R. D. 1978. *Diervilla rivularis* Gattinger. In J. R. Massey and R. D. Whetstone. Final Report for Contract No. 18–606, Threatened and endangered vascular plants of western North Carolina. Southeastern Forest Experiment Station, Asheville, N.C. Unpublished.

White, J. 1978a. Section 4. Natural area categories. Pp. 21–22 in J. White. Illinois natural areas inventory technical report. Vol. 1. Survey methods and results. Illinois Natural Areas Inventory, Urbana.

______. 1978b. Appendix 16. Determining boundaries of natural areas, significant features, and land condition classes. Pp. 236–41 in J. White. Illinois natural areas inventory technical report. Vol. 1. Survey methods and results. Illinois Natural Areas Inventory, Urbana.

______. 1978c. Appendix 21. Some procedures for detecting disturbances. Pp. 267–79 in J. White. Illinois natural areas inventory technical report. Vol. 1. Survey methods and results. Illinois Natural Areas Inventory, Urbana.

______. 1978d. Appendix 22. Grading natural quality. Pp. 280–85 in J. White. Illinois natural areas inventory technical report. Vol. 1. Survey methods and results. Illinois Natural Areas Inventory, Urbana.

Whitson, P. D., and J. R. Massey. Information systems for use in studying the population status of threatened and endangered plants. In L. E. Morse and M. S. Henifin, eds. Geographical data organization for rare plant conservation. New York Botanical Garden, New York. In press.

Whittaker, R. H. 1956. Vegetation of the Great Smoky Mountains. Ecol. Monogr. 26:1–80.

______. 1962. Classification of natural communities. Bot. Rev. (Lancaster) 28:1–239.

______. 1973a. Introduction. Pp. 3–6 in R. H. Whittaker, ed. Handbook of vegetation science. Part 5. Ordination and classification of communities. Dr. W. Junk b.v., Publisher, The Hague.

______. 1973b. Approaches to classifying vegetation. Pp. 323–54 in R. H. Whittaker, ed. Handbook of vegetation science. Part 5. Ordination and classification of communities. Dr. W. Junk b.v., Publisher, The Hague.

______. 1975. Communities and ecosystems. Macmillan Publishing Co., Inc., New York.

Wilbur, H. M., E. W. Tinkle, and J. P. Collins. 1974. Environmental certainty, trophic level, and resource availability in life history evolution. Amer. Naturalist 108:805–17.

Wilderness Act of 1964. Public Law 88–577. 78 Statute 890–96. 88th Congress, S. 4. In United States statutes at large. 1965. Vol. 78. Published under direction of the Administrator of General Services by the Office of the Federal Register, National Archives and Records Service. U.S. Government Printing Office, Washington, D.C.

Willey, G. R. 1966. An introduction to American archaeology. Vol. 1. North and Middle America. Prentice-Hall, Inc., Englewood Cliffs, N.J.

______, and P. Phillips. 1963. Method and theory in American archaeology. University of Chicago Press, Chicago.

Wilson, E. J. 1977. A floristic study of the "savannas" on pine plantations in the Croatan National Forest. M.A. thesis. Department of Botany, University of North Carolina, Chapel Hill.

Winters, R. K., J. A. Putnam, and I. F. Eldredge. 1938. Forest resources of the North Louisiana Delta. Misc. Publ. 308. U.S. Department of Agriculture, Washington, D.C.

Wright, A. H., and A. A. Wright. 1932. The habitats and composition of the vegetation of Okefinokee Swamp, Georgia. Ecol. Monogr. 2:109–232.

Index

Community Type Index

www.ingramcontent.com/pod-product-compliance
Lightning Source LLC
LaVergne TN
LVHW082025200926
848481LV00010B/2922